Computer-Aided Econometrics

STATISTICS: Textbooks and Monographs

1. The Generalized Jackknife Statistic, *H. L. Gray and W. R. Schucany*
2. Multivariate Analysis, *Anant M. Kshirsagar*
3. Statistics and Society, *Walter T. Federer*
4. Multivariate Analysis: A Selected and Abstracted Bibliography, 1957–1972, *Kocherlakota Subrahmaniam and Kathleen Subrahmaniam*
5. Design of Experiments: A Realistic Approach, *Virgil L. Anderson and Robert A. McLean*
6. Statistical and Mathematical Aspects of Pollution Problems, *John W. Pratt*
7. Introduction to Probability and Statistics (in two parts), Part I: Probability; Part II: Statistics, *Narayan C. Giri*
8. Statistical Theory of the Analysis of Experimental Designs, *J. Ogawa*
9. Statistical Techniques in Simulation (in two parts), *Jack P. C. Kleijnen*
10. Data Quality Control and Editing, *Joseph I. Naus*
11. Cost of Living Index Numbers: Practice, Precision, and Theory, *Kali S. Banerjee*
12. Weighing Designs: For Chemistry, Medicine, Economics, Operations Research, Statistics, *Kali S. Banerjee*
13. The Search for Oil: Some Statistical Methods and Techniques, *edited by D. B. Owen*
14. Sample Size Choice: Charts for Experiments with Linear Models, *Robert E. Odeh and Martin Fox*
15. Statistical Methods for Engineers and Scientists, *Robert M. Bethea, Benjamin S. Duran, and Thomas L. Boullion*
16. Statistical Quality Control Methods, *Irving W. Burr*
17. On the History of Statistics and Probability, *edited by D. B. Owen*
18. Econometrics, *Peter Schmidt*
19. Sufficient Statistics: Selected Contributions, *Vasant S. Huzurbazar (edited by Anant M. Kshirsagar)*
20. Handbook of Statistical Distributions, *Jagdish K. Patel, C. H. Kapadia, and D. B. Owen*
21. Case Studies in Sample Design, *A. C. Rosander*
22. Pocket Book of Statistical Tables, *compiled by R. E. Odeh, D. B. Owen, Z. W. Birnbaum, and L. Fisher*
23. The Information in Contingency Tables, *D. V. Gokhale and Solomon Kullback*
24. Statistical Analysis of Reliability and Life-Testing Models: Theory and Methods, *Lee J. Bain*
25. Elementary Statistical Quality Control, *Irving W. Burr*
26. An Introduction to Probability and Statistics Using BASIC, *Richard A. Groeneveld*
27. Basic Applied Statistics, *B. L. Raktoe and J. J. Hubert*
28. A Primer in Probability, *Kathleen Subrahmaniam*
29. Random Processes: A First Look, *R. Syski*
30. Regression Methods: A Tool for Data Analysis, *Rudolf J. Freund and Paul D. Minton*
31. Randomization Tests, *Eugene S. Edgington*
32. Tables for Normal Tolerance Limits, Sampling Plans and Screening, *Robert E. Odeh and D. B. Owen*
33. Statistical Computing, *William J. Kennedy, Jr., and James E. Gentle*
34. Regression Analysis and Its Application: A Data-Oriented Approach, *Richard F. Gunst and Robert L. Mason*
35. Scientific Strategies to Save Your Life, *I. D. J. Bross*
36. Statistics in the Pharmaceutical Industry, *edited by C. Ralph Buncher and Jia-Yeong Tsay*
37. Sampling from a Finite Population, *J. Hajek*
38. Statistical Modeling Techniques, *S. S. Shapiro and A. J. Gross*
39. Statistical Theory and Inference in Research, *T. A. Bancroft and C.-P. Han*
40. Handbook of the Normal Distribution, *Jagdish K. Patel and Campbell B. Read*
41. Recent Advances in Regression Methods, *Hrishikesh D. Vinod and Aman Ullah*
42. Acceptance Sampling in Quality Control, *Edward G. Schilling*
43. The Randomized Clinical Trial and Therapeutic Decisions, *edited by Niels Tygstrup, John M Lachin, and Erik Juhl*

44. Regression Analysis of Survival Data in Cancer Chemotherapy, *Walter H. Carter, Jr., Galen L. Wampler, and Donald M. Stablein*
45. A Course in Linear Models, *Anant M. Kshirsagar*
46. Clinical Trials: Issues and Approaches, *edited by Stanley H. Shapiro and Thomas H. Louis*
47. Statistical Analysis of DNA Sequence Data, *edited by B. S. Weir*
48. Nonlinear Regression Modeling: A Unified Practical Approach, *David A. Ratkowsky*
49. Attribute Sampling Plans, Tables of Tests and Confidence Limits for Proportions, *Robert E. Odeh and D. B. Owen*
50. Experimental Design, Statistical Models, and Genetic Statistics, *edited by Klaus Hinkelmann*
51. Statistical Methods for Cancer Studies, *edited by Richard G. Cornell*
52. Practical Statistical Sampling for Auditors, *Arthur J. Wilburn*
53. Statistical Methods for Cancer Studies, *edited by Edward J. Wegman and James G. Smith*
54. Self-Organizing Methods in Modeling: GMDH Type Algorithms, *edited by Stanley J. Farlow*
55. Applied Factorial and Fractional Designs, *Robert A. McLean and Virgil L. Anderson*
56. Design of Experiments: Ranking and Selection, *edited by Thomas J. Santner and Ajit C. Tamhane*
57. Statistical Methods for Engineers and Scientists: Second Edition, Revised and Expanded, *Robert M. Bethea, Benjamin S. Duran, and Thomas L. Boullion*
58. Ensemble Modeling: Inference from Small-Scale Properties to Large-Scale Systems, *Alan E. Gelfand and Crayton C. Walker*
59. Computer Modeling for Business and Industry, *Bruce L. Bowerman and Richard T. O'Connell*
60. Bayesian Analysis of Linear Models, *Lyle D. Broemeling*
61. Methodological Issues for Health Care Surveys, *Brenda Cox and Steven Cohen*
62. Applied Regression Analysis and Experimental Design, *Richard J. Brook and Gregory C. Arnold*
63. Statpal: A Statistical Package for Microcomputers—PC-DOS Version for the IBM PC and Compatibles, *Bruce J. Chalmer and David G. Whitmore*
64. Statpal: A Statistical Package for Microcomputers—Apple Version for the II, II+, and IIe, *David G. Whitmore and Bruce J. Chalmer*
65. Nonparametric Statistical Inference: Second Edition, Revised and Expanded, *Jean Dickinson Gibbons*
66. Design and Analysis of Experiments, *Roger G. Petersen*
67. Statistical Methods for Pharmaceutical Research Planning, *Sten W. Bergman and John C. Gittins*
68. Goodness-of-Fit Techniques, *edited by Ralph B. D'Agostino and Michael A. Stephens*
69. Statistical Methods in Discrimination Litigation, *edited by D. H. Kaye and Mikel Aickin*
70. Truncated and Censored Samples from Normal Populations, *Helmut Schneider*
71. Robust Inference, *M. L. Tiku, W. Y. Tan, and N. Balakrishnan*
72. Statistical Image Processing and Graphics, *edited by Edward J. Wegman and Douglas J. DePriest*
73. Assignment Methods in Combinatorial Data Analysis, *Lawrence J. Hubert*
74. Econometrics and Structural Change, *Lyle D. Broemeling and Hiroki Tsurumi*
75. Multivariate Interpretation of Clinical Laboratory Data, *Adelin Albert and Eugene K. Harris*
76. Statistical Tools for Simulation Practitioners, *Jack P. C. Kleijnen*
77. Randomization Tests: Second Edition, *Eugene S. Edgington*
78. A Folio of Distributions: A Collection of Theoretical Quantile-Quantile Plots, *Edward B. Fowlkes*
79. Applied Categorical Data Analysis, *Daniel H. Freeman, Jr.*
80. Seemingly Unrelated Regression Equations Models: Estimation and Inference, *Virendra K. Srivastava and David E. A. Giles*

81. Response Surfaces: Designs and Analyses, *Andre I. Khuri and John A. Cornell*
82. Nonlinear Parameter Estimation: An Integrated System in BASIC, *John C. Nash and Mary Walker-Smith*
83. Cancer Modeling, *edited by James R. Thompson and Barry W. Brown*
84. Mixture Models: Inference and Applications to Clustering, *Geoffrey J. McLachlan and Kaye E. Basford*
85. Randomized Response: Theory and Techniques, *Arijit Chaudhuri and Rahul Mukerjee*
86. Biopharmaceutical Statistics for Drug Development, *edited by Karl E. Peace*
87. Parts per Million Values for Estimating Quality Levels, *Robert E. Odeh and D. B. Owen*
88. Lognormal Distributions: Theory and Applications, *edited by Edwin L. Crow and Kunio Shimizu*
89. Properties of Estimators for the Gamma Distribution, *K. O. Bowman and L. R. Shenton*
90. Spline Smoothing and Nonparametric Regression, *Randall L. Eubank*
91. Linear Least Squares Computations, *R. W. Farebrother*
92. Exploring Statistics, *Damaraju Raghavarao*
93. Applied Time Series Analysis for Business and Economic Forecasting, *Sufi M. Nazem*
94. Bayesian Analysis of Time Series and Dynamic Models, *edited by James C. Spall*
95. The Inverse Gaussian Distribution: Theory, Methodology, and Applications, *Raj S. Chhikara and J. Leroy Folks*
96. Parameter Estimation in Reliability and Life Span Models, *A. Clifford Cohen and Betty Jones Whitten*
97. Pooled Cross-Sectional and Time Series Data Analysis, *Terry E. Dielman*
98. Random Processes: A First Look, Second Edition, Revised and Expanded, *R. Syski*
99. Generalized Poisson Distributions: Properties and Applications, *P. C. Consul*
100. Nonlinear L_p-Norm Estimation, *Rene Gonin and Arthur H. Money*
101. Model Discrimination for Nonlinear Regression Models, *Dale S. Borowiak*
102. Applied Regression Analysis in Econometrics, *Howard E. Doran*
103. Continued Fractions in Statistical Applications, *K. O. Bowman and L. R. Shenton*
104. Statistical Methodology in the Pharmaceutical Sciences, *Donald A. Berry*
105. Experimental Design in Biotechnology, *Perry D. Haaland*
106. Statistical Issues in Drug Research and Development, *edited by Karl E. Peace*
107. Handbook of Nonlinear Regression Models, *David A. Ratkowsky*
108. Robust Regression: Analysis and Applications, *edited by Kenneth D. Lawrence and Jeffrey L. Arthur*
109. Statistical Design and Analysis of Industrial Experiments, *edited by Subir Ghosh*
110. *U*-Statistics: Theory and Practice, *A. J. Lee*
111. A Primer in Probability: Second Edition, Revised and Expanded, *Kathleen Subrahmaniam*
112. Data Quality Control: Theory and Pragmatics, *edited by Gunar E. Liepins and V. R. R. Uppuluri*
113. Engineering Quality by Design: Interpreting the Taguchi Approach, *Thomas B. Barker*
114. Survivorship Analysis for Clinical Studies, *Eugene K. Harris and Adelin Albert*
115. Statistical Analysis of Reliability and Life-Testing Models: Second Edition, *Lee J. Bain and Max Engelhardt*
116. Stochastic Models of Carcinogenesis, *Wai-Yuan Tan*
117. Statistics and Society: Data Collection and Interpretation, Second Edition, Revised and Expanded, *Walter T. Federer*
118. Handbook of Sequential Analysis, *B. K. Ghosh and P. K. Sen*
119. Truncated and Censored Samples: Theory and Applications, *A. Clifford Cohen*
120. Survey Sampling Principles, *E. K. Foreman*
121. Applied Engineering Statistics, *Robert M. Bethea and R. Russell Rhinehart*
122. Sample Size Choice: Charts for Experiments with Linear Models: Second Edition, *Robert E. Odeh and Martin Fox*
123. Handbook of the Logistic Distribution, *edited by N. Balakrishnan*
124. Fundamentals of Biostatistical Inference, *Chap T. Le*
125. Correspondence Analysis Handbook, *J.-P. Benzécri*

126. Quadratic Forms in Random Variables: Theory and Applications, *A. M. Mathai and Serge B. Provost*
127. Confidence Intervals on Variance Components, *Richard K. Burdick and Franklin A. Graybill*
128. Biopharmaceutical Sequential Statistical Applications, *edited by Karl E. Peace*
129. Item Response Theory: Parameter Estimation Techniques, *Frank B. Baker*
130. Survey Sampling: Theory and Methods, *Arijit Chaudhuri and Horst Stenger*
131. Nonparametric Statistical Inference: Third Edition, Revised and Expanded, *Jean Dickinson Gibbons and Subhabrata Chakraborti*
132. Bivariate Discrete Distribution, *Subrahmaniam Kocherlakota and Kathleen Kocherlakota*
133. Design and Analysis of Bioavailability and Bioequivalence Studies, *Shein-Chung Chow and Jen-pei Liu*
134. Multiple Comparisons, Selection, and Applications in Biometry, *edited by Fred M. Hoppe*
135. Cross-Over Experiments: Design, Analysis, and Application, *David A. Ratkowsky, Marc A. Evans, and J. Richard Alldredge*
136. Introduction to Probability and Statistics: Second Edition, Revised and Expanded, *Narayan C. Giri*
137. Applied Analysis of Variance in Behavioral Science, *edited by Lynne K. Edwards*
138. Drug Safety Assessment in Clinical Trials, *edited by Gene S. Gilbert*
139. Design of Experiments: A No-Name Approach, *Thomas J. Lorenzen and Virgil L. Anderson*
140. Statistics in the Pharmaceutical Industry: Second Edition, Revised and Expanded, *edited by C. Ralph Buncher and Jia-Yeong Tsay*
141. Advanced Linear Models: Theory and Applications, *Song-Gui Wang and Shein-Chung Chow*
142. Multistage Selection and Ranking Procedures: Second-Order Asymptotics, *Nitis Mukhopadhyay and Tumulesh K. S. Solanky*
143. Statistical Design and Analysis in Pharmaceutical Science: Validation, Process Controls, and Stability, *Shein-Chung Chow and Jen-pei Liu*
144. Statistical Methods for Engineers and Scientists: Third Edition, Revised and Expanded, *Robert M. Bethea, Benjamin S. Duran, and Thomas L. Boullion*
145. Growth Curves, *Anant M. Kshirsagar and William Boyce Smith*
146. Statistical Bases of Reference Values in Laboratory Medicine, *Eugene K. Harris and James C. Boyd*
147. Randomization Tests: Third Edition, Revised and Expanded, *Eugene S. Edgington*
148. Practical Sampling Techniques: Second Edition, Revised and Expanded, *Ranjan K. Som*
149. Multivariate Statistical Analysis, *Narayan C. Giri*
150. Handbook of the Normal Distribution: Second Edition, Revised and Expanded, *Jagdish K. Patel and Campbell B. Read*
151. Bayesian Biostatistics, *edited by Donald A. Berry and Dalene K. Stangl*
152. Response Surfaces: Designs and Analyses, Second Edition, Revised and Expanded, *André I. Khuri and John A. Cornell*
153. Statistics of Quality, *edited by Subir Ghosh, William R. Schucany, and William B. Smith*
154. Linear and Nonlinear Models for the Analysis of Repeated Measurements, *Edward F. Vonesh and Vernon M. Chinchilli*
155. Handbook of Applied Economic Statistics, *Aman Ullah and David E. A. Giles*
156. Improving Efficiency by Shrinkage: The James-Stein and Ridge Regression Estimators, *Marvin H. J. Gruber*
157. Nonparametric Regression and Spline Smoothing: Second Edition, *Randall L. Eubank*
158. Asymptotics, Nonparametrics, and Time Series, *edited by Subir Ghosh*
159. Multivariate Analysis, Design of Experiments, and Survey Sampling, *edited by Subir Ghosh*

160. Statistical Process Monitoring and Control, *edited by Sung H. Park and G. Geoffrey Vining*
161. Statistics for the 21st Century: Methodologies for Applications of the Future, *edited by C. R. Rao and Gábor J. Székely*
162. Probability and Statistical Inference, *Nitis Mukhopadhyay*
163. Handbook of Stochastic Analysis and Applications, *edited by D. Kannan and V. Lakshmikantham*
164. Testing for Normality, *Henry C. Thode, Jr.*
165. Handbook of Applied Econometrics and Statistical Inference, *edited by Aman Ullah, Alan T. K. Wan, and Anoop Chaturvedi*
166. Visualizing Statistical Models and Concepts, *R. W. Farebrother*
167. Financial and Actuarial Statistics: An Introduction, *Dale S. Borowiak*
168. Nonparametric Statistical Inference: Fourth Edition, Revised and Expanded, *Jean Dickinson Gibbons and Subhabrata Chakraborti*
169. Computer-Aided Econometrics, *edited by David E. A. Giles*

Additional Volumes in Preparation

The EM Algorithm and Related Statistical Models, *edited by Michiko Watanabe and Kazunori Yamaguchi*

Multivariate Statistical Analysis, *Narayan C. Giri*

Computer-Aided Econometrics

edited by
David E. A. Giles
University of Victoria
Victoria, British Columbia, Canada

CRC Press
Taylor & Francis Group
Boca Raton London New York

CRC Press is an imprint of the
Taylor & Francis Group, an **informa** business

First published 2003 by Marcel Dekker, Inc.

Published 2020 by CRC Press
Taylor & Francis Group
6000 Broken Sound Parkway NW, Suite 300
Boca Raton, FL 33487-2742

First issued in paperback 2020

ISBN 13: 978-0-367-57849-7 (pbk)
ISBN 13: 978-0-8247-4271-3 (hbk)

Visit the Taylor & Francis Web site at
http://www.taylorandfrancis.com

and the CRC Press Web site at
http://www.crcpress.com

Library of Congress Cataloging-in-Publication Data
A catalog record for this book is available from the Library of Congress.

Preface

Econometrics is an orderly fusion of economists' mathematical models, data that measure economic activities of various types, and the tools of mathematical statistics. It is the empirical side of economic science. In the 21st century it is difficult to think of econometrics as being anything other than "computer-aided." In fact, this description might seem almost redundant. Modern students and practitioners of econometrics find it difficult to conceive of life without almost unlimited, fast, and virtually free computing capacity. Moreover, we have come to expect this capacity to be right there on (or under) our desks. How could we practice our trade without the aid of a computer? Of course, as has been the case in most fields of endeavor in recent years, econometricians have simply expanded their horizons as computers have improved and become more readily accessible.

Econometrics is a relatively "young" discipline, dating from approximately 1930. During the 1940s and 1950s, for example, many new theoretical developments emerged in this fledgling field, ahead of the researchers' ability to put them fully into practice in the context of real economic data. There were many obvious examples of this in the analysis of structural simultaneous equations models, for instance. There was a time when progress in econometrics was constrained by the available computing technology, at least in terms of the application of the theory. At one time, it was acceptable enough for an applied econometrician to comment that although (s)he was aware that a certain procedure should be used in a particular situation, this was not being done because of the unreasonable computational requirements or costs. Thankfully, today this is neither acceptable nor common. In the same vein, the application of conceptually simple statistical tools, such as Monte Carlo simulation, was severely hampered by the high cost and low speed of computing. Consider, for example, the pathbreaking (and "capital intensive") simulations of Summers* relating to the finite-sample properties of various estimators for simultaneous equations models, in which only 50 Monte Carlo replications were used. Compare this with MacKinnon's† response-surface simulations for unit root and cointegration tests, which involved a total of 15 million replications in 600 experiments for cases involving one to five time series, and a total of 12 million replications in 480 experiments in the case of six time series. And then think of the extent to which computing power has exploded in the last decade or so!

This book could, perhaps, have been titled *Computer-Intensive Econometrics*, but this might have created the impression that the methods and applications that are presented in the various chapters that follow are ones that require access to exceptional computing facilities. For the most part, this is not so. It is true that there are many recent and ongoing major developments in econometrics, and in applied statistics generally, that take advantage of parallel processing and other "high end" aspects of computational science. Here, however, we are concerned with the use of readily accessible computing facilities. We are also concerned, though, with the *creative* use of these facilities and the development of algorithms and software that enable us to feel relatively unconstrained in our pursuit of new theoretical results and their application to economic data. Today, econometrics is "computer-aided" to a degree that even our middle-aged peers would never have dreamed possible.

This is a common story, of course, and there is undoubtedly much more to come. At this point in time, however, it is appropriate to reflect on the extent to which econometric techniques and computing methods and hardware are intertwined. Hopefully, the reader will find that the contributions in this volume offer some insights into this matter, and provide a wide range of important new results.

The contributors to this book represent a cross section of the international econometric community. In approaching potential participants I sought to include those who would bring a variety of views, backgrounds, and experiences to this venture. I am convinced that in this respect I have been successful, and I am extremely grateful to all of the contributors for their enthusiastic participation, for their patience, and for the fine chapters

---

*Summers, R., 1965. A capital intensive approach to the small sample properties of various simultaneous equations estimators. *Econometrica* **33**, 1–41.

†MacKinnon, J. G., 1991. Critical values for cointegration tests. In: R. F. Engle and C. W. J. Granger (eds.), *Long-Run Economic Relationships*. Oxford University Press, Oxford, pp. 267–276.

that they have produced. In editing this volume I have benefited from the advice and assistance of a number of people. All of the chapters were anonymously peer-reviewed to ensure a high standard, and I would like to thank all of those who generously gave of their time and expertise in reviewing the draft chapters and providing feedback to me and to the authors. The editorial and production personnel at Marcel Dekker, Inc., were extremely encouraging, skillful, helpful and (above all) patient during the period that this volume was being prepared, and I thank them for making this book possible. I am especially grateful to Maria Allegra and Elissa Ryan at Marcel Dekker, Inc., and to Kelly Hall at Keyword Typesetting Services for all of their help.

David E. A. Giles

Contents

Preface iii
Contributors xi
Introduction xiii

1. Some Methodological Questions Arising from Large Data Sets 1
Clive W.J. Granger

Applications of Simulation Methods

2. Finite-Sample Simulation-Based Tests in Seemingly Unrelated Regressions 11
Jean-Marie Dufour and Lynda Khalaf

3. Finding Optimal Penalties for Model Selection in the Linear Regression Model 37
Maxwell L. King and Gopal K. Bose

4. On Bootstrap Coverage Probability with Dependent Data 69
Janis J. Zvingelis

5. A Comparison of Alternative Causality and Predictive Accuracy Tests in the Presence of Integrated and Cointegrated Economic Variables 91
Norman R. Swanson, Ataman Ozyildirim, and Maria Pisu

6. Finite Sample Performance of the Empirical Likelihood Estimator Under Endogeneity 149
Ron C. Mittelhammer and George G. Judge

7. Testing for Unit Roots in Semiannual Data 175
Sandra G. Feltham and David E. A. Giles

Bayesian and Related Inference

8. Using Simulation Methods for Bayesian Econometric Models 209
John Geweke, William McCausland, and John Stevens

9. Bayesian Inference in the Seemingly Unrelated Regressions Model 263
William E. Griffiths

10. Computationally Intensive Methods for Deriving Optimal Trimming Parameters 291
Marco van Akkeren

Econometric Modeling

11. Estimating and Testing Fundamental Stock Prices: Evidence from Simulated Economies 315
R. Glen Donaldson and Mark J. Kamstra

12. Neural Networks: An Econometric Tool 351
Johan F. Kaashoek and Herman K. van Dijk

13. Real-Time Forecasting with Vector Autoregressions: Spurious Drift, Structural Change, and Intercept Correction 385
Ronald Bewley

14. Econometric Modeling Based on Pattern Recognition via the Fuzzy C-Means Clustering Algorithm 407
David E. A. Giles and Robert Draeseke

Nonparametric and Semiparametric Inference

15. Nonparametric Bootstrap Specification Testing in Econometric Models 451
Tae-Hwy Lee and Aman Ullah

16. The Effect of Economic Growth on Standard of Living: A Semiparametric Analysis 479
Nilanjana Roy

Index *499*

Contributors

Ronald Bewley, Ph.D. Professor, Department of Economics, University of New South Wales, Sydney, Australia

Gopal K. Bose, Ph.D. Statistician, Office of Economic and Statistical Research, Queensland Treasury, Brisbane, Queensland, Australia

R. Glen Donaldson, Ph.D. Finning Junior Professor of Finance, Faculty of Commerce & Business Administration, University of British Columbia, Vancouver, British Columbia, Canada

Robert Draeseke, M.A. Economist, Treaty Negotiations, Ministry of Attorney General, Government of British Columbia, Victoria, British Columbia, Canada

Jean-Marie Dufour, Ph.D. Professor, Département de Sciences Économiques, Université de Montréal, Montreal, Québec, Canada

Sandra G. Feltham, M.A. Senior Benefits Analyst, Ministry of Human Resources, Government of British Columbia, Victoria, British Columbia, Canada

John Geweke, Ph.D. Professor, Economics/Statistics and Harlan E. McGregor Chair in Economic Theory, Department of Economics, University of Iowa, Iowa City, Iowa, U.S.A.

David E. A. Giles, Ph.D. Professor, Department of Economics, University of Victoria, Victoria, British Columbia, Canada

Clive W. J. Granger, Ph.D. Professor, Department of Economics, University of California, San Diego, La Jolla, California, U.S.A.

William E. Griffiths, Ph.D. Professor, Department of Economics, University of Melbourne, Melbourne, Australia

George G. Judge, Ph.D. Professor, Department of Agricultural & Resource Economics, University of California at Berkeley, Berkeley, California, U.S.A.

Johan F. Kaashoek, Ph.D. Professor, Econometric Institute, Erasmus University Rotterdam, Rotterdam, The Netherlands

Mark J. Kamstra, Ph.D. Financial Economist and Associate Policy Advisor, Research Department, Federal Reserve Bank of Atlanta, Atlanta, Georgia, U.S.A.

Lynda Khalaf, Ph.D. Associate Professor, Département d'Économique, Université Laval, Québec City, Québec, Canada

Maxwell L. King, Ph.D. Professor, Department of Econometrics & Business Statistics, and Deputy Dean, Faculty of Business & Economics, Monash University, Melbourne, Australia

Tae-Hwy Lee, Ph.D. Associate Professor, Department of Economics, University of California at Riverside, Riverside, California, U.S.A.

William McCausland, Ph.D. Professor, Département de Sciences Économiques, Université de Montréal, Montréal, Québec, Canada

Ron C. Mittelhammer, Ph.D. Professor, Department of Agricultural & Resource Economics, Washington State University, Pullman, Washington, U.S.A.

Ataman Ozyildirim, Ph.D. Economist, Economics Program, The Conference Board, New York, New York, U.S.A.

Maria Pisu, Ph.D. Assistant Professor, Health Services Administration, University of Alabama, Birmingham, Alabama, U.S.A.

Nilanjana Roy, Ph.D. Assistant Professor, Department of Economics, University of Victoria, Victoria, British Columbia, Canada

John Stevens, Ph.D. Economist, Division of Research & Statistics, Board of Governors of the Federal Reserve System, Washington, D.C., U.S.A.

Norman R. Swanson, Ph.D. Associate Professor, Department of Economics, Rutgers University, New Brunswick, New Jersey, U.S.A.

Aman Ullah, Ph.D. Professor, Department of Economics, University of California at Riverside, Riverside, California, U.S.A.

Marco van Akkeren, Ph.D. Post-Doctoral Fellow, Department of Agricultural & Resource Economics, University of California at Berkeley, Berkeley, California, U.S.A.

Herman K. van Dijk, Ph.D. Professor, Econometric Institute, Erasmus University Rotterdam, and Tinbergen Institute, Rotterdam, The Netherlands

Jānis J. Zvingelis, M.Sc. Ph.D. Student, Department of Finance, University of Iowa, Iowa City, Iowa, U.S.A.

Introduction

This volume comprises sixteen chapters, each of which demonstrates in one way or another the challenges of modern econometric practice, and the role that computer software and hardware play in this field. Without the extraordinary advances that have been made in computing in the last few years, econometrics would have been a very different discipline than the one it is today. Of course, there have been many significant developments in the application of mathematical and statistical tools to the theory of econometrics over this same time period. Moreover, there have been a number of crucial and quite dramatic shifts in the direction or emphasis of econometric methodology. With respect to the latter, some examples that come to mind include the emergence of the tools that we now take for granted in the area of non-stationary time-series analysis; the more central role that Bayesian methods have assumed within our discipline; and the growth in importance of methods for modeling with panel data.

Econometrics is "computer-aided" in many respects, and it is important to note that the relationship is by no means limited to empirical applications. On the contrary, as has been the case in many areas of statistical endeavor, computational methods and abstract theoretical developments often go hand in hand. For instance, the results of an exploratory Monte Carlo study may suggest that the sampling distribution of a new estimator or test statistic has certain desirable properties. This may be just the incentive that is needed for the exact theoretical results to be determined. This volume includes material that will illustrate this idea further. There are, of course many examples of developments in econometric theory that gained real significance only when computational techniques "caught up" to the extent that their implementation was feasible in realistic and relevant situations. The estimation of simultaneous equations models using "full information" methods was an obvious example of this, as was the emergence of integration via importance sampling (and later, the Markov Chain Monte Carlo method) in the context of Bayesian econometrics.

In reality, there is very little activity in econometrics that is *not* "computer-aided" to some degree. However, this volume does not pretend to deal with the impact of computational tools in all areas of econometrics. This would be both pretentious and impractical. Instead, what is offered here is a relatively wide cross section of papers that reflect the role of computers and computational methods in modem econometric theory and applied econometrics. These papers vary considerably in their focus, but as a collection they should provide the interested reader with a clear impression of some of the more important ways in which computational issues are an integral part of econometrics.

The increasing availability of larger and larger sets of data has been both a blessing and a challenge for econometricians. Given that many econometric techniques are justified largely in terms of the desirable "large-sample" (asymptotic) properties that they possess, the limited size of the samples of data that most econometricians have had to work with historically has been problem-ridden and limiting. Indeed, this spawned a vast literature devoted to investigating the "small-sample" properties of these techniques. How relevant is all of this when data constraints begin to disappear? Not only have data sets become larger, but they have also changed in other ways. A good example of this is the availability of vast amounts of very high-frequency (e.g., hourly, by-the-minute, by-the-"tick") data, especially in relation to financial markets. These data have features that do not arise when one has only a few dozen annual observations. This complication has to be balanced against the recognition that sometimes these very features enable one to construct models that are richer than would otherwise be possible, and to test them empirically.

In the opening chapter in this book, Granger looks critically at some of the issues surrounding the availability of very large sets of data for economic analysis. His comments are both timely and thought provoking. What does the availability of *really* large sets of data mean for the bulk of the material that is now taught in traditional econometrics courses? Does a large quantity of data necessarily mean that we will be able to test economic theories more effectively? How will we respond to the fact that with extremely large sample sizes, it is virtually impossible not to reject almost *any* hypothesis using classical methods? Granger reminds us that in this information age, with computing that is basically costless, and data sets that are beyond the wildest dreams that we may have had when we were in graduate school, we have to think afresh about many of the aspects of econometrics that we take for granted. Rightly, he also reminds us that this is an opportunity, not a burden.

Indeed, this sentiment is reflected in the rest of the material that the reader will find in this volume, which has been divided (albeit somewhat loosely, and with many overlaps) into four main parts:

1. Applications of Simulation Methods.
2. Bayesian and Related Inference.
3. Econometric Modeling.
4. Nonparametric and Semiparametric Inference.

The first of these parts includes six chapters, and this in itself speaks to the extent of the impact that simulation techniques have had on econometrics in recent years. The contributions here are quite varied in terms of specifics, but in each case there is a clear reliance on simulation to determine new and important results. Monte Carlo simulation, in particular, receives a considerable amount of attention in this part of the book. In Chapter 2, Dufour and Khalaf consider the problem of constructing tests of quite general hypotheses about the parameters of the well-known and widely used "Seemingly Unrelated Regressions" (SUR) model, first proposed by Zellner (1962). They are especially concerned with the performance of such tests in finite samples, and Monte Carlo methods are used in two ways in their analysis. Exploiting their earlier research into such problems in the context of the multivariate regression model (e.g., Dufour and Khalaf, 1998), they first develop a bounds test that is based on the likelihood ratio criterion. The need for a bounds test, in general, arises from the fact that the exact null distribution for the test statistic involves "nuisance parameters." The distribution of the bounds statistic itself is readily simulated by the Monte Carlo method, thus providing a rather nice example of the blending of formal statistical theory and computer simulation in order to obtain a strong and useful result. Of course, any "bounds test" will not always yield a result – the test statistic may lie in the "inconclusive region." To deal with this possibility, Dufour and Khalaf again resort to Monte Carlo methods to get exact results, one of which effectively involves obtaining the associated p-value for each point in the nuisance parameter space, and then taking the maximum of these p-values. The techniques that the authors describe in this chapter clearly have widespread application to other complex testing problems in econometrics, and they provide some insights into the type of computer-aided testing that we are likely to see much more of in the future.

Model selection, rather than hypothesis testing, is the subject of Chapter 3 by King and Bose. One of the most common problems in applied econometrics, and in regression analysis generally, is the choice of which model to use. More specifically, we are often faced with a number of plausible competing models, but we need to choose one of them as the basis for further analysis. When the competing models are "non-nested," or when their structure precludes the use of standard hypothesis testing for other reasons (e.g., in the selection of the number of "augmentation" terms in a Dickey-Fuller regression), econometricians often turn to "Information Criteria" (IC) as the basis for model selection. Various IC measures are commonly in use, such as those associated with Akaike (1974), Schwarz (1978), Hannan and Quinn (and the Bayesian IC (BIC)). These criteria are likelihood-based, with different penalty functions for the model's lack of parsimony, and for the most part they have only large-sample justification. King and Bose propose the use of simulation methods first to determine the probabilities of correct model selection in finite samples, and second to derive penalty values for IC procedures that are optimal in the sense of maximizing these probabilities on average. Simulated annealing is used in the latter computations, and a Monte Carlo study is conducted to determine the relative merits of the procedures. This study suggests that the new "optimized" procedure proposed by the authors can work very well if the sample size is quite small, but that the BIC is close to being optimal in about two thirds of the situations investigated.

The bootstrap resampling procedure proposed by Efron (1979) has been a key development in statistics, and it has found widespread application in many areas of econometrics. In its basic form, the bootstrap relies on the use of a random sample of data, but it can be modified in various ways if this assumption is not satisfied (as is often the case with economic data). In Chapter 4, Zvingelis takes up this point in the context of the use of the bootstrap to obtain finite-sample critical values for tests (or, equivalently, finite-sample coverage probabilities for confidence intervals). The question that he addresses is whether or not the known advantage of the bootstrap over the use of asymptotic (large n) approximations, in the random sample case, carries over to situations involving dependent data. He uses the non-overlapping block bootstrap to deal with the dependency issue, and uses Edgeworth expansions to prove that in this case the bootstrap provides only a small improvement over the coverage probabilities afforded by the asymptotic approximation. More specifically, he shows that while the errors associated with one-sided and two-sided asymptotic confidence intervals are reduced from $O(n^{-1/2})$ and $O(n^{-1})$, respectively, to $O(n^{-2})$ for a wide class of problems in the random sample case, they are reduced only to $O(n^{-3/4})$ and $O(n^{-4/3})$ in the context of dependent data. This has practical implications for the use of the bootstrap in this role in many areas of econometrics.

In Chapter 5, Swanson, Ozyildirim, and Pisu offer a very extensive set of Monte Carlo results that provides evidence relating to several questions that arise when a vector autoregressive (VAR) model is used as the basis for testing for Granger (non-) causality. Such tests have attracted a great deal of attention in the econometrics literature, and elsewhere, and their implementation has had to be reconsidered in the context of data exhibiting unit roots and/or cointegration. One of the issues that has to be addressed is how to determine the lag lengths for the VAR model that serves as the vehicle for the causality testing. Another issue is how one should deal with the fact that the Wald test, that is usually used when testing for Granger noncausality, does not have its usual distributional properties (even asymptotically) if the data are nonstationary. With regard to the latter matter, the authors' results provide very strong support for the simple modification to the testing procedure that was proposed by Toda and Yamamoto (1995). This support holds whether or not the non-stationary data are cointegrated. They also find that the Schwarz (1979) information criterion (SIC) performs well as a basis for testing Granger causality. Specifically, they favor using the SIC to find the preferred VAR model, and then rejecting the null hypothesis of noncausality if the supposedly causal variable appears appropriately in that model. Swanson et al. provide a large number of results that add significantly to our knowledge of this important topic. In doing so they also remind us that Granger's (1969) original discussion of his concept of causality focused on out-of-sample predictive issues, and they incorporate this important feature of Granger causality into their own study.

Chapter 6, by Mittelhammer and Judge, will no doubt introduce many readers to a type of inference that they have not previously encountered in the econometrics literature. "Empirical Likelihood" (EL) methods of estimation and testing have been pioneered by Owen (1991, 2000) and a handful of other statisticians, but only very recently have these developments been recognized by econometricians. (See Mittelhammer et al., 2000, for an excellent discussion.) Essentially, EL methods lie part way between conventional likelihood methods (such as maximum likelihood estimation, and the likelihood ratio test) and nonparametric methods. One can argue that while the former methods have desirable asymptotic properties, these generally apply only if the underlying data-generating process is specified correctly. That is, if the form of the likelihood function is known. On the other hand, one can avoid the need to specify a parametric likelihood function by adopting nonparametric methods (such as kernel regression), and still retain some weak large-sample results. EL represents a compromise between these two polar positions, and the hope is that one will retain the best of each, rather than the worst of each approach! In general, this hope can be shown to be justifiable. Mittelhammer and Judge apply the EL approach to simultaneous equations models, and compare the small-sample properties of the maximum EL estimator with those of Two Stage Least

Squares (TSLS). EL methods involve computational issues beyond those associated with traditional estimators such as TSLS, so this in itself makes this study particularly relevant for inclusion in this volume. Moreover, the authors use a carefully designed Monte Carlo experiment to simulate the estimator properties – hence the inclusion of this chapter in this part of the book. Mittelhammer and Judge find that the maximum EL estimator fares well in comparison with the TSLS estimator, and they leave us with some interesting avenues for further research in this newly emerging area of econometrics.

As was the case in Chapter 5, nonstationary time-series data are also the focus of the last chapter in the first part of this book. In Chapter 7, Feltham and Giles use both analytical methods and Monte Carlo simulation to investigate the properties of some tests for various types of nonstationarity that can arise with time-series data that are recorded semiannually. Specifically, they look at individual and joint tests for unit roots at the zero and seasonal frequencies, determine their (nonstandard) asymptotic null distributions in terms of functionals of standard Brownian motions, and derive the associated critical values for a range of sample sizes and situations. The latter are obtained from a large Monte Carlo experiment, and the same approach is used to evaluate the power functions of these tests under various scenarios. The results that the authors obtain are linked to other results in the literature on stochastic seasonality, and some simple illustrations using actual data are also provided.

The contributions in the second part of this book are devoted to Bayesian (and related) methods, and these are dealt with in three chapters. The use of the tools of Bayesian statistical inference in the econometric setting dates from the late 1960s, and was energetically promoted from the outset by Zellner (1971) and his students and colleagues. Today, it is widely accepted that the gap that once separated Bayesian econometricians from the majority of their colleagues has all but vanished. I can attest to this with some personal knowledge, having completed my own doctorate in precisely this area of econometrics in 1975! An analogy might be drawn here with another major development in econometrics during roughly the same time period. Thirty or forty years ago there was something of a schism between econometricians who espoused the use of "pure time-series" methods (especially for forecasting purposes), and those who stressed the essential need for complex structural models. When we look at the way in which econometrics has developed, especially since the beginning of the "unit root/cointegration revolution," we see that there has been a healthy blurring of this earlier artificial boundary between these two schools of thought, as can be seen in many of the contributions in this volume. Much the same can be said of the division between Bayesians and "non-Bayesians" in econometrics, and elsewhere. Econometricians are now often quite willing to "change hats" as the situation demands. It is accepted that there are benefits in having access to a range of different tools, so that one can apply the one best suited to a particular task. Even our econometric language is immensely more inclusive, in this respect, than used to be the case.

As was noted in the Preface to this volume, the extensive integration of Bayesian inference into econometrics is a wonderful example of the positive impact that enhanced computing facilities have had on the discipline. Having been developed to a certain stage, Bayesian methods became somewhat difficult to "sell" to the rest of the profession (especially applied researchers) because of the computational difficulties associated with applying them to realistically complex models. Put simply, unless one was prepared to severely limit the way in which one's prior information was expressed, and introduced into the analysis, one was limited to dealing with models with a tiny number of unknown parameters. The introduction of Monte Carlo integration, and later the revolution associated with Markov Chain Monte Carlo (MCMC) methods, changed all of this. Today, we regularly see fully-fledged Bayesian analyses of extremely complex models. Students in econometrics have ready access to these techniques, just one example being through the inclusion of Geweke's (1989) Bayesian regression estimator in the SHAZAM (2001) computer package.

With these remarks in mind, it is most appropriate that this section of the book begins with Chapter 8, by Geweke, McCausland, and Stevens. This chapter begins by providing a clear overview of the basic components of Bayesian inference, and then discusses in some detail the simulation techniques (such as acceptance and importance sampling, and examples of MCMC methods) noted above that have revolutionized this area of econometrics. The use of Geweke's Bayesian econometrics software (the BACC package) is illustrated through some interesting applications, and the reader will be left with a clear impression of the power and flexibility of this software with regard to both estimation and model selection. One could argue that this chapter represents computer-aided econometrics at its best, and those readers who have little inclination in the Bayesian direction will find their lack of faith severely tested!

In Chapter 9, Griffiths provides material that complements the previous chapter extremely well. His concern is with Zellner's (1962) SUR model, discussed above in the context of Chapter 2. As he says, his objective is "... to provide a practical guide to computer-aided Bayesian inference for a variety of problems that arise in applications of the SUR model." Griffiths provides us with insights into some of the practical issues that have to be faced when applying these Bayesian methods, including those that arise when extending the analysis to models that are nonlinear. Three applications serve their purpose admirably, and will leave the reader with a good appreciation of the merits of modern Bayesian econometrics.

One of the key features of the Bayesian methodology is the flexible way in which prior information is "injected" into the analysis, both through prior masses for competing models, and through prior densities to represent the uncertain information about the parameters of these models. This prior information is then combined with the data information, using Bayes' Theorem, and the resulting posterior distributions provide information that may be used (in conjunction with a chosen loss function) to draw inferences about parameters and models. If there is insufficient information in the sample to allow the estimation of all of the individual parameters in a model of interest, then one way to deal with this is to introduce a sufficiently informative prior density, and construct the Bayesian posterior distribution. A case in point is the use of the natural-conjugate prior in the context of a standard regression model whose regressor matrix is rank-deficient. Perfect "multicollinearity" is nothing more than a shortage of information, relative to the proposed task at hand. Often, such "ill-posed" problems can be solved in this way, but there are also alternative ways to proceed, as van Akkeren explores in Chapter 10. Here, he discusses the Data Based Information Theoretic (DBIT) estimator, that has been shown elsewhere to perform well (even in small samples) in the context of ill-conditioned. linear inverse problems. This estimator uses the Kullback-Leibler information measure (e.g., Kullback, 1959) in its objective function, and is a competitor not only of the Bayesian approach, but also other well known techniques such as ridge regression (and shrinkage estimators in general), and principal components regression. Apart from providing an interesting and timely discussion of this important, and relatively new, approach to inference, van Akkeren also illustrates its application to two well known sets of data. One of these is also used by Giles and Draeseke in Chapter 14 of this volume, and the reader may wish to compare these two sets of results. The emergence of the DBIT estimator, and other important related developments in the area of inverse problems, would not have been possible without our modern computing software and hardware. Again, this chapter clearly illustrates yet another direction in which modern econometrics has been "computer-aided."

The third part of this book deals with various modeling techniques that can be applied to economic data, and the four associated chapters include a number of empirical applications as well as proposing some new techniques. One of the areas where econometric techniques have found considerable application in recent times is finance. "Financial Econometrics" is a discernible subfield in its own right, and in Chapter 11

Donaldson and Kamstra provide a very interesting example of work in this area–one that is computationally intensive and creative. They are concerned with an important question that arises in financial economics, namely "are the market prices of stocks in line with the underlying 'fundamentals,' or are stock markets excessively volatile, perhaps containing price 'bubbles'?" Readers who have followed or participated in equity trading in the past five years will need no convincing that this is an interesting question! The methodology that Donaldson and Kamstra adopt is intriguing. They use actual financial data to estimate time-series models for dividend growth and discount rates. These are used to simulate paths for dividend growth and discount rates for various "bubble-free" economies, and fair-value prices for equities in these simulated economies can then be determined. Essentially, they then compare market prices (for S&P 500 stocks) with simulated fundamental prices and look for any discrepancies. Their main findings are that market prices are actually not excessively volatile, and that the tests that other authors have used tend to be biased in the direction "finding" non-existent bubbles and excess volatility. The extent to which this type of analysis is "computer-aided" becomes apparent when one considers that Donaldson and Kamstra simulated approximately one million hypothetical economies, an exercise that involved a month of CPU-time on a SUN UltraSparc workstation.

The remaining three chapters in this part of the book deal with three different approaches to econometric modeling: the use of (artificial) neural networks to model complex nonlinear systems; the use of VAR models for real-time forecasting; and a new technique that adapts ideas from fuzzy set theory and the pattern recognition literature to help estimate nonlinear economic relationships. The first of these topics is covered in Chapter 12, by Kaashoek and van Dijk. Readers who are not familiar with the use of neural network models in econometrics will welcome the very accessible overview that these authors provide in the early parts of this chapter. They then move on to illustrate how neural networks can be used successfully to deal with the complex patterns that are associated with economic data. In particular, Kaashoek and van Dijk show how this type of analysis can be used to uncover the dynamic properties of highly nonlinear systems, and they provide some interesting empirical applications relating to exchange rate data and a "Phillips curve" for the U.S.A.

In Chapter 13, Bewley discusses some important issues that arise when using a VAR model for forecasting purposes. This chapter covers a very wide range of issues–even wider than is suggested by its title. Among the topics that the reader will encounter here are Bayesian VAR models, nonstationary time-series data, structural breaks, and macroeconomic forecasting. Bewley pays careful attention to the role of drift (intercept) terms in the equations of a VAR model, and proposes an automated procedure for ensuring that structural changes are taken into account in a manner that will enhance forecast performance. The merits of this proposal are supported by the results of a Monte Carlo experiment. Notably, Bewley finds that adjustments to the intercept terms of a VAR model estimated in the *differences* of the data, with the intention of bringing the model "back on track," lead to an improvement in forecast mean squared error. Previous work by Clements and Hendry (1996) had suggested that this would be a useful procedure only for VAR models specified in the *levels* of the data. There are aspects of this chapter that will help younger readers appreciate some of the historical connections in the econometrics literature, and econometric practice. By way of example, we are reminded that post-estimation intercept corrections (based on the most recent forecast residuals) were standard practice in the use of the large-scale structural forecasting models of the late 1960s and 1970s. I recall this well from my own experience in the development and use of the Reserve Bank of New Zealand's econometric model at that time. Bewley points also out that the popular Bayesian VAR modeling procedure is really nothing more than a special application of the Theil and Goldberger (1961) "mixed regression" method, using appropriate dummy variables. This is not to say that there is "nothing new under the sun", but it is helpful to have an historical perspective on new developments in any discipline.

Chapter 14, by Giles and Draeseke, completes the third part of this volume. As noted above, it considers some new ways of modeling non-linear economic relationships, by drawing on various concepts from the literature relating to fuzzy sets and fuzzy logic, and that associated with pattern recognition and image processing. The authors show how a fuzzy clustering algorithm can be used to partition a set of data in a flexible manner, and then fuzzy logic can be used to combine separate models estimated from each cluster into one overall model. Because the combining process involves weights that vary continuously through the sample, highly nonlinear characteristics can be captured very successfully, even if the individual component models are themselves linear. Giles and Draeseke illustrate this technique with a range of empirical applications, including simple demand for money and consumption models, trend fitting, and an analysis of the age-earnings data that is also analyzed by van Akkeren in Chapter 10. They find that their "fuzzy regression" approach outperforms nonparametric kernel regression in terms of "goodness of fit." Perhaps even more importantly, they show that this new technique can be applied to multivariate relationships that are beyond the scope of standard nonparametric regression (due to the well-known "curse of dimensionality" associated with the latter).

The final part of the book includes two chapters that reflect the important impact that nonparametric methods have had on econometric theory and practice in recent times, and also fit together remarkably well as a pair. Given that there are substantial penalties (e.g., bias, inconsistency) that are usually incurred if we misspecify the functional form of an econometric model or omit relevant regressors, an attractive alternative may be to use nonparametric kernel regression. In Chapter 15, Lee and Ullah consider various ways of using nonparametric regression not just as an alternative to parametric regression in its own right, but also as a basis for testing for different forms of model misspecification. This testing can be undertaken by using one or more sets of nonparametric regression residual vectors or prediction vectors. Clearly, classical (parametric) hypothesis testing cannot be used, and the authors investigate three different ways of dealing with this problem. In each case, bootstrap methods are used to implement the tests–specifically to determine p-values for finite-sample situations. Lee and Ullah conduct a sizeable Monte Carlo experiment to determine the relative effectiveness of each of these three testing procedures, and they find that the class of tests nonparametric tests proposed by Li and Wang (1998) and by Zheng (1996) generally dominates the other two that are considered. This study provides an insightful "bridge" between parametric and nonparametric regression with respect to an important aspect of econometric analysis. It also illustrates once more the critical role of computational/simulation techniques in modern econometrics.

This volume concludes with an empirical study that (fortuitously, perhaps) makes particular use of the bootstrap specification test of Li and Wang (1998) that emerges so well from the study in the previous chapter. In Chapter 16, Roy returns to a long-standing question from the economics literature: "Does an increase in per capita income lead to an improvement in standard of living?" In this context, "standard of living" has been measured in terms of various individual indicators (such as life expectancy at birth), or composite indices of indicators, by different authors. One of the issues associated with the previous literature on this topic has been the choice of functional form when measuring such a relationship, and with this in mind the use of conventional nonparametric regression has been pursued by other researchers. In this chapter, Roy tests a linear parametric model against a nonparametric specification, using the testing procedure of Li and Wang. She also considers a semiparametric model that includes both per capita income and per capita public health expenditure, the latter entering nonlinearly. This study underscores the importance of undertaking appropriate specification testing, rather than simply abandoning a parametric approach and using standard nonparametric regression. The nonparametric approach that Roy takes also has the merit of at least partly dealing with one of the major limitations of kernel regression–the so-called "curse of dimensionality" that was mentioned above in connection with the research of Giles and Draeseke, reported in Chapter 14.

In many ways it is fitting that this volume ends with this empirical study. Roy's use of an interesting (though rather short) panel of data illustrates some of the data issues that need to be addressed properly in modern econometric applications. Our present level of theory, and the readily available computing facilities, allow one to be flexible as to the choice of technique that is adopted–standard methods can be reconsidered, adapted and applied with comparative ease. Econometrics as a discipline is truly "computer-aided," and the practicing econometrician is freed from many of the constraints that previously limited our horizons. With this freedom comes the opportunity for enhanced and responsible creativity as we push the boundaries of empirical economics.

REFERENCES

1. Akaike, H., 1974. A new look at the statistical model identification. *IEEE Transactions on Automatic Control*, **19**, 716–723.
2. Clements, M. P., Hendry, D. F., 1996. Intercept corrections and structural change. *Journal of Applied Econometrics* **11**, 475–494.
3. Dufour, J.-M., Khalaf, L., 1998. Simulation based finite and large sample tests in multivariate regressions. Technical Report, C.R.D.E., Université de Montréal, Montreal, Quebec.
4. Efron, B., 1979. Bootstrap methods: Another look at the jackknife. *Annals of Statistics* **7**, 1–26.
5. Geweke, J., 1989. Bayesian inference in econometric models using Monte Carlo integration. *Econometrica* **57**, 1317–1340.
6. Granger, C. W. J., 1969. Investigating causal relations by econometric models and cross spectral methods. *Econometrica* **37**, 428–438.
7. Hannan, E. J., Quinn, B. G., 1979. The determination of the order of an autoregression. *Journal of the Royal Statistical Society*, Series B **41**, 190–195.
8. Kullback, S., 1959. *Information Theory and Statistics*. Wiley, New York.
9. Li, Q., Wang, S., 1998. A simple consistent bootstrap test for a parametric regression. *Journal of Econometrics* **87**, 145–165.
10. Mittelhammer, R. C., Judge, G. G., Miller, D. M., 2000. *Econometric Foundations*. Cambridge University Press, Cambridge.
11. Owen, D. B., 1991. Empirical likelihood for linear models. *Annals of Statistics* **19**, 1725–1747.
12. Owen, D. B., 2000. *Empirical Likelihood*. Chapman and Hall, New York.
13. Schwarz, G., 1978. Estimating the dimension of a model. *Annals of Statistics* **6**, 46–464.
14. SHAZAM, 2001. SHAZAM Econometrics Software. User's Reference Manual, Version 9. Northwest Econometrics, Vancouver BC.

15. Theil, H., Goldberger, A. S., 1961. On pure and mixed statistical estimation in economics. *International Economic Review* **2**, 65–78.
16. Toda, H. Y., Yamamoto, T., 1995. Statistical inference in vector autoregressions with possibly integrated processes. *Journal of Econometrics* **66**, 225–250.
17. Zellner, A., 1962. An efficient method of estimating seemingly unrelated regressions and tests of aggregation bias. *Journal of the American Statistical Association* **57**, 500–509.
18. Zellner, A., 1971. *An Introduction to Bayesian Inference in Econometrics.* Wiley, New York.
19. Zheng, J. X., 1996. A consistent test of functional form via nonparametric estimation techniques. *Journal of Econometrics* **75**, 263–289.

Computer-Aided Econometrics

1

Some Methodological Questions Arising from Large Data Sets

Clive W.J. Granger
University of California, San Diego, La Jolla, California, U.S.A.

1 INTRODUCTION

In recent years there have been strong trends in both data availability and computing speed, which will have had dramatic effects on how econometricians approach their tasks. These trends are very likely to continue. To make my discussion easier I will assume that, in the future, computing will be sufficient for all our needs and will be effectively free. These features are almost true now. My home university has a 1 teraflop computer (meaning that it can perform one trillion calculations per second) available for academic use and U.S. government research agencies have a 4.5 teraflop machine available now with a plan for a 100 teraflop machine to come on line in 2004. IBM has already announced work on Blue Gene, claimed to be 500 times faster than any present machine.

Suppose that a statistician or econometrician is faced with the prospect of having enormous quantities of data combined with effectively unlimited computing abilities, what reaction should one expect? Initially there would be simple delight, as one realizes that many of the present constraints on analysis have been removed. However, on reflection, the reaction could be more muted when the fact that it is unclear what techniques should be used on the data. It is also very unclear if the methods traditionally taught to and by econometricians will have any relevance. In the past, the major occupation of econometric theorists was in developing estimation techniques for many different situations and satisfactory tests to be used for inference. As has been pointed out several times before, as in Granger (1998), for a large enough data set it is virtually impossible to not to reject a specific null hypothesis. For example, consider a credit card company that observes one million transactions per day and is interested in the proportion of transactions above a certain monetary amount, say \$200. Suppose that recently this amount has been 41.4% and that yesterday the observed proportion was 0.4152. Denoting this proportion by $\hat{p}$, using the normal approximation for the binomial, the 99.9% confidence interval is $\hat{p} \pm 3.30\sqrt{\hat{p}(1-\hat{p})/n}$ for a sample size n, with $\hat{p} = 0.4152$ and $n =$ one million; this gives a band of 0.4152 ± 0.001, which does not include 0.414. Thus, to a high level of statistical significance, there appears to have been a change in the proportions.

This example, and similar thoughts, lead to the following conclusions that have been discussed in Granger (1998):

1. Limit theorems and asymptotic theory generally will be appropriate with very large samples. See Lehmann (1999) for a recent discussion. Conversely, all the small-sample procedures, which take up so much time in elementary statistics courses, are irrelevant.
2. One can expect that virtually any explicit hypothesis will be rejected, such as H_0: mean $= 3$, but a broader null H_0: mean in range (2.5–3.5) may not be.
3. Statistical significance becomes less relevant compared to economic significance, even though the latter is difficult to quantify in most circumstances.
4. Any consistent estimate will provide satisfactory values; efficiency estimation is not particularly important.
5. It is better to use estimates, and model specifications, that depend on fewer, or looser, assumptions.
6. Bayesian priors become irrelevant unless the number of parameters to estimate is infinite, as in a moving average (MA) (∞) model.
7. The "pooling" assumption of many standard panel methods will be rejected, requiring a joint analysis of many time series from different "agents."

As inference, testing, and estimation have been the life-blood of econometricians for at least the past 50 years, to be reminded that they are of little importance with large data sets may seem to be a major turn-off. There are a number of reactions to this situation; it can be regarded as an exciting opportunity to consider new types of specifications, methods, models, and situations some of which will be considered below. Alternatively, one can ask if something like the old methodology and approaches can be retained and used under particular circumstances.

The answer is obviously yes because however large an information set is, there will usually be sensible ways to subdivide it into subsets that are small enough for classical techniques to be appropriate. Examples would be:

1. A large cross-section of marketing questions involving one million individuals could be divided into many geographical regions, subdivided into income groups, genders, racial, or country of origin groups, and so forth. Eventually, the average number in each subgroup may be fairly small, just a few hundred. What is less clear is if anyone really cares about the average juice consumption of right-handed male Californians originally from Britain with an income in the top 10%, for example, compared to the consumption of some other group. No doubt statistical significances could be found and regressions run but the economic or marketing significance may be slight.
2. A large panel of multinational companies could record monthly sales, changes in stock prices, quantity of shares traded, CAPM beta values, and so forth, over 20 years. The panel members could be considered in various subgroups, depending on size, work force, and country of the home office, and for different subsets of years.
3. A very long single time series, such as daily volume of trading on some exchange, or the percentage of shares increasing over those decreasing, could be recorded from 1927 to 2000, for example, giving over 18,000 daily values. If one uses a running window of, say, 1000 terms and within each window applies standard modeling techniques, a time-varying parameter time series model will be achieved. This model can be tested for significance of changes and interpretations, hopefully of economic relevance, given to the swings in coefficient values or structural changes.

These examples also illustrate a problem with the definition of a large data set. In economic time series, a size of 1000 would be considered large, in fact virtually impossible to achieve in macroeconomics. Only recently in finance have the truly huge data sets become available. In cross-section econometrics, sets involving 10,000 individuals are not unusual, and if one is working with census data, the sample could be in the tens of millions. In this area, large samples are not a novel problem. In panels, the cross-sectional dimension is often quite large, but the time series dimension rather small, though new panels will be large in both dimensions. This area will present the greatest challenges.

To preserve the usefulness of their present human capital, econometricians may concentrate on techniques of great interest but with very slow convergence rates, such as some nonparametric problems and questions involving extreme values, which can have convergence rates of $n^{1/5}$ or less.

Consideration of such possibilities are potentially important in the short-run, but cannot hold back the inevitable development of new techniques.

2 TWO TIME SERIES SCENARIOS

Two hypothetical experiments will be considered:

1. Starting at a fixed point in time, we observe a vector series $\underline{X}_t$ at a fixed observation interval, say monthly, giving a sample length n. Initially, n is quite small, say 50, but after a period it becomes, 200, then 500, and eventually the series consists of 2000 months, say, or possibly more. The question to consider is: what could a time series analyst do differently when n is larger than when it was smaller?
2. The second mind experiment takes a set period of time $(0, T)$ and initially observes it annually, say, giving X_t, $t = 1, \ldots, T$. We then observe it monthly giving $12T$ pieces of data (roughly), then weekly giving $52T$ points, then daily, so that there are now $365T$ points (or $216T$ if we just use weekdays), roughly $2000T$ hours, and so forth.

In each experiment we are getting more numerical information but of a different nature.

For the first experiment, the types of models and techniques used can increase in complexity and sophistication as the sample size increases. For very short series, a simple autoregressive model of low order plus trend is appropriate; for a somewhat larger series, seasonal components, unit root testing, and autoregressive integrated moving average (ARIMA) models come into play. When the series is somewhat larger, one has predictive conditional distributions, quantile estimates, large vector autoregression (VAR) models, complicated nonlinear models, time-varying parameter models, breaks, regime shifts, outliers, and so forth. However, although it is clear that the variety of models available for consideration will increase with sample size, the above discussion is in terms of the viewpoint of an idealized statistician as it is context-free. In practice, as the data set gets larger, one has to be concerned with the possibility that the economy that produced the early terms in the series is quite different from the economy described by the later terms. Economies evolve as the tastes of its agents and the form of its institutions change. A form of time-changing parameter model may capture this, but this essentially shortens the series, as mentioned in Section 1.

In the second experiment, as the observation period is reduced, one obtains more information about the high-frequency part of the spectrum, but no more about the low-frequency components. For example, there is no information about the weekly cycle in data observed at a period of a week or more, but daily data will contain a lot. In a sense, this example illustrates the idea that one can have more data but no more information. If one is looking for exceptionally large rises or drops in stock prices, for example, you do not find more of them by looking at a 5 minute price index rather than an hourly index, and possibly a daily index. The extreme case is the use of tick-by-tick data, called "ultra-high frequency data" by Engle (2000). These are unusual data because it is not clear if the information about the jth price of the day was available to the decision maker(s) who produced the trade for the $j+1$th price, because of the speed required and as there can a queue of as yet unreported trades. That is, the time to decide to make the trade, the time of the trade, and the time the trades reported are by no means kept in the same order. There is certainly a huge amount of tick-by-tick data; its value-added is still being determined. It seems to be clear increasing the frequency of data observation is helpful to some aspects of data analysis, but not others. As an analogy, one could ask if daily data are more useful for studying business cycles than monthly data, except for the fact they are probably issued with a shorter delay.

3 ONWARD AND UPWARD

To reiterate, or expand on, some earlier points, having a lot of data is an opportunity, not a burden. Given the choice of doing the old things better or attempting to do something new, I hope we will chose the latter. Some of the things that we can do with large data sets that are not possible with fewer data include:

1. Consider a wider variety of specifications.
2. Discover more subtle effects.
3. Distinguish between close specifications.
4. Obtain better estimates of parameters.

5. Do more and obtain more useful post- and out-of sample evaluations.
6. Attempt types of analysis that are now not possible.

Details in some cases will be discussed below. There are also a number of general observations that could be made frequently but will just be recorded here:

1. As stated before, standard statistical tests are no help with very large sample sizes, as any simple null hypothesis will almost certainly be rejected. If the null is taken to be the simple case — normality, linearity, stationarity, few explanatory variables — e.g., if these are all rejected one ends up with complicated situations, particularly nonlinear, nonstationary models with many related variables.
2. For panel data, with both many (N) agents and long time (T) dimensions, it follows from (1) that the poolability assumption will be rejected, so that standard panel analysis techniques are largely irrelevant. The data will have to be treated like a vector of time series, presumably inter-related.
3. As there will be many model specifications available, a precise statement about the purpose of the analysis is essential before starting. The purpose is also necessary for the evaluation phase of the modeling process.
4. Evaluation should be in terms of economic significance of alternative models rather than statistical significance. Economic significance can. theoretically at least, be judged in terms of the economic values to decision makers who base their decisions on the outcomes of the models. This perspective is easy in finance and forecasting, but needs further development in other areas of econometrics.
5. As model specifications may well be complicated, presentation of results from these may be difficult. Innovations in presentations of parametric values or outputs of alternative complex models will be needed. The use of mobile trees that increase in complexity about information in a region as one moves around the branches are already successful in containing a great deal of facts in a readable form. Animation of diagrams may also be helpful.
6. The quality of data will not necessarily change as data sets get larger, but running filters to discover missing values, misprints or other data errors, outliers (however defined) will be more difficult to generate in larger multidimensional sets and arbitrary fixes may

be used to remove them. However, using outlier-resistant methods will not affect the efficiency of estimates as there is too much information, unless one is specifically modeling outliers, such as in value-at-risk.

7. With a fairly small data set, the analyst becomes very familiar with its properties and may "find" something unexpected, which occasionally continues to occur out-of-sample. Serendipity has been an important tool in applied statistics but will almost certainly be lost with the advent of very large data sets. Computers can be programmed to search for unexpected events but in a sense they have to be in some way expected to be considered for a search.
8. The answer to the question "Are there too many data?" will depend on the purpose of the analysis and the needs of the decision makers who will use the output. If the purpose is vague, such as "help understanding," the question cannot be answered. For many purposes, having more data does not imply that we have more relevant information, we may just be adding noise. For other purposes, precision may be important and this can increase with sample size.

4 CROSS-SECTION DATA

This area of econometrics already has experience with large data sets. Suppose that we have a vector of m dependent variables Y and a further vector of k explanatory variables X. The objective may be thought of as making statements about the conditional distribution of Y given X. The question is not often tackled in terms of conditional distributions, but rather in conditional moments, such as means, variances, and covariances. The number of specifications to model each is still very high and usually limiting classes are considered, such as linear or quadratic terms of the explanatory variables, based possibly on an assumption that the conditional distribution is Gaussian, even though such an assumption is likely to be rejected by the data. The aim in all approaches is to consider a wide enough class of specifications so that, hopefully, a reasonable approximation to the actual data-generating process will lie with this set. Given enough data it is hoped that one can sort through this collection of models, discard the unsatisfactory ones, and end with one that is suitable for use. The sorting may use the "general to simple" methodology, which has strong theoretical and practical foundations in the class of linear models, but is more difficult to use with nonlinear models, as the number of alternatives is much greater and there are less clear hierarchies. It should be noted that with large data sets there are fewer of the classical data-mining problems, with an inferior model appearing to be the best. Instead, one may obtain several specifications that appear to be different but which have similar statistical properties, and a choice between them in terms of economic benefits may be difficult. The spurious results in small samples are replaced by indifference between specifications when using large samples.

5 TIME SERIES DATA

For time series, methodological problems arise both from the *length* of the series, as discussed above, and also from the *number* of series involved. The problems can be illustrated using a simple linear VAR model with a vector of N series. If a full model, with constant parameters, is constructed with p lags for each variable in each equation, the total number of parameters is N^2p — known as the curse of dimensionality, even without considering estimation of the covariance matrix of the residuals. The value of p is not really a choice, it is the value required to whiten the residuals. For example, if $N=2000$, $p=10$, $N^2p=4\times10^7$. I chose this example because there are roughly 2000 shares quoted in the New York Stock Exchange. A VAR for absolute daily returns would need p larger than 10. The problems that arise concern estimation, presentation, and interpretation of so many parameters plus how to compare alternative specifications. The solution used in building large economic models is to be parsimonious with the inclusion of variables in equations. One might always include the lagged dependent variables, lags of closely associated variables (e.g., from economic theory, or close trading partners for countries, or in some industry for companies), and, most helpfully, a few "common factors" which efficiently mop up inter-relationship between many variables, such as the CAPM model and its extensions. Using such a strategy, which can either be "hands on" or automated, the number of parameters estimated can be reduced to mNp, where m is some integer in the range 5–10, giving a reduction of parameters of over 99% in the example. The economic loss is unclear.

The linear, constant parameter model for the conditional mean, as just considered, is almost certainly not where the future of time series lies. With lots of data, subtle nonlinearities and/or time-varying parameters can be detected (sometimes they may be alternative specifications) and complicated models specified and fitted, probably recursively. Concentration on models for the conditional mean and the conditional variance will evolve into attention being given to other moments and properties of the conditional or predictive distribution, particularly the quantiles. An approximation to the whole predictive distribution will result, not necessarily with an assumption of normality, and with time-varying parameters. The users will concentrate on the quality and usefulness of the output rather than on the specification of the model, which will be very complicated. Alternative models will be compared by the relative usefulness of their outputs for decision making; purely statistical measures are unlikely to have plausible economic interpretations.

6 CONCLUSIONS

In the past 40 years econometrics has progressed considerably because of the increase in computer power and speed, in data availability, and, possibly consequently, in techniques. The overlap in advanced econometrics textbooks written today is quite small compared with those appearing in the early 1960s, except in the first few chapters. The new techniques have allowed economists to view their data in new ways and sometimes to rephrase theories to link with the techniques and to state the properties of the data differently. What is unclear to me is whether or not the increases in data availability has produced solutions to many economic problems? The possibility of testing international trade theory on data between many countries, which is now possible, should certainly be an advantage, but do we learn a lot from having detailed data from 30 stock markets rather than just the big five, say?

In general, having more data is good. For some problems (e.g., dealing with rare events) economists will continue to need more, given that economies evolve through time, but for many problems there will be reducing returns from increased data, particularly when there is little information available about the quality of the new data relative to the old.

REFERENCES

Engle, R. F., 2000. The economics of ultra-high frequency data. *Econometrica* **68**, 1–22.

Granger, C.W.J., 1998. Extracting information from mega-panels and high frequency data. *Statistica Neerlandica* **52**, 258–272.

Lehmann, E. L., 1999. *Elements of Large-Sample Theory*. Springer-Verlag, Berlin.

2

Finite-Sample Simulation-Based Tests in Seemingly Unrelated Regressions

Jean-Marie Dufour
Université de Montréal, Montréal, Québec, Canada
Lynda Khalaf
Université Laval, Québec, Québec, Canada

1 INTRODUCTION

In this chapter, we study the problem of testing general, possibly nonlinear constraints on the coeffcient of the seemingly unrelated regressions (SURE) model introduced by Zellner (1962). The SURE model may be cast as a system of regression equations with contemporaneously correlated disturbances, where the regressors may differ across equations. For a detailed review, the reader may consult Srivastava and Giles (1987).

In connection with the SURE model, very few analytical finite-sample results are available. A rare exception is provided by Harvey and Phillips (1982, Sect. 3) who derived independence tests between the disturbances of an equation and those of the other equations of a SURE model. The tests involve conventional F-statistics and are based on the residuals obtained from regressing each dependent variable on all the independent variables of the system. Of course, this problem is a very special one. In a different vein, Phillips (1985) derived the exact distribution of a two-stage SURE estimator using a fractional matrix calculus. However, the analytical expressions obtained are very complex and, more importantly, involve unknown nuisance parameters, namely, the elements of the error covariance matrix. The latter fact makes the application of Phillips' distributional results to practical hypothesis testing problematic.

Asymptotic Wald, Lagrange multiplier, and likelihood ratio tests are available and commonly employed in empirical applications of the SURE model; see, e.g., Breusch (1979) or Srivastava and Giles (1987). It has been shown, however, that, in finite samples, the asymptotic criteria are seriously biased towards over-rejection, with the problem getting worse as the number of equations grows relative to the sample size; see, for e.g., Laitinen (1978), Meisner (1979), Bera, et al. (1981), Theil and Fiebig (1985), and Dufour and Khalaf (2002). Attempts to improve standard asymptotic tests include, in particular: (1) Bartlett-type corrections, and (2) bootstrap and simulation-based methods. See, e.g., Theil and Fiebig (1985), Theil et al. (1985), Taylor et al. (1986), Theil et al. (1986), Rocke (1989), Rayner (1990), and Rilstone and Veall (1996).

Further results relevant to the SURE model can be found in the statistics and econometrics literature on multivariate linear regressions (MLR). These are relevant because the MLR model can be viewed as a special case of the SURE model where the regressor matrices for the different equations are identical. For reviews and further references on exact and asymptotic inference in MLR models, the reader may consult Rao (1973, Chap. 8), Anderson (1984, Chaps. 8 and 13), Kariya (1985), Stewart (1997), and Dufour and Khalaf (2002). In particular, besides showing the inadequacy of various size-correction procedures (including Bartlett corrections) through simulation, we derived in Dufour and Khalaf (2002) exact bounds on the null distribution of LR test statistics for possibly nonlinear hypotheses on regression coefficients in MLR models. Even though computing these bounds analytically may be difficult, they can be easily evaluated by simulation and implemented as finite-sample *bounds Monte Carlo* (BMC) *tests*. The implications for hypothesis testing are two-fold. First, the finite-sample bounds on the LR criterion easily yield conservative tests, for both linear and nonlinear hypotheses. Second, Monte Carlo (MC) test methods can lead to tests with correct levels. This is related to the fact that LR statistics are *pivotal* or *boundedly pivotal* for quite general hypotheses in the MLR model [see the discussion in Dufour (1997) on boundedly pivotal statistics].

In this chapter, we extend the results presented in Dufour and Khalaf (2002) to the case of SURE systems, under both Gaussian and non-Gaussian disturbance distributions. Indeed the model considered here is an extension of the standard Gaussian SURE model that allows for both Gaussian and non-Gaussian disturbance distribution, as long as the latter is specified up an unknown linear transformation (or contemporaneous covariance matrix). In particular, we discuss two approaches that can be applied on their own or sequentially, namely: (1) a bounds procedure, and (2) MC tests. Practical implementation of both techniques is simple. To obtain the bounds, we exploit the fact that the SURE specification can be viewed as a special case of a properly chosen MLR model constrained by regressor exclusion restrictions on the different equations.

To be more specific, we give at this point a preliminary discussion of the proposed conservative bound, which can be viewed as an extension of an approach described earlier in Dufour (2002) and Dufour and Kiviet (1998). First, we reconsider the testing problem within the framework of an appropriate MLR model, namely the MLR setup of which the model on hand is a restricted form. As pointed out above, this setup allows for Gaussian and non-Gaussian error distributions, provided the latter can be simulated. Second, we introduce, in the relevant MLR framework, a "uniform linear (UL) hypothesis" [Berndt and Savin (1977)], which is a special case of the set of restrictions specified by the null hypothesis. The intuition behind this suggestion follows from the fact that exact nuisance-parameter free critical values for the LR criterion for testing the suggested UL hypothesis conveniently bounds the LR statistic for testing the general constraints.

In addition, we propose alternative MC tests [see Dwass (1957), Barnard (1963), Jöckel (1986), or Dufour (2002)] that can be run whenever the bounds tests are not conclusive. We consider: (1) an asymptotically valid procedure that may be interpreted as a parametric bootstrap, and (2) a method that is exact for any sample size, following Dufour (2002). Further, in situations where maximum likelihood (ML) methods may be computationally expensive, we introduce LR-type test criteria based on non-ML estimators. In particular, we consider two-stage statistics or estimators at any step of the process by which the likelihood is maximized iteratively. We emphasize that parametric bootstrap and bounds tests should be viewed as complementary rather than as alternative procedures.

The chapter is organized as follows. In Section 2, we present the model studied and define the test statistics that will be considered. In Section 3, we describe the proposed bounds and MC test procedures. Simulation results are reported in Section 4. Section 5 illustrates the procedures proposed by applying them to test restrictions on a factor demand model. We conclude in Section 6.

2 FRAMEWORK

We consider here a p-equation SURE system of the form:

$$Y_j = X_j\beta_j + u_j, \quad j = 1, \ldots, p \tag{1}$$

where Y_j is a vector of n observations on a dependent variable, X_j is a full-column rank $n \times k_j$ matrix of regressors, $\beta_j = (\beta_{0j}, \beta_{1j}, \ldots, \beta_{k_j-1,j})'$ is a vector of k_j unknown coeffcients, and $u_j = (u_{1j}, u_{2j}, \ldots, u_{nj})'$ is a $n \times 1$ vector of random disturbances. The system [Eq. (1)] can be rewritten in the stacked form:

$$y = X\beta + u \tag{2}$$

where

$$y = \begin{bmatrix} Y_1 \\ Y_2 \\ \vdots \\ Y_p \end{bmatrix}, \quad X = \begin{bmatrix} X_1 & 0 & \cdots & 0 \\ 0 & X_2 & \cdots & 0 \\ \vdots & \vdots & \ddots & \vdots \\ 0 & 0 & \cdots & X_p \end{bmatrix}, \quad \beta = \begin{bmatrix} \beta_1 \\ \beta_2 \\ \vdots \\ \beta_p \end{bmatrix}, \quad u = \begin{bmatrix} u_1 \\ u_2 \\ \vdots \\ u_p \end{bmatrix} \tag{3}$$

so X is a $(np) \times k$ matrix, y and u each have dimension $(np) \times 1$, and β has dimension $k \times 1$, with $k = \sum_{j=1}^{p} k_j$. Let us also set:

$$U = [u_1, u_2, \ldots, u_p] = \begin{bmatrix} U'_{1.} \\ U'_{2.} \\ \vdots \\ U'_{n.} \end{bmatrix} \tag{4}$$

where $U_{t0} = (u_{t1}, u_{t2}, \ldots, u_{tp})'$ is the disturbance vector for the t-th observation.

In the sequel, we shall also use, when required, some or all of the following assumptions and notations:

$$U_{t1} = JW_t, \quad t = 1, \ldots, n \tag{5}$$

where J is a fixed lower triangular $p \times p$ matrix such that

$$\Sigma \equiv JJ' = [\sigma_{ij}]_{i,j=1,\ldots,p} \text{ is nonsingular} \tag{6}$$

$$W_1, \ldots, W_n \text{ are } p \times 1 \text{ random vectors whose joint distribution is completely specified} \tag{7}$$

and

$$u \text{ is independent of } X \tag{8}$$

Assumption (8) is a strict exogeneity assumption, which clearly holds when X is fixed. The assumptions (5)–(7) mean that the disturbance distribution is completely specified up to an unknown linear transformation that can modify the scaling and dependence properties of the disturbances in the different equations. Note that Eqs. (5)–(7) do not necessarily entail that

Σ is the covariance matrix of $U_{t\cdot}$, because the distribution of $W_1, \ldots, W_n$ is not restricted (e.g., it may not have finite second moments). However, if we make the additional assumption that

$$\begin{aligned} &W_1, \ldots, W_n \text{ are uncorrelated, with} \\ &E(W_t) = 0, \quad E(W_t W_t') = I_p, \quad t = 1, \ldots, n \end{aligned} \tag{9}$$

or the stronger assumption:

$$W_1, \ldots, W_n \overset{i.i.d.}{\sim} N[0, I_p] \tag{10}$$

we have $E(U_{t\cdot} U_{t\cdot}') = \Sigma, t = 1, \ldots, n$, and

$$E(uu') = \Sigma \otimes I_p \tag{11}$$

Assumption (10) yields the Gaussian SURE model. For further reference, we shall write

$$W = [W_1, \ldots, W_n]' = UJ^{-1} \tag{12}$$

In this chapter, we consider the problem of testing general hypotheses of the form:

$$H_0 : A\beta \in \Delta_0 \tag{13}$$

where A is a full row-rank $\nu_0 \times k$ matrix and Δ_0 is a nonempty subset of $\mathbb{R}^{\nu_0}$.

For our subsequent arguments, it will be important to spell out the relation between SURE and MLR models. The MLR model may be defined as a SURE model where the regressors in all the equations are the same ($X_1 = X_2 = \cdots = X_p$). Conversely, a SURE model can be viewed as a restricted MLR system. To be more specific, for each $1 \leq j \leq p$ in the context of Eq. (1), let $\bar{X}_j$ be any matrix such that the columns of $[X_j, \bar{X}_j]$ are linearly independent and span the same space as the columns of the matrix $[X_1, X_2, \ldots, X_p]$. In most practical situations, $\bar{X}_j$ will simply contain the regressors from the matrices $X_k, k \neq j$, which are excluded from the j-th equation. Further, let X_* be any full-column rank $n \times k_*$ matrix whose columns span the same space as those of $[X_1, X_2, \ldots, X_p]$, i.e.,

$$\mathrm{sp}(X_*) = \mathrm{sp}([X_1, X_2, \ldots, X_p]), \quad \det(X_*' X_*) \neq 0, \quad \mathrm{rank}(X_*) = k_* \tag{14}$$

where, for any matrix Z, sp(Z) represents the vector space spanned by the columns of Z. Then, for each j, we can find matrices Sj and $\bar{S}_j$ of dimensions $k_* \times k_j$ and $k_* \times (k_* - k_j)$, respectively, such that $X_* S_j = X_j$, $X_* \bar{S}_j = \bar{X}_j$ and the matrix $T_j = [S_j, \bar{S}_j]$ is invertible. Consequently, Eq. (1) may be rewritten as

$$Y_j = X_* S_j \beta_j + X_* \bar{S}_j \bar{\beta}_j + u_j = X_* \beta_{*j} + u_j, \quad j = 1, \ldots, p \tag{15}$$

where $\beta_{*j} = S_j\beta_j + \bar{S}_j\bar{\beta}_j = T_j(\beta_j', \bar{\beta}_j')'$, with the restrictions:

$$\bar{\beta}_j = 0, j = 1, \ldots, p \tag{16}$$

The latter restrictions may also be expressed in implicit form on β_{*j}, $j = 1, \ldots, p$ as

$$M(S_j)\beta_{*j} = 0, \quad j = 1, \ldots, p \tag{17}$$

where $M(S_j) = [I_{k_*} - S_j(S_j'S_j)^{-1}S_j']$. Clearly Eqs. (15) and (16) define a constrained MLR model. On relaxing the SURE restrictions, Eq. (17), this MLR model can be put in the stacked form:

$$y = (I_p \otimes X_*)\beta_* + u = \bar{X}_*\beta_* + u \tag{18}$$

where $\bar{X}_* = I_p \otimes X_*$ and $\beta_* = (\beta_{*1}', \beta_{*2}', \ldots, \beta_{*p}')'$, or equivalently,

$$Y = X_*B_* + U \tag{19}$$

where $Y = [Y_1, Y_2, \ldots, Y_p]$ and $B_* = [\beta_{*1}, \beta_{*2}, \ldots, \beta_{*p}]$. We shall call the unrestricted MLR model Eq. (18) [or (19)], an *embedding* MLR model for the SURE model, Eq. (1). It is clear any hypothesis on β can be expressed equivalently in terms of β_* within the corresponding embedding MLR model.

In this chapter, we shall emphasize LR-type tests of H_0 derived under the Gaussian distributional assumptions, Eqs. (5)–(10). In this case, the log-likelihood function associated with the SURE model [Eq. (2)] has the form:

$$\mathcal{L}(\beta, \Sigma) = -\frac{np}{2}\ln(2\pi) - \frac{n}{2}\ln(|\Sigma|) - \frac{1}{2}(y - X\beta)'(I_n \otimes \Sigma)^{-1}(y - X\beta) \tag{20}$$

Then, provided that the relevant optima do exist and are unique, the "unconstrained" maximized value of $\mathcal{L}(\beta, \Sigma)$ can be written as

$$\begin{aligned} L(H_S) &= \sup\{\mathcal{L}(\beta, \Sigma): \beta \in \mathbb{R}^k \text{ and } \Sigma \text{ is p.d.}\} \\ &= -\frac{np}{2}[\ln(2\pi) + 1] - \frac{n}{2}\ln(|\hat{\Sigma}_S|) \end{aligned} \tag{21}$$

while its constrained maximized value subject to H_0 is

$$\begin{aligned} L(H_0) &= \sup\{\mathcal{L}(\beta, \Sigma): A\beta \in \Delta_0 \text{ and } \Sigma \text{ is p.d.}\} \\ &= -\frac{np}{2}[\ln(2\pi) + 1] - \frac{n}{2}\ln(|\hat{\Sigma}_0|) \end{aligned} \tag{22}$$

where $\hat{\Sigma}_0$ and $\hat{\Sigma}_S$ are the restricted and unrestricted ML estimates of Σ, assuming that Σ is positive definite (p.d.). Thus, the Gaussian LR statistic for testing H_0 against the unrestricted SURE model [or, equivalently, against the embedding MLR model, Eq. (18), with the restrictions, Eq. (17)] is given by:

$$LR(H_0) = 2[L(H_S) - L(H_0)] = n \ln(\Lambda_S), \quad \Lambda_s = |\hat{\Sigma}_0|/|\hat{\Sigma}_S| \tag{23}$$

Note that the exclusion SURE restrictions are imposed under both the null and the alternative hypotheses. In the statistics literature, Λ_S^{-1} is known as the Wilks criterion.

Similarly, the log-likelihood function associated with the MLR model, Eq. (18), taken jointly with the Gaussian distributional assumptions, Eqs. (5)–(10), is

$$\begin{aligned}\mathcal{L}_*(\beta_*, \Sigma) &= -\frac{np}{2}\ln(2\pi) - \frac{n}{2}\ln(|\Sigma|) - \frac{1}{2}(y - \bar{X}_*\beta_*)'(\text{In} \otimes \Sigma)^{-1}(y - \bar{X}_*\beta_*) \\ &= -\frac{np}{2}\ln(2\pi) - \frac{n}{2}\ln(|\Sigma|) - \frac{1}{2}\text{tr}[\Sigma(Y - X_*B_*)(Y - X_*B_*)']\end{aligned} \tag{24}$$

where $\beta_* = \text{vec}(B_*)$. It is clear that

$$\mathcal{L}_*(\beta_*, \Sigma) = \mathcal{L}(\beta, \Sigma), \text{ when } \bar{\beta}_j = 0, \quad j = 1, \ldots, p \tag{25}$$

Let

$$\hat{B}_* = (X_*'X_*)^{-1}X_*'Y, \quad \hat{U} = Y - X_*\hat{B}_* = M(X_*)U, \quad \hat{\Sigma}_M = \frac{1}{n}\hat{U}'\hat{U} \tag{26}$$

where $M(X_*) \equiv I_n - X_*(X_*'X_*)X_*'$. Then, provided that $\hat{U}$ has full-column rank (which requires $n \geq k_* + p$), $\mathcal{L}_*(\beta_*, \Sigma)$ attains a unique (unconstrained) finite maximum at $B_* = \hat{B}_*$ and $\Sigma = \hat{\Sigma}_M$ [see Anderson (1984, Chap. 3)], yielding the maximal value:

$$\begin{aligned}L(H_M) &= \sup\{\mathcal{L}_*(\beta_*, \Sigma) : \beta_* \in \mathbb{R}^{k*} \text{ and } \Sigma \text{ is p.d.}\} \\ &= -\frac{np}{2}[\ln(2\pi) + 1] - \frac{n}{2}\ln(|\hat{\Sigma}_M|)\end{aligned} \tag{27}$$

Thus, $\hat{B}_*$ and $\hat{\Sigma}_M$ are the ML estimators of the parameters of the unrestricted embedding MLR model associated with the SURE model, Eq. (2). Given the assumptions (5)–(10), a necessary and sufficient condition for $\hat{U}$ to have full-column rank with probability one is

$$P[\text{rank}[M(X_*)W] = p] = 1$$

The latter will hold, e.g., if rank $(X_*) = k_* \leq n - p$ and vec(W) follows an absolutely continuous distribution on $\mathbb{R}^{np}$. In view of the relation between the SURE model and an embedding MLR model, we shall also consider the LR statistic for testing H_0 against the completely unrestricted MLR model [Eq. (18) without the SURE restrictions, Eq. (17)]:

$$LR_{\mathcal{M}}(H_0) = 2[L(H_{\mathcal{M}}) - L(H_0)] = n\ln(\Lambda_{\mathcal{M}}), \quad \Lambda_{\mathcal{M}} = |\hat{\Sigma}_0|/|\hat{\Sigma}_{\mathcal{M}}| \tag{28}$$

3 TEST PROCEDURES

We will now show how one can obtain finite-sample LR-based tests in the context of SURE models as defined above. For that purpose, we shall exploit special features of so-called *uniform linear* restrictions for which LR test statistics have nuisance-parameter null distributions in the context of MLR models (which entails that they are pivotal statistics under the null hypothesis). In the MLR case (where $X_1 = \cdots = X_p \equiv X_*$ and $k_1 = \cdots = k_p \equiv k_*$), the hypothesis H_0 in Eq. (13) is uniform linear if it can be expressed in the form:

$$H_{\mathcal{UL}} : (C' \otimes R)\beta = \text{vec}(D_0) \tag{29}$$

or, equivalently,

$$H_{\mathcal{UL}} : RBC = D_0 \tag{30}$$

where B is the $k_* \times p$ matrix such that $\beta = \text{vec}(B)$, R is a known $r \times k_*$ matrix of rank r, C is a known $p \times c$ matrix of rank c, and D_0 is a known $r \times c$ matrix. In Dufour and Khalaf (2002), it is shown that the null distribution of the Gaussian LR statistic [derived under the assumptions (5)–(10)] for testing $H_{\mathcal{UL}}$ (against the unrestricted MLR model) does not involve any nuisance parameter under the weaker assumptions (1)–(8) — which allow for non-normal disturbances — and may easily be simulated. In particular, the parameters of the covariance matrix $\Sigma = JJ'$ do not appear in the distribution. Beyond this specific hypothesis class, it is well known that the LR statistic is not pivotal, even if the null hypothesis is linear. For further discussion of uniform linear hypotheses in MLR models, the reader may consult Berndt and Savin (1977), Stewart (1997), and Dufour and Khalaf (2002).

Let us now turn to the SURE model. In the context of the embedding MLR model, Eq. (18), the null hypothesis H_0 in Eq. (13) is equivalent to the conjunction of $A\beta \in \Delta_0$ with the SURE restrictions, Eq. (16):

$$H_0^*: AF\beta_* \in \Delta_0 \text{ and } M(S_j)\beta_{*j} = 0, \qquad j = 1, \ldots, p \tag{31}$$

where $\beta_*, S_j, M(S_j)$, and k_* are defined as in Eqs. (14)–(18), and

$$F = \begin{bmatrix} (S_1'S_1)^{-1}S_1' & 0 & \cdots & 0 \\ 0 & (S_2'S_2)^{-1}S_2' & \cdots & 0 \\ \vdots & \vdots & \ddots & \vdots \\ 0 & 0 & \cdots & (S_p'S_p)^{-1}S_p' \end{bmatrix} \tag{32}$$

It is clear we can find a full row-rank matrix A_* of dimension $\nu_{0*} \times (pk_*)$ such that H_0^* can be re-expressed in terms of β_* according to a form similar to H_0 in Eq. (13):

$$H_0^* : A_*\beta_* \in \Delta_{0*} \tag{33}$$

where $\nu_{0*} \geq \nu_0$ and Δ_{0*} is a nonempty subset of $\mathbb{R}^{\nu_{0*}}$.

We now state our main result on the distribution of LR statistics in SURE models.

Theorem 1 Bound on LR statistics in Sure models. *Suppose the assumptions* (1)–(8) *hold, with*

$$P[\text{rank}[M(X_*)W] = p] = 1 \tag{34}$$

where X_* is defined as in Eq. (14), *and let $H_{\mathcal{UL}}^* : RB_*C = D_0$ be a set of uniform linear restrictions on Eq.* (19) *such that $H_{\mathcal{UL}}^*$ entails H_0^*, where the matrices R, B_*, C, D_0, and the hypothesis H_0^* are defined as in Eqs.* (30) *and* (31). *The following inequalities then hold:*

$$\Lambda_s \leq \Lambda_{\mathcal{M}} \leq \Lambda_{\mathcal{UL}} \tag{35}$$

where Λ_s and $\Lambda_{\mathcal{M}}$ are defined as in Eqs. (23) *and* (28), $\Lambda_{\mathcal{UL}} = |\hat{\Sigma}_{\mathcal{UL}}|/|\hat{\Sigma}_{\mathcal{M}}|$ *and $\hat{\Sigma}_{\mathcal{UL}}$ is a ML estimator of Σ obtained under the uniform linear restrictions $H_{\mathcal{UL}}^*$. Furthermore, under $H_{\mathcal{UL}}^*$, the distribution of $\Lambda_{\mathcal{UL}}$ (conditional on X) does not depend on the unknown parameter matrices B and Σ nor on the values of the constants in D_0.*

Proof. The proof is based on observing that the hypotheses involved in the definitions of the statistics Λ_S, $\Lambda_{\mathcal{M}}$, and $\Lambda_{\mathcal{UL}}$ can be viewed as special cases of the embedding MLR model, Eq. (19). First, we note that, under the assumptions (1)–(8) and (34), the log-likelihood $\mathcal{L}_*(\beta_*, \Sigma)$ has (with probability one) a unique maximum given by $L(H_{\mathcal{M}})$ in Eq. (27). This entails that the supremum of $\mathcal{L}_*(\beta_*, \Sigma)$ under any set of restrictions on β_* must be finite. In particular, the supremum (with respect to β_* and Σ) of $\mathcal{L}_*(\beta_*, \Sigma)$ under $H_{\mathcal{UL}}^*$ can be written

$$\begin{aligned} L(H_{\mathcal{UL}}^*) &= \sup\{\mathcal{L}_*(\beta_*, \Sigma): RB_*C = D_0 \text{ and } \Sigma \text{ is p.d.}\} \\ &= -\frac{np}{2}[\ln(2\pi) + 1] - \frac{n}{2}\ln(|\hat{\Sigma}_{\mathcal{UL}}|) \end{aligned} \tag{36}$$

where $\hat{\Sigma}_{\mathcal{UL}}$ is the ML estimator of Σ under $H^*_{\mathcal{UL}}$. Thus, the Gaussian LR statistic for testing $H^*_{\mathcal{UL}}$ against the embedding MLR model [i.e., Eq. (18) jointly with Eqs. (5)–(10)] is

$$LR_{\mathcal{M}}(H^*_{\mathcal{UL}}) = 2[L(H_{\mathcal{M}}) - L(H^*_{\mathcal{UL}})] = n \ln(\Lambda_{\mathcal{UL}}),$$
$$\Lambda_{\mathcal{UL}} = |\hat{\Sigma}_{\mathcal{UL}}|/|\hat{\Sigma}_{\mathcal{M}}| \tag{37}$$

From Theorem 3.1 in Dufour and Khalaf (2002), the exact distribution of $\Lambda_{\mathcal{UL}}$ under $H^*_{\mathcal{UL}}$ only depends on the distribution of W and the known matrices X, R, and C, but not on D_0 nor on the otherwise unknown parameters in B_* and Σ.

Now, by the definition of the embedding MLR model, Eq. (25), and the equivalence between H_0 and H^*_0, we see that $L(H_S)$ and $L(H_0)$ in Eqs. (21)–(22) can also be expressed in terms of $\mathcal{L}_*(\beta_*, \Sigma)$:

$$L(H_S) = \sup\{\mathcal{L}_*(\beta_*, \Sigma)\colon M(S_j)\beta_{*j} = 0, \quad j = 1, \ldots, p, \text{ and } \Sigma \text{ is p.d.}\} \tag{38}$$

$$L(H_0) = \sup\{\mathcal{L}_*(\beta_*, \Sigma)\colon \beta_* \text{ satisfies } H^*_0 \text{ and } \Sigma \text{ is p.d.}\} \tag{39}$$

Since $H^*_{\mathcal{UL}}$ entails H^*_0, which in turn is a restricted form of the SURE model (H_S), and since H_S can be obtained by imposing linear restrictions on the embedding MLR model ($H_{\mathcal{M}}$), it follows that

$$L(H^*_{\mathcal{UL}}) \le L(H_0) \le L(H_S) \le L(H_{\mathcal{M}}) \tag{40}$$

hence

$$L(H_S) - L(H_0) \le L(H_{\mathcal{M}}) - L(H_0) \le L(H_{\mathcal{M}}) - L(H^*_{\mathcal{UL}}) \tag{41}$$

and

$$\Lambda_S \le \Lambda_{\mathcal{M}} \le \Lambda_{\mathcal{UL}} \tag{42}$$

This completes the proof of the theorem.

It follows from the latter theorem that

$$P[\Lambda_S \ge \lambda_{\mathcal{UL}}(\alpha)] \le P[\Lambda_{\mathcal{M}} \ge \lambda_{\mathcal{UL}}(\alpha)] \le \alpha \tag{43}$$

under H_0, where $\lambda_{\mathcal{UL}}(\alpha)$ is determined such that $P[\Lambda_{\mathcal{UL}} \ge \lambda_{\mathcal{UL}}(\alpha)] = \alpha$ (or, at least, $P[\Lambda_{\mathcal{UL}} \ge \lambda_{\mathcal{UL}}(\alpha)] \le \alpha$) and $0 \le \alpha \le 1$. It is important to note here that the inequality (35) holds for *any* set of uniform linear restrictions which entails H_0. In particular, on taking $R = I_{k_*}$, $C = I$, and $D_0 = B_*$ (the true value of B_*), it is clear such a set of restrictions does always exist, although other choices (for R, C, and D_0) may be available in view of the form of H_0 and lead to a tighter bound. Further, the distribution of $\Lambda_{\mathcal{UL}}$ (when $RB_*C = D_0$) only depends on X, R, and C, but not on D_0, so one can use any possible value of D_0 (such as $D_0 = 0$) in order to compute or simulate this distribution.

Using well known results from Anderson (1984) and Rao (1973) for Gaussian MLR models, it is also possible to show that the bounding statistic $\Lambda_{\mathcal{UL}}$ is distributed like the product of the inverse of p beta variables with degrees of freedom that depend only on r, c, p, and k_* [see Dufour (1997) and Dufour and Khalaf (2002)]. The latter result is, however, hardly useful for practical applications of the proposed bound. Hence, we do not restate our conclusions here for this specific Gaussian case: it is more convenient to derive $\lambda_{\mathcal{UL}}(\alpha)$ by simulation as shown below and using Theorem 1, under any distributional assumptions that satisfy Eq. (5) including the normal case. Finally, we note that the same bound applies to both criteria Λ_s and $\Lambda_{\mathcal{M}}$. Since $\Lambda_{\mathcal{M}} \geq \Lambda_{\mathcal{S}}$, it will thus be preferable to apply the bound to $\Lambda_{\mathcal{M}}$ rather than $\Lambda_{\mathcal{S}}$, since this will yield a more powerful test.

Theorem 1 has further implications for LR-based hypothesis tests. The fact that the null distribution of the LR statistic can be bounded (in a nontrivial way) by a statistic whose distribution can be simulated fairly easily entails that MC test techniques may be used to obtain finite-sample p-values based on the LR-based statistics when the bounds test is not conclusive. In earlier work, we have discussed in detail how such procedures can be implemented; see Dufour and Kiviet (1996, 1998), Dufour (2002), Dufour et al. (1998), and Dufour and Khalaf (2001, 2002). These include techniques for the construction of: (1) a (parametric) bootstrap-type p-value which we denote a *local* MC p-value (LMC) to account for the fact that the underlying simulation routine is implemented given a specific nuisance parameter estimate, and (2) an exact randomized p-value that corresponds to the largest MC p-value over the relevant nuisance parameter space; conformably, we call the latter a *maximized* Monte Carlo (MMC) p-value [Dufour (2002)]. Both procedures are summarized below. Although the LMC p-value is only valid asymptotically, nonrejections are conclusive from a finite-sample perspective, in the following sense. Indeed, for all $0 \leq \alpha \leq 1$, if the LMC p-value exceeds α, we can be sure that the maximum p-value also exceeds α. We emphasize the fact that the MMC test can be implemented in complementarity with the above defined bounds tests. Indeed, if the bounds test rejects the null then the MMC test is certainly significant. For a more detailed discussion of the justification and implementation of such simulation-based procedures, we refer the reader to the papers just cited.

To illustrate how the above results may be used in the context of a SURE model, we will now discuss an illustrative example.

Example 1 THREE-EQUATION SURE MODEL. In the SURE model, Eq. (1), with Gaussian errors and $p=3$, $k_i=2$, $X_i=[\iota_n, x_i]$ where ι_n denotes a vector of n 1's, consider the problem of testing

$$H_0: \beta_{11} = \beta_{22} = \beta_{33} \tag{44}$$

We suppose also that the matrix $X_* = [\iota_n, x_1, x_2, x_3]$ has full-column rank $k_* = 4 \leq n-3$. Then, it is easy to see that this problem is equivalent to testing

$$H_0^*: b_{11} = b_{22} = b_{33} \quad \text{and} \quad b_{12} = b_{13} = b_{21} = b_{23} = b_{31} = b_{32} = 0$$

in the framework of the MLR model:

$$Y = X_* B_* + U \tag{45}$$

with $Y=[Y_1, Y_2, Y_3]$, $U=[U_1, U_2, U_3]$, $B_* = [b_1, b_2, b_3]$, and $b_j = (b_{0j}, b_{1j}, b_{2j}, b_{3j})$, $j=1, 2, 3$. In order to use the above results on the conservative bound, we need to construct a set of UL restrictions on the coefficients of the later MLR model that satisfy the hypothesis in question. It is easy to see that constraints setting the values of the coefficients $b_{ij}, i, j = 1, \ldots, 3$ to specific values (9 restrictions on 12 coefficients) meet this purpose:

$$H_{\mathcal{UL}}^*: \begin{bmatrix} 0 & 1 & 0 & 0 \\ 0 & 0 & 1 & 0 \\ 0 & 0 & 0 & 1 \end{bmatrix} B_* = D_0 \tag{46}$$

All that remains is to calculate the LR, as defined in Eq. (23), and use the critical value associated with the uniform linear restriction, Eq. (46).

A bounds MC test may then be applied as described in Dufour and Khalaf (2002) for testing general possibly nonlinear restrictions in MLR models. The procedure can be described as follows for the example just considered. For further reference, and conformably with Dufour and Khalaf (2002), we call the latter bounds test, a BMC test.

1. Denote $\Lambda^{(0)}$ the observed test statistic, which could be the statistic $\Lambda_{\mathcal{S}}$ in Eq. (23) or $\Lambda_{\mathcal{M}}$ in Eq. (28).
2. By MC methods, draw N simulated samples (conditional on the right-hand-side regressors) from model (18) in $H_{\mathcal{UL}}^*$. For instance, in the above example, one way to do this is to draw the simulated samples from the base model, Eq. (45), with parameters set to their constrained (imposing H_0^*) SURE estimates on which the additional restrictions underlying $H_{\mathcal{UL}}^*$ have been imposed (simply setting to fixed values the relevant coefficients of B_*). To be more specific, let us denote by $\tilde{b}_{ij}$ the SURE estimate of β_{ij}, $0 \leq i \leq 3, 1 \leq j \leq 3$, several of which should be zero to account for the SURE exclusion restrictions. Then, if we choose to draw

from model (45) with coefficients $\tilde{b}_{ij}, i = 0, \ldots, 3,\ j = 1, \ldots, 3$ and the conformable covariance matrix estimate, then $H^*_{\mathcal{UL}}$ could be of the form: $b_{ij} = \tilde{b}_{ij}, i, j = 1, \ldots, 3$.* Furthermore, any error distribution that satisfies Eqs. (5)–(8) may be considered.

3. From each simulated sample, compute the bounding statistic $\Lambda_{\mathcal{UL}}$, which corresponds to the LR-based statistic associated with $H^*_{\mathcal{UL}}$: this yields $\Lambda^{(h)}_{\mathcal{UL}}, h = 1, \ldots, N$. As emphasized above, it is important to make sure that the regression coefficients selected to generate the MC drawings correspond to the restrictions implied by $H^*_{\mathcal{UL}}$. In other words, the simulated values of the bounding statistic should satisfy the null hypothesis.
4. Compute the simulated p-value $\hat{p}_N(\Lambda^{(0)})$ where

$$\hat{p}_N(x) = \left\{ 1 + \sum_{h=1}^{N} I[\Lambda^{(h)}_{\mathcal{UL}} - x] \right\} / (N+1) \tag{47}$$

$I[z] = 1$ if $z \geq 0$ and $I[z] = 0$ if $z < 0$.

5. The procedure rejects at level α *if* $\hat{p}_N(\Lambda^{(0)}) \leq \alpha$.

Using the same arguments as in Dufour and Kiviet (1996, 1998), Dufour (2002), and Dufour and Khalaf (2002), it is easy to see that

$$P[\hat{p}_N(\Lambda^{(0)}) \leq \alpha] \leq \alpha \text{ under } H_0$$

so the critical region $\hat{p}_N(\Lambda^{(0)}) \leq \alpha$ has level α. If the above procedure is implemented replacing $\Lambda^{(h)}_{\mathcal{UL}}, j = 1, \ldots, N$ with $\hat{\Lambda}^{(h)}, h = 1, \ldots, N$, which refer to realized values of the LR criterion associated with H^*_0 and the simulated samples, then Eq. (47) yields a parametric bootstrap or an LMC p-value. In this case, the p-value in question depends on the choice of the intervening nuisance parameters. The MMC p-value corresponds to the largest MC p-value overall nuisance parameters compatible with the null hypothesis. A global optimizing algorithm is required to maximize the MC p-value. In this chapter, we used simulated annealing [see Goffe et al. (1994)]. The reader may consult Dufour and Kiviet (1996, 1998), Dufour (2002), Dufour et al. (1998), and Dufour and Khalaf (2001) for further discussion of MC tests in econometrics.

---

*It is also possible, and perhaps more efficient, to rewrite the bounding LR statistic as a pivotal quantity, using the results in Dufour and Khalaf (2002). The procedure just presented exploits the pivotal property of the statistic *implicitly*. Yet it is quite intuitive and relates to the familiar parametric bootstrap.

4 SIMULATION STUDY

In this section, we present simulation results illustrating the performance of the above proposed procedures along with the one of more traditional asymptotically justified methods. In particular, we examine the performance of BMC and MMC tests in SURE contexts. In the linear case, we also consider LMC tests based on standard Wald-type criteria and several alternative statistics justified on the basis of computational cost as opposed to those relying on full maximum likelihood estimation.

4.1 Design

We studied two Gaussian SURE designs, similar to that of Example 1. In the first one (D1), we considered the problem of testing linear cross-equation constraints, while in the second one (D2), we studied a nonlinear constraint.

D1. *SURE system, cross-equation constraints*

Model (1); $k_j = 2, \quad j = 1, \ldots, p; \; p = 3, 5; \; n = 25;$

$$H_0: \beta_{jj} = \beta_{11}, \quad j = 2, \ldots, p.$$

D2. *SURE system, non-linear constraints*

Model (1); $k_j = 3, \quad j = 1, \ldots, p; \; p = 7; \; n = 25;$

the regressor which corresponds to β_{1j} is common to all equations;

$$H_o: \beta_{1j} = \gamma\beta_{2j}, \quad j = 1, \ldots, p, \; \gamma \text{ unknown.}$$

For each model, a constant regressor was included and the other regressors were independently drawn (once) from a normal distribution; the errors were independently generated as *i.i.d.* $N(0, \Sigma)$ with $\Sigma = JJ'$ and the elements of J drawn (once) from a normal distribution. The coefficients are reported in Table 1.

The statistics examined for D1 include the relevant LR criteria defined by Eqs. (23) and (28), as well as three other types of statistics: (1) quasi-LR statistics based on incompletely maximized likelihood functions;

TABLE 1 Coefficient Values used in the Simulation Experiments

D1	$\beta_{(3EQ)} = (1.2, 0.1, 0.8, 0.1, -1.1, 0.1)'$
	$\beta_{(5EQ)} = (1.2, 0.1, 0.8, 0.1, -1.1, 0.1, 1.9, 0.1, -0.2, 0.1)'$
D2	$\gamma = 0.009$ and $\beta_{0j}, \beta_{2j}, j = 1 \ldots, p$, drawn (once) as *i.i.d.* $N(0, 0.16)$

(2) test statistics similar to those suggested by Theil et al. (1985), and (3) a number of Wald-type criteria. To be more precise, the latter are defined as follows:

1. The quasi-LR (QLR) statistics are

$$QLR_{(l)} = n\ln(\Lambda_{(l)}),\ \Lambda_{(l)} = |\tilde{\Sigma}_{0(l)}|/|\tilde{\Sigma}_{(l)}| \tag{48}$$

   where $\tilde{\Sigma}_{0(l)}$ and $\tilde{\Sigma}_{(l)}$ denote the constrained and unconstrained iterative estimators of Σ and the subscript l refers to the number of iterations involved. Though we did not analytically establish the asymptotic distribution of the latter criteria, we assessed their asymptotic significance using the χ^2 reference distribution for the usual LR statistic. We append the subscript *LMC* to the notation for the QLR test to refer to the corresponding LMC test.
2. The test statistics suggested by Theil et al. (1985) may be interpreted as unscaled Wald-type statistics, whose level is controlled by a MC (or bootstrap) method. We consider these here mainly for historical reasons, because they are really the first simulation-based test procedures proposed in the SURE setup. For the model with three equations, we considered:

$$\mu_{31} = |\hat{\beta}_{11} - \hat{\beta}_{22}| + |\hat{\beta}_{22} - \hat{\beta}_{33}|$$
$$\mu_{32} = |\hat{\beta}_{11} - \hat{\beta}_{33}| + |\hat{\beta}_{22} - \hat{\beta}_{33}|$$
$$\mu_{33} = |\hat{\beta}_{11} - \hat{\beta}_{22}| + |\hat{\beta}_{11} - \hat{\beta}_{33}|$$

   In the five-equation case, the following were selected among many possible choices:

$$\mu_{51} = |\hat{\beta}_{11} - \hat{\beta}_{22}| + |\hat{\beta}_{22} - \hat{\beta}_{33}| + |\hat{\beta}_{33} - \hat{\beta}_{44}| + |\hat{\beta}_{44} - \hat{\beta}_{55}|$$
$$\mu_{52} = |\hat{\beta}_{22} - \hat{\beta}_{33}| + |\hat{\beta}_{33} - \hat{\beta}_{44}| + |\hat{\beta}_{44} - \hat{\beta}_{55}| + |\hat{\beta}_{55} - \hat{\beta}_{11}|$$
$$\mu_{53} = |\hat{\beta}_{33} - \hat{\beta}_{44}| + |\hat{\beta}_{44} - \hat{\beta}_{55}| + |\hat{\beta}_{55} - \hat{\beta}_{11}| + |\hat{\beta}_{11} - \hat{\beta}_{22}|$$
$$\mu_{54} = |\hat{\beta}_{44} - \hat{\beta}_{55}| + |\hat{\beta}_{55} - \hat{\beta}_{11}| + |\hat{\beta}_{11} - \hat{\beta}_{22}| + |\hat{\beta}_{22} - \hat{\beta}_{33}|$$
$$\mu_{55} = |\hat{\beta}_{55} - \hat{\beta}_{11}| + |\hat{\beta}_{11} - \hat{\beta}_{22}| + |\hat{\beta}_{22} - \hat{\beta}_{33}| + |\hat{\beta}_{33} - \hat{\beta}_{44}|$$

   For the purpose of this experiment, we used for $\hat{\beta}$ the Gaussian ML estimator of β.
3. The Wald-type criteria are based on feasible generalized least squares (GLS) parameter estimates. Specifically, we considered the statistic suggested in Srivastava and Giles (1987, Chap. 10) for

an hypothesis of the form $A\beta - r = 0$:

$$\mathcal{W} = \left(\frac{\nu_1}{\nu_0}\right) \frac{(A\hat{\beta} - r)'[A(X'(S^{-1} \otimes I_n)X)^{-1}A']^{-1}(A\hat{\beta} - r)}{(y - X\hat{\beta})'(S^{-1} \otimes I_n)(y - X\hat{\beta})} \tag{49}$$

where A is a $\nu_0 \times k$ full-row rank fixed matrix, $\nu_1 = np - k$, $k = \sum_{j=1}^{p} k_j$, while S and $\hat{\beta}$ are feasible generalized least squares parameter estimates.* Under the null hypothesis (and standard regularity conditions), $\nu_0 \mathcal{W}$ has a $\chi^2(\nu_0)$ asymptotic distribution. Theil (1971, Chap. 6) suggests that the $F(\nu_0, \nu_1)$ distribution better captures the finite-sample distribution of the statistics. Yet this claim is not supported by either analytical or simulation evidence. Maximum likelihood estimators may also be substituted for $\hat{\beta}$ and S in the formulas for the Wald criterion. Here, we have considered both the standard feasible estimator of β [using the estimate of S based on least squares residuals applied to each one of the regressions in Eq. (1)] as well as the ML estimators (iterated to convergence) of β and Σ. For further reference, the GLS and ML based $\mathcal{W}$ tests will be denoted $\mathcal{W}[GLS, j]$ and $\mathcal{W}[ML, j]$, respectively, where $j \in \{\chi^2, F\}$ indicates whether the critical value was obtained from the χ^2 asymptotic distribution or from the $F(\nu_0, \nu_1)$ distribution. The local MC (parametric bootstrap) counterparts will be denoted $\mathcal{W}[GLS, LMC]$ and $\mathcal{W}[ML, LMC]$.

In both D1 and D2, we computed empirical frequencies of type I errors, based on a nominal size of 5% and 1000 replications. In D1, the powers of the tests were investigated by simulating the model with the same parameter values except for β_{11} respectively.† The LMC and BMC tests were applied with 19 and 99 replications. Because of the computational cost involved, the MMC test was only applied with $p = 3$ and $N = 19$. The BMC test was performed based on the bounding statistic as described in Example 1. For each test statistic, the *LMC* randomized procedure was based on simulations that use a restricted estimator similar to the estimator(s) involved in the corresponding test statistic: a restricted ML (or quasi-ML) estimator for LR or Wald-type tests based on ML (quasi-ML) estimators, and restricted feasible GLS estimators for tests based on GLS estimators. All the experiments were conducted using Gauss-386i VM version 3.1.

---

*The statistic $\mathcal{W}$ above corresponds to the z statistic in Eq. (10.11) of Srivastava and Giles (1987, Chap. 10).

†For the purpose of power comparisons, the asymptotic tests were size corrected using an independent simulation.

4.2 Results and Discussion

The results of the limited size study in D2 reveal the following: the observed empirical frequency of type I errors for the LR statistic was 12.5% whereas the bounds test (2.6%) satisfied the 5% level constraint. The results of experiment D1 are summarized in Tables 2–5. The subscripts *asy*, *BMC*, *LMC*, and *MMC*, which appear in these tables, refer, respectively, to the standard asymptotic tests, MC bounds tests, local MC tests (parametric bootstrap), and maximized MC tests. *LR*[*asy*],

TABLE 2 Empirical Levels of Various Tests: Experiment D1

Asymptotic tests			MC tests					
Test	3EQ	5EQ	Test	3EQ	5EQ	Test	3EQ	5EQ
$\mathcal{W}[GLS, \chi^2]$	0.061	0.130	$\mathcal{W}[GLS_{LMC}]$	0.049	0.047	μ_{31}	0.058	-
$\mathcal{W}[ML, \chi^2]$	0.124	0.254	$\mathcal{W}[ML_{LMC}]$	0.047	0.049	μ_{32}	0.051	-
$\mathcal{W}[GLS, F]$	0.052	0.121	$LR[LMC]$	0.047	0.043	μ_{33}	0.055	-
$\mathcal{W}[ML, F]$	0.111	0.242	$LR[MMC]$	0.038		μ_{51}	-	0.027
$LR[asy]$	0.094	0.143	$LR_{\mathcal{M}}[MMC]$	0.036		μ_{52}	-	0.026
$QLR_{(0)}$	0.068	0.077	$LR[BMC]$	0.036	0.029	μ_{53}	-	0.025
$QLR_{(1)}$	0.088	0.131	$QLR_{(0)}[LMC]$	0.045	0.052	μ_{54}	-	0.011
$QLR_{(2)}$	0.094	0.143	$QLR_{(1)}[LMC]$	0.048	0.052	μ_{55}	-	0.025
			$QLR_{(2)}[LMC]$	0.047	0.044			

TABLE 3 Power of the Bounds Tests: Experiment D1 ($H_0 : \beta_{11} = 0.1$)

		Three equations					Five equations				
N	β_{11}	0.3	0.5	0.7	0.9	1	0.3	0.5	0.7	0.9	1.0
19	p_1	0.065	0.383	0.791	0.963	0.987	0.082	0.416	0.792	0.958	0.995
	p_2	0.171	0.324	0.171	0.034	0.013	0.249	0.497	0.207	0.042	0.005
	p_3	0.030	0.021	0.008	0.000	0.000	0.038	0.011	0.00	0.000	0.000
	p_4	0.734	0.272	0.030	0.003	0.000	0.631	0.076	0.001	0.000	0.000
99	p_1	0.077	0.434	0.858	0.986	0.999	0.075	0.474	0.877	0.990	1.000
	p_2	0.204	0.372	0.127	0.014	0.001	0.256	0.439	0.122	0.010	0.000
	p_3	0.022	0.007	0.003	0.000	0.000	0.035	0.010	0.000	0.000	0.000
	p_4	0.697	0.187	0.012	0.000	0.000	0.634	0.077	0.001	0.000	0.000

Note: p_1 is the empirical probability that $LR[LMC]$ and $LR[BMC]$ reject, p_2 measures the probability that $LR[BMC]$ fails to reject and $LR[LMC]$ rejects, p_3 measures the probability that $LR[BMC]$ rejects and $LR[LMC]$ fails to reject, and p_4 is the empirical probability that both tests fail to reject.

TABLE 4 Power of Various Tests: Experiment D1, Three equations ($H_0 : \beta_{11} = 0.1$)

	19 Replications					99 Replications				
β_{11}	0.3	0.5	0.7	0.9	1.0	0.3	0.5	0.7	0.9	1.0
$\mathcal{W}[GLS, \chi^2]$	0.192	0.647	0.939	0.993	0.999	0.192	0.647	0.939	0.993	0.999
$\mathcal{W}[ML, \chi^2]$	0.264	0.787	0.984	1.000	1.000	0.264	0.787	0.984	1.000	1.000
$LR[asy]$	0.281	0.806	0.985	1.000	1.000	0.281	0.806	0.985	1.000	1.000
$\mathcal{W}[GLS, LMC]$	0.185	0.579	0.884	0.974	0.986	0.202	0.640	0.934	0.990	0.998
$\mathcal{W}[ML, LMC]$	0.225	0.704	0.958	0.997	1.000	0.260	0.774	0.985	1.000	1.000
$LR[LMC]$	0.236	0.707	0.962	0.997	1.000	0.262	0.779	0.985	1.000	1.000
$QLR_{(0)}$	0.227	0.689	0.950	0.993	0.988	0.256	0.762	0.977	0.997	0.999
$QLR_{(1)}$	0.238	0.709	0.961	0.997	1.000	0.259	0.776	0.986	1.000	1.000
$QLR_{(2)}$	0.236	0.707	0.962	0.997	1.000	0.262	0.776	0.985	1.000	1.000
$LR_M[MMC]$	0.095	0.404	0.799	0.963	0.987	0.099	0.441	0.861	0.986	0.999
$LR[MMC]$	0.054	0.388	0.804	0.978	0.993	-	-	-	-	-
$LR[BMC]$	0.095	0.404	0.799	0.963	0.987	0.099	0.441	0.861	0.986	0.999
μ_{31}	0.076	0.108	0.148	0.216	0.259	0.064	0.108	0.165	0.219	0.268
μ_{32}	0.197	0.552	0.869	0.974	0.992	0.210	0.641	0.935	0.995	0.998
μ_{33}	0.093	0.183	0.307	0.432	0.489	0.088	0.184	0.328	0.503	0.601

TABLE 5 Power of Various Tests: Experiment D1, Five Equations ($H_0 : \beta_{11} = 0.1$)

	19 Replications					99 Replications				
β_{11}	0.3	0.5	0.7	0.9	1.1	0.3	0.5	0.7	0.9	1.1
$\mathcal{W}[GLS, \chi^2]$	0.200	0.703	0.961	0.994	0.999	0.200	0.703	0.961	0.994	0.999
$\mathcal{W}[ML, \chi^2]$	0.317	0.918	1.000	1.000	1.000	0.317	0.918	1.000	1.000	1.000
LR_{asy}	0.331	0.913	0.999	1.000	1.000	0.331	0.913	0.999	1.000	1.000
$\mathcal{W}[GLS, LMC]$	0.162	0.619	0.918	0.982	0.998	0.186	0.684	0.946	0.990	0.999
$\mathcal{W}[ML, LMC]$	0.265	0.832	0.991	0.999	1.000	0.297	0.903	1.000	1.000	1.000
$LR[LMC]$	0.286	0.841	0.999	0.999	1.000	0.328	0.908	0.998	1.000	1.000
$QLR_{(0)}$	0.265	0.806	0.971	0.998	1.000	0.316	0.864	0.983	0.999	1.000
$QLR_{(1)}$	0.290	0.849	0.988	0.998	1.000	0.334	0.900	0.997	1.000	1.000
$QLR_{(2)}$	0.287	0.842	0.991	0.999	1.000	0.331	0.908	0.997	1.000	1.000
$LR[BMC]$	0.120	0.427	0.792	0.958	0.995	0.110	0.484	0.877	0.990	1.000
μ_{51}	0.029	0.034	0.038	0.041	0.048	0.032	0.036	0.039	0.041	0.044
μ_{52}	0.031	0.036	0.039	0.042	0.045	0.031	0.034	0.038	0.040	0.041
μ_{53}	0.042	0.085	0.154	0.258	0.359	0.035	0.077	0.152	0.241	0.397
μ_{54}	0.023	0.071	0.159	0.289	0.456	0.025	0.067	0.175	0.302	0.512
μ_{55}	0.031	0.050	0.071	0.118	0.170	0.033	0.056	0.092	0.128	0.180

$LR[BMC]$, $LR[LMC]$, and $LR[MMC]$ refer to the corresponding LR tests, and similarly to LR_M and QLR. Our results show the following:

1. The asymptotic criteria have an upward bias in size; as can be seen in Table 2, rejection of the null is repeatedly many times larger than what it should be. The bias clearly worsens in the five equation example (5EQ). Across the cases examined, the Wald-type statistics have larger sizes when based on their asymptotic χ^2 critical values. Although the F approximation seems to correct the problem in the 3EQ model, it clearly fails to do so in the 5EQ case. The nonlinear LR test examined in D2 is also oversized.
2. The BMC test was found to be well behaved. Power gains are possible in other test problems where a tighter critical bound is available. Indeed, we have observed reasonable power even if we have experimented with the worst scenario, in the sense that bounding test statistics correspond to a null hypothesis that fixes the values of all regression coefficients (except the intercept). Furthermore, we found that the BMC and the MMC tests based on LR_M yield equivalent decisions for all cases examined; the MMC test based on LR performs marginally better. This illustrates the value of the conservative bounds test as a tool to be used in conjunction with LMC test methods and not necessarily as an alternative to those methods. As emphasized earlier, the bounds procedure is computationally inexpensive and exact. In addition, whenever the bounds test rejects, inference may be made without further appeal to randomized tests.
3. There is no indication of over-rejection for the *LMC* tests considered. While the critical values used, conditional on the particular choice of consistent estimator for the error covariance matrix, were only asymptotically justified, the procedure was remarkably effective in correcting the bias. Whether this conclusion would carry to larger systems remains an open question. In this regard, note that available simulation evidence on the SURE model, specifically the experiment in Rocke (1989) on *large* systems, is limited to three equations at best.
4. While they did exhibit adequate sizes, the statistics inspired by Theil et al. (1985) did not fare well in terms of power. For the 3EQ model, the performance was dramatically poor for μ_{32} and μ_{33} but less so in the case of μ_{31}. Even then, as compared to the randomized LR, the performance is less than satisfactory.
5. The *LMC* tests performed noticeably well in terms of power in all instances, even when the number of replications was as low as 19.

It is worth noting, however, that simulation evidence does not favor the randomized usual LR tests over those based on $\Lambda_{(l)}$ typically involving fewer iterations, although we are uncertain as to the asymptotic equivalence of both procedures. This observation has an important bearing on empirical practice. The simplicity of the method based on $\Lambda_{(l)}$ has much to recommend it for larger models in which statistics requiring full ML may be quite expensive to randomize.

5 EMPIRICAL ILLUSTRATION

In this section, we present an empirical application that illustrates the results presented in this paper. We consider testing restrictions on the parameters of a generalized Leontief cost function. We used the data from Berndt and Wood (1975) and the factor demand system from Berndt (1991, pp. 460–462). The model imposes constant returns to scale and linear homogeneity in prices, and includes four inputs: capital (K), labor (L), energy (E), and nonenergy intermediate materials (M). If we denote the output by Y and the input prices by P_j, j, = K, L, E, M, the stochastic cost minimizing input–output KLEM equations are

$$\mathrm{K/Y} = d_{\mathrm{KK}} + d_{\mathrm{KL}}(P_{\mathrm{L}}/P_{\mathrm{K}})^{1/2} + d_{\mathrm{KE}}(P_{\mathrm{E}}/P_{\mathrm{K}})^{1/2} + d_{\mathrm{KM}}(P_{\mathrm{M}}/P_{\mathrm{K}})^{1/2} + e_{\mathrm{K}} \tag{50}$$

$$\mathrm{L/Y} = d_{\mathrm{LL}} + d_{\mathrm{LK}}(P_{\mathrm{K}}/P_{\mathrm{L}})^{1/2} + d_{\mathrm{LE}}(P_{\mathrm{E}}/P_{\mathrm{L}})^{1/2} + d_{\mathrm{LM}}(P_{\mathrm{M}}/P_{\mathrm{L}})^{1/2} + e_{\mathrm{L}} \tag{51}$$

$$\mathrm{E/Y} = d_{\mathrm{EE}} + d_{\mathrm{EK}}(P_{\mathrm{K}}/P_{\mathrm{E}})^{1/2} + d_{\mathrm{EL}}(P_{\mathrm{L}}/P_{\mathrm{E}})^{1/2} + d_{\mathrm{EM}}(P_{\mathrm{M}}/P_{\mathrm{E}})^{1/2} + e_{\mathrm{E}} \tag{52}$$

$$\mathrm{M/Y} = d_{\mathrm{MM}} + d_{\mathrm{MK}}(P_{\mathrm{K}}/P_{\mathrm{M}})^{1/2} + d_{\mathrm{ML}}(P_{\mathrm{L}}/P_{\mathrm{M}})^{1/2} + d_{\mathrm{ME}}(P_{\mathrm{E}}/P_{\mathrm{M}})^{1/2} + e_{\mathrm{M}} \tag{53}$$

where the error terms e_{K}, e_{L}, e_{E}, e_{M} satisfy the distributional assumptions, Eq. (10). We focus on testing the symmetry restrictions entailed by microeconomic theory, i.e.,

$$H_{01}\colon \begin{cases} d_{\mathrm{KL}} = d_{\mathrm{LK}}, & d_{\mathrm{KM}} = d_{\mathrm{MK}} \\ d_{\mathrm{KE}} = d_{\mathrm{EK}}, & d_{\mathrm{LM}} = d_{\mathrm{ML}} \\ d_{\mathrm{LE}} = d_{\mathrm{EL}}, & d_{\mathrm{EM}} = d_{\mathrm{ME}} \end{cases}$$

as well as a subset of these constraints:

$$H_{02}\colon d_{\mathrm{EM}} = d_{\mathrm{ME}}, \quad d_{\mathrm{KM}} = d_{\mathrm{MK}}$$

Confirming with the procedures described above, we reconsider the testing problem in the context of the MLR model of which the KLEM system is a restricted form. The individual equations of the latter model include the 32 price ratios $(P_i/P_j)^{1/2}$, $i, j = \mathrm{K, L, E, M}$ as regressors. The unrestricted MLE SURE estimates using the data provided in Berndt (1991) on the manufacturing sector of the U.S. economy over the period 1947–1971 are given below (with asymptotic standard errors in parentheses):

$$\mathrm{K/Y} = \underset{(0.0143)}{0.0263} + \underset{0.0088}{0.0036}(P_\mathrm{L}/P_\mathrm{K})^{1/2} + \underset{(0.0301)}{0.0649}(P_\mathrm{E}/P_\mathrm{K})^{1/2} - \underset{(0.0426)}{0.0443}(P_\mathrm{M}/P_\mathrm{K})^{1/2} + \hat{e}_\mathrm{K} \tag{54}$$

$$\mathrm{L/Y} = \underset{(0.0157)}{0.0719} + \underset{(0.0245)}{0.0517}(P_\mathrm{K}/P_\mathrm{L})^{1/2} + \underset{(0.0476)}{0.2200}(P_\mathrm{E}/P_\mathrm{L})^{1/2} + \underset{(0.0676)}{0.0264}(P_\mathrm{M}/P_\mathrm{L})^{1/2} + \hat{e}_\mathrm{L} \tag{55}$$

$$\mathrm{E/Y} = \underset{(0.0183)}{0.0403} - \underset{(0.0088)}{0.0111}(P_\mathrm{K}/P_\mathrm{E})^{1/2} - \underset{(0.0053)}{0.0048}(P_\mathrm{L}/P_\mathrm{E})^{1/2} + \underset{(0.0259)}{0.0150}(P_\mathrm{M}/P_\mathrm{E})^{1/2} + \hat{e}_\mathrm{E} \tag{56}$$

$$\mathrm{M/Y} = \underset{(0.1214)}{0.7401} - \underset{(0.0420)}{0.0542}(P_\mathrm{K}/P_\mathrm{M})^{1/2} - \underset{(0.0258)}{0.1374}(P_\mathrm{L}/P_\mathrm{M})^{1/2} + \underset{(0.0855)}{0.0399}(P_\mathrm{E}/P_\mathrm{M})^{1/2} + \hat{e}_\mathrm{M} \tag{57}$$

For both hypotheses, we computed the GLS and ML-based Wald statistics, Eq. (49) the LR and $LR_{\mathcal{M}}$ criteria as defined in Eqs. (23) and (28) and the QLR statistics, Eq. (48). In the case of the Wald and QLR test, we obtained the asymptotic χ^2 and LMC p values using 19 and 99 simulated samples. The exact BMC and MMC p values were also obtained for the LR criteria. The bounding statistic $LR_{\mathcal{UL}} = n\ln(\Lambda_{\mathcal{UL}})$ corresponds to the UL hypothesis that sets all the coefficients of the MLR model (except the intercepts) to specific values. As stated in Section 3, the BMC procedure based on $LR_{\mathcal{M}}$ yields tighter bounds [see inequality (42)]. Our results are summarized in Tables 6 and 7.

From these results, we see that the symmetry hypothesis H_{01} is rejected using all asymptotic and exact tests. In the case of H_{02}, all tests against the unconstrained SURE specification are not significant. However, the asymptotic χ^2 and LMC tests $LR_{\mathcal{M}}$ are significant at the 5% level. Although the bounds p-value is larger than 0.05, the MMC test is significant at the 5% level, even with 19 simulated samples. It is worth noting that the QLR and the LR LMC tests yield equivalent decisions for both testing problems. Moreover, all MC tests based on 19 and 99 replications also yield similar decisions.

TABLE 6 Generalized Leontief Factor Demands: Cross-Equation Symmetry Tests

$$H_{01}{}^{a}: \begin{cases} d_{KL} = d_{LK}, & d_{KM} = d_{MK}, & d_{KE} = d_{EK}, \\ d_{LM} = d_{ML}, & d_{LE} = d_{EL}, & d_{EM} = d_{ME} \end{cases}$$

		$LR_{\mathcal{M}}$	LR	$QLR_{(0)}$	$QLR_{(1)}$	$QLR_{(2)}$	Wald_{GLS}	Wald_{ML}
Statistic		176.582	74.159	75.545	75.140	74.911	239.597	238.777
Asymptotic p-value		0.000	0.000	0.000	0.000	0.000	0.000	0.000
Reps[b]	MC p-value							
19	BMC	0.05	0.70	-	-	-	-	-
99		0.01	0.67	-	-	-	-	-
19	MMC	0.05	0.05	-	-	-	-	-
99		0.01	0.01	-	-	-	-	-
19	LMC	0.05	0.05	0.05	0.05	0.05	0.05	0.05
99		0.01	0.01	0.01	0.01	0.01	0.01	0.01

[a]Under $H_{01}, LR_{\mathcal{M}} \overset{asy}{\sim} \chi^2(42)$ while the other statistics have asymptotic χ^2 (6) distributions. The $LR_{\mathcal{M}}$ statistic tests the symmetry restrictions (6 constraints) jointly with the SURE exclusion restrictions (36 constraints)—a total of 42 restrictions—against the unrestricted MLR model.
[b]Replications.

TABLE 7 Generalized Leontief Factor Demands: Partial Cross-Equation Symmetry Tests

$$H_{02}{}^{a}: d_{EM} = d_{ME}, \quad d_{KM} = d_{MK}$$

		$LR_{\mathcal{M}}$	LR	$QLR_{(0)}$	$QLR_{(1)}$	$QLR_{(2)}$	Wald_{GLS}	Wald_{ML}
Statistic		102.574	0.15179	0.15180	0.15179	0.15179	0.1283	0.1279
Asymptotic p-value		0.000	0.927	0.927	0.927	0.927	0.937	0.938
Reps[b]	MC p-value							
19	BMC	0.15	1.0	-	-	-	-	-
99		0.17	1.0	-	-	-	-	-
19	MMC	0.05	1.0	-	-	-	-	-
99		0.05	1.0	-	-	-	-	-
19	LMC	0.05	0.90	0.90	0.90	0.90	0.90	0.90
99		0.04	0.94	0.94	0.94	0.94	0.94	0.94

[a]Under H_{02}, $LR_{\mathcal{M}} \overset{asy}{\sim} \chi^2(38)$ while the other statistics have asymptotic $\chi^2(2)$ distributions. $LR_{\mathcal{M}}$ tests a subset of symmetry restrictions (2 constraints) jointly with the SURE exclusion restrictions (36 constraints)—38 restrictions in all—against the unrestricted MLR model.
[b]Replications.

6 CONCLUSION

In this chapter we have extended the MLR-based LR test procedure to the SURE framework. We have combined the bounds and Monte Carlo test approaches to provide *p*-values for test statistics that can yield provably exact tests in finite samples even for nonlinear hypotheses as well as more reliable large sample tests. The feasibility of the test strategy was also illustrated with an extensive Monte Carlo experiment and an empirical application. We have found that standard asymptotic tests exhibit serious errors in level, particularly in larger systems. In contrast, the various tests we have proposed displayed excellent size and power properties.

ACKNOWLEDGMENTS

This work was supported by the Canadian Network of Centres of Excellence [program on Mathematics of Information Technology and Complex Systems (MITACS)], the Canada Council for the Arts (Killam Fellowship), the Natural Sciences and Engineering Research Council of Canada, the Social Sciences and Humanities Research Council of Canada, and the Fonds FCAR (Government of Quebec). This chapter is a revised version of an earlier working paper by Dufour and Khalaf (1998). The authors thank Emanuela Cardia, Marcel Dagenais, John Galbraith, Eric Ghysels, James MacKinnon, Christophe Muller, Olivier Torrès, and Michael Veall for useful comments. Earlier versions of the work were presented at the North American Meetings of the Econometric Society, the Third International Conference on Computing and Finance (Hoover Institute, Stanford University), the Annual Meetings of the American Statistical Association, the Canadian Econometric Study group, the Canadian Economic Association, and at Ohio State University (Economics).

REFERENCES

Anderson, T.W., 1984. *An Introduction to Multivariate Statistical Analysis*, 2nd ed. Wiley, New York.

Barnard, G.A., 1963. Comment on "The spectral analysis of point processes" by M. S. Bartlett. *Journal of the Royal Statistical Society, Series B* **25**, 294.

Bera, A.K., Byron, R.P., and Jarque, C.M., 1981. Further evidence on asymptotic tests for homogeneity and symmetry in large demand systems. *Economics Letters* **8**, 101–105.

Berndt, E.R., 1991. *The Practice of Econometrics: Classic and Contemporary.* Addison-Wesley, Reading, MA.

Berndt, E.R. and Savin, N.E., 1977. Conflict among criteria for testing hypotheses in the multivariate linear regression model. *Econometrica* **45**, 1263–1277.

Berndt, E.R. and Wood, D.O., 1975. Technology, prices and the derived demand for energy. *Review of Economics and Statistics* **53**, 259–268.

Breusch, T.S., 1979. Conflict among criteria for testing hypotheses: Extensions and comments. *Econometrica* **47**, 203–207.

Dufour, J.-M., 1989. Nonlinear hypotheses, inequality restrictions, and non-nested hypotheses: Exact simultaneous tests in linear regressions. *Econometrica* **57**, 335–355.

Dufour, J.-M., 1997. Some impossibility theorems in econometrics, with applications to structural and dynamic models. *Econometrica* **65**, 1365–1389.

Dufour, J.-M., 2002. Monte Carlo tests with nuisance parameters: A general approach to finite-sample inference and nonstandard asymptotics in econometrics. Technical report, C.R.D.E., Université de Montréal.

Dufour, J.-M. and Khalaf, L., 1998. Simulation based finite and large sample inference methods in multivariate regressions and seemingly unrelated regressions. Technical report, C.R.D.E., Université de Montréal.

Dufour, J.-M. and Khalaf, L., 1998b. Simulation based finite and large sample tests in multivariate regressions. *Journal of Econometrics*, **111**, 303–322.

Dufour, J.-M. and Khalaf, L., 2001. Monte Carlo test methods in econometrics. In B. Baltagi, ed., *Companion to Theoretical Econometrics.* Blackwell, Oxford, pp. 494–519.

Dufour, J.-M. and Kiviet, J.F., 1996. Exact tests for structural change in first-order dynamic models. *Journal of Econometrics* **70**, 39–68.

Dufour, J.-M. and Kiviet, J.F., 1998. Exact inference methods for first-order autoregressive distributed lag models. *Econometrica* **66**, 79–104.

Dufour, J.-M., Farhat, A., Gardiol, L., and Khalaf, L., 1998. Simulation-based finite sample normality tests in linear regressions. *The Econometrics Journal* **1**, 154–173.

Dwass, M., 1957. Modified randomization tests for nonparametric hypotheses. *Annals of Mathematical Statistics* **28**, 181–187.

Goffe, W.L., Ferrier, G.D., and Rogers, J., 1994. Global optimization of statistical functions with simulated annealing. *Journal of Econometrics* **60**, 65–99.

Harvey, A.C. and Phillips, G.D.A., 1982. Testing for contemporaneous correlation of disturbances in systems of regression equations. *Bulletin of Economic Research* **34**, 79–81.

Jöckel, K.-H. 1986. Finite sample properties and asymptotic efficiency of Monte Carlo tests. *The Annals of Statistics* **14**, 336–347.

Kariya, T., 1985. *Testing in the Multivariate General Linear Model.* No. 22 in "Economic Research Series, The Institute of Economic Research, Hitotsubashi University, Japan," Kinokuniya Co., Tokyo.

Laitinen, K., 1978. Why is demand homogeneity so often rejected? *Economics Letters* **1**, 187–191.

Meisner, J.F., 1979. The sad fate of the asymptotic Slutsky symmetry test for large systems. *Economics Letters* **2**, 231–233.

Phillips, P.C.B., 1985. The exact distribution of the SUR estimator. *Econometrica* **53**, 745–756.

Rao, C.R., 1973. *Linear Statistical Inference and its Applications*, 2nd ed. Wiley, New York.

Rayner, R.K., 1990. Bootstrapping *p*-values and power in the first-order autoregression: A Monte Carlo study. *Journal of Business and Economics Statistics* **8**, 251–263.

Rilstone, P. and Veall, M., 1996. Using bootstrapped confidence intervals for improved inferences with seemingly unrelated regression equations. *Econometric Theory* **12**, 569–580.

Rocke, D.M., 1989. Bootstrap Bartlett adjustment in seemingly unrelated regressions. *Journal of the American Statistical Association* **84**, 598–601.

Srivastava, V.K. and Giles, D.E.A., 1987. *Seemingly Unrelated Regression Equations Models. Estimation and Inference*. Marcel Dekker, New York.

Stewart, K.G., 1997. Exact testing in multivariate regression. *Econometric Reviews* **16**, 321–352.

Taylor, T.G., Shonkwiler, J.S., and Theil, H., 1986. Monte Carlo and bootstrap testing of demand homogeneity. *Economics Letters* **20**, 55–57.

Theil, H., 1971. *Principles of Econometrics*. Wiley, New York.

Theil, H. and Fiebig, D.E., 1985. Small sample and large equation systems. In E.J. Hannan, P.R. Krishnaiah, and M.M. Rao, eds. *Handbook of Statistics 5: Time Series in the Time Domain*. North-Holland, Amsterdam, pp. 451–480.

Theil, H., Shonkwiler, J.S. and Taylor, T.G., 1985. A Monte Carlo test of Slutsky symmetry. *Economics Letters* **19**, 331–332.

Theil, H., Taylor, T. and Shonkwiler, J.S., 1986. Monte Carlo testing in systems of equations. In D.J. Slottje and G.F. Rhodes Jr., eds. *Advances in Econometrics*, Vol. 5, *Innovations in Quantitative Economics: Essays in Honor of Robert L. Basmann*, Vol. 5, JAI Press, Greenwich, CT, pp. 227–239.

Zellner, A., 1962. An efficient method for estimating seemingly unrelated regressions and tests for aggregate bias. *Journal of the American Statistical Association* **57**, 348–368.

3

Finding Optimal Penalties for Model Selection in the Linear Regression Model

Maxwell L. King
Monash University, Melbourne, Australia
Gopal K. Bose
Queensland Treasury, Brisbane, Queensland, Australia

1 INTRODUCTION

In econometrics and statistics, we often need to make a choice between a number of alternative statistical models. In some cases, economic theory or our understanding of the phenomenon we are trying to model can help us make an appropriate decision. However, there are many situations in which we are forced to ask which model best fits the data. The problem of choosing between a limited range of alternative models using only the available data is known as the model selection problem.

There is a large literature on this problem with a range of methods or strategies being suggested as solutions. A popular method of deciding on a final model using the data has been to use a series of pairwise hypothesis tests. Unfortunately, as noted by Granger et al. (1995), this approach has many limitations. At each step, one model is chosen as the null hypothesis and can be unfairly favored because the probability of wrongly rejecting it is set to a small value like 0.05 or 0.01. This can be particularly troublesome in situations where the test being used is not particularly powerful and therefore a choice of the null hypothesis model is the most likely outcome. There is also the well known problem of pretest bias (see, e.g., Wallace, 1977; Giles and Giles, 1993). Finally different researchers working on the same model selection problem could easily end up with different final models purely because they performed the tests in a different order or used different significance levels.

The most widely accepted class of model selection procedures is the use of information criteria (IC) based on choosing the model with the largest maximized log-likelihood function minus a penalty that is an increasing function of the number of free parameters included in the model. Examples of these procedures include Akaike's (1973, 1974) IC (AIC), Schwarz's (1978) Bayesian IC (BIC), and Hannan and Quinn's (1979) procedure denoted HQ. Fox (1995) has also demonstrated how a number of other model selection procedures, particularly those developed for choosing from a set of linear regression models, can be thought of in the IC framework. Those of interest in this chapter include the use of Theil's (1961) $\bar{R}^2$ criterion, Mallows' (1964) C_p procedure, denoted MCP, Schmidt's (1975) generalised cross-validation (GCV) procedure, and Hocking's (1976) criterion, denoted HOC. As Granger et al. (1995) and others have noted, the IC approach has the advantages that:

1. No particular model is favored because it has been chosen to be a null hypothesis.
2. The order of computation is irrelevant.
3. Pretesting bias is not an issue.
4. If the IC procedure is asymptotically consistent, the correct model is chosen with probability one asymptotically.
5. There is no need to choose an arbitrary level of significance although there is the related issue of which penalty function is appropriate.

The last is a major issue that has featured in the literature. There is currently little agreement about what the form of the penalty function should be. The early literature focused on asymptotic arguments to justify various choices of penalty functions (see, e.g., Akaike, 1973; Schwarz, 1978; Hannan and Quinn, 1979). Since then we have seen a number of Monte Carlo studies of the small sample properties of different model selection procedures, which have shown a number of problems. They suggest that asymptotic properties are no guarantee of acceptable small sample properties. For example, Grose and King (1994), in the context of choosing between first-order autoregressive and first-order moving average disturbances in the linear regression model, have shown that a particular model can be unfairly favored because of the functional form of its log-likelihood function. They also found that the presence of nuisance parameters can adversely affect the probabilities of correct selection.

There are clear parallels between the hypothesis testing literature and the model selection literature, although it does appear that the latter lags the former. The computer revolution has meant that we can now ask what kind of testing procedures would we like to use rather than what kind of testing procedure is convenient to use (see, e.g., King, 1987). We should be asking similar questions for model selection, in the context of finite samples. We now regularly use simulation to find critical values for hypothesis tests. Can we use similar methods to find penalty values for IC procedures that are in some sense optimal?

In this chapter we investigate a new approach to IC model selection in the context of the classical problem of choosing between different linear regression models. Our approach involves the use of the simulation method to estimate probabilities of correct selection and choosing penalties that optimize these probabilities on average. Maximizing simulated probabilities can be a difficult optimization problem. We use a relatively new optimization algorithm called simulated annealing to overcome any difficulties in this regard.

A similar approach to model selection has been investigated by Kwek (1999) and Kwek and King (1999), in the context of selecting between different conditional heteroscedastic processes. Billah and King (2000a,b) have also considered a similar method for choosing between different time-series processes for linear regression disturbances. Each of these studies has involved choosing between variance–covariance matrix functions with restricted parameter spaces. In this chapter we consider model selection of different mean processes with unrestricted parameter spaces.

The plan of this chapter is as follows. The new approach is outlined in abstract terms in Section 2. The main practical issues of applying this approach to choosing between m linear regression models are discussed in Section 3. These include the choice of weighting function, the issue of how best to estimate mean average probabilities via simulation and the choice of optimization method. The design of our Monte Carlo study to compare the new approach with some existing IC methods is given in Section 4 and the results of this study are discussed in Section 5. The chapter closes with some concluding remarks.

2 FRAMEWORK FOR OPTIMAL PENALTY SELECTION

Our problem of interest is one of selecting a model from m alternative models that we will denote by $M_1, M_2, \ldots, M_m$, for a given set of n

observations on the dependent variable denoted by the $n \times 1$ vector y. Let model M_j, $j=1,\ldots,m$, be represented by

$$y = f(X_j, \beta_j, u) \tag{1}$$

where f is a known $n \times 1$ vector function, X_j is an $n \times q_j$ matrix of observations on q_j fixed variables, β_j is a $q_j \times 1$ vector of unknown structural parameters and u is an $n \times 1$ vector of random disturbances distributed as $N(0, \sigma_j^2 I_n)$ where σ_j^2 is a further unknown parameter. Let $\theta_j = (\beta_j', \sigma_j^2)'$ denote the $k_j \times 1$ vector of unknown parameters in this model where $k_j = q_j + 1$. Let $L_j(\theta_j)$ denote the log-likelihood function of model M_j and let $L_j(\hat{\theta}_j)$ denote its maximized value where $\hat{\theta}_j$ is the maximum likelihood estimator (MLE) of θ_j.

Almost all IC-based model selection procedures are based on choosing the model with the largest value of

$$I_j = L_j(\hat{\theta}) - p_j \tag{2}$$

for $j=1,\ldots,m$, where p_j is the penalty for model M_j. Examples of values for p_j include k_j for AIC, $k_j \log(n)/2$ for BIC, $k_j \log[\log(n)]$ for HQ, $-n\log(n - k_j)/2$ for $\bar{R}^2$, $-n\log[(n-k_j)/n]$ for GCV, $-n\log\{(n - k_j)\times(n - k_j - 1)\}/2$ for HOC, and $n\log[(1 + 2k_j/(n - k^*)]/2$ for MCP where k^* is the number of free parameters in the smallest model that includes all models under consideration as special cases.

The main question we wish to address is how do we find optimal penalties, $p_1, \ldots, p_m$, for Eq. (2)? We can look to the hypothesis testing literature for some guidance. There we make a choice of test based on maximizing power which is the probability of correctly rejecting the null hypothesis. It therefore seems reasonable to want to choose p_j to somehow maximize the probability of correctly selecting the true model. Unfortunately, these probabilities depend on the true model and the value of θ_j for this model. Our suggestion of a criterion for optimally choosing $p_1, \ldots, p_m$ is as follows. For any given set of penalty values (i.e., p_j, $j=1,\ldots,m$) and for each model under consideration, we estimate the average probability of correctly selecting this model when it is indeed the true model. For the same penalty set, we then sum these probabilities of correct selection, thus treating all models equally. The penalties are chosen to maximize this sum.

In order to understand more closely what is involved, let CS_j denote the event of correct selection of M_j when M_j is the true model and let

$$P(CS_j \mid \theta_j, p_1, \ldots, p_m)$$

denote the probability of correct selection when M_j is the true model with parameter vector θ_j and $p_1,\ldots,p_m$ are used as the penalties in Eq. (2). Then the average probability of correct selection for the jth model with penalties $p_1,\ldots,p_m$ can then be written as

$$APCS_j(p_1,\ldots,p_m) = \int P(CS_j|\theta_j,p_1,\ldots,p_m)g_j(\theta_j)\,d\theta_j \tag{3}$$

where $g_j(\theta_j)$ is a weighting density function rather like a prior density function used in Bayesian methods. Its purpose is to weight different parameter vector values when calculating the average probability of correct selection. The objective function we wish to choose $p_1,\ldots,p_m$ to maximize is, therefore,

$$MAPCS(p_1,\ldots,p_m) = \frac{1}{m}\sum_{j=1}^{m} APCS_j(p_1,\ldots,p_m) \tag{4}$$

For our approach to be operational, we need a method of estimating Eq. (3). Given that $g_j(\theta_j)$ is a joint density function, the method of Monte Carlo estimation of Eq. (3) suggests itself. This allows Eq. (3) to be estimated by first taking a large sample of drawings from the distribution implied by $g_j(\theta_j)$. We denote these independent drawings by $\theta_j(i)$, $i=1,\ldots,r$, where r is the number of drawings. The estimate of Eq. (3) is then obtained by calculating the sample mean:

$$\frac{1}{r}\sum_{i=1}^{r} P(CS_j|\theta_j(i),p_1,\ldots,p_m) \tag{5}$$

This in turn requires us to estimate

$$P(CS_j|\theta_j(i),p_1,\ldots,p_m) \tag{6}$$

for each value of $\theta_j(i)$, $p_1,\ldots p_m$, $i=1,\ldots,r$, which can be achieved in the normal manner by a standard Monte Carlo simulation of N replications from the model represented by Eq. (1).

This still leaves a number of issues for resolution before the method can be implemented. These are the choice of $g_j(\theta_j)$ and how we might sample from its implied distribution, the choice of values for r and N, and the method for maximizing the estimated value of the $MAPCS(p_1,\ldots,p_m)$. These are addressed in Section 3 in the context of choosing between different linear regression models.

3 SOME PRACTICAL ISSUES

We now turn our attention to the specific problem of choosing between m regression models of the form:

$$y = X_j\beta_j + u \tag{7}$$

where y, X_j, β_j, and u are defined as in Eq. (1).

3.1 Choice of Weighting Function

In choosing a weighting function for $\theta_j = (\beta_j', \sigma_j^2)'$, we need to balance the need not to favor unduly a particular part of the θ_j space with the need for $APCS_j(p_1, \ldots, p_m)$ to take a value between 0 and 1 so that it is reasonably sensitive to different values of $p_1, \ldots, p_m$. For example, if Eq. (3) was always unity because of our choice of $g_j(\theta_j)$, it would be almost impossible to find unique values of $p_1, \ldots, p_m$ that maximize Eq. (4) as required. This calls for the use of a publicly respected prior for θ_j that can be tightened around the origin should Eq. (4) prove to get too close to unity.

For the standard linear regression model, Eq. (7), Zellner (1971) suggests the use of an inverted gamma distribution for σ_j^2 with scale parameter s_e^2 and a two-step prior for β_j which involves drawing β_j from the multivariate normal distribution:

$$N(0, \tilde{\sigma}_j^2 I)$$

where $\tilde{\sigma}_j^2$ itself is a separate drawing from an inverted gamma distribution with scale parameter s_b^2.

The required $\theta_j(i) = [\beta_j(i)', \sigma_j^2(i)]'$, $i = 1, \ldots, r$, can be generated as follows:

Step 1: To generate $\sigma_j^2(i)$, draw a simple random sample of size $n - q_j$ denoted $z_1, \ldots, z_{n-q_j}$ from the $N(0, 1)$ distribution and then calculate

$$\sigma_j^2(i) = (n - q_j)s_e^2 \left[\sum_{i=1}^{n-q_j} z_i^2 \right]^{-1}$$

where s_e^2 is the predetermined scale parameter which is held constant for all r drawings.

Step 2: Generate $\tilde{\sigma}_j^2$ with a new independent sample as outlined in *Step 1* but with s_e^2 replaced by s_b^2, another predetermined scale parameter that is held constant for all r drawings.

Step 3: To generate $\beta_j(i)$, draw a simple random sample of size q_j denoted $w_j = (w_1, \ldots, w_{q_j})'$ from the $N(0,1)$ distribution and calculate

$$\beta_j(i) = w_j \tilde{\sigma}_j$$

where $\tilde{\sigma}_j$ is the square root of $\tilde{\sigma}_j^2$ calculated in *Step 2*.

Observe that increasing (decreasing) s_e^2 and/or s_b^2 makes it easier (harder) to select the correct model and hence increases (decreases) the probability of correct selection and also of the objective function (4).

3.2 Simulation Experiment to Determine the Best Choice of *N* and *r* Values

In this subsection, we investigate different choices of values for N, the number of Monte Carlo replications used to estimate the probability of correctly choosing model j when it is the true model with parameter vector $\theta_j(i)$ and for r, the number of drawings used to estimate Eq. (3) by Eq. (5). The total number of replications in the whole process of estimating Eq. (3) is, therefore, rN. Our aim is to find the best values of r and N for fixed rN. In Subsection 3.2.1, we outline the simulation experiments conducted for this purpose and discuss the results of these experiments in Subsection 3.2.2.

3.2.1 Experimental Design

We conducted the simulation experiment in the context of choosing between the following four linear regression models:

$$\begin{array}{lll} M_1: & y_t = \beta_{10} + u_{1t} & u_{1t} \sim IN(0, \sigma_1^2) \\ M_2: & y_t = \beta_{20} + x_{1t}\beta_{21} + u_{2t} & u_{2t} \sim IN(0, \sigma_2^2) \\ M_3: & y_t = \beta_{30} + x_{2t}\beta_{31} + u_{3t} & u_{3t} \sim IN(0, \sigma_3^2) \\ M_4: & y_t = \beta_{40} + x_{1t}\beta_{41} + x_{2t}\beta_{42} + u_{4t} & u_{4t} \sim IN(0, \sigma_4^2) \end{array}$$

where y_t is the tth observation of the dependent variable, x_{it} is the tth observation on the ith nonconstant regressor, the β's are the regression coefficients and the u_{jt} are the random disturbances for $t = 1, \ldots, n$. The following two data sets were used as design matrices for these regression models:

X1: x_{1t} is the quarterly Australian retail trade as recorded by the Australian Bureau of Statistics and x_{2t} is the same series lagged by one quarter.

X2: x_{1t} is the real per capita GDP and x_{2t} is the investment of the tth country. We used the annual data from Summers and Heston (1991) revised version 5.6 and World Bank world tables.

The experiments were conducted for $n=20$, 50, and 96 for *X1*, and $n=20$, 50, and 100 for *X2*. For each model in turn, parameter values were randomly generated as outlined in Subsection 3.1 with $s_e^2 = 55$ and $s_b^2 = 6$ for *X1* and $s_e^2 = 2.5$ and $s_b^2 = 0.002$ for *X2*. These values were chosen after some experimentation in order to obtain the expected probabilities of correct selection around the middle value of 0.5. As the aim was to find the best choice of r and N values for fixed rN, the following 15 combinations of N and r were used that give $rN=2000$:

N	2000	1000	500	250	200	100	50	40	20	10	8	5	4	2	1
r	1	2	4	8	10	20	40	50	100	200	250	400	500	1000	2000

For each experiment, the values of the four penalties (one for each model) were held fixed. We used seven sets of penalties, namely those for AIC, BIC, HQ, $\bar{R}^2$, GCV, MCP, and HOC noted in Section 2.

Each simulation experiment was repeated with different random numbers 20 times so that the standard deviation of the estimated mean probabilities of correct selection [our estimate of Eq. (3)] can be calculated with each model being the true model in turn. We also computed the average of these standard deviations over the four competing models for each set of penalties with the aim of looking for N and r combinations where this average standard deviation (ASD) is lowest (and hence our estimates of Eq. (3), on average, have the least variation).

3.2.2 Results

Selected results of the simulation experiments are presented in Tables 1 and 2. From these tables, we see a very clear tendency for the value of ASD to decrease as r increases, particularly for small r. In almost all cases, the lowest ASD is obtained when the model parameters are generated 2000 times ($r=2000$) and only one replication ($N=1$) is used to estimate $P(CS_j \mid \theta_j(i), p_1, \ldots, p_m)$. There are some exceptions to these general trends that may be due in part to sampling error.

We therefore decided to estimate the mathematical relationship between ASD and r. Plots of the data suggested

$$\log(ASD_i) = a + b\log(r_i) + u_i$$

where a and b are parameters and u_i is an error term. Ordinary least squares estimation of these models for different penalty functions and values of n resulted in strong evidence of positive autocorrelation, suggesting functional misspecification. Further experimentation led to the model:

$$\log(\mathrm{ASD}_i) = a + b\log(r_i) + c(\log(r_i))^2 + u_i \tag{8}$$

where $u_i \sim IN(0, \sigma^2)$.

We estimated this model for each set of penalties, values of n, and set of design matrices using our 15 observations on ASD_i and r_i. The resulting estimates, values of $\bar{R}^2$, and the Durbin–Watson (DW) statistics for *X1* and *X2* are given in Tables 3 and 4, respectively.

Our best guess of the relationship between ASD and r is, therefore, given by

$$\log(\mathrm{ASD}) = \hat{a} + \hat{b}\log(r) + \hat{c}[\log(r)]^2 \tag{9}$$

We can differentiate Eq. (9) and set the derivative equal to zero in order to find which value of log(ASD) (and hence ASD) is a minimum. This value turns out to be

$$\hat{r} = e^{-(\hat{b}/(2\hat{c}))} \tag{10}$$

Tables 3 and 4 also give the resulting values of $\hat{r}$ calculated from Eq. (10). All values of $\hat{r}$ for *X1* are greater than 2000. This is also true for *X2* when $n = 50$ and 100. Only for *X2* with $n = 20$ do we get $\hat{r}$ values below 2000. The minimum would seem to be around $r = 1000$ in this particular case.

3.2.3 Conclusions

Taken as a whole, these results lead us to use $r = 2000$ and $N = 1$ when estimating Eq. (3). The evidence is particularly compelling for *X1* for all n values and for *X2* with $n = 50$ and 100. It certainly gets stronger as n increases.

3.3 Issues of Optimization

Our basic problem is one of estimating Eq. (4) and then maximizing it with respect to $p_1, \ldots, p_m$. Without loss of generality, we can set $p_1 = 0$ because, from Eq. (2), decisions about which model is chosen depends on differences between penalties, not on their overall level. Because our objective function is made up of averaged probabilities which in turn have been estimated by the Monte Carlo method, it is essentially a step function. These types of functions are difficult to optimize using standard numerical methods. They have many plateaus that will cause standard optimization techniques great difficulties.

There are two approaches to this problem that we recommend. The first is grid-search which can be very time consuming when m is large. We therefore recommend the use of the simulated annealing (SA) algorithm. From the work of Kirkpatrick et al. (1983), Romeo et al. (1984), White

TABLE 1 Average Over Four Models of Mean Over 20 Iterations of Estimated Mean Probability of Correct Selection (Denoted AOM) and the Average Standard Deviations for These 20 Iterations Averaged Over Four Models (Denoted ASD) for *X1*

	AIC		BIC		GCV		HOC		HQ		$\bar{R}^2$		MCP	
r	AOM	ASD	AOM	ASD	AOM	ASD	AOM	ASD	AOM	ASD	AOM	ASD	AOM	ASD
							$n=20$							
1	0.456	0.1527	0.455	0.1647	0.458	0.1565	0.458	0.1577	0.458	0.1563	0.434	0.1232	0.457	0.1532
4	0.423	0.0904	0.417	0.0947	0.423	0.0920	0.423	0.0926	0.423	0.0920	0.408	0.0779	0.423	0.0905
8	0.429	0.0601	0.422	0.0638	0.430	0.0617	0.430	0.0620	0.430	0.0616	0.412	0.0494	0.430	0.0604
20	0.430	0.0343	0.422	0.0373	0.429	0.0350	0.429	0.0353	0.429	0.0351	0.413	0.0289	0.430	0.0343
50	0.431	0.0288	0.425	0.0333	0.431	0.0300	0.431	0.0300	0.431	0.0299	0.412	0.0253	0.431	0.0290
100	0.431	0.0196	0.424	0.0214	0.431	0.0203	0.431	0.0206	0.431	0.0205	0.416	0.0169	0.431	0.0200
200	0.432	0.0140	0.427	0.0150	0.432	0.0142	0.432	0.0139	0.432	0.0142	0.414	0.0137	0.432	0.0139
500	0.433	0.0139	0.427	0.0138	0.433	0.0139	0.433	0.0138	0.433	0.0139	0.415	0.0107	0.433	0.0139
1000	0.431	0.0111	0.426	0.0107	0.431	0.0113	0.431	0.0111	0.431	0.0113	0.413	0.0111	0.431	0.0112
2000	0.433	0.0111	0.426	0.0099	0.433	0.0107	0.433	0.0105	0.433	0.0105	0.417	0.0127	0.433	0.0108
							$n=50$							
1	0.677	0.2022	0.704	0.2571	0.680	0.2060	0.680	0.2068	0.698	0.2303	0.600	0.1457	0.677	0.2023
4	0.665	0.1060	0.703	0.1311	0.669	0.1077	0.669	0.1083	0.691	0.1182	0.585	0.0785	0.665	0.1060

8	0.681	0.0776	0.721	0.0941	0.685	0.0790	0.685	0.0793	0.708	0.0870	0.600	0.0570	0.682	0.0777
20	0.676	0.0409	0.713	0.0495	0.680	0.0416	0.680	0.0417	0.701	0.0447	0.595	0.0314	0.676	0.0410
50	0.673	0.0258	0.709	0.0296	0.677	0.0261	0.677	0.0263	0.698	0.0277	0.594	0.0228	0.673	0.0259
100	0.674	0.0260	0.711	0.0292	0.678	0.0258	0.678	0.0256	0.700	0.0271	0.596	0.0221	0.674	0.0261
200	0.672	0.0157	0.709	0.0184	0.677	0.0161	0.677	0.0162	0.698	0.0174	0.593	0.0150	0.673	0.0157
500	0.675	0.0143	0.713	0.0138	0.679	0.0143	0.679	0.0140	0.701	0.0140	0.594	0.0126	0.675	0.0143
1000	0.673	0.0129	0.709	0.0118	0.677	0.0130	0.677	0.0130	0.698	0.0127	0.593	0.0127	0.673	0.0130
2000	0.670	0.0109	0.707	0.0099	0.675	0.0109	0.675	0.0109	0.696	0.0104	0.591	0.0109	0.671	0.0109
							$n=96$							
1	0.834	0.0801	0.936	0.1002	0.838	0.0809	0.838	0.0811	0.897	0.0922	0.706	0.0608	0.834	0.0800
4	0.822	0.0544	0.913	0.0719	0.826	0.0551	0.826	0.0552	0.881	0.0635	0.699	0.0388	0.822	0.0545
8	0.811	0.0430	0.902	0.0569	0.815	0.0433	0.815	0.0433	0.869	0.0504	0.692	0.0332	0.811	0.0430
20	0.817	0.0236	0.908	0.0286	0.821	0.0240	0.821	0.0240	0.874	0.0259	0.694	0.0203	0.817	0.0236
50	0.811	0.0173	0.900	0.0208	0.814	0.0172	0.814	0.0171	0.868	0.0186	0.687	0.0151	0.811	0.0172
100	0.815	0.0140	0.906	0.0155	0.818	0.0141	0.818	0.0141	0.873	0.0148	0.691	0.0137	0.815	0.0140
200	0.812	0.0131	0.902	0.0138	0.816	0.0133	0.816	0.0132	0.869	0.0138	0.690	0.0119	0.812	0.0131
500	0.812	0.0109	0.903	0.0087	0.815	0.0112	0.815	0.0112	0.869	0.0095	0.689	0.0112	0.812	0.0109
1000	0.812	0.0094	0.902	0.0071	0.816	0.0092	0.816	0.0092	0.870	0.0087	0.690	0.0106	0.812	0.0094
2000	0.812	0.0091	0.902	0.0067	0.815	0.0089	0.815	0.0089	0.869	0.0077	0.690	0.0098	0.812	0.0090

TABLE 2 Average Over Four Models of Mean over 20 Iterations of Estimated Mean Probability of Correct Selection (Denoted AOM) and the Average Standard Deviations for These 20 Iterations Averaged Over Four Models (Denoted ASD) for *X2*

	AIC		BIC		GCV		HOC		HQ		$\bar{R}^2$		MCP	
r	AOM	ASD	AOM	ASD	AOM	ASD	AOM	ASD	AOM	ASD	AOM	ASD	AOM	ASD
							$n=20$							
1	0.501	0.0633	0.518	0.0618	0.506	0.0634	0.508	0.0632	0.506	0.0634	0.461	0.0564	0.501	0.0632
4	0.490	0.0306	0.507	0.0282	0.495	0.0298	0.496	0.0293	0.495	0.0299	0.451	0.0303	0.491	0.0306
8	0.493	0.0223	0.509	0.0218	0.497	0.0228	0.499	0.0228	0.497	0.0227	0.455	0.0220	0.493	0.0224
20	0.492	0.0165	0.507	0.0164	0.496	0.0162	0.497	0.0163	0.496	0.0161	0.453	0.0153	0.492	0.0166
50	0.488	0.0148	0.505	0.0144	0.492	0.0149	0.494	0.0149	0.493	0.0148	0.449	0.0154	0.489	0.0147
100	0.492	0.0136	0.509	0.0107	0.497	0.0131	0.498	0.0129	0.497	0.0132	0.454	0.0130	0.493	0.0136
200	0.492	0.0105	0.508	0.0099	0.497	0.0103	0.498	0.0099	0.497	0.0104	0.452	0.0114	0.492	0.0102
500	0.492	0.0118	0.509	0.0092	0.496	0.0115	0.498	0.0114	0.496	0.0115	0.454	0.0113	0.492	0.0116
1000	0.494	0.0108	0.509	0.0090	0.498	0.0105	0.499	0.0104	0.498	0.0104	0.456	0.0107	0.495	0.0107
2000	0.491	0.0102	0.509	0.0083	0.496	0.0092	0.497	0.0088	0.496	0.0093	0.452	0.0094	0.491	0.0101
							$n=50$							
1	0.545	0.0952	0.557	0.0924	0.546	0.0956	0.546	0.0956	0.557	0.0974	0.500	0.0809	0.545	0.0952
4	0.545	0.0676	0.558	0.0670	0.547	0.0680	0.548	0.0679	0.557	0.0699	0.497	0.0544	0.545	0.0676

8	0.554	0.0368	0.567	0.0339	0.556	0.0369	0.556	0.0369	0.565	0.0378	0.504	0.0324	0.554	0.0368
20	0.551	0.0265	0.564	0.0247	0.552	0.0269	0.553	0.0268	0.563	0.0258	0.503	0.0233	0.550	0.0266
50	0.550	0.0198	0.563	0.0185	0.552	0.0196	0.553	0.0196	0.562	0.0193	0.500	0.0186	0.550	0.0199
100	0.547	0.0164	0.560	0.0140	0.549	0.0166	0.549	0.0166	0.559	0.0163	0.499	0.0146	0.547	0.0165
200	0.552	0.0118	0.567	0.0102	0.554	0.0120	0.554	0.0121	0.565	0.0107	0.504	0.0113	0.552	0.0119
500	0.549	0.0108	0.561	0.0088	0.551	0.0108	0.551	0.0105	0.561	0.0100	0.502	0.0105	0.549	0.0108
1000	0.547	0.0113	0.563	0.0088	0.549	0.0112	0.550	0.0111	0.560	0.0115	0.499	0.0125	0.547	0.0113
2000	0.549	0.0099	0.562	0.0067	0.551	0.0097	0.551	0.0096	0.561	0.0082	0.499	0.0108	0.549	0.0099
							$n = 100$							
1	0.607	0.1313	0.621	0.1449	0.608	0.1323	0.608	0.1324	0.626	0.1442	0.545	0.1036	0.607	0.1314
4	0.603	0.0599	0.613	0.0626	0.604	0.0601	0.605	0.0601	0.619	0.0634	0.542	0.0493	0.603	0.0599
8	0.604	0.0530	0.614	0.0578	0.605	0.0532	0.606	0.0534	0.621	0.0569	0.544	0.0410	0.604	0.0530
20	0.622	0.0412	0.634	0.0428	0.624	0.0413	0.624	0.0413	0.641	0.0439	0.557	0.0335	0.622	0.0412
50	0.621	0.0260	0.633	0.0254	0.622	0.0264	0.622	0.0264	0.640	0.0267	0.556	0.0221	0.621	0.0260
100	0.619	0.0216	0.632	0.0195	0.620	0.0216	0.620	0.0216	0.637	0.0207	0.554	0.0171	0.619	0.0217
200	0.617	0.0148	0.632	0.0131	0.618	0.0148	0.619	0.0148	0.637	0.0137	0.551	0.0134	0.617	0.0148
500	0.619	0.0125	0.632	0.0104	0.620	0.0125	0.621	0.0125	0.638	0.0115	0.554	0.0121	0.619	0.0126
1000	0.619	0.0097	0.633	0.0081	0.620	0.0099	0.621	0.0099	0.638	0.0095	0.555	0.0112	0.619	0.0097
2000	0.619	0.0111	0.634	0.0071	0.620	0.0113	0.620	0.0113	0.638	0.0093	0.555	0.0120	0.619	0.0111

TABLE 3 Ordinary Least Squares (OLS) Parameter Estimates and Diagnostics for the Model Eq. (8) Together with the $\hat{r}$ Value that Minimizes $\hat{\log}$(ASD) for *X1*

n	Penalty function	$\hat{a}$	$\hat{b}$	$\hat{c}$	$\bar{R}^2$	DW	$\hat{r}$
20	AIC	−1.800***	−0.594***	0.0301***	0.988***	1.709^{ns}	19,286
	BIC	−1.754***	−0.569***	0.0242**	0.987***	1.720^{ns}	127,542
	GCV	−1.788***	−0.582***	0.0280***	0.989***	1.691^{ns}	32,626
	HOC	−1.778***	−0.583***	0.0276***	0.988***	1.571^{ns}	38,623
	HQ	−1.788***	−0.582***	0.0278***	0.989***	1.593^{ns}	35,159
	RBAR	−1.979***	−0.603***	0.0353***	0.979***	1.682^{ns}	5,121
	MCP	−1.801***	−0.590***	0.0295***	0.988***	1.665^{ns}	22,026
50	AIC	−1.478***	−0.565***	0.0333***	0.987***	1.692^{ns}	4,834
	BIC	−1.280***	−0.633***	0.0249***	0.993***	1.817^{ns}	331,321
	GCV	−1.459***	−0.659***	0.0333***	0.987***	1.583^{ns}	19,829
	HOC	−1.451***	−0.661***	0.0335***	0.987***	1.542^{ns}	19,258
	HQ	−1.369***	−0.649***	0.0297***	0.991***	1.622^{ns}	55,599
	RBAR	−1.807***	−0.630***	0.0359***	0.989***	1.898^{ns}	6,466
	MCP	−1.478***	−0.655***	0.0332***	0.987***	1.695^{ns}	19,234
96	AIC	−2.264***	−0.551***	0.0293*	0.972***	2.243^{ns}	12,121
	BIC	−2.029***	−0.507***	0.0132^{ns}	0.977***	2.159^{ns}	218,991,934
	GCV	−2.254***	−0.551***	0.0290*	0.972***	2.251^{ns}	13,360
	HOC	−2.254***	−0.550***	0.0288*	0.973***	2.249^{ns}	14,025
	HQ	−2.120***	−0.539***	0.0225*	0.974***	2.211^{ns}	159,178
	RBAR	−2.573***	−0.528***	0.0338***	0.969***	2.495^{ns}	2,467
	MCP	−2.263***	−0.552***	0.0294*	0.972***	2.237^{ns}	11,941

***, **, *Denote significant at the 0.1, 1 and 5% levels, respectively, "ns" denotes not significant, and "nd" denotes an inconclusive test.

(1984) and Goffe et al. (1994), it does seem that this algorithm performs well at finding the global maxima in the presence of local maxima and for functions like ours which have plateaus and other ill-behaviour. The algorithm works well because it accepts both uphill and downhill moves in a random but systematic manner thus allowing the algorithm to by-pass local maxima and plateaus.

Much has been written on the SA algorithm. See in particular Corana et al. (1987), Goffe et al. (1994), Billah and King (2000b).

4 EXPERIMENTAL DESIGN OF THE SIMULATION STUDY

In order to investigate the small sample properties of our proposed model selection procedure and also to check its feasibility, we conducted a

TABLE 4 OLS Parameter Estimates and Diagnostics for the Model Eq. (8) Together with the $\hat{r}$ Value that Minimizes $\log\hat{g}$(ASD) for *X2*

n	Penalty Function	$\hat{a}$	$\hat{b}$	$\hat{c}$	$\bar{R}^2$	DW	$\hat{r}$
20	AIC	−2.681***	−0.568***	0.0434***	0.962***	2.335^{ns}	695
	BIC	−2.682***	−0.587***	0.0417***	0.964***	2.553nd	1,140
	GCV	−2.693***	−0.554***	0.0404***	0.955***	2.300^{ns}	950
	HOC	−2.692***	−0.556***	0.0400***	0.955***	2.328^{ns}	1,043
	HQ	−2.692***	−0.556***	0.0408***	0.957***	2.302^{ns}	910
	RBAR	−2.829***	−0.512***	0.0375***	0.973***	1.843^{ns}	922
	MCP	−2.682***	−0.566***	0.0429***	0.961***	2.418^{ns}	733
50	AIC	−2.212***	−0.575***	0.0334***	0.981***	2.172^{ns}	5,474
	BIC	−2.236***	−0.583***	0.0292**	0.981***	2.513^{ns}	21,653
	GCV	−2.211***	−0.570***	0.0325***	0.982***	2.205^{ns}	6,433
	HOC	−2.215***	−0.566***	0.0319***	0.982***	2.277^{ns}	7,126
	HQ	−2.186***	−0.582***	0.0321**	0.974***	2.417^{ns}	8,651
	RBAR	−2.400***	−0.562***	0.0369***	0.981***	2.219^{ns}	2,029
	MCP	−2.215***	−0.573***	0.0332***	0.981***	2.196^{ns}	5,594
100	AIC	−2.035***	−0.491***	0.0192*	0.987***	2.235^{ns}	357,345
	BIC	−1.926***	−0.502***	0.0125*	0.992***	2.537^{ns}	525,573,182
	GCV	−2.023***	−0.499***	0.0202**	0.986***	2.123^{ns}	231,303
	HOC	−2.022***	−0.499***	0.0202*	0.987***	2.129nd	231,303
	HQ	−1.938***	−0.505***	0.0171*	0.989***	2.285^{ns}	2,587,187
	RBAR	−2.243***	−0.528***	0.0295***	0.986***	1.449^{ns}	7,701
	MCP	−2.035***	−0.491***	0.0192*	0.987***	2.244^{ns}	357,345

***, **, *Denotes significant at the 0.1, 1, and 5% levels, respectively, "ns" denotes not significant, and "nd" denotes an inconclusive test.

simulation study comparing probabilities of correct selection of eight model selection procedures in a range of settings. The procedures included in the study are the new method outlined above involving the maximization of the estimated value of MAPCS($p_1, \ldots, p_m$) with respect to $p_2, \ldots, p_m$ (when $p_1 = 0$), which we will denote by MEM, and AIC, BIC, HQ, $\bar{R}^2$, GCV, HOC, and MCP.

The following four linear regression models along with the four models described in Subsection 3.2.1 were used in the study:

$$
\begin{aligned}
&M_5\colon y_t = \beta_{50} + x_{3t}\beta_{51} + u_{5t} && u_{5t} \sim IN(0, \sigma_5^2)\\
&M_6\colon y_t = \beta_{60} + x_{1t}\beta_{61} + x_{3t}\beta_{62} + u_{6t} && u_{6t} \sim IN(0, \sigma_6^2)\\
&M_7\colon y_t = \beta_{70} + x_{2t}\beta_{71} + x_{3t}\beta_{72} + u_{7t} && u_{7t} \sim IN(0, \sigma_7^2)\\
&M_8\colon y_t = \beta_{80} + x_{1t}\beta_{81} + x_{2t}\beta_{82} + x_{3t}\beta_{83} + u_{8t} && u_{8t} \sim IN(0, \sigma_8^2)
\end{aligned}
$$

where, as for $M_1 - M_4$, y_t is the tth observation of the dependent variable, x_{it} is the tth observation on the ith nonconstant regressor, the β's are regression coefficients, and the u_{jt} are the random disturbance terms for $t=1,\ldots,n$.

In addition to *X1* and *X2* described in Subsection 3.2.1, the following data sets were used to construct design matrices as required:

X3: x_{1t} and x_{2t} are randomly and independently generated values from the $N(0,1)$ distribution.

X4: This is an extension of *X1* where the additional regressor x_{3t} is x_{1t} lagged two quarters.

X5: This is an extension of *X2* where the additional regressor x_{3t} is the price level of consumption in the tth country.

X6: This is an extension of *X3* where the additional regressor x_{3t} is a further set of independently generated values from the $N(0,1)$ distribution.

The simulation experiments involved estimating the probabilities of correct selection when choosing between M_1, M_2, M_3, and M_4 in the context of *X1*, *X2*, and *X3*, and choosing between $M_1, M_2, \ldots, M_8$ in the context of *X4*, *X5*, and *X6*. Two thousand replications were used throughout and in the case of *X3* and *X6*, the randomly generated regressors were generated once and then held constant across the 2000 replications. Sample sizes of $n=20$, 50, and 100 were used with *X2*, *X3*, *X5*, and *X6* while $n=20$, 50, and 96 were used for *X1* and *X4*. MEM was applied with $s_e^2 = 55$ and $s_b^2 = 6$ for *X1*, $s_e^2 = 0.1$ and $s_b^2 = 0.5$ for *X2*, $s_e^2 = 0.12$ and $s_b^2 = 0.08$ for *X3* and *X6*, $s_e^2 = 0.1$ and $s_b^2 = 0.15$ for *X4*, and $s_e^2 = 0.35$ and $s_b^2 = 0.02$ for *X5*; $r=2000$ and $N=1$ were used to estimate Eq. (4) as described in Sections 2 and 3.

MEM was applied using the SA algorithm. In applying this method of maximizing the estimated value of Eq. (4), we experimented with different sets of starting values, upper and lower boundaries, and temperature reduction values, which are a requirement of the SA algorithm. Because AIC and BIC are so widely used in econometrics and statistics, we used the relative penalties (namely, 0, $p_2 - p_1$, $p_3 - p_1, \ldots, p_m - p_1$) of AIC and BIC as starting values. We also used zeros as starting values. A condition of the SA algorithm is that the starting values are not outside the boundary values of the parameters. The parameters in our case are $p_2, \ldots, p_m$ with $p_1 = 0$ and we selected boundary values with this in mind. When zeros or the relative penalties of AIC were used as the starting values, we used the maximum number of free parameters, k, among the competing models, the number of competing models, m, as well as km and the arbitrary value 10 as upper boundary values. For these starting values we used two types of lower boundary values, namely zeros and the negative of each of the upper boundaries. When the relative penalties of BIC were used as starting values, we used $k\log(n)/2$, $m\log(n)/2$, km, and the arbitrary value 10 as upper boundaries together with zeros and the negative of k, m, 10, and km.

In all cases, we used four temperature reduction values, namely, 0.1, 0.01, 0.001 and 0.0001. This means that in total, we experimented with 96 combinations of start-up values for the application of SA to the MEM method.

5 THE RESULTS

The estimated average probabilities of choosing the true model (APCS) for each of the models in turn plus the estimated mean average probabilities of correct selection (MAPCS) for each design matrix are given in Tables 5–10. We have also included the standard deviation (SD) of the m individual APCSs that make up the MAPCS. These give some idea of the relative variability in APCS for each model. The smaller the SD is, the less variation in the APCSs, which indicates a greater evenness in the APCS values which does seem desirable. We would prefer model selection procedures that do not unduly favor particular types of models.

Although we performed 96 sets of optimizations using SA for the MEM approach, we only give the results for four cases. These are for the penalties that give the largest MAPCS over all 96 optimizations, as well as the results for the smallest SD. In many cases, the different optimizations often result in exactly the same APCS and MAPCS so we give the results for the modal value of MAPCS together with those for the median MAPCS. As can be seen from the tables, it is not unusual for the largest MAPCS also to be the modal and the median MAPCS. Finally, for *X1* to *X3*, we give the relative penalties for each of the IC procedures.

5.1 Results when Choosing Between Four Models

We will first discuss the results for the existing IC methods (AIC, BIC, GCV, HOC, HQ, $\bar{R}^2$, and MCP) before turning to the results for the MEM approach.

In each case, BIC has the highest MAPCS of the existing IC procedures, followed by HOC for $n=20$ and HQ for larger sample sizes. BIC has the largest APCS for the minimal model M_1 and the lowest APCS for the largest model M_4. In contrast, $\bar{R}^2$ has the lowest MAPCS by a very clear margin and has the lowest APCS for M_1 and the largest APCS for M_4. The remaining existing IC procedures behave in a reasonably similar

TABLE 5 Estimated Average Probabilities (APCS), Mean Average Probabilities (MAPCS), and Standard Deviations (SD) of Average Probabilities of Correct Selection for *X1* Together with Relative Penalty Values

	APCS						Relative penalties		
Criteria	M_1	M_2	M_3	M_4	MAPCS	SD	p_2	p_3	p_4
					$n=20$				
AIC	0.6690	0.6545	0.3850	0.2605	0.4923	0.2023	1.0000	1.0000	2.0000
BIC	0.7970	0.6850	0.3485	0.1700	0.5001	0.2911	1.4979	1.4979	2.9957
GCV	0.6985	0.6710	0.3900	0.2295	0.4973	0.2265	1.0813	1.0813	2.2245
HOC	0.7095	0.6745	0.3885	0.2245	0.4992	0.2329	1.1123	1.1123	2.2901
HQ	0.7035	0.6635	0.3835	0.2370	0.4969	0.2242	1.0972	1.0972	2.1944
$\bar{R}^2$	0.4505	0.5775	0.4190	0.3785	0.4564	0.0860	0.5407	0.5407	1.1123
MCP	0.6715	0.6565	0.3850	0.2545	0.4919	0.2059	1.0059	1.0059	2.0220
MEM with largest MAPCS	0.8015	0.7255	0.3975	0.1095	0.5085	0.3186	1.6079	1.3140	3.6832
MEM with smallest SD	0.7940	0.6075	0.3635	0.2465	0.5029	0.2455	1.5974	1.3141	2.5162
MEM with modal MAPCS	0.8015	0.7255	0.3975	0.1095	0.5085	0.3186	1.6079	1.3140	3.6832
MEM with median MAPCS	0.8015	0.7255	0.3975	0.1095	0.5085	0.3186	1.6079	1.3140	3.6832
					$n=50$				
AIC	0.7190	0.7780	0.6655	0.5925	0.6887	0.0789	1.0000	1.0000	2.0000
BIC	0.9095	0.8830	0.7100	0.4530	0.7389	0.2101	1.9560	1.9560	3.9120
GCV	0.7290	0.7920	0.6715	0.5840	0.6941	0.0884	1.0310	1.0310	2.0836

HOC	0.7325	0.7940	0.6720	0.5815	0.6950	0.0906	1.0418	1.0418	2.1058
HQ	0.8125	0.8335	0.6925	0.5315	0.7175	0.1387	1.3641	1.3641	2.7281
$\bar{R}^2$	0.4900	0.6390	0.5800	0.6915	0.6001	0.0864	0.5155	0.5155	1.0418
MCP	0.7190	0.7785	0.6660	0.5915	0.6887	0.0795	1.0009	1.0009	2.0035
MEM with largest MAPCS	0.9745	0.8650	0.7185	0.4210	0.7448	0.2400	3.3296	2.8476	5.2849
MEM with smallest SD	0.8690	0.8165	0.6840	0.5460	0.7289	0.1446	1.7732	1.6949	2.9996
MEM with modal MAPCS	0.9745	0.8650	0.7185	0.4210	0.7448	0.2400	3.3296	2.8476	5.2849
MEM with median MAPCS	0.9745	0.8650	0.7185	0.4210	0.7448	0.2400	3.3296	2.8476	5.2849
					$n=96$				
AIC	0.7415	0.8155	0.8035	0.8830	0.8109	0.0580	1.0000	1.0000	2.0000
BIC	0.9390	0.9445	0.9160	0.8255	0.9063	0.0552	2.2822	2.2822	4.5643
GCV	0.7455	0.8195	0.8080	0.8815	0.8136	0.0557	1.0159	1.0159	2.0426
HOC	0.7480	0.8205	0.8085	0.8815	0.8146	0.0547	1.0213	1.0213	2.0536
HQ	0.8665	0.8905	0.8740	0.8595	0.8726	0.0133	1.5183	1.5183	3.0366
$\bar{R}^2$	0.5290	0.6600	0.6610	0.9155	0.6914	0.1618	0.5079	0.5079	1.0213
MCP	0.7415	0.8155	0.8035	0.8830	0.8109	0.0580	1.0003	1.0003	2.0009
MEM with largest MAPCS	0.9940	0.9530	0.9550	0.7920	0.9235	0.0897	4.7801	4.1070	7.6732
MEM with smallest SD	0.8650	0.8805	0.8820	0.8600	0.8719	0.0110	1.5598	1.4537	2.9997
MEM with modal MAPCS	0.9940	0.9530	0.9550	0.7920	0.9235	0.0897	4.7801	4.1070	7.6732
MEM with median MAPCS	0.9940	0.9530	0.9550	0.7920	0.9235	0.0897	4.7801	4.1070	7.6732

TABLE 6 Estimated Average Probabilities (APCS), Mean Average Probabilities (MAPCS), and Standard Deviations (SD) of Average Probabilities of Correct Selection for *X2* Together with Relative Penalty Values

	APCS						Relative penalties		
Criteria	M_1	M_2	M_3	M_4	MAPCS	SD	p_2	p_3	p_4
					$n=20$				
AIC	0.6800	0.8060	0.1580	0.2220	0.4665	0.3244	1.0000	1.0000	2.0000
BIC	0.8225	0.8815	0.1150	0.1345	0.4884	0.4206	1.4979	1.4979	2.9957
GCV	0.7125	0.8305	0.1495	0.1865	0.4698	0.3521	1.0813	1.0813	2.2245
HOC	0.7195	0.8395	0.1450	0.1820	0.4715	0.3593	1.1123	1.1123	2.2901
HQ	0.7135	0.8205	0.1455	0.1960	0.4689	0.3476	1.0972	1.0972	2.1944
$\bar{R}^2$	0.4710	0.6790	0.2195	0.3695	0.4347	0.1928	0.5407	0.5407	1.1123
MCP	0.6810	0.8080	0.1565	0.2165	0.4655	0.3272	1.0059	1.0059	2.0220
MEM with largest MAPCS	0.6245	0.5435	0.4550	0.5080	0.5328	0.0712	13.0759	0.4639	13.4040
MEM with smallest SD	0.5535	0.5440	0.4845	0.5080	0.5225	0.0320	2.6281	0.3685	2.9563
MEM with modal MAPCS	0.6255	0.5435	0.4535	0.5080	0.5326	0.0721	8.2473	0.4660	8.5757
MEM with median MAPCS	0.6490	0.5460	0.4265	0.5055	0.5318	0.0926	5.1629	0.5032	5.4936
					$n=50$				
AIC	0.7560	0.8230	0.3165	0.2080	0.5259	0.3088	1.0000	1.0000	2.0000
BIC	0.9260	0.9450	0.2840	0.0800	0.5587	0.4430	1.9560	1.9560	3.9120
GCV	0.7630	0.8320	0.3155	0.2010	0.5279	0.3161	1.0310	1.0310	2.0836

HOC	0.7670	0.8340	0.3150	0.1990	0.5287	0.3185	1.0418	1.0418	2.1058
HQ	0.8450	0.8865	0.3085	0.1430	0.5457	0.3760	1.3641	1.3641	2.7281
$\bar{R}^2$	0.5145	0.6660	0.3070	0.3660	0.4634	0.1608	0.5155	0.5155	1.0418
MCP	0.7560	0.8240	0.3165	0.2075	0.5260	0.3093	1.0009	1.0009	2.0035
MEM with largest MAPCS	0.8770	0.5170	0.5935	0.5285	0.6290	0.1687	8.9402	1.2482	9.2081
MEM with smallest SD	0.7765	0.5180	0.6275	0.5275	0.6124	0.1201	2.6381	0.8357	2.9068
MEM with modal MAPCS	0.8770	0.5170	0.5935	0.5285	0.6290	0.1687	8.9402	1.2482	9.2081
MEM with median MAPCS	0.8745	0.5170	0.5920	0.5285	0.6280	0.1676	5.4455	1.2475	5.7132
					$n = 100$				
AIC	0.7615	0.8450	0.5215	0.4040	0.6330	0.2052	1.0000	1.0000	2.0000
BIC	0.9495	0.9580	0.5190	0.2095	0.6590	0.3631	2.3026	2.3026	4.6052
GCV	0.7645	0.8520	0.5225	0.3970	0.6340	0.2107	1.0152	1.0152	2.0409
HOC	0.7665	0.8520	0.5220	0.3970	0.6344	0.2112	1.0204	1.0204	2.0514
HQ	0.8745	0.9170	0.5305	0.2990	0.6552	0.2939	1.5272	1.5272	3.0544
$\bar{R}^2$	0.5380	0.6935	0.4825	0.5435	0.5644	0.0904	0.5076	0.5076	1.0204
MCP	0.7615	0.8450	0.5215	0.4035	0.6329	0.2054	1.0002	1.0002	2.0009
MEM with largest MAPCS	0.9340	0.7990	0.7380	0.4635	0.7336	0.1978	15.2409	1.7396	16.0419
MEM with smallest SD	0.8450	0.7530	0.7405	0.5040	0.7106	0.1454	2.3308	1.1201	2.9778
MEM with modal MAPCS	0.9340	0.7990	0.7375	0.4635	0.7335	0.1978	7.4871	1.7373	8.2880
MEM with median MAPCS	0.9340	0.7990	0.7355	0.4635	0.7330	0.1978	6.1524	1.7391	6.9534

Table 7 Estimated Average Probabilities (APCS), Mean Average Probabilities (MAPCS) and Standard Deviations (SD) of Average Probabilities of Correct Selection for *X3* Together with Relative Penalty Values

	APCS						Relative penalties		
Criteria	M_1	M_2	M_3	M_4	MAPCS	SD	p_2	p_3	p_4
					$n=20$				
AIC	0.6355	0.5995	0.5725	0.5085	0.5790	0.0536	1.0000	1.0000	2.0000
BIC	0.7815	0.6210	0.5750	0.4205	0.5995	0.1486	1.4979	1.4979	2.9957
GCV	0.6650	0.6155	0.5815	0.4810	0.5857	0.0778	1.0813	1.0813	2.2245
HOC	0.6755	0.6165	0.5825	0.4740	0.5871	0.0846	1.1123	1.1123	2.2901
HQ	0.6695	0.6090	0.5735	0.4900	0.5855	0.0750	1.0972	1.0972	2.1944
$\bar{R}^2$	0.4305	0.5365	0.5320	0.6100	0.5272	0.0737	0.5407	0.5407	1.1123
MCP	0.6400	0.6025	0.5750	0.5075	0.5813	0.0559	1.0059	1.0059	2.0220
MEM with largest MAPCS	0.8740	0.5745	0.5590	0.4110	0.6046	0.1941	2.1992	1.8608	3.6479
MEM with smallest SD	0.8000	0.6100	0.5330	0.4590	0.6005	0.1466	1.5052	1.6540	2.8529
MEM with modal MAPCS	0.8740	0.5745	0.5590	0.4110	0.6046	0.1941	2.1992	1.8608	3.6479
MEM with median MAPCS	0.8740	0.5745	0.5590	0.4110	0.6046	0.1941	2.1992	1.8608	3.6479
					$n=50$				
AIC	0.7050	0.7140	0.6610	0.6610	0.6853	0.0282	1.0000	1.0000	2.0000
BIC	0.8985	0.7660	0.6980	0.5605	0.7308	0.1408	1.9560	1.9560	3.9120
GCV	0.7140	0.7235	0.6720	0.6565	0.6915	0.0323	1.0310	1.0310	2.0836

HOC	0.7165	0.7265	0.6740	0.6545	0.6929	0.0342	1.0418	1.0418	2.1058
HQ	0.7945	0.7460	0.6875	0.6140	0.7105	0.0778	1.3641	1.3641	2.7281
$\bar{R}^2$	0.4735	0.6145	0.5735	0.7455	0.6018	0.1127	0.5155	0.5155	1.0418
MCP	0.7050	0.7145	0.6615	0.6605	0.6854	0.0284	1.0009	1.0009	2.0035
MEM with largest MAPCS	0.9060	0.7665	0.7035	0.5560	0.7330	0.1452	2.0229	1.9573	4.0287
MEM with smallest SD	0.8005	0.7780	0.7010	0.5930	0.7181	0.0937	1.3051	1.4416	2.9445
MEM with modal MAPCS	0.9060	0.7665	0.7035	0.5560	0.7330	0.1452	2.0229	1.9573	4.0287
MEM with median MAPCS	0.9060	0.7665	0.7035	0.5560	0.7330	0.1452	2.0229	1.9573	4.0287
					$n=100$				
AIC	0.7025	0.7665	0.7125	0.7515	0.7332	0.0306	1.0000	1.0000	2.0000
BIC	0.9315	0.8325	0.7580	0.6375	0.7899	0.1240	2.3026	2.3026	4.6052
GCV	0.7090	0.7710	0.7170	0.7475	0.7361	0.0286	1.0152	1.0152	2.0409
HOC	0.7090	0.7710	0.7175	0.7470	0.7361	0.0284	1.0204	1.0204	2.0514
HQ	0.8450	0.8095	0.7605	0.6980	0.7782	0.0637	1.5272	1.5272	3.0544
$\bar{R}^2$	0.4575	0.6440	0.6055	0.8275	0.6336	0.1522	0.5076	0.5076	1.0204
MCP	0.7025	0.7665	0.7125	0.7515	0.7332	0.0306	1.0002	1.0002	2.0009
MEM with largest MAPCS	0.8905	0.8330	0.8165	0.6305	0.7926	0.1126	2.1758	1.5387	4.2676
MEM with smallest SD	0.8485	0.7945	0.7700	0.7065	0.7799	0.0589	1.6293	1.4306	2.9826
MEM with modal MAPCS	0.8905	0.8330	0.8160	0.6305	0.7925	0.1126	2.1811	1.5461	4.2745
MEM with median MAPCS	0.8905	0.8330	0.8160	0.6305	0.7925	0.1126	2.1811	1.5461	4.2745

Table 8 Estimated Average Probabilities (APCS), Mean Average Probabilities (MAPCS), and Standard Deviations (SD) of Average Probabilities of Correct Selection for *X4*

	APCS									
Criteria	M_1	M_2	M_3	M_4	M_5	M_6	M_7	M_8	MAPCS	SD
					$n=20$					
AIC	0.5320	0.3800	0.3635	0.2760	0.3470	0.2625	0.2560	0.2585	0.3344	0.0946
BIC	0.7105	0.4060	0.3755	0.2280	0.3655	0.2185	0.2160	0.1905	0.3388	0.1726
GCV	0.5730	0.3955	0.3745	0.2670	0.3625	0.2590	0.2475	0.2210	0.3375	0.1156
HOC	0.5880	0.4000	0.3760	0.2655	0.3635	0.2570	0.2435	0.2160	0.3387	0.1219
HQ	0.5760	0.3865	0.3630	0.2675	0.3555	0.2585	0.2470	0.2380	0.3365	0.1131
$\bar{R}^2$	0.2990	0.2990	0.3220	0.3155	0.2910	0.3010	0.2905	0.3590	0.3096	0.0228
MCP	0.5325	0.3820	0.3640	0.2740	0.3460	0.2615	0.2525	0.2505	0.3329	0.0965
MEM with largest MAPCS	0.6320	0.4095	0.4280	0.2650	0.3425	0.2550	0.1870	0.2305	0.3437	0.1444
MEM with smallest SD	0.5695	0.3850	0.4620	0.2410	0.3745	0.2810	0.2320	0.1955	0.3426	0.1290
MEM with modal MAPCS	0.6255	0.4100	0.4115	0.2685	0.3345	0.2555	0.2465	0.1950	0.3434	0.1381
MEM with median MAPCS	0.6245	0.4170	0.4170	0.2650	0.3395	0.2555	0.2345	0.1930	0.3433	0.1405
					$n=50$					
AIC	0.6240	0.5700	0.5905	0.5540	0.5945	0.5290	0.5635	0.5725	0.5747	0.0286
BIC	0.8770	0.7125	0.7325	0.5280	0.7300	0.5115	0.5400	0.4565	0.6360	0.1467
GCV	0.6450	0.5870	0.6040	0.5530	0.6085	0.5305	0.5675	0.5580	0.5817	0.0368

HOC	0.6505	0.5900	0.6050	0.5530	0.6130	0.5315	0.5685	0.5575	0.5836	0.0385
HQ	0.7530	0.6470	0.6640	0.5535	0.6765	0.5300	0.5690	0.5200	0.6141	0.0831
$\bar{R}^2$	0.3620	0.4010	0.4220	0.5010	0.4090	0.4765	0.5230	0.6650	0.4699	0.0958
MCP	0.6255	0.5710	0.5910	0.5540	0.5955	0.5295	0.5635	0.5720	0.5753	0.0290
MEM with largest MAPCS	0.9255	0.6885	0.7245	0.5625	0.7180	0.5330	0.5300	0.4530	0.6419	0.1519
MEM with smallest SD	0.7720	0.6845	0.6485	0.5685	0.6845	0.4940	0.5335	0.5475	0.6166	0.0952
MEM with modal MAPCS	0.9135	0.6940	0.7250	0.5665	0.7275	0.5320	0.5160	0.4580	0.6416	0.1503
MEM with median MAPCS	0.9135	0.6940	0.7250	0.5665	0.7275	0.5320	0.5160	0.4580	0.6416	0.1503
					$n = 96$					
AIC	0.6895	0.6565	0.6860	0.7480	0.6630	0.7490	0.7210	0.8735	0.7233	0.0702
BIC	0.9525	0.8850	0.8905	0.8340	0.8840	0.8030	0.8055	0.8135	0.8585	0.0531
GCV	0.6990	0.6660	0.6935	0.7535	0.6715	0.7535	0.7235	0.8720	0.7291	0.0667
HOC	0.7025	0.6685	0.6945	0.7555	0.6750	0.7545	0.7250	0.8720	0.7309	0.0657
HQ	0.8525	0.7895	0.8130	0.8065	0.7945	0.7845	0.7780	0.8460	0.8081	0.0278
$\bar{R}^2$	0.3975	0.4445	0.4600	0.6270	0.4485	0.6345	0.6070	0.9040	0.5654	0.1658
MCP	0.6895	0.6575	0.6860	0.7480	0.6635	0.7490	0.7210	0.8735	0.7235	0.0700
MEM with largest MAPCS	0.9970	0.9230	0.9395	0.8380	0.9300	0.8150	0.7870	0.7690	0.8748	0.0831
MEM with smallest SD	0.7145	0.8480	0.8470	0.7920	0.8285	0.7465	0.7720	0.8475	0.7995	0.0515
MEM with modal MAPCS	0.9970	0.9230	0.9395	0.8380	0.9300	0.8150	0.7870	0.7690	0.8748	0.0831
MEM with median MAPCS	0.9970	0.9240	0.9350	0.8390	0.9310	0.8140	0.7825	0.7745	0.8746	0.0825

Table 9 Estimated Average Probabilities (APCS), Mean Average Probabilities (MAPCS), and Standard Deviations (SD) of Average Probabilities of Correct Selection for *X5*

	APCS									
Criteria	M_1	M_2	M_3	M_4	M_5	M_6	M_7	M_8	MAPCS	SD
					$n=20$					
AIC	0.5445	0.6280	0.2395	0.2985	0.5015	0.6275	0.2335	0.3330	0.4257	0.1681
BIC	0.7240	0.7735	0.2325	0.2300	0.5660	0.6555	0.1855	0.2420	0.4511	0.2519
GCV	0.5865	0.6825	0.2485	0.2765	0.5325	0.6440	0.2250	0.2915	0.4359	0.1935
HOC	0.6000	0.6910	0.2515	0.2780	0.5390	0.6485	0.2210	0.2860	0.4394	0.1983
HQ	0.5885	0.6630	0.2400	0.2825	0.5205	0.6295	0.2265	0.3140	0.4331	0.1852
$\bar{R}^2$	0.3175	0.4295	0.2200	0.3450	0.3775	0.5635	0.2735	0.4380	0.3706	0.1075
MCP	0.5465	0.6340	0.2415	0.2955	0.5055	0.6290	0.2320	0.3280	0.4265	0.1706
MEM with largest MAPCS	0.6815	0.7115	0.4390	0.3670	0.5095	0.5375	0.3600	0.3455	0.4939	0.1433
MEM with smallest SD	0.5190	0.6290	0.4905	0.4040	0.5150	0.5270	0.3115	0.3915	0.4734	0.0992
MEM with modal MAPCS	0.6830	0.7120	0.4415	0.3675	0.5045	0.5375	0.3550	0.3455	0.4933	0.1441
MEM with median MAPCS	0.6770	0.7015	0.4350	0.3625	0.5070	0.5490	0.3570	0.3490	0.4922	0.1416
					$n=50$					
AIC	0.5775	0.6915	0.3595	0.3875	0.6210	0.7475	0.3975	0.4700	0.5315	0.1485
BIC	0.8535	0.8860	0.3410	0.3020	0.7820	0.8140	0.3325	0.3125	0.5779	0.2755
GCV	0.5910	0.7100	0.3660	0.3815	0.6340	0.7605	0.3955	0.4595	0.5373	0.1566

HOC	0.5965	0.7165	0.3675	0.3790	0.6365	0.7600	0.3950	0.4565	0.5384	0.1584
HQ	0.7120	0.7950	0.3710	0.3505	0.7015	0.7900	0.3725	0.4045	0.5621	0.2036
$\bar{R}^2$	0.3215	0.4665	0.2890	0.3965	0.4230	0.6115	0.4105	0.5760	0.4368	0.1124
MCP	0.5795	0.6925	0.3595	0.3875	0.6210	0.7485	0.3970	0.4690	0.5318	0.1491
MEM with largest MAPCS	0.7875	0.7685	0.5530	0.5260	0.7260	0.7060	0.4620	0.4450	0.6218	0.1402
MEM with smallest SD	0.5950	0.6810	0.5380	0.5035	0.6675	0.7055	0.4875	0.4905	0.5836	0.0910
MEM with modal MAPCS	0.7955	0.7585	0.5560	0.5200	0.7090	0.7165	0.4650	0.4485	0.6211	0.1387
MEM with median MAPCS	0.7845	0.7660	0.5505	0.5220	0.7245	0.7095	0.4565	0.4465	0.6200	0.1407
					$n = 100$					
AIC	0.5980	0.7005	0.4750	0.5105	0.6650	0.7875	0.5170	0.6320	0.6107	0.1070
BIC	0.9100	0.9230	0.5005	0.4215	0.8325	0.8740	0.4630	0.4610	0.6732	0.2288
GCV	0.6030	0.7100	0.4810	0.5125	0.6735	0.7950	0.5170	0.6250	0.6146	0.1090
HOC	0.6050	0.7110	0.4820	0.5130	0.6750	0.7970	0.5175	0.6245	0.6156	0.1094
HQ	0.7755	0.8345	0.5230	0.4920	0.7650	0.8420	0.5030	0.5560	0.6614	0.1560
$\bar{R}^2$	0.3435	0.4670	0.3640	0.4860	0.4550	0.6520	0.4860	0.7355	0.4986	0.1337
MCP	0.5980	0.7010	0.4750	0.5110	0.6650	0.7875	0.5170	0.6310	0.6107	0.1070
MEM with largest MAPCS	0.8955	0.8545	0.6525	0.5735	0.7545	0.7690	0.5975	0.6070	0.7130	0.1230
MEM with smallest SD	0.6770	0.7735	0.6115	0.5480	0.7580	0.7165	0.5820	0.6775	0.6680	0.0817
MEM with modal MAPCS	0.8925	0.8705	0.6560	0.5690	0.7540	0.7580	0.5980	0.6040	0.7127	0.1253
MEM with median MAPCS	0.8755	0.8635	0.6440	0.5650	0.7560	0.7760	0.6210	0.5910	0.7115	0.1224

TABLE 10 Estimated Average Probabilities (APCS), Mean Average Probabilities (MAPCS), and Standard Deviations (SD) of Average Probabilities of Correct Selection for *X6*

	APCS									
Criteria	M_1	M_2	M_3	M_4	M_5	M_6	M_7	M_8	MAPCS	SD
					$n=20$					
AIC	0.5310	0.3275	0.5425	0.3325	0.5705	0.3515	0.5650	0.3740	0.4493	0.1116
BIC	0.7065	0.3285	0.6260	0.2885	0.6865	0.3190	0.5770	0.3025	0.4793	0.1858
GCV	0.5655	0.3445	0.5730	0.3255	0.6125	0.3530	0.5775	0.3385	0.4613	0.1302
HOC	0.5775	0.3460	0.5835	0.3220	0.6195	0.3525	0.5775	0.3330	0.4639	0.1352
HQ	0.5690	0.3345	0.5610	0.3280	0.6010	0.3460	0.5660	0.3525	0.4572	0.1258
$\bar{R}^2$	0.3070	0.2810	0.4020	0.3530	0.3940	0.3840	0.5090	0.4890	0.3899	0.0795
MCP	0.5325	0.3300	0.5455	0.3335	0.5750	0.3550	0.5680	0.3675	0.4509	0.1129
MEM with largest MAPCS	0.7615	0.4425	0.6540	0.3435	0.6940	0.3455	0.4360	0.3675	0.5056	0.1702
MEM with smallest SD	0.5170	0.5160	0.6735	0.3215	0.6520	0.3730	0.3760	0.4835	0.4891	0.1290
MEM with modal MAPCS	0.7520	0.4450	0.6540	0.3450	0.6855	0.3595	0.4355	0.3665	0.5054	0.1648
MEM with median MAPCS	0.7580	0.4515	0.6545	0.3440	0.6625	0.3720	0.4350	0.3605	0.5047	0.1619
					$n=50$					
AIC	0.5815	0.4510	0.6105	0.4730	0.5990	0.4900	0.6810	0.5505	0.5546	0.0787
BIC	0.8675	0.4735	0.7660	0.4165	0.7895	0.4375	0.7155	0.4190	0.6106	0.1913
GCV	0.5940	0.4575	0.6270	0.4785	0.6190	0.4945	0.6905	0.5400	0.5626	0.0828

HOC	0.6010	0.4600	0.6295	0.4790	0.6225	0.4925	0.6895	0.5370	0.5639	0.0834
HQ	0.7165	0.4770	0.6960	0.4605	0.7055	0.4795	0.6985	0.4955	0.5911	0.1213
$\bar{R}^2$	0.3105	0.3460	0.4265	0.4730	0.4105	0.4705	0.5695	0.6570	0.4579	0.1134
MCP	0.5815	0.4510	0.6125	0.4730	0.5995	0.4905	0.6810	0.5500	0.5549	0.0789
MEM with largest MAPCS	0.8560	0.5595	0.7680	0.4675	0.7210	0.5220	0.6500	0.4485	0.6241	0.1486
MEM with smallest SD	0.6155	0.6325	0.7045	0.4880	0.7010	0.5030	0.5965	0.5635	0.6006	0.0808
MEM with modal MAPCS	0.8225	0.6085	0.7420	0.4655	0.7125	0.5320	0.6575	0.4480	0.6236	0.1347
MEM with median MAPCS	0.8390	0.5735	0.7660	0.4640	0.7255	0.5195	0.6510	0.4490	0.6234	0.1450
						$n=100$				
AIC	0.5860	0.5310	0.6470	0.5245	0.6660	0.5740	0.7245	0.6465	0.6124	0.0700
BIC	0.8865	0.5875	0.8220	0.4905	0.8540	0.5390	0.7685	0.5065	0.6818	0.1670
GCV	0.5905	0.5385	0.6515	0.5265	0.6755	0.5770	0.7300	0.6390	0.6161	0.0702
HOC	0.5925	0.5390	0.6550	0.5265	0.6755	0.5765	0.7315	0.6385	0.6169	0.0707
HQ	0.7540	0.5855	0.7570	0.5295	0.7785	0.5825	0.7590	0.5840	0.6663	0.1043
$\bar{R}^2$	0.3105	0.3870	0.4345	0.4880	0.4430	0.5110	0.6160	0.7415	0.4914	0.1351
MCP	0.5860	0.5310	0.6470	0.5245	0.6670	0.5740	0.7245	0.6465	0.6126	0.0701
MEM with largest MAPCS	0.8700	0.6725	0.8025	0.5630	0.7955	0.6235	0.7065	0.5500	0.6979	0.1174
MEM with smallest SD	0.6715	0.6365	0.7400	0.5590	0.7920	0.5820	0.6535	0.6600	0.6618	0.0764
MEM with modal MAPCS	0.8490	0.6855	0.8055	0.5630	0.8010	0.6295	0.6955	0.5520	0.6976	0.1130
MEM with median MAPCS	0.8705	0.6725	0.7970	0.5585	0.7785	0.6155	0.7265	0.5605	0.6974	0.1151

manner with perhaps GCV being the next best. They are closer to HQ and HOC in their performance and clearly superior to $\bar{R}^2$. Interestingly, $\bar{R}^2$ has the smallest SD for *X1* and *X2* but not for *X3*.

For *X2*, MEM has a higher MAPCS than all existing IC. The improvement in MAPCS is clear cut and seems to be achieved by a greater balance in the individual APCSs. On the other hand, MEM has only a slightly higher MAPCS than BIC for *X1* and *X3* whether it be the MEM with the largest, modal, or median MAPCS. We also found that for the larger sample sizes, the MEM with the smallest SD has an MAPCS below that of BIC.

MEM has much lower APCSs for the smaller M_1 model than BIC for *X2* and larger APCSs for the larger M_4 model. It seems that its clear advantage over BIC comes from being able to adjust the penalties so as not to overfavor M_1 with the result that there are larger gains in the APCSs for M_4. In the case of *X1* and *X3*, it appears BIC has reasonably optimal penalties so little can be done by way of changing these penalties to improve the MAPCS.

5.2 Results when Choosing Between Eight Models

The relative patterns for *X4–X6* are very similar to those for *X1–X3*, particularly with regard to MAPCS values. Again, BIC is the best of the existing IC procedures with $\bar{R}^2$ being clearly the worst. BIC favors models with fewer parameters while $\bar{R}^2$ favors models with more parameters. HQ is the second best existing procedure, especially for the larger sample sizes. The largest, median, and modal MAPCS for MEM is always higher than that of BIC, particularly for *X5*. This time we see that MEM and BIC and to a lesser extent HQ are the best procedures. MEM seems to be better placed to exploit any advantage for $n=20$, but it does appear that as the sample size increases, it is a little harder for MEM to clearly dominate BIC in terms of MAPCS.

6 CONCLUDING REMARKS

In this chapter we have outlined a new approach to IC model selection. It is based on the use of simulation to estimate average probabilities of correct selection and choosing penalties that optimize the mean of these average probabilities over the range of models under consideration. It requires the use of the SA algorithm to find these penalties, which can be computationally demanding.

We conducted a Monte Carlo experiment designed to investigate and compare the small sample properties of our new procedure with seven existing IC procedures. The results of this experiment suggest that often BIC (two out of three cases on average in our limited experience) has close to optimal penalties. Occasionally, and particularly for smaller samples, the new procedure clearly has better small sample properties than BIC. For the best results, care must be taken with the optimization. It may be desirable to try different starting values and not allow boundary values to be restrictive. On the basis of our experiment, we do not recommend the use of $\bar{R}^2$ for selecting regression models.

Finally, it is worth observing that the approach outlined in this chapter need not be restricted to penalizing maximized likelihood functions. The latter require restrictive distributional assumptions about the models being investigated. It could be that some other estimation criterion such as least squares or least absolute deviations needs to be employed. Clearly, the model selection procedure outlined in this chapter could be readily adapted to such an estimation criterion.

ACKNOWLEDGMENTS

This work was supported by an ARC grant. The authors are grateful to Kwek Kian Teng and Baki Billah for helpful discussions and to Sivagowry Sriananthakumar for research assistance.

REFERENCES

Akaike, H., 1973. Information theory and an extension of the maximum likelihood principle. In B.N. Petrov and F. Csaki (eds.) *Second International Symposium of Information Theory*. Akademiai Kiado, Budapest, pp. 267–281.

Akaike, H., 1974. A new look at the statistical model identification. *IEEE Transactions on Automatic Control*, **AC-19**, 716–723.

Billah, M.B. and King, M.L., 2000a. Time series model selection and forecasting via optimal penalty estimation. *Pakistan Journal of Statistics* **16**, 126–145.

Billah, M.B. and King, M.L., 2000b. Using simulated annealing to estimate penalty functions for time series model selection. Unpublished manuscript, Monash University, Melbourne, Australia.

Corana, A., Marchesi, M., Martini, C., and Ridella, S., 1987. Minimising multimodal functions of continuous variables with the simulated annealing algorithm. *ACM Transactions of Mathematical Software* **13**, 262–280.

Fox, K.J., 1995. Model selection criteria: A reference source. Unpublished manuscript, University of British Columbia, Vancouver, Canada, and University of New South Wales, Sydney, Australia.

Giles, J.A. and Giles, D.E.A., 1993. Pre-test estimation and testing in econometrics: Recent developments. *Journal of Economic Surveys* **7**, 145–197.

Goffe, W.L., Ferrier, G.D., and Rogers, J., 1994. Global optimization of statistical functions with simulated annealing. *Journal of Econometrics* **60**, 65–99.

Granger, C.J.W., King, M.L., and White, H., 1995. Comments on testing economic theories and the use of model selection criteria. *Journal of Econometrics* **67**, 173–187.

Grose, S.D., and King, M.L., 1994. The use of information criteria for model selection between models with equal numbers of parameters. Unpublished manuscript, Monash University, Melbourne, Australia.

Hannan, E.J. and Quinn, B.G., 1979. The determination of the order of an autoregression. *Journal of the Royal Statistical Society Series* B **41**, 190–195.

Hocking, R.R., 1976. The analysis and selection of variables in linear regression. *Biometrics* **32**, 1–49.

King, M.L., 1987. Towards a theory of point-optimal testing. *Econometric Reviews* **6**, 169–218.

Kirkpatrick, S., Gelatt, C.D., and Vecchi, M.P., 1983. Optimization by simulated annealing. *Science* **220**, 671–680.

Kwek, K.T., 1999. Model selection for a class of conditional heteroscedastic processes. Unpublished Ph.D. thesis, Monash University, Melbourne, Australia.

Kwek, K.T., King, M.L., 1999. Model selection and optimal penalty function estimation for conditional heteroskedastic processes. Unpublished manuscript, Monash University, Melbourne, Australia.

Mallows, C.L., 1964. Choosing variables in a linear regression: A graphical aid. Presented at the Central Regional Meeting of the Institute of Mathematical Statistics, Manhattan, KS.

Romeo, F., Sangiovanni, V. A., and Sechen, C., 1984. Research on simulated annealing at Berkeley. In *Proceedings of the IEEE International Conference on Computer Design, ICCD 84*. IEEE, New York, pp. 652–657.

Schmidt, P., 1975. Choosing among alternative linear regression models: A correction and some further results. *Atlantic Economic Journal* **3**, 61–63.

Schwarz, G., 1978. Estimating the dimension of a model. *Annals of Statistics* **6**, 461–464.

Summers, R., and Heston, A., 1991. The Penn World Table (Mark 5): An expanded set of International comparisons, 1950–1988. *Quarterly Journal of Economics* **106**, 327–368.

Theil, H., 1961. *Economic Forecasts and Policy*, 2nd ed. North Holland, Amsterdam.

Wallace, T.D., 1977. Pre-test estimation in regression: A survey. *American Journal of Agricultural Economics* **59**, 431–443.

White, S.R., 1984. Concepts of scale in simulated annealing. In *Proceedings of the IEEE International Conference on Computer Design, ICCD 84*. IEEE, New York, 646–651.

Zellner, A., 1971. *An Introduction to Bayesian Inference in Econometrics*. Wiley, New York.

4

On Bootstrap Coverage Probability with Dependent Data

Janis J. Zvingelis
University of Iowa, Iowa City, Iowa, U.S.A.

1 INTRODUCTION

The bootstrap, introduced by Efron (1979), is a statistical procedure for estimating the distribution of a given estimator. The distinguishing feature of the bootstrap is that it replaces the unknown population distribution of the data by an estimate of it, which is formed by resampling the original sample randomly with replacement.

When the observed data form a random sample, the bootstrap often provides more accurate critical values for the tests than asymptotic theory (e.g., Singh, 1981; Hall, 1986, 1992; Beran, 1988). Hall (1988) proves that the error in coverage probability made by symmetric two-sided confidence intervals, when bootstrap critical values are used in the IID case, is of order $\mathcal{O}(n^{-2})$. This amounts to sizable refinement for the precision of asymptotic confidence intervals, since the errors made by one- and symmetric two-sided asymptotic confidence intervals are of orders $\mathcal{O}(n^{-1/2})$ and $\mathcal{O}(n^{-1})$, respectively.* Monte Carlo experiments support the predictions of the theory, sometimes producing spectacular results (Horowitz, 1994).

In the case of dependent data the bootstrap procedure must be designed in a way that suitably captures the dependence structure of the original sample. Several different sampling procedures have been invented to tackle this task. Carlstein (1986) proposes to divide the original data set in nonoverlapping blocks and then sample these blocks randomly with replacement. Künsch (1989) proceeds similarly, except that he divides the original sample in overlapping blocks. Hall (1985) also suggested these techniques in the context of spatial data. Despite the blocking, the dependence structure of the original sample is not replicated exactly in the bootstrap sample. For example, if nonoverlapping blocks are used, the observations from different blocks in the bootstrap sample are independent with respect to the probability measure induced by bootstrap sampling. Furthermore, observations from the same block are deterministically related. Lastly, the block bootstrap sample is nonstationary even if the original sample is stationary. This dependence structure is unlikely to be present in the original sample. As a result bootstrap performance deteriorates. Hall and Horowitz (1996) give conditions under with Carlstein's block bootstrap provides asymptotic refinements through $\mathcal{O}(n^{-1})$ for coverage probabilities, when bootstrap critical values are used to construct symmetric two-sided confidence intervals for generalized method of moments (GMM) estimators.

The random variables of interest in this chapter are standardized and studentized smooth functions of sample moments of $\tilde{X}$ or sample moments of functions of $\tilde{X}$, where $\tilde{X}$ denotes the sample. For this broad class of random variables we have established the following result: the errors made in the coverage probabilities by one- and symmetric two-sided nonoverlapping block bootstrap confidence intervals are of orders $\mathcal{O}(n^{-3/4})$ and $\mathcal{O}(n^{-4/3})$, respectively, when optimal block lengths are used. The optimal block lengths are equal to $C_1 n^{1/4}$ and $C_2 n^{1/3}$ for one- and symmetric two-sided confidence intervals, respectively.† Note, however, that the improvement from using the bootstrap over the asymptotics leaves much

*By "refinement through $O(n^{-r})$" we mean that the estimated parameter of interest is correct up to and including the term of order $O(n^{-r})$, and the estimation error is of size $o(n^{-r})$.

†For convenience and simplicity we will employ C and/or C_i, $i = 1, 2, \ldots$ to denote some finite constants that depend on the specifics of the data-generation process and the functional form of the estimator, but not on sample size, n. These constants may assume different values at each appearance. Since the constants do not depend on the sample size, they are not of direct interest given the goals of this paper.

to be desired, especially in the symmetric two-sided confidence interval case. The lackluster performance of the nonoverlapping block bootstrap, as demonstrated in the Monte Carlo experiments of Hall and Horowitz (1996) and Hansen (1999), among others, is consistent with the theory established in the present paper.

To achieve asymptotic refinement, the Edgeworth expansions of the statistic of interest and its bootstrap equivalent have to have the same structure apart from replacing bootstrap cumulants with sample cumulants in the bootstrap expansion.* Lahiri (1992) and Hall and Horowitz (1996) proposed "corrected" bootstrap estimators that achieve asymptotic refinement and partially account for the change in the dependence structure in the bootstrap sample. The corrected versions of bootstrap test statistics are also used in this chapter. The point of the correction factor is to make the exact variance of the leading term of the Taylor series expansion of the bootstrap test statistic equal to one and to do this without introducing new (bootstrap) stochastic terms that would affect the structure of the Edgeworth expansion.

An enlightening fact to note is that one does not need correction factors in the one-sided confidence interval case to achieve asymptotic refinement through $\mathcal{O}(n^{-1/2})$ (see, e.g., Lahiri, 1992; Davison and Hall, 1993; Götze and Künsch, 1996; Lahiri, 1996). The reason for this is that the differences between the population and bootstrap variances of higher order terms of the Taylor series expansions of the random variable of interest are of order smaller than $\mathcal{O}(n^{-1/2})$.

The solution method used in this chapter was introduced in Hall (1988). A crucial prerequisite for the use of this technique is the existence of the Edgeworth expansions for the random variable of interest. Götze and Hipp (1983, 1994) and Götze and Künsch (1996) give regularity conditions under which the Edgeworth expansions exist for smooth functions of sample averages in the dependent data setting. However, the method of solution of this chapter involves terms that are not smooth functions of sample moments. These terms are of the form $(1/b)^m \sum_{i=1}^{b}(\sum_{j=1}^{\ell} X_{ij}/l)^m$, where m is some positive integer, l is the bootstrap block length, b is the number of blocks, and X_{ij} is the jth observation in the ith block of the bootstrap sample.

---

*An *Edgeworth expansions* is an approximation to the distribution function of a random variable. Under certain assumptions Edgeworth expansion takes on the form of a power series in n^{-r}, where the first term is the standard normal distribution function and r depends on the type of a random variable. The power series form of an Edgeworth expansion makes it a convenient tool for determining the size of the error made by an estimator of a finite sample distribution function. See Hall (1992) for a detailed discussion on Edgeworth expansions.

At present, there are no Edgeworth expansion results in the literature that apply to the statistics of the above type in the dependent data setting. Thus, the techniques used and results obtained in this chapter are heuristic.

The chapter is organized as follows: Section 2 introduces the test statistics of interest, Section 3 lays out the main theoretical results, and Section 4 concludes. This is followed by an Appendix containing the relevant mathematical derivations.

2 TEST STATISTICS

The notation will largely follow that used in Hall et al. (1995) and Hall and Horowitz (1996).

2.1 Sample

The test statistics of interest in this chapter are either normalized or studentized smooth functions of sample moments of $\tilde{X}$ or sample moments of functions of $\tilde{X}$.* Most of the test statistics and estimators in economics are smooth functions of sample averages or can be approximated by such with negligible error. Test statistics based on GMM estimators constitute an example of the latter case (Hall and Horowitz, 1996, Proposition 1, 2).

Denote the data by $\chi = (X_1, \ldots, X_{n_{\text{full}}})$, where $X_i \in \mathcal{R}^d$ is a $d \times 1$ random variable. Assume that $\{X_i\}$ is an ergodic stationary strongly mixing stochastic process and that $EX_iX_j' = 0$ if $|i-j| > k$ for some integer $k < \infty$.† We need to make this assumption so that the estimator of the covariance necessary for studentization would be a smooth function of sample moments. Andrews (1991) proposes a nonparametric covariance matrix estimator, but it is not a smooth function of sample moments and convergence at a rate that is slower than $n^{-1/2}$. Existing theory on Edgeworth expansions with dependent data (Götze and Hipp, 1983, 1994) applies only to smooth functions of sample moments. Set $n = n_{\text{full}} - k$. Define the samples as $\tilde{\chi} \equiv \{\tilde{X}_i : i = 1, \ldots, n\}$, where $\tilde{X}_i = \{X_i', \ldots, X_{i+k}'\}'$. We need to make this redefinition of the sample so that the consistent estimator of the asymptotic variance is a smooth function of sample moments, because the cross-product components of the covariance estimator are not smooth functions of sample moments.

---

*By "smooth" we mean that the appropriate number of derivatives exist.

†Note that the restriction on the expectations of the products is not equivalent to the assumption of m dependence. Also, this assumption is not as restrictive as it might seem, because X_i, for example, can be equal to the GMM moment function. In that case the data can be dependent with an infinite lag, given that some assumptions are satisfied (see Hall and Horowitz, 1996).

2.2 Carlstein's Blocking Rule

Let b and l denote integers such that $n = bl$. Carlstein's rule divides the sample $\tilde{\chi}$ in b disjoint blocks, where the kth block is $\mathcal{B}_k = (\tilde{X}_{(k-1)l+1}, \ldots, \tilde{X}_{kl})$ for $1 \leq k \leq b$. According to the Carlstein rule, bootstrap sample $\tilde{\chi}^*$ is formed by choosing b blocks randomly with replacement out of the set of blocks formed from the original sample and laying the chosen blocks side by side in the order that they are chosen. Bootstrap sample $\tilde{\chi}^*$ then consists of $\{\tilde{X}_i^*\} = \{(X_i^{*'}, \ldots, X_{i+k}^{*'})' : i = 1, \ldots, n\}$.

2.3 Normalized Statistic

Let us denote the random variable of interest by $U_N = (\hat{\theta} - \theta)/s$, where $\hat{\theta} = f(\bar{X})$, $\theta = f[E(X)]$, $s = [V(\hat{\theta} - \theta)]^{1/2}$, $V(\cdot)$ is an exact variance, and $f(\cdot)$: $\mathcal{R}^d \rightarrow \mathcal{R}$ is a smooth function of sample moments of $\tilde{\chi}$ or sample moments of functions of $\tilde{\chi}$.

Let U_N^* denote the bootstrap equivalent of U_N, where $U_N^* = (\hat{\theta}^* - \hat{\theta})/\tilde{s}$, $\hat{\theta}^* = f(\bar{X}^*)$, and $\bar{X}^* = n^{-1} \sum X_i^*$ is the resample mean. Define $\tilde{s} = (V^*[\hat{\theta}^* - \hat{\theta}])^{1/2}$, where $V^*[\hat{\theta}^* - \hat{\theta}] = E^*(\hat{\theta}^* - E'[\hat{\theta}^*])^2$. Here, $E^*[\cdot]$ denotes the expectation induced by the bootstrap sampling, conditional on the sample, $\tilde{\chi}$. Thus, bootstrap expectation $E^*[\cdot]$, and therefore $\tilde{s}$, can be calculated, since we know the probability distribution induced by bootstrap sampling.

Next, we define the Edgeworth expansions of U_N and U_N^*:

$$P(U_N < x) - \Phi(x) - n^{-1/2} p_1(x) - n^{-1} p_2(x) = o(n^{-1})$$

where $p_1(x)$ and $p_2(x)$ are even and odd functions, respectively; both of the functions are polynomials with coefficients depending on cumulants of U_N and both are of order $\mathcal{O}(1)$.*

$$P^*(U_N^* < x) - \Phi(x) - n^{-1/2} \hat{p}_1(x) - n^{-1} \hat{p}_2(x) = o(n^{-1})$$

except, possibly, if χ is contained in a set of probability $o(n^{-1})$. Here $\hat{p}_1(x)$ and $\hat{p}_2(x)$ are the same polynomials as above only the population cumulants of U_N are replaced by sample cumulants of U_N^*, and $P^*(\cdot)$ is a

---

*Cumulants are defined as the coefficients of $(1/j!)(it)^j$ terms in a power series expansion of log $\chi(t)$, where $\chi(t)$ is the characteristic function of a random variable and $\chi(t) = \exp(k_1 it + (1/2)k_2(it)^2 + \cdots + (1/j!)k_j(it)^j + \cdots +)$.

probability measure (conditional on the sample, $\tilde{\chi}$) induced by the bootstrap sampling. Let k_i denote the ith cumulant of U_N. Then,

$$n^{-1/2}p_1(x) = -k_1 - \frac{k_3}{6}(x^2 - 1)$$

$$n^{-1}p_2(x) = -\frac{1}{2}k_1^2x + \left(\frac{k_4}{24} + \frac{k_1k_3}{6}\right)(3x - x^3) - \frac{k_3^2}{72}(x^5 - 10x^3 + 15x)$$

The first four cumulants of U_N have the following form (see Appendix, Result 2):

$$k_1 \equiv E(U_N) = \frac{k_{1,2}}{n^{1/2}} + \frac{k_{1,3}}{n^{3/2}} + \mathcal{O}(n^{-5/2})$$

$$k_2 \equiv E(U_N - E(U_N))^2 = 1$$

$$k_3 \equiv E(U_N - E(U_N))^3 = \frac{k_{3,1}}{n^{1/2}} + \frac{k_{3,2}}{n^{3/2}} + \mathcal{O}(n^{-5/2})$$

$$k_4 \equiv E(U_N - E(U_N))^4 - 3[V(U_N)]^2 = \frac{k_{4,1}}{n} + \mathcal{O}(n^{-2}),$$

where $k_{i,j}$'s are constants that do not depend on n and $E(U_N^2) = \mathcal{O}(1)+\mathcal{O}(n^{-1})$.

Define u_α as $P(U_N < u_\alpha) = \alpha$. Inverting the Edgeworth expansion produces the Cornish–Fisher expansion (see Hall, 1992):

$$u_\alpha - z_\alpha - n^{-1/2}p_{11}(z_\alpha) - n^{-1}p_{21}(z_\alpha) = o(n^{-1})$$

The expansions are to be interpreted as asymptotic series, and in that sense are available uniformly in

$\varepsilon < \alpha < 1 - \varepsilon$ for any $0 < \varepsilon < 1/2$*

Similarly, define $\hat{u}_\alpha$ as $P^*(U_N^* < \hat{u}_\alpha) = \alpha$.† Then, uniformly in $\varepsilon < \alpha < 1 - \varepsilon$ for each $\varepsilon > 0$.

$$\hat{u}_\alpha - z_\alpha - n^{-1/2}\hat{p}_{11}(z_\alpha) - n^{-1}\hat{p}_{21}(z_\alpha) = o(n^{-1})$$

---

*In the notation $p_{ij}(\cdot)$ [and later $q_{ij}(\cdot)$], i denotes the term in the Cornish–Fisher expansion and j is equal to 1, if u_α is a percentile of a one-sided distribution, and 2, if it is a percentile of a two-sided distribution.

†A way to obtain an empirical estimate of $\hat{u}_\alpha$ is to carry out a Monte Carlo experiment that consists of resampling the original sample $\tilde{\chi}$, calculating the bootstrap test statistic U_N^*, and forming the empirical distribution of U_N^* with the desired level of accuracy. The αth quantile of the empirical distribution of the bootstrap test statistic is the empirical estimate of $\hat{u}_\alpha$. See Hall (1992) for regularity conditions and details of implementation.

except, possibly, if χ is contained in a set of probability $o(n^{-1})$. Here,

$$n^{-1/2}p_{11}(x) = -n^{-1/2}p_1(x)$$

$$n^{-1}p_{21}(x) = n^{-1/2}p_1(x)n^{-1/2}p_1'(x) - \tfrac{1}{2}xn^{-1}p_1(x)^2 - n^{-1}p_2(x) \tag{1}$$

with obvious modifications for $\hat{p}_{11}(x)$ and $\hat{p}_{21}(x)$. Here, $p_1'(\cdot)$ denotes the derivatives of $p_1(\cdot)$ with respect to x.

Let us also introduce some notation for the two-sided distribution function of the normalized test statistic. Noting that $P(|U_N| < x) = P(U_N < x) - P(U_N < -x)$ and that $p_1(x)$ is an even polynomial, the Edgeworth expansions for $|U_N|$ and $|U_N^*|$ take on the following form:

$$P(|U_N| < x) - 2\Phi(x) + 1 - 2n^{-1}p_2(x) = o(n^{-2})$$

$$P^*(|U_N^*| < x) - 2\Phi(x) + 1 - 2n^{-1}\hat{p}_2(x) = o(n^{-2})$$

where the latter equality holds except, possibly, if χ is contained in a set of probability $o(n^{-2})$.

Define $\xi = 1/2\times(1+\alpha)$, $P(|U_N| < w_\alpha) = \alpha$, and $n^{-1}p_{12}(\cdot) = -n^{-1}p_2(\cdot)$. Inverting the population Edgeworth expansion we obtain the following Cornish–Fisher expansion:

$$w_\alpha - z_\xi - n^{-1}p_{12}(z_\xi) = o(n^{-2})$$

uniformly in $\varepsilon < \alpha < 1 - \varepsilon$ for each $\varepsilon > 0$. Equivalently, define $P^*(|U_N^*| < \hat{w}_\alpha) = \alpha$ and $n^{-1}\hat{p}_{12}(\cdot) = -n^{-1}\hat{p}_2(\cdot)$, where $\hat{p}_2(\cdot)$ is as $p_2(\cdot)$ with population moments replaced by their sample equivalents. Then, uniformly in $\varepsilon < \alpha < 1 - \varepsilon$ for each $\varepsilon > 0$,

$$\hat{w}_\alpha - z_\xi - n^{-1}\hat{p}_{12}(z_\xi) = o(n^{-2})$$

except, possibly, if χ is contained in a set of probability $o(n^{-2})$.

2.4 Studentized Statistic

The random variable of interest here is $U_S = (\hat{\theta} - \theta)/\hat{s}$, where $\hat{s}^2$ is a consistent estimate of s^2. The functional forms of s^2 and $\hat{s}^2$ are*

$$s^2 \sim \sum_{i=1}^{d} C_i^2\, V(\bar{X}_i)$$

where $V(\bar{X}_i) = \gamma(0)/n + (2/n)\sum_{j=1}^{k} \gamma(j)(1 - n^{-1}j)$, $\gamma(j)$ is the jth autocovariance of X, k is the highest lag for nonzero covariance, $f(\cdot): \mathcal{R}^d \to \mathcal{R}$,

*The following variances are expressed in terms of being asymptotically equivalent to something, because, in general, the function $f(\cdot)$ in the random variable of interest, U_S, is not a linear function. To be able to evaluate the variance of the random variable of interest, we have to linearize $f(\cdot)$ using Taylor's theorem.

and C_i's are constants that depend of function $f(\cdot)$, but not on n. Also, $\bar{X}_i$ is a sample mean of the ith argument of the function $f(\cdot)$. A consistent estimator of s^2 is given by

$$\hat{s}^2 \sim \sum_{i=1}^{d} C_i^2 \hat{V}(\bar{X}_i)$$

where $\hat{V}(\bar{X}_i) = n^{-2}\sum_{j=1}^{n}(X_{ij} - \bar{X}_i)^2 + (2/n)\sum_{m=1}^{k}(1 - n^{-1}m)\sum_{j=1}^{n-m}(X_{ij} - \bar{X}_i)(X_{i,j+m} - \bar{X}_i)/n$ and X_{ij} is the jth element of the sample from the ith argument.

The corrected bootstrap test statistic is $U_S^* = (\hat{s}/\tilde{s})(\hat{\theta}^* - \hat{\theta})/\hat{s}^*$, where $\hat{s}^{*2}$ is the bootstrap equivalent of $\hat{s}^2$:

$$\hat{s}^{*2} \sim \sum_{i=1}^{d} C_i^2 \hat{V}(\bar{X}_i^*)$$

Here, $\hat{V}(\bar{X}_i^*) = n^{-2}\sum_{j=1}^{n}(X_{ij}^* - \bar{X}_i^*)^2 + (2/n)\sum_{m=1}^{k}(1 - n^{-1}m)\sum_{j=1}^{n-m}(X_{ij}^* - \bar{X}_i^*)(X_{i,j+m}^* - \bar{X}_i^*)/n$, X_{ij}^* is the jth observation of the ith argument of $f(\cdot)$ in the block bootstrap sample, and $\bar{X}_i^*$ is a sample mean of the block bootstrap sample for the ith argument. The exact bootstrap variance of $\hat{\theta}^* - \hat{\theta}$ is denoted by $\tilde{s}^2$:

$$\begin{aligned}
\tilde{s}^2 &\sim \sum_{i=1}^{d} C_i^2 \cdot V^*(\bar{X}_i^* - \bar{X}_i) \\
&= \sum_{i=1}^{d} C_i^2 \frac{1}{b}\sum_{j=1}^{b}\frac{(\bar{X}_{ij} - \bar{X}_i)^2}{b} \\
&= \sum_{i=1}^{d} C_i^2 \frac{1}{n^2}\sum_{j=1}^{b}\sum_{k_1=1}^{l}\sum_{k_2=1}^{l}(X_{ijk_1} - \bar{X})(X_{ijk_2} - \bar{X})
\end{aligned}$$

where $V^*(\cdot)$ is the variance induced by block bootstrap sampling, $\bar{X}_{ij}$ is the sample mean of the jth block of the ith argument, and X_{ijk_m} is the k_mth observation in the jth block of the ith argument.

Note that the Taylor series expansions of U_S and U_S^* have the following forms.*

$$\begin{aligned}
U_S &= U_N \times \left[1 - \frac{\hat{s}^2 - s^2}{2s^2} + \frac{3}{8}\left(\frac{\hat{s}^2 - s^2}{s^2}\right) + o_p(n^{-1})\right] \\
U_S^* &= U_N^* \times \left[1 - \frac{\hat{s}^{*2} - \tilde{s}^2}{2\tilde{s}^2} + \frac{3}{8}\left(\frac{\hat{s}^{*2} - \tilde{s}^2}{\tilde{s}^2}\right) + o_p(b^{-1})\right]
\end{aligned} \tag{2}$$

*The expansion of U_S, of course, is a theoretical construct, since in the studentized case it is assumed that we do not know s^2.

where the error in the second expansion holds conditional on the sample χ. The exact variances of U_N and U_N^* in the above two equations are equal to one. Furthermore, first four cumulants of U_S have the same expansions and rates as the cumulants of U_N above with an exception of the second cumulant. The second cumulant of U_S is equal to $1 + \mathcal{O}(n^{-1})$. With this change in the variance, the Edgeworth expansion, say, for U_S, is

$$P(U_S < x) - \Phi(x) - n^{-1/2}q_1(x)\phi(x) - n^{-1}q_2(x)\phi(x) = o(n^{-1})$$

where

$$\begin{aligned} n^{-1/2}q_1(x) &= -k_1^U - \frac{k_3^U}{6}(x^2 - 1) \\ n^{-1}q_2(x) &= \left(-\frac{k_{2,2}^U}{2} - \frac{k_1^{U2}}{2}\right)x + \left(\frac{k_4^U}{24} + \frac{k_1^U k_3^U}{6}\right)(3x - x^3) \\ &\quad - \frac{k_3^{U2}}{72}(x^5 - 10x^3 + 15x) \end{aligned} \tag{3}$$

where k_i^U is the ith population cumulant of U_S, $k_2^U = 1 + k_{2,2}^U/n + o(n^{-1})$, $\Phi(\cdot)$ is the standard normal cumulative distribution function, and $\phi(\cdot)$ is the standard normal density function. The functional forms of the first two components of one-sided Cornish–Fisher expansion are defined as

$$\begin{aligned} n^{-1/2}q_{11}(x) &= -n^{-1/2}q_1(x) \\ n^{-1}q_{21}(x) &= n^{-1/2}q_1(x)n^{-1/2}q_1'(x) - \tfrac{1}{2}xn^{-1}q_1(x)^2 - n^{-1}q_2(x) \end{aligned} \tag{4}$$

The functional forms of $n^{-1/2}\hat{q}_{11}(\cdot)$ and $n^{-1}\hat{q}_{21}(\cdot)$ are the same as those of $n^{-1/2}q_{11}(\cdot)$ and $n^{-1}q_{21}(\cdot)$, respectively, with population moments of U_S replaced by the sample cumulants of U_S^*. Also, note that the following equality holds for the first polynomial, $n^{-1}q_{12}(\cdot)$, in the Cornish–Fisher expansion of the αth quantile of the two-sided population distribution function of the studentized test statistic: $n^{-1}q_{12}(\cdot) = -n^{-1}q_2(\cdot)$, with the obvious equivalent for the bootstrap case.

Note the difference between $n^{-1}p_2(\cdot)$ (introduced earlier) and $n^{-1}q_2(\cdot)$. Although the functional forms of the polynomials in the Edgeworth and Cornish–Fisher expansions for the standardized and the studentized statistics are the same (as functions of cumulants), some cancellations happen in the normalized case, when we replace the second cumulant with its expansion. In the normalized case the second cumulant is exactly equal to one, whereas it is equal to $1 + \mathcal{O}(n^{-1})$ in the studentized case.

3 MAIN RESULTS

In this section we will establish the optimal nonoverlapping bootstrap block lengths by minimizing the error in the coverage probabilities of one- and symmetric two-sided block bootstrap confidence intervals of normalized and studentized smooth functions of sample averages. Solution methods to the problems involving normalized and studentized statistics are very similar. Section 3.1 deals with the normalized statistic, while the details of the studentized statistic are discussed in Section 3.2. Algebraic details of the important calculations can be found in the Appendix.

3.1 Normalized Statistic

3.1.1 One-sided Confidence Intervals

Here, we find the block length ℓ that satisfies the following expression:

$$l^* = \arg\min_{i \in \mathcal{L}} |P(U_N < \hat{u}_\alpha) - \alpha|$$

where $\mathcal{L}$ is the set of block lengths that are no larger than n and that go to infinity as the sample size n goes to infinity.

Intuitively, the above probability should equal α plus some terms that disappear asymptotically and are functions of l. The goal, therefore, is to find these approximating terms. We start out by expanding the objective function from the above minimization problem:

$$P(U_N < \hat{u}_\alpha) = P\Bigg[U_N - n^{-1/2}(\hat{p}_{11}(z_\alpha) - p_{11}(z_\alpha)) - n^{-1}(\hat{p}_{21}(z_\alpha) - p_{21}(z_\alpha)) \leq \sum_{j=1}^{2} n^{-j/2} p_{j1}(z_\alpha) + z_\alpha + r_N\Bigg]$$

where $r_N = o(n^{-1})$, except, possibly, if χ is contained in a set of probability $o(n^{-1})$. Let us denote $n^{-1/2}\Delta_N \equiv n^{-1/2}(\hat{p}_{11}(z_\alpha) - p_{11}(z_\alpha))$, $S_N \equiv U_N - n^{-1/2}(\hat{p}_{11}(z_\alpha) - p_{11}(z_\alpha)) - n^{-1}(p_{21}(z_\alpha) - p_{21}(z_\alpha))$, and $p_{ij}(\cdot)$'s are defined in Eq. (1). By application of the Delta method (see Appendix, Result 3):

$$P(U_N < \hat{u}_\alpha) = P(S_N < x) + o(n^{-1})$$

where $x = \sum_{j=1}^{2} n^{-j/2} p_{j1}(z_\alpha) + z_\alpha$.

Now the objective is to develop the first four cumulants of S_N as functions of cumulants of U_N. Then, using cumulants of S_N, to derive an Edgeworth expansion of S_N as an Edgeworth expansion of U_N plus some error terms.* Lastly, evaluate the resulting expression at $x = \sum_{j=1}^{2} n^{-j/2} \times p_{j1}(z_\alpha) + z_\alpha + z_\alpha$.

Denote the cumulants of S_N by k_i^S. Then (see Appendix, Result 2 for details)

$$\begin{aligned}
k_1^S &= k_1 - n^{-1/2} E(\Delta_N) + o(n^{-1}) \\
k_2^S &= k_2 - 2n^{-1/2} E(U_N \Delta_N) + o(n^{-1}) \\
k_3^S &= k_3 - 3n^{-1/2} E(U_N^2 \Delta_N) + 3n^{-1/2} E(U_N^2) E(\Delta_N) + o(n^{-1}) \\
k_4^S &= k_4 - 4n^{-1/2} E(U_N^3 \Delta_N) + 12n^{-1/2} E(U_N^2) E(U_N \Delta_N) + o(n^{-1})
\end{aligned}$$

where we have used the result that $U_N = \mathcal{O}_p(1)$, $n^{-1/2}\Delta_N = \mathcal{O}_p(A_1^{1/2})$ and $n^{-1}(\hat{p}_{21}(z_\alpha) - p_{21}(z_\alpha)) = \mathcal{O}_p(A_2^{1/2})$ (see Appendix, Result 1), and $A_1 = C_1 n^{-1} l^{-2} + C_2 n^{-2} l^2$ and $A_2 = C_2 n^{-2} l^{-2} + C_4 n^{-2} l^2$. The rates of A_1 and A_2 follow from Hall et al. (1995). Next, substitute these cumulants in the Edgeworth expansion of S_N. The resulting equation is

$$\begin{aligned}
P(U_N \le \hat{u}_\alpha) &= P(S_N \le x) + o(n^{-1}) \\
&= P(U_N \le x) + n^{-1/2} E(\Delta_N)\phi(x) + n^{-1/2} E(U_N \Delta_N) x\phi(x) \\
&\quad + \left(\tfrac{1}{2} n^{-1} E(U_N^2 \Delta_N) - \tfrac{1}{2} n^{-1/2} E(U_N^2)(\Delta_N)\right)(x^2 - 1)\phi(x) \\
&\quad + \left(\tfrac{1}{2} n^{-1/2} E(U_N^2) E(U_N \Delta_N) - \tfrac{1}{6} n^{-1/2} E(U_N^3 \Delta_N)\right) \\
&\quad \times (3x - x^3)\phi(x) + o(n^{-1})
\end{aligned}$$

Evaluating the above equation at $x = \sum_{j=1}^{2} n^{-j/2} p_{j1}(z_\alpha) + z_\alpha$ and noting that $P(U_N \le x) = \alpha + \mathcal{O}(n^{-1})$ does not depend on the block length, l, gives us the following objective function:

$$\begin{aligned}
&n^{-1/2} \sum_{i=0}^{3} E(U_N^i \Delta_N) C_i + n^{-1/2} E(U_N^2) E(\Delta_N) C_4 \\
&\quad + n^{-1/2} E(U_N^2) E(U_N \Delta_N) C_5 + o(n^{-1})
\end{aligned}$$

where $n^{-1/2} E(U_N^i \Delta_N) \sim n^{-1/2} E(U_N^2) E(U_N^i \Delta_N)$, $\{i = 0, 1\}$. Thus, we are left with four terms: $n^{-1/2} E(U_N^i \Delta_N)$, $\{i = 0, \ldots, 3\}$. Result 4 in the Appendix

---

*In this chapter we have not derived the regularity conditions under which this expansion exists.

shows that these terms have the following orders:

$$n^{-1/2}E(\Delta_N) = O(n^{-3/2}l^{3/2}) + O(n^{-1/2}l^{-1})$$
$$n^{-1/2}E(U_N\Delta_N) = O(n^{-1}l) + O(n^{-1}l^{-1})$$
$$n^{-1/2}E(U_N^2\Delta_N) = O(n^{-3/2}l^{3/2}) + O(n^{-1/2}l^{-1})$$
$$n^{-1/2}E(U_N^3\Delta_N) = O(n^{-1}l^{1/2}) + O(n^{-1}l^{-1})$$

Therefore, the error in the bootstrap coverage probability of a one-sided block bootstrap confidence interval is: $O(n^{-1}l) + O(n^{-1/2}l^{-1})$. The block length, l, that minimizes this quantity is proportional to $n^{1/4}$. Furthermore, the size of the coverage error is $(n^{-3/4})$, when block lengths proportional to $n^{1/4}$ are used.

3.1.2 Symmetric Two-Sided Confidence Intervals

The solution methods for one- and symmetric two-sided confidence interval cases are very similar. Again, we are looking for the block length, l, that satisfies the following equation:

$$l^* = \min_{l \in \mathcal{L}} |P(|U_N| < \hat{w}_\alpha) - \alpha|$$

Note that $\hat{w}_\alpha = w_\alpha = n^{-1}\Delta_N^A + o(n^{-2})$, except, if χ is contained in a set of probability $o(n^{-2})$, where $n^{-1}\Delta_N^A = n^{-1}(\hat{p}_{12}(z_\xi) - p_{12}(z_\xi))$ and $n^{-1}p_{12}(\cdot) = -n^{-1}p_2(\cdot)$. One can show (see Appendix, Result 1) that $n^{-1}\Delta_N^A = O_p(A_2^{1/2})$, where $A_2 = C_1 n^{-2}l^{-2} + C_1 n^{-2}l^{-2} + C_2 n^{-3}l^3$ and the rate of A_2 follows from Hall et al. (1995).

Then:

$$\begin{aligned} P(|U_N| < \hat{w}_\alpha) &= P(|U_N| < w_\alpha + n^{-1}\Delta_N^A + r_N^A) \\ &= P(|U_N| < w_\alpha + n^{-1}\Delta_N^A) + o(n^{-2}) \\ &= P(U_N < w_\alpha + n^{-1}\Delta_N^A) - P(U_N < w_\alpha + n^{-1}\Delta_N^A) \\ &\quad + o(n^{-2}) \end{aligned}$$

where $r_N^A = o(n^{-2})$, except, possibly, if χ is contained in a set of probability $o(n^{-2})$ and the second equality follows by the Delta method (see Appendix, Result 3). The next task is to develop cumulants of $U_N - n^{-1}\Delta_N^A$ and $U_N + n^{-1}\Delta_N^A$ and substitute them in the Edgeworth expansion of $P(U_N - n^{-1}\Delta_N^A < w_\alpha) - P(U_N + n^{-1}\Delta_N^A < -w_\alpha)$. Following the steps of the solution method for the one-sided confidence interval case, it is straightforward to show that the relevant error terms are: $n^{-1}E(\Delta_N^A)$,

$n^{-1}E(U_N^2\Delta_N^A)$, and $n^{-1}E(U_N^2\Delta_N^A)E(U_N)\tilde{n}^{-3}E(U_N\Delta_N^A)$. The above terms are of the following orders (see Appendix, Result 4 for the methods used):

$$n^{-1}E(\Delta_N^A) = \mathcal{O}(n^{-2}l^2) + \mathcal{O}(n^{-1}l^1)$$
$$n^{-3/2}E(U_N\Delta_N^A) = \mathcal{O}(n^{-2}l^{3/2}) + \mathcal{O}(n^{-2}l^{-1})$$
$$n^{-1}E(U_N^2\Delta_N^A) = \mathcal{O}(n^{-2}l^2) + \mathcal{O}(n^{-1}l^{-1})$$

Thus, the error in the coverage probability of a symmetric two-sided block bootstrap confidence interval is of order $\mathcal{O}(n^{-2}l^2) + \mathcal{O}(n^{-1}l^{-1})$. The block length, l, that minimizes this error is proportional to $n^{1/3}$. The error then is of size $\mathcal{O}(n^{-4/3})$.

3.2 Studentized Statistic

It is intuitively clear that the error rates of the coverage probability in the studentized case should be the same in the normalized case. The reason for this is that the Taylor series expansion of the studentized test statistic is equal to normalized test statistic plus some higher order error terms [see Eq. (2)].

The solution method for the studentized statistic case is very similar to that of the normalized statistic. The derivation of the error terms is identical to the normalized statistic case for both one- and symmetric two-sided confidence intervals. The dominant error terms are: $n^{-1}E(U_S^i\Delta_S)$, $\{i=0,\ldots,3\}$ for the one-sided case and $n^{-1}E(\Delta_S^A)$, $n^{-1}E(U_S^2\Delta_S^A)$, and $n^{-1}E(U_S\Delta_S^A)E(U_S)$ for the two-sided case, where $n^{-1/2}\Delta_s = n^{-1/2}(\hat{q}_{11}(z_\alpha) - q_{11}(z_\alpha))$ and $n^{-1}\Delta_S^A = n^{-1}(\hat{q}_{12}(z_\xi) - q_{12}(z_\xi))$ [see Equation (4)].*

Let k_i^U and $\hat{k}_i^U$ denote the population and bootstrap cumulants of U_S and U_S^*, respectively. Given the structure of the polynomials $q_1(\cdot)$ and $q_2(\cdot)$ in Eq. (3) we see that the following errors terms have to be bounded: for the one-sided case, $E[U_S^i(\hat{k}_1^U - k_1^U)]$ and $E[U_S^i(\hat{k}_3^U - k_3^U)]$, $\{i=0,\ldots,3\}$; for the symmetric two-sided case: $E[U_S^j(\hat{k}_2^U - k_2^U)]$, $E[U_S^j(\hat{k}_1^{U2} - k_1^{U2})]$, $E[U_S^j(\hat{k}_4^U - k_4^U)]$, $E[U_S^j(\hat{k}_1^U\hat{k}_3^U - k_1^Uk_3^U)]$, and $E[U_S^i(\hat{k}_3^{U2} - k_3^{U2})]$, $\{j=0,\ldots,2\}$.

Notice that the above terms are dominated by their normalized statistic equivalents. This is easy to see from Eq. (2), where we break down U_S and U_S^* in U_N and U_N^*, respectively, times something that is asymptotically equal to one. The only exception occurs in the case of the terms $E[U_S^j(\hat{k}_2^U - k_2^U)]$, $\{j=0,\ldots,2\}$. In the normalized statistic case the exact variances of U_N and U_N^* are both equal to one. Thus, the leading terms of k_2^U and $\hat{k}_2^U$ both cancel,

---

*Note that all the error terms above are nonstochastic. Thus, the word "dominance" is used in its usual limit sense. When the terms involved are stochastic, a specific reference to stochastic dominance is made.

and $E[U_S^j(\hat{k}_2^U - k_2^U)]$ is dominated by the population and the bootstrap variances of the second brackets in Eq. (2). However, one can show (see Appendix) that $E[U_S^j(\hat{k}_2^U - k_2^U)]$ terms are either equal to or dominated by $\mathcal{O}(n^{-1}l^{-1})$. If follows that the error rates in the coverage probabilities of one- and symmetric two-sided block bootstrap confidence intervals of the studentized statistics are $\mathcal{O}(n^{-1}l)+ \mathcal{O}(n^{-1/2}l^{-1})$ and $\mathcal{O}(n^{-2}l^{-1}) + \mathcal{O}(n^{-1}l^{-1})$, respectively. Thus, the optimal block lengths and the coverage error rates are the same for both studentized and normalized cases, where the former are proportional to $n^{1/4}$ and $n^{1/3}$ for one- and symmetric two-sided confidence intervals, respectively.

4 CONCLUSIONS

In this chapter we have established that the minimum error rates in coverage probabilities of one- and symmetric two-sided nonoverlapping block bootstrap confidence intervals are of orders $\mathcal{O}(n^{-3/4})$ and $\mathcal{O}(n^{-4/3})$, respectively, for normalized and studentized smooth functions of sample moments. These rates are attained when the blocks for one- and symmetric two-sided block bookstrap confidence intervals are proportional to $n^{1/4}$ and $n^{1/3}$, respectively. The above rates in the coverage errors are consistent with the Monte Carlo evidence of Hall and Horowitz (1996) and Hansen (1999). The reason for such slight refinement over the asymptotic case is clear: the blocking damages the dependence structure of the original sample, and even under the optimal blocking rate, the block bookstrap does not recover this dependence sufficiently well.

Bühlman (1997, 1998) has suggested a promising alternative to blocking called "sieve bootstrap," which seems to work well under certain restrictions.

ACKNOWLEDGMENT

I thank Professor Joel Horowitz for his indispensable comments and continued support. All of the errors are my own.

REFERENCES

Andrews, D.W.K., 1991. Heteroskedasticity and autocorrelation consistent covariance matrix estimation. *Econometrica*, **59**, 817–858.

Beran, R., 1988. Prepivoting test statistics: a bootstrap view of asymptotic refinements. *Journal of the American Statistical Association*, **83**, 687–697.

Bühlman, P., 1997. Sieve bootstrap for time series. *Bernoulli*, **3**, 123–148.

Bühlman, P., 1998. Sieve bootstrap for smoothing in nonstationary time series. *Annals of Statistics*, **26**, 48–83.

Carlstein, E., 1986. The use of subseries methods for estimating the variance of a general statistic from a stationary time series. *Annals of Statistics*, **14**, 1171–1179.

Davison, A.C. and Hall, P., 1993. On studentizing and blocking methods for implementing the bootstrap with dependent data. *Australian Journal of Statistics*, **35**, 215–224.

Doukhan, P., 1994. *Mixing: Properties and Examples*. Springer-Verlag, New York.

Efron, B., 1979. Bootstrap methods: another look at the jackknife. *Annals of Statistics*, **7**, 1–26.

Götze, F. and Hipp, C., 1983. Asymptotic expansions for sums of weakly dependent random vectors. *Zeitschrift für Wahrscheinlichkeitstheorie und verwandte Gebiete*, **64**, 211–239.

Götze, F. and Hipp, C., 1994. Asymptotic distributions of statistics in time series. *Annals of Statistics*, **22**, 2062–2088.

Götze, F. and Künsch, H.R., 1996. Second-order correctness of the blockwise bootstrap for stationary observations. *Annals of Statistics*, **24**, 1914–1933.

Hall, P., 1985. Resampling a coverage process. *Stochastic Processes and Applications*, **20**, 231–246.

Hall, P., 1986. On the bootstrap and confidence intervals. *Annals of Statistics*, **14**, 1431–1452.

Hall, P., 1988. On symmetric bootstrap confidence intervals. *Journal of Royal Statistical Society*, **50**, 35–45.

Hall, P., 1992. *The Bootstrap and Edgeworth Expansions*. Springer-verlag.

Hall, P. and Heyde, C.C., 1980. *Martingale Limit Theory and Its Applications*. Academic Press, New York.

Hall, P. and Horowitz, J.L., 1996. Bootstrap critical values for tests based on generalized-method-of-moments estimators. *Econometrica*, **64**, 891–916.

Hall, P., Horowitz, J.L. and Jing, B.Y., 1995. On blocking rules for the bootstrap with dependent data. *Biometrika*, **82**, 561–574.

Hansen, B., 1999. Non-parametric dependent data bootstrap for conditional moment models. Working paper, Dept. of Economics, University of Wisconsin, Madison, Wisconsin.

Horowitz, J.L., 1994. Bootstrap-based critical values for the information matrix test. *Journal of Econometrics*, **61**, 395–411.

Künsch, H.R., 1989. The jackknife and the bootstrap for general stationary observations. *Annals of Statistics*, **17**, 1217–1241.

Lahiri, S.N., 1992. Edgeworth correction by 'moving block' bootstrap for stationary and nonstationary data. In: *Exploring the Limits of Bootstrap*, R. Lepage and L. Billard (eds.) 183–214, Wiley, New York, pp. 183–214.

Lahiri, S.N., 1996. On Edgeworth expansion and moving block bootstrap for studentized M-estimators in multiple linear regression models. *Journal of Multivariate Analysis*, **56**, 42–59.

Politis, D. and Romano, J., 1994. Large sample confidence regions based on subsamples under minimal assumptions. *Annals of Statistics*, **22**, 2031–2050.

Singh, K., 1981. On the asymptotic accuracy of Efron's bootstrap. *Annals of Statistics*, **9**, 1187–1195.
Yokoyama, R., 1980. Moment bounds for stationary mixing sequences. *Zeitschrift für Wahrscheinlichkeitstheorie und verwandte Gebiete*, **52**, 45–57.

APPENDIX

Result 1 *Derivation of the probability bounds for $n^{-1/2}\Delta_N$, $n^{-1}\Delta_N^A$ and $n^{-1}(\hat{p}_{21}(x) - p_{21}(x))$.*

From Hall et al. (1995), we know that $n^{-1}E(\hat{p}_1(x) - p_1(x)^2) = \mathcal{O}(A_1)$, where $A_1 = C_1 n^{-1} l^{-2} + C_2 n^{-2} l^2$. Also, note that the probability rate of $n^{-1/2}(\hat{p}_1(x) - p_1(x))$ is the samel as that of $n^{-1/2}(\hat{p}_{11}(x) - p_{11}(x))$, since $n^{-1/2}p_{11}(\cdot) = -n^{-1/2}p_1(\cdot)$ with the obvious modifications for $n^{-1/2}/\hat{p}_{11}(\cdot)$. Then, by Chebyshev's inequality:

$$P(A_1^{-1/2}n^{-1/2}|\Delta_N| > M_\varepsilon) < \frac{n^{-1}E(\Delta_N)^2}{A_1 M_\varepsilon^2} \equiv \varepsilon^*$$

where $M_\varepsilon < \infty$ and ε^* can be made arbitrarily small. The latter statement is true, because $n^{-1}E(\Delta_N)^2/A_1 = \mathcal{O}(1)$. Thus, $A_1^{-1}n^{-1/2}\Delta_N = \mathcal{O}_p(1)$, i.e., it is bounded in probability.

To find the probability bound for $n^{-1}\Delta_N^A$ we use the result from Hall et al. (1995): $n^{-2}E(\hat{p}_2(x) - p_2(x))^2 = \mathcal{O}(A_2)$, where $A_2 = C_1 n^{-2} l^{-2} + C^2 n^{-3} l^3$. Also, the probability rate of $n^{-1}(\hat{p}_2(x) - p_2(x))$ is the same as that of $n^{-1}(\hat{p}_{12}(x) - p_{12}(x))$, since $n^{-1}p_{12}(\cdot) = -n^{-1}p_2(\cdot)$ with the obvious modifications for $n^{-1}\hat{p}_{12}(\cdot)$. We then follow the steps above to establish that $n^{-1}\Delta_N^A = \mathcal{O}_p(A_2^{1/2})$.

Lastly, to establish the probability bound of $n^{-1}(\hat{p}_{21}(x) - p_{21}(x))$, we note that the probability rate of $n^{-1}(\hat{p}_2(x) - p_2(x))$ is the same as that of $n^{-1}(\hat{p}_{21}(x) - p_{21}(x))$ (this is not hard to show), and then proceed as in the case above.

Result 2 *Derivation of the cumulants of U_N, S_N, and $U_N \pm n^{-1}\Delta_N^A$*

The derivation of the cumulants of U_N depend on applying the Taylor series expansion to the random variable of interest. We know that

$$U_N = \frac{n^{1/2}(f(\bar{X}) - f(\mu))}{\sqrt{V[n^{1/2}(f(\bar{X}) - f(\mu))]}}$$

Note that $V[n^{1/2}(f(\bar{X}) - f(\mu))] = \mathcal{O}(1)$. Then, using the Taylor expansion with respect to $\bar{X}$ around μ:

$$n^{1/2}(f(\bar{X}) - f(\mu)) = \sum_{i=1}^{d}(D_i f)(\mu)n^{1/2}(\bar{X}_i - \mu_i) + \tfrac{1}{2}\sum_{i=1}^{d}\sum_{j=1}^{d}(D_i D_j f) \times (\mu)n^{1/2}(\bar{X}_i - \mu_i)(\bar{X}_j - \mu_j) + o_p(n^{-1/2})$$

where the notation is as in Eq. (2). Then

$$E(U_N) = \frac{k_{1,2}}{n^{1/2}} + \frac{k_{1,3}}{n^{3/2}} + \mathcal{O}(n^{-5/2})$$

where $k_{i,j}$ are constants that do not depend on n. Here, we have used the following equalities from Hall et al. (1995):

$$E(\bar{X}_i - \mu_i)^2 = \frac{\gamma(0)_i}{n} + \frac{2}{n}\sum_{j=1}^{k}(1 - n^{-1}j)\gamma(j)_i \tag{5}$$

$$\begin{aligned} E(\bar{X}_i - \mu_i)^3 = {} & \frac{E(X_{i1} - \mu_i)^3}{n^2} + \frac{3}{n^2}\sum_{j_1=1}^{k}(1 - n^{-1}j_1)E((X_{i0} - \mu_i)(X_{ij_1} - \mu_i)^2 \\ & + (X_{i0} - \mu_i)^2(X_{ij_1} - \mu_i)) + \frac{6}{n^2}\sum_{j_2,j_3\geq 1;}\sum_{j_2+j_3\leq k}(1 - n^{-1}(j_2 + j_3)) \\ & \times E((X_{i0} - \mu_i)(X_{ij_2} - \mu_i)(X_{i,j_2+j_3} - \mu_i) \end{aligned} \tag{6}$$

where X_{ij} is the jth observation of the ith element of the vector X, $\gamma(j)_i$ is the lag-j covariance of the ith element of the vector X. Also, we used the moment inequalities of Yokoyama (1980), Doukhan (1994) (Remark 2, p. 30), Hölder, and a consequence of Hölder and Burkholder inequalities (Hall and Heyde, 1980, eq. 3.67, p. 87). These tools with be used repeatedly when we bound the error terms in the coverage probability (see Results 4 and 5).

To derive higher order cumulants, use the Taylor series expansion, taken to the appropriate power.

The method of derivation of cumulants of S_N and $U_N \pm n^{-1}\Delta_N^A$ is to derive them as sums of cumulants of U_N plus an error that is asymptotically equal to zero. Let us demonstrate this for the second cumulant of S_N:

$$\begin{aligned} k_2^S &= E(S_N)^2 - E^2(S_N) \\ &= E(U_N - n^{-1/2}/\Delta_N - n^{-1}(\hat{p}_{21}(z_\alpha) - p_{21}(z_\alpha)))^2 \\ &\quad - E^2(U_N - n^{-1/2}/\Delta_N - n^{-1}(\hat{p}_{21}(z_\alpha) - p_{21}(z_\alpha))) \\ &= k_2 - 2n^{-1/2}E(U_N\Delta_N) + n^{-1}E(\Delta_N^2) \\ &\quad + 2n^{-1/2}E(U_N)E(\Delta_N) - n^{-1}E^2(\Delta_N) + o(n^{-1}) \end{aligned}$$

Using this method it is straightforward to derive cumulants of higher orders.

Result 3 *Derivations involving the Delta method*

Here, we will demonstrate the derivation of equality $P(U_N < \hat{u}_\alpha) = P(S_N < x) + o(n^{-1})$. The derivation of other equalities involving applications of Delta method are similar.

$$P(U_N < \hat{u}_\alpha) = P(S_N < x + r_N)$$

where $r_N = o(n^{-1})$, except, possibly, if $\mathcal{Y}$ is contained in a set of probability $o(n^{-1})$. That is, $P(r_N \neq o(n^{-1})) = o(n^{-1})$. Therefore, as $n \to \infty$, $P(m|r_N| \geq \varepsilon$, for some $m \geq n) = o(n^{-1})$, for all $\varepsilon > 0$. This is equivalent to $P(n \cdot |r_N| \geq \varepsilon) = o(n^{-1})$ for all $\varepsilon > 0$, as $n \to \infty$. Then,

$$\begin{aligned} P(S_N < x + r_N) &= P\left(S_n x + r_N, \bigcup_{n=1}^{\infty}\bigcap_{m=n}^{\infty}\{\omega : m \cdot |r_N(\omega)| < \varepsilon\}\right) \\ &\quad + P\left(S_N < x + r_N, \bigcup_{n=1}^{\infty}\bigcap_{m=n}^{\infty}\{\omega : m \cdot |r_N(\omega)| \geq \varepsilon\}\right) \\ &\leq P\left(S_N < x + r_N, \bigcup_{n=1}^{\infty}\bigcap_{m=n}^{\infty}\{\omega : m \cdot |r_N(\omega)| < \varepsilon\}\right) \\ &\quad + P\left(\bigcup_{n=1}^{\infty}\bigcap_{m=n}^{\infty}\{\omega : m \cdot |r_N(\omega)| \geq \varepsilon\}\right) \\ &\leq \lim_{n\to\infty} P(S_N < x + r_N, m \cdot |r_N| \leq \varepsilon, \forall m \geq n) \\ &\quad + P(\{\omega : r_N(\omega) \neq o(n^{-1})\}) \\ &\leq P(S_N < x) + o(n^{-1}) \end{aligned}$$

Similarly, we can show that $P(S_N < x) + r_N) \geq P(S_N < x) + o(n^{-1})$.

Result 4 *Bounding of* $n^{-1/2}E(U_N^i \Delta_N)$, $\{i = 0, \dots, 3\}$

$$\begin{aligned} n^{-1/2}E(\Delta_N) &= n^{-1/2}E(\hat{p}_1(x) - p_1(x)) \\ &= E(\hat{k}_1 - k_1) + C \cdot E(\hat{k}_3 - k_3). \end{aligned}$$

Start with $E(\hat{k}_1 - k_1)$ and define $\hat{\beta} \equiv E^*(f(\bar{X}^*) - f(\bar{X}))$, $\beta \equiv E(f(\bar{X}) - f(\mu))$, and $\hat{k}_1 = \hat{\beta}/\tilde{s}$. Then,

$$\hat{k}_1 = \frac{\hat{\beta} - \beta + \beta}{s}\left(1 - \frac{\tilde{s}^2 - s^2}{2s^2} + \frac{3}{8}\left(\frac{\tilde{s}^2 - s^2}{s^2}\right)^2 + \cdots\right)$$

$$= k_1 + \frac{\hat{\beta} - \beta}{s} - \frac{\beta}{s}\frac{\tilde{s}^2 - s^2}{2s^2} - \frac{\hat{\beta} - \beta}{s}\frac{\tilde{s}^2 - s^2}{2s^2} + \frac{\beta}{s}\frac{3}{8}\left(\frac{\tilde{s}^2 - s^2}{s^2}\right)^2 \mathcal{O}_p(n^{3/2}A_0)$$

where $A_0 = C_1 n^{-2} l^{-2} + C_2 n^{-3} l$, $s^2 = \mathcal{O}(n^{-1})$, $\beta = \mathcal{O}(n^{-1})$, $\hat{\beta} - \beta = \mathcal{O}_p(A_0^{1/2})$, $\tilde{s}^2 - s^2 = \mathcal{O}_p(A_0^{1/2})$, $E(\hat{\beta} - \beta) \sim C_1 n^{-1} l^{-1} + C_2 n^{-2} l \sim E(\tilde{s}^2 - s^2)$, $E(\hat{\beta} - \beta)^2 \sim C_1 n^{-2} l^{-2} + C_2 n^{-3} l \sim E(\tilde{s}^2 - s^2)^2$.

The last six bounds are from Hall et al. (1995). Also, note that $\hat{\beta} - \beta \sim (f''(\bar{X})/2)E^*(\bar{X}_i^* - \bar{X}_i)^2 - (f''(\mu)/2)E(\bar{X}_i - \mu_i)^2$ and $\tilde{s}^2 - s^2 \sim (f'(\bar{X}))^2 \times E^*(\bar{X}_i^* - \bar{X}_i)^2 - (f'(\mu))^2 E(\bar{X}_i - \mu_i)^2$. Then, $E(\hat{k}_1 - k_1) \sim C_1 n^{-1/2} l^{-1} + C_2 n^{-3/2} l$.

Next, bound $E(\hat{k}_3 - k_3)$:

$$\hat{k}_3 - k_3 = E^*\left(\frac{f(\bar{X}^*) - f(\bar{X})}{\sqrt{V^*(f(\bar{X}^*))}} - E^*\left(\frac{f(\bar{X}^*) - f(\bar{X})}{\sqrt{V^*(f(\bar{X}^*))}}\right)\right)^3$$

$$- E\left(\frac{f(\bar{X}) - f(\mu)}{\sqrt{V(f(\bar{X}))}} - E\left(\frac{f(\bar{X}) - f(\mu)}{\sqrt{V(f(\bar{X}))}}\right)\right)^3$$

$$\sim C\left(\frac{E^*(\bar{X}_i^* - \bar{X}_i)^3}{\sqrt{V^*(\bar{X}_i^* - \bar{X}_i))^{3/2}}} - \left(\frac{E(\bar{X}_i - \mu_i)^3}{(V(\bar{X}_i - \mu_i))^{3/2}}\right)\right)$$

Note that $\bar{X}$ is a vector random variable and $\bar{X}_i$ is a scalar random variable. Here, we have used Taylor's theorem for vector-valued functions. By Hall et al. (1995):

$$E\left(\frac{E^*(\bar{X}_i^* - \bar{X}_i)^3}{\sqrt{V^*(\bar{X}_i^* - \bar{X}_i))^{3/2}}} - \left(\frac{E(\bar{X}_i - \mu_i)^3}{(V(\bar{X}_i - \mu_i))^{3/2}}\right)\right)$$
$$= \mathcal{O}(n^{-1/2} l^{-1}) + \mathcal{O}(n^{-3/2} l^{3/2})$$

Therefore, $n^{-1/2}E(\Delta_N) \sim C_1 n^{-3/2} l^{3/2} + C_2 n^{-1/2} l^{-1}$.

$$n^{-1/2}E(U_N \Delta_N) = E(U_N \hat{k}_1) - k_1^2 + C \cdot (E(U_N \hat{k}_3) - k_1 k_3)$$

Since $\hat{k}_1 = k_1 + (\hat{\beta} - \beta)/s - (\beta/s)((\hat{s}^2 - s^2)/2s^2) + \mathcal{O}_p(n^{3/2}A_0)$:

$$E(U_N\hat{k}_1) = k_1^2 + \frac{1}{s^2}E((f(\bar{X}) - f(\mu))(\hat{\beta} - \beta)) + E\left(U_N\frac{\beta}{s}\frac{\hat{s}^2 - s^2}{2s^2}\right) + o(A_3)$$

where $A_3 = C_1n^{-1} + C_2n^{-1}l^{-1}$. The rate of the error $o(A_3)$ stems from the following two considerations. First, the terms covered by the error are farther out in the Taylor series expansion of $\hat{k}_1$ than the terms left in the expansion, and therefore their rates are smaller than those of the terms left in the expansion. Second, the error of the term $n^{-1/2}E(U_N\Delta_N)$ turns out to be $\mathcal{O}(A_3)$. Next, define $Y(X - \mu$, i.e., Y is the demeaned random vector X. Some algebra:

$$\begin{aligned}(f(\bar{X}) - f(\mu))(\hat{\beta} - \beta) &\sim f'(\mu)(\bar{X}_i - \mu_i) \\ &\quad \times \left[\frac{f''(\mu)}{2b}\frac{\sum_{j=1}^{b}(X_{ij} - \bar{X}_i)^2}{b} - \frac{f''(\mu)}{2}E(\bar{X}_i - \mu_i)^2\right] \\ &= C\left(\bar{Y}_i\frac{1}{b}(\bar{Y}_i^{(2)} - \bar{Y}_i^2) - \bar{Y}_i\frac{1}{b}E(\bar{Y}_i^{(2)})\right) \\ &\quad + C\left(\bar{Y}_i\frac{1}{b}E(\bar{Y}_i^{(2)}) - \bar{Y}_iE(\bar{Y}_i^{(2)})\right)\end{aligned}$$

where $X_{ij} = (1/l)\sum_{k=1}^{l}X_{i,(j-1)l+k}$, $\bar{Y}_i^{(k)} = (1/b)\sum_{j=1}^{b}(X_{ij} - \mu_i)^k$, and $C = f'(\mu) \times f(\mu)/2$.

$$\begin{aligned}E[(f(\bar{X}) - f(\mu))(\hat{\beta} - \beta) &\sim \frac{C}{b}E[\bar{Y}_i(\bar{Y}_i^{(2)} - \bar{Y}_i^2) - \bar{Y}_iE(Y_i^{(2)})] \\ &= \frac{C}{b}E[\bar{Y}_i(\bar{Y}_i^{(2)} - E(\bar{Y}_i^2)) - \bar{Y}_i^3] \\ &= \mathcal{O}(n^{-2}) + \mathcal{O}(n^{-3}l)\end{aligned}$$

where the second equality follows from Taylor's theorem and the last equality follows from the application in equalities of Yokoyama (1980), Doukhan (1994) (Remark 2, p. 30), and Eq. (6). By noting that $E[U_N(\hat{\beta} - \beta)/s] \sim E[U_N((\hat{s}^2 - s^2)/2s^2)]$, it follows then that $E(U_N\hat{k}_1) - k_1^2 = \mathcal{O}(n^{-1}) + \mathcal{O}(n^{-2}l)$.

Let us examine $E(U_N\hat{k}_3) - k_1k_3$. From Hall et al. (1995), $\hat{k}_3 = k_3 + (l^{1/2}/n^{1/2})k_3^l - k_3 + O_p(A_4^{1/2})$, where $A_4 = C_1n^{-1}l^{-2} + C_2n^{-2}l^2$ and k_3^l is the third cumulant for a sample with l observations. Then,

$$
\begin{aligned}
E(U_N\hat{k}_3) - k_1k_3 &\sim k_1\left(\frac{l^{1/2}}{n^{1/2}}k_3^l - k_3\right) + E(U_NR_N) \\
&= \frac{k_1}{n^{1/2}} \cdot O(l^{-1}) + O(n^{-1}l) \\
&= O(n^{-1}l^{-1}) + O(n^{-1}l)
\end{aligned}
$$

where $R_N = O(n^{-1/2}l^2)(\bar{Y}_i^{(3)} - E(\bar{Y}_i^{(3)})) + O(n^{-1/2}l)\bar{Y}_i + O(n^{-1/2}l)(\bar{Y}_i^{(2)} - E(\bar{Y}_i^{(2)}))$. The first part of the second line of the above equation follows from Hall et al. (1995), and the second part of the second line follows from inequalities of Yokoyama (1980), Doukhan (1994) (Remark 2, p. 30), and Eqs. (5) and (6).

Thus, $n^{-1/2}E(U_N\Delta_N) = O(n^{-1}l) + O(n^{-1}l^{-1})$. Following methods developed above we can establish that $n^{-1/2}E(U_N^2\Delta_N) = O(n^{-3/2}l^{3/2}) + O(n^{-1/2}l^{-1})$ and $n^{-1/2}E(U_N^3\Delta_N) = O(n^{-1}l^{1/2}) + O(n^{-1}l^{-1})$. Therefore, the error in coverage probability of the one-sided confidence interval is equal to $O(n^{-1/2}l^{-1}) + O(n^{-1}l)$.

To obtain the rate of the error in coverage probability for the symmetric two-sided confidence interval, we have to bound the following terms: $n^{-1}E(\Delta_N^4)$, $n^{-3/2}E(U_N\Delta_N^4)$, and $n^{-1}E(U_N^2\Delta_N^4)$. Applying the methods above we can show that $n^{-1}E(\Delta_N^4) = O(n^{-2}l^2) + O(n^{-1}l^{-1})$, $n^{-3/2}E(U_N\Delta_N^4) = O(n^{-2}l) + O(n^{-2}l^{-1})$, and $n^{-1}E(U_N^2\Delta_N^4) = O(n^{-2}l^2) + O(n^{-1}l^{-1})$.

Result 5 *Bounding of* $E[U_S^j(V^*(U_S^*) - V(U_S))]$, $\{j=0,\ldots,2\}$

$$
\begin{aligned}
E[V^*(U_S^*) - V(U_S)] &\sim CE\left\{E^*\left(\left(\frac{f'(\bar{X})(\bar{X}_i^* - \bar{X}_i)}{\tilde{s}}\right)\frac{\hat{s}^{*2} - \hat{s}^2}{2\hat{s}^2}\right)\right. \\
&\qquad \left. -E\left(\left(\frac{f'(\mu)(\bar{X}_i - \mu_i)}{s}\right)^2\frac{\hat{s}^2 - s^2}{2s^2}\right)\right\} \\
&\sim O(n)E\{E^*((\bar{X}_i^* - \bar{X}_i)^2(\bar{X}_i^{*(2)} - \bar{X}^{(2)})) \\
&\quad - E((\bar{X}_i - \mu_i)^2(\bar{X}_i^{(2)} - E(X_i^2)))\} \\
&\quad + O(n)E\{E^*((\bar{X}_i^* - \bar{X}_i)^2(\bar{X}_i^{*(2)} - \bar{X}_i^2)) \\
&\quad - E((\bar{X}_i - \mu_i)^2(\bar{X}_i^2 - \mu_i^2))\}
\end{aligned}
$$

$$
\begin{aligned}
&\sim \mathcal{O}(n)E\{E^*(\bar{X}_i^* - \bar{X}_i)^3 E(\bar{X}_i - \mu_i)^3\} \\
&\quad + \mathcal{O}(n)E\left\{\frac{l}{b^3}E(\bar{X}_{ij} - \bar{X}_i)^4 - E(\bar{X}_i - \mu_i)^4\right\} \\
&\sim C\,E\left\{\frac{l^2}{n}\sum_{j=1}^{b}\frac{(\bar{X}_{ij} - \mu_i)^3}{b} - nE(\bar{X}_i - \mu_i)^3\right\} \\
&= C\left(\frac{l^2}{n}E(\bar{X}_{ij} - \mu_i)^3 - nE(\bar{X}_i - \mu_i)^3\right)
\end{aligned}
$$

where $E^*(\bar{X}^* - \bar{X})^3 = (1/b^2)\sum_{i=1}^{b}(\bar{X}_i - \bar{X})^3/b$, $s^2 = \mathcal{O}(n^{-1})$, and the rest of the notation is as in Result 4. Then, by Eq. (6), we have that $l^2/nE(X_{ij} - \mu_i)^3 - nE(\bar{X}_i - \mu_i)^3 = \mathcal{O}(n^{-1}l^{-1})$. Using the above methodology it is straightforward to show that the terms $E[U_S(\hat{k}_2^U - k_2^U)]$ and $E[U_S^2(\hat{k}_2^U - k_2^U)]$ are dominated by the term $E[\hat{k}_2^U - k_2]$.

5

A Comparison of Alternative Causality and Predictive Accuracy Tests in the Presence of Integrated and Cointegrated Economic Variables

Norman R. Swanson
Rutgers University, New Brunswick, New Jersey, U.S.A.

Ataman Ozyildirim
The Conference Board, New York, New York, U.S.A.

Maria Pisu
University of Alabama, Birmingham, Alabama, U.S.A.

1 INTRODUCTION

In a 1983 paper, Geweke et al. compared eight alternative tests of the absence of causal ordering, in the context of stationary variables. All were versions of the tests proposed by Granger (1969) and Sims (1972). Their unambiguous finding was that Wald variants of the tests are preferred. This chapter is meant to extend the results of Geweke et al. (1983) in two directions. First, cases where the variables are not $I(0)$ [in the sense of Engle and Granger (1987)], but also integrated of higher order and/or cointegrated, are considered. Second, in addition to classical in-sample tests of Granger noncausality, various model selection procedures for checking causal ordering are outlined and examined.

Our results, based on a series of Monte Carlo experiments, indicate that surplus lag Wald-type tests due to Toda and Yamamoto (1995) and Dolado and Lütkepohi (1996) (hereafter the TYDL test) perform well, regardless of integratedness and/or cointegratedness properties. However, when cointegrated data are examined, the sequential Wald-type tests due Toda and Phillips (1993, 1994) come in a close second, and in some cases are superior. In addition, we find that model selection procedures based on the use of Schwarz information criteria as well as on the *ex ante* forecasting approach suggested in Diebold and Mariano (1995) and Swanson (1998) perform surprisingly well in a variety of different contexts.

It should perhaps be noted that out-of-sample predictive ability type tests such as the model selection-based *ex ante* procedure that we consider here are receiving considerable attention in the current causality literature. One reason for this may be an aspect of Granger's original definition which had not previously received much attention in the literature. In particular, at issue is whether or not standard in-sample implementations of Granger's definition are wholly in the spirit originally intended by Granger, and whether out-of-sample implementations might also be useful. Arguments in favor of using out-of-sample implementations are given in Granger (1980, 1988), and are summarized nicely in Ashley et al. (1980), where it is stated on page 1149 that: "... a sound and natural approach to such tests [Granger causality tests] must rely primarily on the out-of-sample forecasting performance of models relating the original (non-prewhitened) series of interest." Recent papers that develop and discuss predictive ability-type causality tests include Clark and McCracken (1999), McCracken (1999), Chao et al. (2000), and Corradi and Swanson (2000b). Other important papers in the burgeoning literature on predictive ability tests include Diebold and Mariano (1995), West (1996), Harvey et al. (1997), West and McCracken (1998), Clark (2000), Corradi and Swanson (2000a) and White (2000).

In addition to straightforward evaluation of the finite sample properties of various procedures, we address a number of additional empirical issues, such as:

1. Which model selection criteria perform best for choosing lags in finite sample scenarios, when the objective is to construct classical hypothesis tests of noncausality?
2. What is the relative impact of over- versus under-specification of the true lag order and/or the cointegration rank?
3. DGPs used include simple VAR(1) models and more complex VAR(3) models [possibly with moving average (MA) errors].

This allows us to examine not only which tests perform well in the context of simple models, but also in the context of more complex models.

We ask the question: is it true a procedure which performs "best" in the context of a simple VAR(1) model may be dominated by another test when more complex DGPs describe the data? Further, we are able to examine the rate of increase in test size distortion as model complexity, as well as sample size, increases.

The rest of the chapter is organized as follows: In Section 2, a variety of different procedures for ascertaining causal ordering are summarized and discussed. In Section 3, our experimental setup is discussed. Monte Carlo results are summarized in Section 4, and Section 5 concludes and offers a number of recommendation. Throughout, $\to_d$ signifies convergence in distribution.

2 SUMMARY OF ALTERNATIVE TESTS AND MODEL SELECTION PROCEDURES

For simplicity, we start by assuming that an n-vector time series, y_t, is generated by a Kth order VAR model:

$$y_t = J(L)y_{t-1} + u_t, \qquad t = -K+1, \ldots, T \tag{1}$$

where $J(L) = \sum_{i=1}^{K} J_i L^{i-1}$. For simplicity of exposition, we leave the discussion of additional deterministic components in Eq. (1) until later. Assume also that

A.1. $u_t = (u_{1t}, \ldots, u_{nt})$ is an i.i.d. sequence of n-dimensional random vectors with mean zero and convariance matrix $\Sigma_u > 0$ such that $E|u_{it}|^{2+\delta} < \infty$ infinity for some $\delta > 0$.

As we consider both integrated (integration of order 0 or higher) and cointegrated [by cointegrated we always refer to linear combinations of $I(1)$ variables, which are $I(0)$ as discussed in Engle and Granger (1987)] economic variables, a number of alternative assumptions on the J matrices are in order. the cases below where error correction models are estimated, we assumed that:

A.2. $|I_n - J(z)z| = 0 => |z| > 1$ or $z = 1$.

A.3. $J(1) - I_n = \Gamma A'$ where Γ and A are $n \times r$ matrices of full column rank r, $0 \leq r \leq n-1$.

A.4. $\Gamma'_P(J^*(1) - I_n)A_P$ is non-singular, where Γ_P and A_P are $n \times (n-r)$ matrices of full column rank such that $\Gamma_P \Gamma = 0 = A_P A$.

Assumptions A.2–A.4 are standard, are taken from Toda and Phillips (1994), and ensure that Eq. (1) can be written in error-correction form:

$$\Delta y_t = J^*(L)\Delta y_{t-1} + \Gamma A' y_{t-1} + u_t \tag{2}$$

where $J^*(L) = \sum_{i=1}^{K-1} J_i^* L^{i-1}$. Throughout, we assume also that the covariance matrix of the stationary component in the system is greater than zero. Finally, in the following discussion assume that we are interested in testing the hypothesis:

$$H_0 : f(B) = 0 \quad \text{against} \quad H_A : f(B) \neq 0$$

where $f(\cdot)$ is an m-vector valued function satisfying standard assumptions. To simplify, we focus on the case where $f(\cdot) = C\beta$, and C is a standard $R \times (n^2 \times K)$ linear restriction matrix of rank R, where R is the number of restrictions [see Lütkepohl (1991) for further details].

2.1 Wald-Type Tests

2.1.1 All Variables Are Stationary

Assuming that y_t is generated by a stationary, stable VAR(K) process [see Lütkepohl (1991) for one definition of stability, e.g., assume A.1 and A.2 where $|z| > 1$], a frequently used version of the Wald test has

$$\lambda_{ST} = C\hat{\beta}'(C(ZZ')^{-1} \otimes \hat{\Sigma}_u C')^{-1} C\hat{\beta} \rightarrow_d \chi^2_R$$

where $\hat{\Sigma}_u$ is a consistent estimator of the convariance matrix of u (e.g., $\hat{\Sigma}_u = T^{-1}\hat{u}\hat{u}'$ or $\hat{\Sigma}_u = (T - nK)^{-1}\hat{u}\hat{u}'$), and $\hat{\beta}$ is the least squares estimate of vec(B) from levels regressions of the form:

$$Y = BZ + U$$

where $Y \equiv ((y_1, \ldots, y_T)$ is the $K \times T$ matrix of data, $B \equiv ((J_1, \ldots, J_K)$, $Z_t \equiv ((y_t, \ldots, y_{t-K+1})'$, $Z((Z_0, \ldots, Z_{T-1})$, $\beta \equiv$ (vec(B), and $U \equiv ((u_1, \ldots, u_T)$ is an $n \times T$ matrix of errors. Of course, it has been known for quite some time that the χ^2_R distribution is not generally appropriate under the null hypothesis, when variables are integrated and/or cointegrated, so that in our experiments we do not expect particularly good results from this statistic when data are generated under assumptions of difference stationarity and/or cointegration. Indeed, the focus of our Monte Carlo experiments is on cases where variables are difference stationary and/or cointegrated. Finite sample properties of causal order tests under stationary data are well known, and are thus not included here. However, full results for stationary variable cases are available in an earlier working paper version (Swanson et al., 1996). In addition, the reader is referred to Geweke et al. (1983).

2.1.2 All Variables Are Difference Stationary

Assuming that A.1–A.4 hold and that there are n unit roots in the system (in which case we set $A_P = I_n = \Gamma_P$, and $J(1) = I_n$] it is reasonable to use λ_{ST} to test the null hypothesis, but with differenced data, and $K - 1$ lags. We will call this statistic λ_{DS}.

2.1.3 Variables Are Integrated of any Order, and Possibly Cointegrated

Assuming that A.1 holds, consider the case where A.2–A.4 may hold. Alternatively, given A.1, assume that each element of $\{y_t\}$ may be integrated of any order [e.g., $I(0)$ or $I(1)$], and that any combination of variables in $\{y_t\}$ may be cointegrated. Further, assume that we do not know, are unable to determine, or do not care what the orders of integratedness and/or cointegratedness are. Interestingly, in this case, Toda and Yamamoto 1995), and in independent work Dolado and Lütkepohl (1996), show that standard Wald tests can be applied to the coefficient matrices up to the correct lag order, when the system in levels [Eq. (1)] is artificially augmented by including at least as many extra lags of each variable as the highest order of integration of any of the variables in the system. (This test is hereafter referred to as the TYDL test.) In particular, it turns out that asymptotic chi-squared tests of causality restrictions can be applied to the submatrix of coefficients up to the correct lag order. The reason for this property is that all nonstandard asymptotics are essentially confined to the coefficient matrices beyond the correct lag order, and standard asymptotics apply with respect to the coefficient matrices up to the correct lag order. [Choi (1993) uses similar arguments to show that standard asymptotics can be applied to unit root tests which are suitably augmented.] Overall, the method of including surplus lags provides a very tractable, and hence appealing means with which to deal with many of the nonstandard distribution problems that characterize so many economic variables. However, it should be pointed out that the method is inefficient in the sense that extra coefficient matrices are estimated, so that the methods' power may be affected in finite samples. Indeed, it remains to see what the local power properties of such tests are. As the test is a standard "levels regression" type Wald test, it is clearly very easy to apply, and has the form:

$$\lambda_{LW} = C\hat{\beta}'(C((ZQZ')^{-1} \otimes \hat{\Sigma}_\varepsilon)C')^{-1}C\hat{\beta} \to_d \chi^2_R$$

where all regressions are the same standard least squares regressions estimated using levels data (i.e., as used to construct λ_{ST}), except that extra lags of the regressors are added as surplus explanatory variables. That is,

$$Y = \hat{B}^*Z + \hat{\Psi}W + \hat{\varepsilon} \qquad (3)$$

where $\hat{\Sigma}_\varepsilon = T^{-1}\hat{\varepsilon}\hat{\varepsilon}'$. Note that $\hat{B}^*$, which is estimated using Eq. (3), is not the same as the estimator of B used in λ_{ST} and λ_{DS}, as W extra variables are added to the standard levels model, where $W_t \equiv (y_{t-K}, \ldots, y_{t-K-d+1})'$, and $W \equiv (W_0, \ldots, W_{T-1})$; Ψ is the coefficient matrix associated with these extra lags, and d is the highest order of integration of any of the variables in the system. The matrix Q is defined in the standard way as: $Q = Q_\tau - Q_\tau W'(WQ_\tau W')^{-1}WQ_\tau$, where $Q_\tau = I_T - S'(SS')^{-1}S$, where S is a matrix that contains constant and/or deterministic trend variables. In our case, we have set S to zero. However, if S is nonzero, then the variables can be added in the obvious way as additional regressors in Eq. (3), without affecting the distributional result under the null hypothesis.

One feature of the TYDL method that deserves further mention is that *any* form of Granger causal relation is captured by the test, regardless of whether it is "long run", in the sense that error-correction terms are constructed using the variable(s) whose causal effect is being examined, or "short run", in the sense that lagged differences of the relevant variable(s) appear in the error-correction model, or both. Alternatively, the sequential Wald tests discussed below have the feature that both "short-run noncausality" and "long-run noncausality" are sequentially tested for. However, as we will see below, if we incorrectly specify a difference stationary model (when the true model is cointegrated), then the estimated coefficient matrices become so "contaminated" that standard Wald tests (i.e., λ_{DS}) suffer from severe size distortion, often rendering the results of tests of short-run noncausality" meaningless. Thus, it is not sufficient simply to model differenced data if all that we are interested in is "short-run noncausality". The reader is referred to Granger and Lin (1995) and Toda and Phillips (1994) for further discussion of short- and long-run causality.

2.1.4 Variables Are Integrated, and Possibly Cointegrated *I*

Toda and Phillips (1993) provide a theory for Wald tests of Granger noncausality in levels vector autoregressions that allow for stochastic and deterministic trends as well as arbitrary degrees of cointegration. The theory extends earlier work by Sims et al. (1990) on trivariate VARs and suggests a tractable strategy for implementing a variety of Wald-type tests. An important feature of this approach is that tests are based on estimated versions of Eq. (2), the error correction model. Thus, while potentially offering gains in efficiency when estimating the system (relative to the TYDL approach of estimating the system in levels) the sequential Wald test adds a layer of estimation to the problem, in the sense that cointegrating ranks must be first estimated. Nevertheless, as shown in Toda and Phillips (1994), the sequential Wald tests appear to have good size properties, for a number of alternative specifications. To be more precise about the null hypothesis of interest for sequential Wald tests, we here consider variations of the joint null:

$$H_0^*: J_{1,13}^* = \cdots = J_{K-1,13}^* = 0 \quad \text{and} \quad \Gamma_1 A_3' = 0$$

where $J^*_{13}(L) = \sum_{i=1}^{K-1} L^{i-1}$ is the $n_1 \times n_3$ upper right submatrix of $J^*(L)$, Γ_1 is the first n_1 rows of the loading coefficient matrix Γ, and $y_t = (y_{1t}, y_{2t}, y_{3t})'$ has been partitioned into three vectors of length n_1, n_2, and n_3, respectively.

Before discussing the sequential Wald tests any further, it is worth reiterating an important point that is made (and elaborated on) in Toda and Phillips (1994) [and also to some extent in Sims et al. (1990)]. In the case of cointegrated economic variables, standard levels regression based causality tests are valid asymptotically as χ^2 criteria when there is sufficient cointegration among the variables whose causal effects are being tested. However, the rank condition associated with this criterion suffers from simultaneous equations bias when levels VARs are estimated, suggesting that there is no valid statistical basis in VARs for testing if the condition holds. Furthermore, when the rank condition fails, there is no valid statistical basis for mounting a standard Wald test of causality. One interpretation of this point is simply that if we are using the "correct" specification of Eq. (1) (e.g., we have specified the correct number of lags) and there is cointegration among the variables, then standard Wald tests on levels are not directly feasible, and sequential Wald tests that are based on error-correction models are. Of final note is that, if there is no cointegration, the standard levels regression-based Wald statistic (i.e., λ_{ST}) for causality has a nuisance parameter-free nonstandard limit distribution, and so can conceivably be used. These points are explained in some detail in Toda and Phillips (1994). We now state the three sequential procedures, and the testable hypotheses associated with them. The actual test statistics are given in Section 2 of Toda and Phillips (1994). For simplicity, assume that Γ_1 and A_3 are r, or $\hat{r}$-dimensional row vectors, and that $K > 1$ (with obvious modifications otherwise).

Then, the tests are based on the null hypotheses:

$$H_1^*: J^*_{1,13} = \cdots = J^*_{K-1,13} = 0$$
$$H_1^*: J^*_{1,13} = \cdots = J^*_{K-1,13} = 0$$
$$H_2^*: \Gamma_1 = 0$$
$$H_3^*: A_3 = 0$$
$$H_4^*: \Gamma_3 A_3' = 0$$

and as above:

$$H_0^*: J^*_{1,13} = \cdots = J^*_{K-1,13} = 0 \quad \text{and} \quad \Gamma_1 A_3' = 0$$

The sequential testing procedures are:

P1. Test H_2^*. If H_2^* is rejected, test H_0^* using a $\chi^2_{n_3 K}$ critical value. Otherwise, test H_1^*.

P2. Test H_3^*. If H_3^* is rejected, test H_0^* using a $\chi^2_{n_1 K}$ critical value. Otherwise, test H_1^*.

P3. Test H_1^*. If H_1^* is rejected, reject the null hypothesis of non-casuality. Otherwise, test H_2^* and H_3^*. If both are rejected, test H_4^* if $\hat{r} > 1$, or reject the null if $\hat{r} = 1$. Otherwise, accept the null of noncausality. These tests are easy to apply once standard procedures have been used to estimate the rank of the cointegrating space, and the number of cointegrating vectors in the system, although it should be noted that exact control of the overall size of these causality tests is not feasible. In what follows, we will refer to the test statistics associated with P1–P3 as λ_{P1}, λ_{P2}, and λ_{P3}, respectively.

In closing, note that while the sequential Wald discussed above is essentially designed to deal with $I(1)$ variables, which may or may not be cointegrated [of order (1,1) using the notation of Engle and Granger (1987)], the TYDL method allows for any mixtures of $I(0)$, $I(1)$, $I(2)$, etc., variables, which may or may not be cointegrated. Finally, the FM–VAR test (see below) also requires that A.1–A.4 hold [e.g., $|I_n - J(L)L| = 0$ has roots on or outside the unit circle], with the modification that assumption A.3 can be weakened so that $0 \leq r \leq n$, allowing for the possibility that all of the variables are stationary, for example. Of course, one reason why the sequential Wald test has $r < n$ stems from the fact that it is assumed that the variables are pretested to determine their order of integration, and if all are found to be $I(0)$ then standard chi-squared asymptotics apply, and sequential Wald tests are not called for.

2.1.5 Variables Are Integrated, and Possibly Cointegrated II

Another type of Wald causality test, which is in keeping with the unrestricted levels VAR method of TYDL, is that based on fully modified vector autoregression (FM-VAR) (Phillips, 1995). Phillips proposes making corrections to standard OLS–VAR regression formulas, which account for serial correlation effects as well as for regressor "endogeneity" that results from the existence of cointegrating relationship(s). The endogeneity in the context of cointegrated models arises for the following reason. Assume that Eq. (1) can be partitioned into two subsystems (1) $y_{1t} = Dy_{2t-1} + u_{1t}$, and (2) $y_{2t} = y_{2t-1} + u_{2t}$. The standard procedure is to treat y_{2t-1} as predetermined, and regard the model as a reduced form. However, even though $E(u_{1t}y_{2t-1}) = 0$, the sample covariance, $T^{-1}\sum_{t=1}^{T} u_{1t}y'_{2t-1}$, does not converge to zero, and instead converges to a function of Brownian motions, which are generally correlated in the limit.

The FM–VAR method is largely a generalization of work by Phillips and Hansen (1990), and has the feature that advance knowledge of which variables are stationary, nonstationary, and/or cointegrated is not required to achieve consistency of the fully modified VAR estimator. Given a kernel condition (Phillips, 1995, p. 1031), a bandwidth expansion rate condition (Phillips, 1995, p. 1032), and a VAR condition (Phillips 1995, p. 1044), which is essentially the same as assumptions A.1–A.4 above, the relevant Wald-type test statistic is

$$\lambda_{FM} = C\hat{\beta}^{+'}(C((XX')^{-1} \otimes \hat{\Sigma}_{\varepsilon})C')^{-1}C\hat{\beta}^{+}$$

where X is a matrix of observations with the same informational content of Z above, except that it is comprised of $K-1$ differences of the variables, as well as the lagged level of the variables; $\hat{\Sigma}_{\varepsilon}$ is the usual OLS estimate of the error covariance matrix from a regression of Y on X. The test used here is actually distributed as a weighted sum of χ^2 values, whose weights are difficult to estimate. However, as pointed out by Phillips (1995), an asymptotically conservative test can by constructed by using the χ^2_R distribution as the limiting distribution for λ_{FM}. We adopt that approach in the sequel. Although the reader is referred to Phillips (1995) for further details, note that the FM–VAR estimator, $\hat{\beta}^{+} = \text{vec}(\hat{F}^{+})$, is

$$\hat{F}^{+} = [YX_D'|YY_{-1}' - \hat{\Omega}_{\varepsilon y}\hat{\Omega}_{yy}^{-1}(\Delta Y_{-1}Y_{-1}' - T\hat{\Delta}_{\delta y \Delta y})](XX')^{-1}$$

where X_D is the same as X, except that the levels variables are removed, Y_{-1} is Y lagged one period, and the Ω and Δ estimates are long-run covariance matrix estimates, and one-sided long-run covariance matrix estimates calculated in the usual way as the weighted sum of estimated auto-covariance terms, using some appropriate kernel function as the weighting scheme [see Phillips (1995) for details]. Deterministic components can also be added to the FM–VAR regression model by augmenting the matrix of explanatory data used in the construction of λ_{FM} in the usual way.

2.2 Model Selection-Type Causality Procedures

The next two procedures that can be used to examine causal ordering are not classical, in the sense that critical values are not used. In practice, these types of procedures involve choosing a 'best" model. If that "best" model contains the variable(s) whose causal effect is being examined, then we "choose" the model that is not consistent with a null hypothesis of noncausality. The "best" model may be a single equation, or a model consisting of many equations, depending on how many variables are in y_1, when we are testing whether y_3 is noncausal for y_1. One advantage of model selection, e.g., based on complexity penalized likelihood criteria, is that the approach does not require specification of a correct model for its valid application. Another desirable feature is that the probability of selecting the truly best model approaches one as the sample size increases, if the model selection approach is properly designed. This is contrary to the standard practice of fixing a test size, and rejecting the null hypothesis at that fixed size, regardless of sample size. An interesting result of our simulations, which is in part a consequence of these features (see below), is that the empirical rejection frequencies based on both Akaike and Schwarz information criteria (AIC and SIC, respectively) become vanishingly small (in many cases) for sample sizes of only 300 observations, regardless as to whether the "true" model is among those in the "selection" set.

As mentioned above, Granger causality tests have a natural interpretation as tests of predictive ability. In this sense, it may seem natural to use the results of causality tests when constructing forecasting models. However, it is commonly argued that any good forecasting model should be "tested" using some sort of *ex ante* forecasting experiment before being implemented in any practical setting. Along these lines, a reasonable alternative* procedure for checking causal ordering involves comparing competing forecasting models, both with and without the variable whose causal effect is being examined. These "competing" models could then be subjected to *ex ante* forecasting analysis and the "best" model chosen using some sort of out-of-sample loss function (e.g., mean square prediction error), or by constructing an out-of-sample predictive ability test. If the "best" model contains the variable of interest, noncausality is rejected. In general, there are many model selection criteria to choose from, including forecast error summary measures, directional forecast accuracy measures, and profit measures, for example. Choice of criterion is naturally dependent on the objective function of the practitioner. Diebold and Lopez (1995) and Swanson and White (1997a,b) discuss these and related issues, and contain other related references.

2.3 Complexity-Based Likelihood Criteria

Granger et al. (1995) suggest that although standard hypothesis testing has a role to play in terms of testing economic theories, it is more difficult to justify using standard hypothesis tests for choosing between two competing models. One reason for their concern is that one model must be selected as the null, and this model is often the more parsimonious. However, it is often difficult to distinguish between the two models (because of multicollinearity, near-identification, etc.), so that the null hypothesis may be unfairly favored. For example, it is far from clear that pretest significance levels of 5 and 1% are optimal, as pointed out by

*In our experiments, we use the predictive ability test due to Diebold and Mariano (1995) to "compare" our alternative models.

Fomby and Guilkey (1978) in the context of Durbin–Watson tests. The use of model selection criteria neatly avoids related sticky issues associated with how to test theories and how arbitrarily to choose significance levels. Further, as discussed above, model selection procedure have a number of optimal properties is used to "select" among competing models, and seem a natural alternative to standard hypothesis testing when considering issues of Granger causality and/or predictive ability.

Sin and White (1995) consider the use of penalized likelihood criteria for selecting models of dependent processes. In the context of strictly nested, overlapping or non-nested, linear or nonlinear, and correctly specified or mis-specified models they provide sufficient conditions on the penalty to ensure that the model selected attains the lower average Kullback–Leibler information criterion, with probability (approaching) one. Further, as special cases, their results describe the AIC and SIC.

Our experiments into the use of model selection criteria focus on their ability to identify "correctly" and unravel the causal relationships among stationary, integrated, and cointegrated time series variables. If it turns out that these model selection criteria perform well, then we have direct evidence that model selection criteria offer a valid alternative to standard hypothesis testing in the context of uncovering causal relationships. Further, as model selection criteria are very easy to calculate, regardless of the properties of the data, evidence of the usefulness of AIC and SIC might be of some interest to empirical economists. The two model selection criteria which we examine below are the AIC (Akaike, 1973, 1974):

$$AIC = T\log|\hat{\Sigma}|2f$$

and the SIC (Rissanen, 1978; Schwarz, 1978):

$$SIC = T\log|\hat{\Sigma}| + f\log(T)$$

where we define f to be the total number of parameters in the system [if we are measuring the causal effect of some variable(s) on a *group* of more than one other variable], or f is the number of parameters in the single equation of interest (when the *group* being examined is a single variable). Similarly, $\hat{\Sigma}$ is some standard estimate of the error covariance matrix, which is scalar if only one equation in the system is being examined.

2.4 *Ex Ante* Forecast Comparison

Although AIC and SIC may prove to be valuable when testing the null hypothesis of non causality, note that they are calculated "in sample," as are the standard Wald-type tests discussed in Section 2.1. However, if, as is often the case, the goal of our analysis is to construct an "optimal" forecasting model, then causality might be better tested for in an *ex ante* rather than *ex post* setting, as discussed above. After all, such procedures would allow us to tackle directly the issue of predictive ability. In addition to those papers cited above, Engle and Brown (1986), Fair and Shiller (1990), and Swanson (1998) discuss *ex ante* model selection. However, these papers, as well as Diebold and Mariano (1995), West (1996), Swanson (1998), White (2000), and Corradi and Swanson (2000a) consider the application of predictive accuracy tests, which are not shown to be valid in the context of nested models (i.e, when forming Granger causality-type predictive ability tests). For tests that are valid in the context of nested models, see Clark and McCracken (1999), McCracken (1999), Chao et al. (2000), and Corradi and Swanson (2000b). The approach that we adopt involves constructing sequences of real-time one-step ahead forecasts of the variable of interest using a variety of models (levels VARs, differences VARs, and VECs) as well as a variety of different lag structures, cointegration assumptions, etc. The resulting sequences of forecasts are then used to construct three measures of out-of-sample forecast performance. The measures are mean squared forecast error ($MSE = \sum_{t=1}^{T^*} \hat{fe}_t^2 / T^*$), and the well known related statistics: mean absolute forecast error deviation (MAD) and mean absolute forecast percentage error (MAPE), where T^* is the out-of-sample period (which here varies with sample size), while $\hat{fe}_t$ is the appropriate forecast error associated with the particular econometric model in question. The "best" model is not selected based on direct examination of MSE, MAD, and MAPE criteria, however. Rather, we construct predictive accuracy tests described in Diebold and Mariano (1995), and implemented in Swanson and White (1997a,b). These tests statistics are constructed by first forming the mean ($\bar{d}$) of a given loss differential series, d_t (e.g., for MSE we construct $d_t = \hat{fe}_{1,t}^2 - \hat{fe}_{2,t}^2$, where the subscript 1 corresponds to the null model, and subscript 2 denotes the same model as subscript 1, but with additional regressors corresponding to the variable whose causal effect is being examined). The loss differential test statistic is then constructed by dividing ($\bar{d}$) by an estimate of the standard error of d_t, say $\hat{s}_d$. We use the parametric convariance matrix estimation procedure of Den Haan and Levin (1995) to estimate s_d, in part because the method is very easy to implement, and in part because these authors show that their procedure compares favorably with a number of nonparametric covariance matrix estimation techniques. Further details of these tests are given in Swanson and White (1995, 1997). McCracken (1999) gives critical values that are valid when Diebold–Mariano tests are used to compare nested alternatives.

In our simulations we are able to implement a truly *ex ante* forecasting procedure. However, as is well known, the use of real-time forecasting methods becomes suspect when the actual economic data that are used have been subjected to two-sided moving average filtering, periodic rebasing, and periodic revision, for example. Thus, we stress that these and related problems may make valid implementation of these types of tests quite difficult in practice. In Section 3 we discuss the design of a number of Monte Carlo experiments that were used to examine the finite sample properties of the above tests.

3 EXPERIMENTAL DESIGN

Our Monte Carlo experiments are all based on 5000 simulations, and start with various difference stationary and cointegrated versions of the following two VAR models:

$$y_t = J_1 y_{t-1} + u_t, t = 0, \ldots, T \tag{4}$$

$$y_t = J_1 y_{t-1} + J_2 y_{t-2} + J_3 y_{t-3} + u_t, t = -2, \ldots, T \tag{5}$$

We use a VAR(1) and a VAR(3) model because it is well known that causality tests often result in quite different conclusions, depending on how many lags are used to estimate a model. Clearly, one reason why this problem arises is that models which are misspecified in the sense that surplus lags are included, may suffer decreased test power. However, another possibility is that increasing the number of lags estimated results in what we will call "least squares confusion" (i.e., the tendency for least squares to yield poor parameter estimates), even if the "true" data are generated by a process with many lags. In practice, this type of situation may arise because of multicollinearity, and would probably manifest itself in the form of an increased incidence of rejecting the null hypothesis of Granger noncausality, even when the null is true. If this is the case, then we would expect experimental results based solely on VAR(1) processes to be conservative relative to those based on higher order VARs. As empirical models often contain many lags, "least squares confusion" may be damaging in the sense that we may later construct tests under the assumption that their size is close to the nominal test size, when in actuality the size is much higher than the nominal size. Further, the tests that perform "best" might change as we increase the number of lags in the system. In fact, we will see later that these issues are relevant, and have a non-negligible impact on our findings.

Our data generating processes can be summarized as follows:

Cointegrated models (cointegrating rank $=1$; $I_3 - \sum J_i = \Gamma A'$)

$$\text{(VAR1}-\text{CI1)}\quad J_1 = \begin{bmatrix} 0.5 & 0.5 & 0.0 \\ 0.0 & 1.0 & 0.0 \\ -1.0 & 1.0 & 1.0 \end{bmatrix} \quad \Gamma = \begin{bmatrix} 0.5 \\ 0.0 \\ 1.0 \end{bmatrix} \quad A = \begin{bmatrix} 1.0 \\ -1.0 \\ 0.0 \end{bmatrix}$$

$$(\text{VAR3}-\text{CI1})\ J_1 = J_2 = \begin{bmatrix} 0.17 & 0.17 & 0.0 \\ 0.0 & 0.4 & 0.0 \\ -0.4 & 0.4 & 0.4 \end{bmatrix} \quad J_3 = \begin{bmatrix} 0.16 & 0.16 & 0.0 \\ 0.0 & 0.2 & 0.0 \\ -0.2 & 0.2 & 0.2 \end{bmatrix}$$

$$\Gamma = \begin{bmatrix} 0.5 \\ 0.0 \\ 1.0 \end{bmatrix} \quad A = \begin{bmatrix} 1.0 \\ -1.0 \\ 0.0 \end{bmatrix}$$

$$(\text{VAR1}-\text{CI2})\ J_1 = \begin{bmatrix} 1.0 & -1.0 & -1.0 \\ 0.0 & 0.5 & -0.5 \\ 0.0 & -1.0 & 0.0 \end{bmatrix} \quad \Gamma = \begin{bmatrix} 1.0 \\ 0.5 \\ 1.0 \end{bmatrix} \quad A = \begin{bmatrix} 0.0 \\ 1.0 \\ 1.0 \end{bmatrix}$$

$$(\text{VAR3}-\text{CI2})\ J_1 = J_2 = \begin{bmatrix} 0.4 & -0.4 & -0.4 \\ 0.0 & 0.17 & -0.17 \\ 0.0 & -0.4 & 0.0 \end{bmatrix}$$

$$J_3 = \begin{bmatrix} 0.2 & -0.2 & -0.2 \\ 0.0 & 0.16 & -0.16 \\ 0.0 & -0.2 & 0.0 \end{bmatrix} \quad \Gamma = \begin{bmatrix} 1.0 \\ 0.5 \\ 1.0 \end{bmatrix} \quad A = \begin{bmatrix} 0.0 \\ 1.0 \\ 1.0 \end{bmatrix}$$

Difference stationary models

$$(\text{VAR1} - \text{DS1})\quad J_1 = I_3$$

$$(\text{VAR1} - \text{DS1})\quad J_1 = \begin{bmatrix} 0.2 & 0.1 & 0.0 \\ 0.0 & 0.4 & 0.0 \\ 0.4 & 0.5 & 0.5 \end{bmatrix} \quad J_2 = \begin{bmatrix} 0.2 & 0.1 & 0.0 \\ 0.0 & 0.4 & 0.0 \\ 0.2 & 0.1 & 0.3 \end{bmatrix}$$

$$J_3 = \begin{bmatrix} 0.6 & -0.2 & 0.0 \\ 0.0 & 0.2 & 0.0 \\ -0.6 & -0.6 & 0.2 \end{bmatrix}$$

It should perhaps be noted that the above parameterizations are ad hoc in the sense that they are not calibrated by examining actual economic data, for example. In addition, economic data at most frequencies are usually modelled by including more than three lags. In this sense, the above DGPs are too simplistic. For these reasons, the Monte Carlo results reported below are meant only to be a guide to the sorts of relationships among the various causality tests, which might be observed in practice. In addition, the following structure is placed on the errors of the above DGPs, where $u_t = \eta_t - \theta\eta_{t-1}$, and η_t is a unit normal random vector:

$$(MA1)\,\theta = 0$$

$$(MA2)\,\theta = \begin{bmatrix} 0.5 & 0.0 & 0.0 \\ 0.0 & 0.5 & 0.0 \\ 1.0 & 0.0 & 0.5 \end{bmatrix}$$

These values of θ are the same as those used by Toda and Phillips (1994). As discussed in Toda and Phillips (1994), even though (MA2) appears to be inconsistent with A.1, u_t is an invertible MA process so that VAR(1) and VAR(3) can be rewritten as infinite order VARs. As the higher lags are exponentially decaying in the parameters, the actual number of lags needed for a "reasonable" approximation is low. Further, it will be interesting to see how well our various statistics perform when faced with inherently mis-specified models (in the MA1 models, the true specification is always included within the set of models that are fit). Finally, the addition of MA2 to our experimental setup ensures that some of our models [i.e., the VAR(3) models] should ideally contain more than three lags when estimated.

In most cases we are testing the null hypothesis that y_3 is Granger noncausal for y_1 (we use the notation $y_3 \not\rightarrow y_1$ for cases such as this), and thus we focus somewhat on the empirical size of the seven tests. The exception is CI2, in which case we test the null hypothesis that $y_1 \not\rightarrow y_2$. In all experiments, whenever a test of the null that $y_i \not\rightarrow y_j$ is carried out, we also perform a test of the null that $y_j \not\rightarrow y_i$. This allows us to tabulate power at the same time that we tabulate empirical size. It should be noted that in the case of our model selection procedures, "empirical size" is not the appropriate measure. Instead, we report the rejection frequency of the correct model in favor of the "incorrect" model. As the tests that we examine all have good power properties (albeit rejection frequencies under alternative hypotheses are not "size adjusted"), these findings leave little to choose between the tests. Not surprisingly, then, may analyses related to ours focus on empirical size. We follow suit, and report only empirical size (and its analog for our model selection procedures). Complete finding are available on request from the authors.

In each of the experiments we use 100, 200, and 300 observations. Assuming that the true number of lags in each case is p, then the number of lags used in each estimation is set at $\{p-1, p, p+1, p_{AIC}, p_{SIC}\}$, where p_{AIC} and p_{SIC} use the number of lags selected by implementation of the AIC and SIC model selection criteria, where the model selection criteria are constructed using estimates of levels VARs. Also, all lags are reported in terms of a VAR in levels, so that results for difference VAR Wald-type tests (λ_{DS}) for one lag are not given. This is because a level VAR with one lag translates to a difference VAR with no lags.

In all experiments we calculate many difference stationary, stationary, and VEC systems, regardless of the true DGP. This is because the seven procedures considered variously involve estimation of all three types of VARs. For example, a cointegration model is estimated, and sequential Wald tests are constructed, even when the true data-generating process is stationary. One reason for an extensive analysis of this type is that absolute size distortions associated with each test type are not necessarily as important as the size distortion of one test relative to the size distortion of other tests, regardless as to whether either test is theoretically valid. As an example of why this issue is perhaps important, note that Gonzalo and Lee (1996) show that, while it is well known that augmented Dickey–Fuller tests suffer from low power when processes have near unit roots, and suffer from size distortions in other cases, it is less widely acknowledged that standard t-statistics also have similar power and size problems, even when the null hypothesis is that $\beta = 0.5$ and the true process is a stationary AR(1) with slope coefficient β. Extending this argument, it should be of interest to determine whether size and power distortions associated with using standard χ^2 asymptotics when they are not valid are large enough to warrant the use of more complex (albeit asymptotically valid) testing procedures which themselves may suffer from size and power distortions of other types. In particular, even though asymptotically valid, the finite sample performance of some of the more complex test statistics considered here is not certain.

We also used F-test versions of all Wald-type tests, as it is well known that these often have better size properties in finite samples. To be more specific, we construct $F = \lambda/R$, where R is the number of restrictions. These statistics are assumed to be distributed as F, with $T - f$ degrees of freedom in the denominator, where f is (1) the number of parameters in the equation being examined, and f is also defined as (2) the number of parameters estimated in the entire system. Interestingly, the properties of these versions of the Wald tests were qualitatively the same as when the χ^2 versions of the tests were used. For this reason, these results are not reported in the tables, although complete results are available in an earlier working paper version of this study.

Whenever VEC models are estimated, we use the maximum likelihood procedures developed by Johansen (1988, 1991) and Johansen and Juselius (1990). In particular, assume that the true cointegrating rank is r. Then, the cointegrating ranks used are usually $\{r-1, r, r+1, r_{JO1}, r_{JO2}\}$.

The last two are rank estimates based on the well known likelihood ratio trace statistic using a 5% size, and with and without a constant in the estimated system, respectively. Results based on the related maximum statistic are not reported for the sake of brevity. The exception to this rule are cases where the true cointegrating rank is zero. In these cases, we use $\{r+1, r, r+2, r_{JO1}, r_{JO2}\}$, where r_{JO1} and r_{JO2} may be zero, in which case VEC models are not estimated, and corresponding sequential Wald test results are not reported. With the exception of the test statistic based on fully modified VAR estimates, all other estimations done in the experiments are multivariate least squares.

Overall, we calculate all Wald-type tests whenever possible. However, it should be noted that there are some cases where Wald-type tests are not applicable. As mentioned above, difference stationary versions of the standard Wald test are not calculated when the lag is one (as a one lag VAR in levels is a zero lag VAR in differences). Also, there are some lag structures and cointegrating rank cases (e.g., $r=n$) for which the sequential Wald-type tests are not applicable, and are not calculated. [The reader is referred to Toda and Phillips (1994) for more details.] In such cases, the reported simulation results are calculated using possibly fewer than 5000 simulation trials. Instances where this situation arises are reported on in the next section.

The AIC and SIC type tests are implemented as follows. For all lags, cointegrating ranks, and alternative models (i.e., levels VAR, differences VAR, an VEC) AIC and SIC statistics are calculated. In particular, the statistics are calculated for versions of the models both with and without the variable(s) whose causal effect is being examined. As a summary test, the AIC and SIC best models are chosen. If the "best" model contains any of the relevant variable(s), then the null hypothesis of Granger noncausality is rejected. We also track with *type* of model is chosen as "best" in each simulation. These types of tests are individually reported on for each lag and each model (e.g., cointegrating model with cointegrating space rank of unity), as well as across all models. The latter reporting style yields what we will call "global" tests, while the former is more in keeping with a comparison of only "two" alternative models when testing the null hypothesis of noncausality.

Ex ante forecasting tests are carried out the same way as the AIC and SIC type tests, with two exceptions. First, the statistics calculated are MSE, MAD, and MAPE. Second the statistics are constructed from sequences of one-step ahead forecast errors, and these forecast errors are in turn used to construct predictive accuracy test statistics (as discussed above). The procedure is implemented in the following way. A starting sample of $0.7\,T$ observations is used to construct a one-step ahead forecast, using all different model types, lags, etc., and for $T=100$, 200, or 300. From this, the first forecast error is saved. One more observation is then added to the sample, and another set of models are estimated, forecasts are constructed, and forecast errors are saved. This procedure continuous until the entire sample is exhausted, and the resultant *ex ante* sequence of $(0.3T)-1$ forecast errors is used to construct the model selection criteria. It should be noted that one case arises where the actual number of forecast errors used to construct the model selection criteria is not $(0.3\,T)-1$. For the VEC models that are estimated using cointegrating rank, $r = r_{JO1}$ or $r = r_{JO2}$, there may be some simulations where the estimated cointegrating rank is 0 or n (in which case the model is not estimated). This situation arises most frequently when the actual DGPs are stationary or difference stationary VARs. In these cases, we will report the average number of observations used in the construction of the selection criteria. In addition, it should be noted that, due to possible model misspecification (see, e.g., Corradi and Swanson, 2000b), finite sample results may not be the same for the one-step ahead versus multistep ahead approach to *ex ante* forecasting implementations of Granger causality analysis.

4 RESULTS

A summary of our experimental findings is given in Tables 1–15. Complete experimental results are contained in an earlier working paper version of this chapter (i.e., see Swanson et al. and Pisu, 1996). Our findings from the Monte Carlo experiments are fairly clear cut, and can be summarized as follows:

1. In all cases (i.e., regardless of lag and model specification) the Toda and Yamamoto type Wald tests have empirical sizes close to nominal (nominal size is set at 5% throughout).
2. Sequential Wald tests, and standard Wald tests based on the use of differenced data only perform the best when correctly specified models (i.e., correct lags, cointegrating rank, data transformation) are estimated.
3. SIC type approaches, and to a lesser degree those based on the AIC perform quite well, even when the DGP has moving average (MA) error components (i.e., when the true lag order of the model is infinite and is hence only approximated using AIC and SIC measures). In many cases and for samples of moderate and large size (200 and 300 observations), the SIC-type causality test has rejection frequency approaching zero very rapidly, and power close to unity.
4. As expected, test performance (with the exception of all of our model selection type causality "tests") worsens substantially when model complexity is increased (i.e., going from one to three lags and from no MA to an MA component in the true DGP).
5. Lag selection based on the AIC and SIC does very well. In particular, the lag structure chosen using the SIC almost always yields the "best" results when the DGP is a VAR(1) with no MA component. However, for other DGPs, the lag structure given by the AIC generally results in better test performance.
6. *Ex ante* forecasting type approaches tend to perform well [see Chao et al. (2000) and McCracken (1999) for further evidence]. In particular, when

TABLE 1 Summary of Wald Test Monte Carlo Results for λ_{ST}, λ_{DS}, λ_{LW}, and λ_{FM} Empirical Size;[a] Data Generating-Processes are (CI-1), (CI-2), and (DS1)

		(CI-1)				(CI-2)				(DSI)			
DGP/sample	Lag	λ_{ST}	λ_{DS}	λ_{LW}	λ_{FM}	λ_{ST}	λ_{DS}	λ_{LW}	λ_{FM}	λ_{ST}	λ_{DS}	λ_{LW}	λ_{FM}
VAR(1)-(MA1)	1	0.167	—	0.065	—	0.146	—	0.059	—	0.173	—	0.061	—
$T=100$	2	0.142	0.999	0.074	0.279	0.127	0.651	0.068	0.800	0.152	0.054	0.070	0.329
	AIC	0.175	0.990	0.064	0.408	0.154	0.571	0.059	0.660	0.186	0.234	0.062	0.335
	SIC	0.167	—	0.065	—	0.146	—	0.059	—	0.173	1.00	0.061	—
	1	0.147	—	0.058	—	0.156	—	0.057	—	0.177	—	0.053	—
$T=200$	2	0.133	1.00	0.064	0.247	0.129	0.911	0.064	0.938	0.146	0.051	0.061	0.299
	AIC	0.155	1.00	0.057	0.336	0.160	0.788	0.057	0.788	0.185	0.211	0.053	0.338
	SIC	0.147	—	0.058	—	0.156	—	0.057	—	0.177	—	0.053	—
$T=300$	1	0.162	—	0.048	—	0.149	—	0.051	—	0.172	—	0.052	—
	2	0.128	1.00	0.058	0.256	0.119	0.979	0.053	0.975	0.137	0.052	0.058	0.295
	AIC	0.169	1.00	0.049	0.366	0.155	0.912	0.052	0.837	0.180	0.259	0.054	0.384
	SIC	0.162	—	0.048	—	0.149	—	0.051	—	0.172	—	0.052	—
VAR(1)-(MA2)	1	0.244	—	0.154	—	0.407	—	0.071	—	0.703	—	0.135	—
$T=100$	2	0.217	1.00	0.108	0.784	0.215	0.964	0.074	0.948	0.344	0.298	0.108	0.821
	AIC	0.215	0.996	0.114	0.353	0.207	0.941	0.116	0.507	0.282	0.169	0.135	0.483
	SIC	0.238	1.00	0.104	0.741	0.213	0.967	0.075	0.779	0.582	0.328	0.096	0.718
	1	0.258	—	0.215	—	0.433	—	0.057	—	0.798	—	0.242	—

(continued)

Table 1 (Continued)

DGP/sample	Lag	(CI-1) λ_{ST}	λ_{DS}	λ_{LW}	λ_{FM}	(CI-2) λ_{ST}	λ_{DS}	λ_{LW}	λ_{FM}	(DSI) λ_{ST}	λ_{DS}	λ_{LW}	λ_{FM}
$T=200$	2	0.274	1.00	0.106	0.861	0.217	1.00	0.063	0.996	0.504	0.505	0.134	0.876
	AIC	0.169	1.00	0.079	0.260	0.159	0.994	0.082	0.719	0.188	0.117	0.104	0.485
	SIC	0.251	1.00	0.093	0.724	0.170	1.00	0.069	0.406	0.413	0.363	0.105	0.710
	1	0.258	—	0.285	—	0.447	—	0.055	—	0.813	—	0.400	—
$T=300$	2	0.324	1.00	0.119	0.908	0.226	1.00	0.059	1.00	0.656	0.681	0.165	0.918
	AIC	0.140	1.00	0.066	0.266	0.140	1.00	0.073	0.687	0.158	0.099	0.084	0.587
	SIC	0.217	1.00	0.078	0.560	0.151	1.00	0.061	0.384	0.321	0.278	0.090	0.656
VAR(3)-(MA1)	2	0.185	0.530	0.075	0.554	0.162	0.236	0.074	0.605	0.371	0.099	0.086	0.914
	3	0.139	0.874	0.085	0.229	0.132	0.276	0.088	0.165	0.145	0.066	0.103	0.408
$T=100$	4	0.147	0.895	0.100	0.276	0.137	0.256	0.100	0.175	0.150	0.077	0.123	0.296
	AIC	0.170	0.797	0.089	0.308	0.162	0.280	0.093	0.261	0.171	0.089	0.114	0.406
	SIC	0.192	0.547	0.076	0.536	0.168	0.243	0.075	0.575	0.145	0.066	0.103	0.408
	2	0.169	0.821	0.061	0.565	0.160	0.399	0.060	0.814	0.382	0.106	0.070	0.924
	3	0.110	0.996	0.064	0.198	0.113	0.466	0.067	0.143	0.132	0.057	0.076	0.473
$T=200$	4	0.107	0.997	0.070	0.285	0.112	0.426	0.077	0.152	0.132	0.065	0.086	0.271
	AIC	0.117	0.993	0.066	0.209	0.118	0.466	0.070	0.155	0.142	0.066	0.078	0.465
	SIC	0.162	0.866	0.060	0.480	0.148	0.422	0.060	0.638	0.132	0.057	0.076	0.473
	2	0.181	0.941	0.053	0.602	0.147	0.533	0.056	0.918	0.394	0.097	0.062	0.931
	3	0.107	1.00	0.062	0.213	0.110	0.653	0.063	0.141	0.128	0.054	0.067	0.555

$T=300$	4	0.108	1.00	0.064	0.335	0.109	0.596	0.065	0.149	0.130	0.055	0.067	0.268
	AIC	0.114	1.00	0.062	0.219	0.118	0.651	0.061	0.144	0.136	0.061	0.068	0.545
	SIC	0.147	0.974	0.056	0.398	0.127	0.613	0.060	0.458	0.128	0.054	0.067	0.555
VAR(3)-(MA2)	2	0.448	0.932	0.197	0.978	0.464	0.149	0.114	0.626	0.865	0.493	0.227	0.963
	3	0.242	0.977	0.143	0.516	0.221	0.457	0.105	0.411	0.431	0.110	0.154	0.799
$T=100$	4	0.187	0.985	0.129	0.564	0.177	0.812	0.114	0.480	0.226	0.152	0.144	0.815
	AIC	0.254	0.964	0.161	0.585	0.239	0.835	0.161	0.600	0.326	0.178	0.205	0.846
	SIC	0.295	0.971	0.142	0.613	0.221	0.570	0.109	0.441	0.437	0.123	0.151	0.810
	2	0.603	0.998	0.266	0.999	0.517	0.194	0.097	0.738	0.946	0.738	0.259	0.983
	3	0.293	1.00	0.146	0.598	0.229	0.733	0.082	0.477	0.553	0.131	0.194	0.881
$T=200$	4	0.191	1.00	0.099	0.814	0.158	0.988	0.085	0.578	0.290	0.212	0.123	0.908
	AIC	0.166	1.00	0.101	0.678	0.165	0.987	0.101	0.744	0.182	0.117	0.123	0.871
	SIC	0.251	1.00	0.115	0.693	0.165	0.968	0.086	0.589	0.461	0.194	0.139	0.921
	2	0.718	1.00	0.352	1.00	0.539	0.244	0.095	0.792	0.969	0.858	0.268	0.989
	3	0.369	1.00	0.159	0.671	0.230	0.900	0.080	0.535	0.613	0.166	0.237	0.927
$T=300$	4	0.200	1.00	0.090	0.953	0.155	1.00	0.073	0.672	0.372	0.302	0.139	0.946
	AIC	0.136	1.00	0.075	0.714	0.149	0.998	0.092	0.835	0.169	0.110	0.117	0.887
	SIC	0.223	1.00	0.094	0.899	0.152	1.00	0.074	0.733	0.349	0.220	0.130	0.974

[a]The tests examined are: λ_{ST} (Wald test based on levels OLS regression); λ_{DS} (Wald test based on differences OLS regression); λ_{LW} (Wald test based on levels OLS regression with surplus lags added to isolate nonstandard asymptotics); and λ_{FM} (Wald test based on fully modified vector autoregression). For each test, 5% χ^2 critical values are used. The lags used are listed, as is the number of observations in the sample, and the DGP used to generate the data. All results are based on 5000 Monte Carlo trials. Entries in the table correspond to the proportion of rejections of the null hypothesis of noncasuality. For example, for the (CI-1)—VAR(1)—(MA1) with $T=100$ observations, the upper left entry of the table is 0.167. Thus, a standard Wald test when applied to the data in levels has an empirical rejection rate of 16.7%.

TABLE 2 Summary of Sequential Wald Test Monte Carlo Results, Empirical Size;[a] Data-Generating Process is (CI-1)

Lag/sample	CI rank	λ_{p1} VAR(1) (MA1)	λ_{p1} VAR(1) (MA2)	λ_{p1} VAR(3) (MA1)	λ_{p1} VAR(3) (MA2)	λ_{p2} VAR(1) (MA1)	λ_{p2} VAR(1) (MA2)	λ_{p2} VAR(3) (MA1)	λ_{p2} VAR(3) (MA2)	λ_{p3} VAR(1) (MA1)	λ_{p3} VAR(1) (MA2)	λ_{p3} VAR(3) (MA1)	λ_{p3} VAR(3) (MA2)
	1	0.067	0.069			0.887	0.948			—	—		
Lag = 1	2	0.162	0.237			0.230	0.328			—	—		
$T=100$	j_{o1}	0.067	0.069			0.889	0.947			—	—		
	j_{o2}	0.068	0.067			0.889	0.946			—	—		
	1	0.095	0.124	0.063	0.338	0.112	0.151	0.094	0.387	0.170	0.172	0.125	0.419
Lag = 2	2	0.150	0.208	0.181	0.448	0.155	0.227	0.191	0.468	0.229	0.285	0.272	0.495
$T=100$	j_{o1}	0.095	0.123	0.061	0.330	0.112	0.150	0.093	0.381	0.169	0.172	0.124	0.411
	j_{o2}	0.094	0.123	0.060	0.328	0.111	0.150	0.091	0.380	0.168	0.172	0.122	0.410
	1			0.102	0.213			0.118	0.218			0.191	0.268
Lag = 3	2			0.140	0.242			0.147	0.251			0.235	0.324
$T=100$	j_{o1}			0.102	0.212			0.118	0.217			0.191	0.265
	j_{o2}			0.102	0.212			0.117	0.216			0.190	0.265
	1			0.118	0.145			0.146	0.155			0.217	0.201
Lag = 4	2			0.152	0.190			0.157	0.198			0.253	0.285
$T=100$	j_{o1}			0.119	0.144			0.148	0.154			0.222	0.200
	j_{o2}			0.117	0.143			0.145	0.153			0.217	0.199
	1	0.082	0.144	0.114	0.201	0.628	0.174	0.137	0.241	0.376	0.187	0.202	0.268
Lag = AIC	2	0.174	0.215	0.168	0.255	0.241	0.225	0.175	0.264	0.411	0.299	0.264	0.348
$T=100$	j_{o1}	0.082	0.141	0.113	0.194	0.631	0.170	0.134	0.234	0.370	0.184	0.200	0.262
	j_{o2}	0.082	0.141	0.112	0.195	0.636	0.170	0.134	0.235	0.379	0.184	0.199	0.262

	1	0.067	0.126	0.069	0.243	0.887	0.197	0.101	0.259	—	0.192	0.135	0.302
Lag = SIC	2	0.162	0.231	0.187	0.293	0.230	0.263	0.196	0.304	—	0.299	0.273	0.367
$T = 100$	j_{o1}	0.067	0.126	0.068	0.236	0.889	0.197	0.100	0.252	—	0.192	0.134	0.294
	j_{02}	0.068	0.125	0.066	0.234	0.889	0.196	0.098	0.250	—	0.191	0.132	0.291
	1	0.054	0.058			0.940	0.964				—		
Lag = 1	2	0.150	0.246			0.214	0.343				—		
$T = 200$	j_{o1}	0.054	0.058			0.940	0.963				—		
	j_{02}	0.054	0.056			0.940	0.961				—		
	1	0.069	0.147	0.047	0.489	0.088	0.196	0.078	0.568	0.127	0.207	0.103	0.581
Lag = 2	2	0.131	0.267	0.169	0.595	0.142	0.288	0.176	0.619	0.202	0.334	0.250	0.641
$T = 200$	j_{o1}	0.069	0.147	0.047	0.487	0.088	0.196	0.078	0.567	0.128	0.207	0.103	0.578
	j_{02}	0.069	0.146	0.047	0.486	0.087	0.196	0.077	0.566	0.127	0.207	0.103	0.577
	1			0.075	0.231			0.088	0.254			0.137	0.278
Lag = 3	2			0.114	0.287			0.122	0.303			0.203	0.353
$T = 200$	j_{o1}			0.075	0.230			0.087	0.253			0.136	0.277
	j_{02}			0.075	0.228			0.087	0.252			0.136	0.276
	1			0.087	0.128			0.095	0.134			0.154	0.162
Lag = 4	2			0.110	0.182			0.115	0.190			0.207	0.279
$T = 200$	j_{o1}			0.087	0.127			0.094	0.133			0.154	0.161
	j_{02}			0.087	0.127			0.094	0.133			0.154	0.161

(*continued*)

TABLE 2 (Continued)

		λ_{p1}				λ_{p2}				λ_{p3}			
Lag/sample	CI rank	VAR(1) (MA1)	VAR(1) (MA2)	VAR(3) (MA1)	VAR(3) (MA2)	VAR(1) (MA1)	VAR(1) (MA2)	VAR(3) (MA1)	VAR(3) (MA2)	VAR(1) (MA1)	VAR(1) (MA2)	VAR(3) (MA1)	VAR(3) (MA2)
	1	0.062	0.103	0.083	0.120	0.649	0.118	0.096	0.137	0.341	0.144	0.144	0.173
Lag = AIC	2	0.155	0.172	0.122	0.168	0.220	0.181	0.130	0.175	0.403	0.277	0.209	0.275
$T = 200$	j_{o1}	0.062	0.102	0.082	0.119	0.650	0.117	0.096	0.136	0.342	0.143	0.144	0.173
	j_{o2}	0.062	0.102	0.082	0.118	0.650	0.117	0.095	0.135	0.342	0.143	0.143	0.171
	1	0.054	0.133	0.059	0.188	0.940	0.174	0.086	0.205	—	0.186	0.119	0.232
Lag = SIC	2	0.150	0.243	0.162	0.246	0.214	0.260	0.168	0.258	—	0.320	0.241	0.325
$T = 200$	j_{o1}	0.054	0.132	0.058	0.187	0.940	0.174	0.086	0.204	—	0.185	0.118	0.231
	j_{o2}	0.054	0.132	0.058	0.186	0.940	0.174	0.085	0.203	—	0.185	0.118	0.230
	1	0.057	0.049			0.953	0.957			—	—		
Lag = 1	2	0.160	0.245			0.225	0.344			—	—		
$T = 300$	j_{o1}	0.057	0.047			0.953	0.955			—	—		
	j_{o2}	0.057	0.047			0.953	0.954			—	—		
	1	0.061	0.197	0.041	0.617	0.079	0.263	0.072	0.701	0.117	0.273	0.087	0.707
Lag = 2	2	0.123	0.311	0.178	0.716	0.130	0.342	0.185	0.741	0.201	0.389	0.260	0.763
$T = 300$	j_{o1}	0.061	0.196	0.041	0.614	0.079	0.263	0.072	0.700	0.117	0.272	0.087	0.705
	j_{o2}	0.061	0.196	0.041	0.617	0.079	0.263	0.071	0.703	0.116	0.272	0.087	0.709

	1			0.055	0.291			0.072	0.329			0.120	0.349
Lag = 3	2			0.107	0.361			0.112	0.381			0.203	0.423
$T = 300$	j_{o1}			0.055	0.290			0.072	0.329			0.120	0.349
	j_{02}			0.055	0.290			0.072	0.329			0.120	0.348
	1			0.071	0.137			0.084	0.145			0.137	0.164
Lag = 4	2			0.111	0.194			0.116	0.203			0.212	0.285
$T = 300$	j_{o1}			0.071	0.137			0.084	0.145			0.137	0.164
	j_{02}			0.071	0.137			0.084	0.145			0.137	0.164
	1	0.064	0.082	0.062	0.099	0.648	0.093	0.080	0.112	0.291	0.119	0.127	0.149
Lag = AIC	2	0.167	0.139	0.113	0.143	0.231	0.146	0.120	0.147	0.330	0.241	0.209	0.257
$T = 300$	j_{o1}	0.064	0.082	0.062	0.099	0.646	0.093	0.080	0.112	0.288	0.119	0.127	0.149
	j_{02}	0.064	0.082	0.062	0.099	0.649	0.093	0.080	0.112	0.290	0.119	0.127	0.149
	1	0.057	0.116	0.054	0.157	0.953	0.155	0.077	0.169	—	0.166	0.111	0.189
Lag = SIC	2	0.160	0.209	0.146	0.217	0.225	0.226	0.152	0.227	—	0.292	0.234	0.306
$T = 300$	j_{o1}	0.057	0.115	0.054	0.157	0.953	0.155	0.077	0.169	—	0.166	0.111	0.189
	j_{02}	0.057	0.115	0.053	0.156	0.953	0.155	0.076	0.169	—	0.166	0.110	0.189

[a]See footnotes to Table 1. The tests examined are: λ_{P1}–λ_{P3} (sequential Wald tests based on VEC models estimated using Johansen's maximum likelihood procedure).

TABLE 3 Summary of Squential Wald Test Monte Carlo Results, Empirical Size;[a] Data-Generating Process is (CI-2)

Lag/sample	CI rank	λ_{p1} VAR(1) (MA1)	λ_{p1} VAR(1) (MA2)	λ_{p1} VAR(3) (MA1)	λ_{p1} VAR(3) (MA2)	λ_{p2} VAR(1) (MA1)	λ_{p2} VAR(1) (MA2)	λ_{p2} VAR(3) (MA1)	λ_{p2} VAR(3) (MA2)	λ_{p3} VAR(1) (MA1)	λ_{p3} VAR(1) (MA2)	λ_{p3} VAR(3) (MA1)	λ_{p3} VAR(3) (MA2)
	1	0.061	0.013	—	—	0.908	0.970	—	—	—	—	—	—
Lag = 1	2	0.147	0.409	—	—	0.213	0.530	—	—	—	—	—	—
$T = 100$	j_{o1}	0.061	0.013	—	—	0.908	0.969	—	—	—	—	—	—
	j_{o2}	0.061	0.012	—	—	0.910	0.968	—	—	—	—	—	—
	1	0.065	0.037	0.059	0.105	0.083	0.069	0.089	0.162	0.140	0.076	0.118	0.174
Lag = 2	2	0.137	0.211	0.163	0.470	0.141	0.221	0.174	0.480	0.206	0.295	0.253	0.536
$T = 100$	j_{o1}	0.065	0.037	0.058	0.087	0.082	0.069	0.088	0.148	0.139	0.075	0.117	0.153
	j_{o2}	0.065	0.037	0.058	0.087	0.082	0.068	0.088	0.148	0.139	0.075	0.117	0.152
	1	—	—	0.079	0.070	—	—	0.096	0.098	—	—	0.175	0.109
Lag = 3	2	—	—	0.137	0.224	—	—	0.146	0.232	—	—	0.227	0.330
$T = 100$	j_{o1}	—	—	0.078	0.067	—	—	0.096	0.096	—	—	0.175	0.106
	j_{o2}	—	—	0.078	0.067	—	—	0.096	0.095	—	—	0.174	0.104
	1	—	—	0.089	0.084	—	—	0.105	0.103	—	—	0.178	0.126
Lag = 4	2	—	—	0.144	0.180	—	—	0.151	0.187	—	—	0.236	0.291
$T = 100$	j_{o1}	—	—	0.088	0.083	—	—	0.105	0.103	—	—	0.180	0.126
	j_{o2}	—	—	0.089	0.082	—	—	0.105	0.101	—	—	0.179	0.124
	1	0.071	0.115	0.093	0.163	0.603	0.151	0.119	0.203	0.316	0.171	0.186	0.222
Lag = AIC	2	0.157	0.210	0.165	0.243	0.221	0.216	0.175	0.251	0.395	0.296	0.254	0.339
$T = 100$	j_{o1}	0.070	0.112	0.090	0.156	0.604	0.148	0.115	0.195	0.312	0.168	0.183	0.214
	j_{o2}	0.070	0.112	0.090	0.157	0.603	0.149	0.115	0.196	0.311	0.168	0.183	0.215

Lag = SIC $T = 100$	1	0.061	0.048	0.061	0.085	0.908	0.078	0.094	0.110	—	0.086	0.127	0.126
	2	0.147	0.208	0.172	0.221	0.213	0.215	0.183	0.228	—	0.286	0.253	0.328
	j_{o1}	0.061	0.048	0.061	0.081	0.908	0.078	0.093	0.106	—	0.085	0.127	0.121
	j_{o2}	0.061	0.047	0.061	0.081	0.910	0.077	0.093	0.106	—	0.084	0.126	0.119
Lag = 1 $T = 200$	1	0.055	0.008	—	—	0.917	0.953	—	—	—	—	—	—
	2	0.157	0.428	—	—	0.225	0.562	—	—	—	—	—	—
	j_{o1}	0.056	0.008	—	—	0.917	0.952	—	—	—	—	—	—
	j_{o2}	0.056	0.008	—	—	0.917	0.950	—	—	—	—	—	—
Lag = 2 $T = 200$	1	0.061	0.025	0.045	0.111	0.082	0.054	0.077	0.190	0.129	0.056	0.095	0.195
	2	0.129	0.209	0.157	0.525	0.137	0.218	0.296	0.539	0.207	0.299	1.00	0.590
	j_{o1}	0.062	0.025	0.045	0.105	0.082	0.054	0.076	0.191	0.129	0.056	0.095	0.192
	j_{o2}	0.062	0.025	0.044	0.104	0.082	0.054	0.076	0.191	0.130	0.056	0.094	0.192
Lag = 3 $T = 200$	1	—	—	0.054	0.048	—	—	0.072	0.084	—	—	0.134	0.089
	2	—	—	0.112	0.228	—	—	0.119	0.236	—	—	0.206	0.336
	j_{o1}	—	—	0.054	0.047	—	—	0.072	0.083	—	—	0.134	0.086
	j_{o2}	—	—	0.054	0.047	—	—	0.071	0.083	—	—	0.133	0.086
Lag = 4 $T = 200$	1	—	—	0.063	0.053	—	—	0.078	0.070	—	—	0.145	0.077
	2	—	—	0.116	0.155	—	—	0.118	0.161	—	—	0.207	0.271
	j_{o1}	—	—	0.063	0.053	—	—	0.078	0.069	—	—	0.145	0.076
	j_{o2}	—	—	0.064	0.052	—	—	0.078	0.069	—	—	0.145	0.076
Lag = AIC $T = 200$	1	0.062	0.084	0.061	0.098	0.610	0.117	0.081	0.126	0.306	0.135	0.142	0.150
	2	0.163	0.160	0.118	0.167	0.229	0.166	0.127	0.172	0.322	0.266	0.212	0.271
	j_{o1}	0.063	0.084	0.061	0.098	0.612	0.117	0.081	0.125	0.309	0.135	0.141	0.150
	j_{o2}	0.063	0.084	0.061	0.097	0.611	0.117	0.081	0.124	0.309	0.135	0.141	0.149

(continued)

Table 3 (Continued)

		λ_{p1}				λ_{p2}				λ_{p3}			
Lag/sample	CI rank	VAR(1) (MA1)	VAR(1) (MA2)	VAR(3) (MA1)	VAR(3) (MA2)	VAR(1) (MA1)	VAR(1) (MA2)	VAR(3) (MA1)	VAR(3) (MA2)	VAR(1) (MA1)	VAR(1) (MA2)	VAR(3) (MA1)	VAR(3) (MA2)
	1	0.055	0.042	0.052	0.059	0.917	0.069	0.083	0.075	—	0.076	0.116	0.083
Lag = SIC	2	0.157	0.172	0.151	0.167	0.225	0.179	0.163	0.174	—	0.271	0.239	0.279
$T=200$	j_{o1}	0.056	0.043	0.052	0.058	0.917	0.069	0.082	0.075	—	0.076	0.115	0.083
	j_{o2}	0.056	0.042	0.051	0.057	0.917	0.069	0.082	0.073	—	0.076	0.115	0.083
	1	0.048	0.006	—	—	0.938	0.912	—	—	—	—	—	—
Lag = 1	2	0.147	0.447	—	—	0.209	0.581	—	—	—	—	—	—
$T=300$	j_{o1}	0.049	0.006	—	—	0.938	0.912	—	—	—	—	—	—
	j_{o2}	0.049	0.006	—	—	0.941	0.912	—	—	—	—	—	—
	1	0.055	0.024	0.042	0.134	0.075	0.055	0.074	0.227	0.109	0.059	0.089	0.229
Lag = 2	2	0.122	0.220	0.156	0.549	0.130	0.228	0.165	0.567	0.195	0.313	0.238	0.608
$T=300$	j_{o1}	0.055	0.024	0.042	0.130	0.075	0.056	0.074	0.227	0.109	0.060	0.090	0.227
	j_{o2}	0.055	0.023	0.042	0.131	0.075	0.055	0.074	0.227	0.109	0.059	0.089	0.227
	1	—	—	0.058	0.046	—	—	0.072	0.079	—	—	0.118	0.080
Lag = 3	2	—	—	0.113	0.230	—	—	0.119	0.238	—	—	0.204	0.343
$T=300$	j_{o1}	—	—	0.058	0.045	—	—	0.071	0.079	—	—	0.118	0.081
	j_{o2}	—	—	0.058	0.046	—	—	0.071	0.080	—	—	0.118	0.080

	1	—	—	0.062	0.045	—	—	0.076	0.070	—	—	0.132	0.075
Lag = 4	2	—	—	0.113	0.151	—	—	0.116	0.158	—	—	0.214	0.277
$T = 300$	j_{o1}	—	—	0.062	0.045	—	—	0.076	0.071	—	—	0.132	0.075
	j_{o2}	—	—	0.062	0.045	—	—	0.076	0.071	—	—	0.132	0.075
	1	0.054	0.074	0.065	0.090	0.602	0.098	0.079	0.114	0.286	0.116	0.126	0.136
Lag = AIC	2	0.153	0.140	0.120	0.147	0.215	0.146	0.127	0.153	0.335	0.255	0.212	0.268
$T = 300$	j_{o1}	0.054	0.074	0.065	0.090	0.601	0.097	0.079	0.114	0.283	0.116	0.126	0.135
	j_{o2}	0.054	0.075	0.065	0.090	0.601	0.098	0.079	0.114	0.283	0.116	0.126	0.135
	1	0.048	0.042	0.052	0.054	0.938	0.066	0.077	0.076	—	0.071	0.111	0.082
Lag = SIC	2	0.147	0.148	0.133	0.147	0.209	0.156	0.141	0.154	—	0.261	0.219	0.272
$T = 300$	j_{o1}	0.049	0.042	0.052	0.053	0.938	0.066	0.076	0.076	—	0.071	0.111	0.082
	j_{o2}	0.049	0.042	0.052	0.054	0.941	0.066	0.076	0.076	—	0.070	0.110	0.082

[a]See footnotes to Table 2.

TABLE 4 Summary of Sequential Wald Test Monte Carlo Results, Empirical Size;[a] Data-Generating Process is (DS1)

Lag/ sample	CI rank	λ_{p1}				λ_{p2}				λ_{p3}			
		VAR(1) (MA1)	VAR(1) (MA2)	VAR(3) (MA1)	VAR(3) (MA2)	VAR(1) (MA1)	VAR(1) (MA2)	VAR(3) (MA1)	VAR(3) (MA2)	VAR(1) (MA1)	VAR(1) (MA2)	VAR(3) (MA1)	VAR(3) (MA2)
	1	0.498	0.960	—	—	0.246	0.856	—	—	—	—	—	—
Lag = 1	2	0.452	0.799	—	—	0.195	0.724	—	—	—	—	—	—
$T=100$	j_{o1}	0.846	0.997	—	—	0.688	0.959	—	—	—	—	—	—
	j_{o2}	0.759	0.995	—	—	0.537	0.947	—	—	—	—	—	—
	1	0.165	0.438	0.491	0.935	0.164	0.413	0.484	0.936	0.290	0.503	0.593	0.946
Lag = 2	2	0.157	0.364	0.382	0.870	0.157	0.346	0.382	0.869	0.199	0.390	0.441	0.895
$T=100$	j_{o1}	0.500	0.499	0.696	0.949	0.500	0.489	0.690	0.949	0.688	0.551	0.754	0.958
	j_{o2}	0.407	0.495	0.668	0.950	0.407	0.482	0.661	0.950	0.574	0.546	0.727	0.959
	1	—	—	0.164	0.552	—	—	0.162	0.554	—	—	0.305	0.679
Lag = 3	2	—	—	0.155	0.432	—	—	0.154	0.436	—	—	0.214	0.521
$T=100$	j_{o1}	—	—	0.500	0.660	—	—	0.500	0.664	—	—	0.500	0.775
	j_{o2}	—	—	0.425	0.633	—	—	0.375	0.637	—	—	0.525	0.754
	1	—	—	0.174	0.286	—	—	0.175	0.272	—	—	0.331	0.398
Lag = 4	2	—	—	0.166	0.237	—	—	0.166	0.233	—	—	0.237	0.292
$T=100$	j_{o1}	—	—	0.250	0.350	—	—	0.300	0.338	—	—	0.400	0.447
	j_{o2}	—	—	0.269	0.339	—	—	0.288	0.324	—	—	0.500	0.435
	1	0.478	0.329	0.187	0.382	0.259	0.322	0.186	0.383	0.487	0.430	0.325	0.513
Lag = AIC	2	0.441	0.292	0.179	0.331	0.208	0.291	0.179	0.335	0.405	0.332	0.237	0.401
$T=100$	j_{o1}	0.750	0.524	0.579	0.634	0.667	0.515	0.553	0.626	0.600	0.546	0.605	0.730
	j_{o2}	0.605	0.507	0.438	0.592	0.500	0.491	0.391	0.587	0.500	0.539	0.547	0.689

	1	0.498	0.687	0.164	0.553	0.246	0.653	0.162	0.557	1.00	0.485	0.305	0.674
Lag = SIC	2	0.453	0.611	0.155	0.438	0.195	0.588	0.154	0.444	1.00	0.388	0.214	0.522
$T = 100$	j_{o1}	0.860	0.897	0.500	0.669	0.688	0.869	0.500	0.673	0.00	0.567	0.500	0.780
	j_{o2}	0.759	0.869	0.425	0.646	0.537	0.837	0.375	0.650	0.00	0.546	0.525	0.762
	1	0.490	0.968			0.267	0.908	—	—	—	—	—	—
Lag = 1	2	0.451	0.875			0.203	0.816	—	—	—	—	—	—
$T = 200$	j_{o1}	1.00	0.998			0.833	0.981	—	—	—	—	—	—
	j_{o2}	0.818	0.996			0.500	0.974	—	—	—	—	—	—
	1	0.165	0.629	0.503	0.974	0.162	0.599	0.499	0.976	0.290	0.664	0.611	0.978
Lag = 2	2	0.154	0.547	0.398	0.949	0.157	0.515	0.393	0.949	0.194	0.570	0.460	0.958
$T = 200$	j_{o1}	0.750	0.678	0.712	0.978	0.750	0.660	0.708	0.979	0.750	0.703	0.770	0.982
	j_{o2}	0.500	0.670	0.684	0.981	0.533	0.648	0.679	0.981	0.700	0.696	0.751	0.984
	1	—	—	0.147	0.663	—	—	0.144	0.671	—	—	0.299	0.765
Lag = 3	2	—	—	0.136	0.552	—	—	0.140	0.560	—	—	0.205	0.624
$T = 200$	j_{o1}	—	—	0.200	0.750	—	—	0.200	0.758	—	—	0.500	0.842
	j_{o2}	—	—	0.387	0.733	—	—	0.355	0.741	—	—	0.548	0.827
	1	—	—	0.140	0.354	—	—	0.140	0.344	—	—	0.311	0.458
Lag = 4	2	—	—	0.135	0.299	—	—	0.137	0.298	—	—	0.212	0.356
$T = 200$	j_{o1}	—	—	0.267	0.415	—	—	0.267	0.401	—	—	0.467	0.511
	j_{o2}	—	—	0.293	0.406	—	—	0.293	0.391	—	—	0.488	0.499
	1	0.465	0.213	0.158	0.218	0.269	0.213	0.154	0.213	0.383	0.364	0.308	0.387
Lag = AIC	2	0.431	0.193	0.148	0.195	0.211	0.191	0.151	0.195	0.350	0.260	0.214	0.273
$T = 200$	j_{o1}	0.833	0.373	0.182	0.447	0.714	0.373	0.182	0.433	0.000	0.452	0.545	0.538
	j_{o2}	0.643	0.371	0.400	0.375	0.450	0.374	0.371	0.365	0.250	0.446	0.571	0.485

(continued)

TABLE 4 (Continued)

Lag/ sample	CI rank	λ_{p1} VAR(1) (MA1)	λ_{p1} VAR(1) (MA2)	λ_{p1} VAR(3) (MA1)	λ_{p1} VAR(3) (MA2)	λ_{p2} VAR(1) (MA1)	λ_{p2} VAR(1) (MA2)	λ_{p2} VAR(3) (MA1)	λ_{p2} VAR(3) (MA2)	λ_{p3} VAR(1) (MA1)	λ_{p3} VAR(1) (MA2)	λ_{p3} VAR(3) (MA1)	λ_{p3} VAR(3) (MA2)
	1	0.490	0.491	0.147	0.524	0.267	0.473	0.144	0.525	—	0.548	0.299	0.619
Lag = SIC	2	0.451	0.432	0.136	0.456	0.203	0.419	0.140	0.463	—	0.449	0.205	0.515
$T=200$	j_{o1}	1.00	0.695	0.200	0.784	0.839	0.680	0.200	0.793	—	0.701	0.500	0.860
	j_{02}	0.818	0.678	0.387	0.751	0.500	0.658	0.355	0.758	—	0.688	0.548	0.825
	1	0.469	0.971	—	—	0.248	0.912	—	—	—	—	—	—
Lag = 1	2	0.458	0.886	—	—	0.202	0.834	—	—	—	—	—	—
$T=300$	j_{o1}	0.600	1.00	—	—	0.600	0.985	—	—	—	—	—	—
	j_{02}	0.696	0.997	—	—	0.533	0.979	—	—	—	—	—	—
	1	0.151	0.757	0.514	0.985	0.148	0.729	0.508	0.985	0.270	0.777	0.623	0.988
Lag = 2	2	0.145	0.696	0.408	0.970	0.146	0.664	0.407	0.970	0.185	0.716	0.464	0.977
$T=300$	j_{o1}	0.667	0.776	0.732	0.987	0.667	0.760	0.727	0.987	0.833	0.790	0.790	0.989
	j_{02}	0.364	0.776	0.699	0.987	0.333	0.756	0.695	0.987	0.576	0.790	0.765	0.989
	1	—	—	0.144	0.716	—	—	0.146	0.722	—	—	0.284	0.802
Lag = 3	2	—	—	0.135	0.613	—	—	0.138	0.620	—	—	0.190	0.675
$T=300$	j_{o1}	—	—	0.333	0.803	—	—	0.333	0.808	—	—	0.444	0.872
	j_{02}	—	—	0.296	0.783	—	—	0.296	0.789	—	—	0.370	0.858

	1	—	—	0.133	0.447	—	—	0.134	0.428	—	—	0.290	0.531
Lag = 4	2	—	—	0.125	0.381	—	—	0.127	0.377	—	—	0.198	0.429
$T = 300$	j_{o1}	—	—	0.417	0.536	—	—	0.417	0.497	—	—	0.417	0.610
	j_{02}	—	—	0.355	0.519	—	—	0.387	0.485	—	—	0.452	0.592
	1	0.453	0.179	0.152	0.202	0.253	0.177	0.153	0.198	0.482	0.338	0.289	0.373
Lag = AIC	2	0.447	0.166	0.142	0.180	0.209	0.165	0.146	0.177	0.397	0.239	0.196	0.264
$T = 300$	j_{o1}	0.571	0.371	0.333	0.328	0.667	0.371	0.333	0.310	0.667	0.427	0.444	0.474
	j_{02}	0.654	0.382	0.310	0.306	0.531	0.387	0.310	0.301	0.500	0.462	0.414	0.451
	1	0.469	0.374	0.144	0.404	0.248	0.366	0.146	0.390	—	0.459	0.284	0.505
Lag = SIC	2	0.458	0.336	0.135	0.357	0.202	0.329	0.138	0.356	—	0.373	0.190	0.408
$T = 300$	j_{o1}	0.600	0.651	0.333	0.678	0.600	0.653	0.333	0.646	—	0.685	0.444	0.725
	j_{02}	0.696	0.593	0.296	0.641	0.533	0.587	0.296	0.613	—	0.637	0.370	0.691

[a]See footnotes to Table 2.

TABLE 5 Summary of SIC and AIC Criteria Approach to Causal Analysis;[a] Data-Generating Process is (CI-1)

				Model Selected—AIC			Model Selected—SIC		
DGP/sample	Lag	AIC	SIC	CI	Levels	Difference	CI	Levels	Difference
VAR(1)-(MA1)	1	0.958	0.995	0.953	0.047	—	0.995	0.005	—
$T=100$	2	0.358	0.113	0.889	0.105	0.006	0.929	0.052	0.019
	AIC	0.945	0.960	0.946	0.048	0.006	0.984	0.008	0.008
	SIC	0.958	0.995	0.953	0.047	—	0.995	0.005	—
	1	0.963	0.995	0.962	0.038	—	0.995	0.005	—
$T=200$	2	0.380	0.072	0.907	0.093	0.000	0.966	0.034	0.000
	AIC	0.945	0.955	0.957	0.042	0.001	0.991	0.008	0.001
	SIC	0.963	0.995	0.962	0.038	—	0.995	0.005	—
	1	0.973	0.999	0.973	0.027	—	0.999	0.001	—
$T=300$	2	0.343	0.057	0.901	0.099	0.000	0.977	0.023	0.000
	AIC	0.950	0.957	0.967	0.033	0.000	0.998	0.002	0.000
	SIC	0.973	0.999	0.973	0.027	—	0.999	0.001	—
VAR(1)-(MA2)	1	0.962	0.994	0.948	0.052	—	0.994	0.006	—
$T=100$	2	0.450	0.143	0.919	0.080	0.001	0.972	0.023	0.005
	AIC	0.366	0.051	0.829	0.134	0.037	0.920	0.063	0.017
	SIC	0.583	0.359	0.922	0.077	0.001	0.974	0.021	0.005
	1	0.967	0.994	0.955	0.045	—	0.994	0.006	—
$T=200$	2	0.505	0.136	0.936	0.064	0.000	0.986	0.014	0.000
	AIC	0.261	0.015	0.866	0.117	0.017	0.955	0.039	0.006
	SIC	0.459	0.106	0.934	0.066	0.000	0.981	0.017	0.002

	1	0.980	0.998	0.965	0.035	—	0.998	0.002	—
$T=300$	2	0.586	0.162	0.966	0.034	0.000	0.994	0.006	0.000
	AIC	0.239	0.003	0.897	0.098	0.005	0.966	0.033	0.001
	SIC	0.425	0.056	0.943	0.057	0.000	0.985	0.015	0.000
VAR(3)-(MA1)	2	0.371	0.096	0.920	0.080	0.000	0.982	0.018	0.000
$T=100$	3	0.275	0.047	0.835	0.163	0.002	0.907	0.089	0.004
	4	0.250	0.027	0.815	0.172	0.013	0.870	0.123	0.007
	AIC	0.326	0.063	0.853	0.144	0.003	0.905	0.091	0.004
	SIC	0.379	0.114	0.913	0.087	0.000	0.973	0.027	0.000
	2	0.383	0.058	0.933	0.067	0.000	0.986	0.014	0.000
$T=200$	3	0.278	0.024	0.883	0.117	0.000	0.936	0.064	0.000
	4	0.232	0.013	0.860	0.139	0.001	0.918	0.082	0.000
	AIC	0.290	0.029	0.886	0.114	0.000	0.934	0.065	0.001
	SIC	0.381	0.054	0.929	0.071	0.000	0.970	0.030	0.000
	2	0.390	0.066	0.925	0.075	0.000	0.996	0.004	0.000
$T=300$	3	0.260	0.018	0.873	0.127	0.000	0.950	0.050	0.000
	4	0.224	0.002	0.850	0.150	0.000	0.936	0.064	0.000
	AIC	0.264	0.016	0.876	0.124	0.000	0.953	0.047	0.000
	SIC	0.336	0.043	0.897	0.103	0.000	0.967	0.033	0.000
VAR(3)-(MA2)	2	0.669	0.364	0.956	0.044	0.000	0.975	0.025	0.000
$T=100$	3	0.433	0.091	0.915	0.082	0.003	0.927	0.052	0.021
	4	0.304	0.032	0.854	0.119	0.027	0.925	0.056	0.019
	AIC	0.323	0.022	0.814	0.150	0.036	0.910	0.080	0.010
	SIC	0.504	0.147	0.925	0.068	0.007	0.943	0.043	0.014

(*continued*)

TABLE 5 (Continued)

				Model Selected—AIC			Model Selected—SIC		
DGP/sample	Lag	AIC	SIC	CI	Levels	Difference	CI	Levels	Difference
	2	0.828	0.483	0.960	0.040	0.000	0.988	0.012	0.000
$T=200$	3	0.485	0.062	0.936	0.064	0.000	0.960	0.039	0.001
	4	0.312	0.006	0.915	0.085	0.000	0.964	0.034	0.002
	AIC	0.225	0.001	0.854	0.134	0.012	0.955	0.045	0.000
	SIC	0.418	0.029	0.936	0.064	0.000	0.960	0.038	0.002
	2	0.895	0.385	0.976	0.024	0.000	0.992	0.008	0.000
$T=300$	3	0.562	0.082	0.948	0.052	0.000	0.968	0.032	0.000
	4	0.312	0.001	0.922	0.078	0.000	0.971	0.029	0.000
	AIC	0.185	0.000	0.860	0.131	0.009	0.952	0.048	0.000
	SIC	0.350	0.011	0.928	0.072	0.000	0.975	0.025	0.000

[a]Results based on the model selection criteria-type approaches discussed above are reported in this table. The lags used are listed, as is the number of observations in the sample, and the DGP used to generate the data. The proportion of "wins" accruing to each type of estimated model (VEC model (CI)), stationary levels model (levels), and difference stationary model (difference) are listed in columns 5–10.

Table 6 SIC and AIC Approach Rejection Frequencies;[a] Data-Generating Process is (CI-1)[b]

Lag/sample	CI rank	VAR(1)—(MA1)		VAR(1)—(MA2)		VAR(3)—(MA1)		VAR(3)—(MA2)	
		AIC	SIC	AIC	SIC	AIC	SIC	AIC	SIC
Lag = 1	1	0.076	0.076	0.023	0.023	—	—	—	—
$T = 100$	2	1.000	1.000	1.000	1.000	—	—	—	—
Lag = 2	1	0.241	0.115	0.289	0.121	0.203	0.072	0.541	0.315
$T = 100$	2	1.000	1.000	1.000	1.000	1.000	0.999	1.000	1.000
Lag = 3	1	—	—	—	—	0.233	0.096	0.337	0.104
$T = 100$	2	—	—	—	—	1.000	0.974	0.999	0.994
Lag = 4	1	—	—	—	—	0.231	0.097	0.205	0.040
$T = 100$	2	—	—	—	—	0.999	0.724	0.998	0.914
Lag = AIC	1	0.104	0.091	0.260	0.059	0.252	0.108	0.240	0.041
$T = 100$	2	1.000	0.999	0.999	0.926	0.999	0.954	0.991	0.675
Lag = SIC	1	0.076	0.076	0.248	0.113	0.206	0.078	0.388	0.141
$T = 100$	2	1.000	1.000	1.000	1.000	1.000	0.997	0.998	0.991
Lag = 1	1	0.061	0.061	0.023	0.023	—	—	—	—
$T = 200$	2	1.000	1.000	1.000	1.000	—	—	—	—
Lag = 2	1	0.228	0.086	0.372	0.129	0.212	0.047	0.729	0.434
$T = 200$	2	1.000	1.000	1.000	1.000	1.000	1.000	1.000	1.000
Lag = 3	1	—	—	—	—	0.212	0.070	0.403	0.082
$T = 200$	2	—	—	—	—	1.000	1.000	1.000	1.000
Lag = 4	1	—	—	—	—	0.189	0.075	0.220	0.021
$T = 200$	2	—	—	—	—	1.000	0.996	1.000	1.000
Lag = AIC	1	0.084	0.067	0.177	0.026	0.223	0.074	0.154	0.024

(continued)

Table 6 (Continued)

Lag/sample	CI rank	VAR(1)—(MA1)		VAR(1)—(MA2)		VAR(3)—(MA1)		VAR(3)—(MA2)	
		AIC	SIC	AIC	SIC	AIC	SIC	AIC	SIC
$T=200$	2	1.000	1.000	1.000	0.982	1.000	0.999	1.000	0.887
Lag = SIC	1	0.061	0.061	0.339	0.101	0.233	0.059	0.322	0.048
$T=200$	2	1.000	1.000	1.000	1.000	1.000	1.000	1.000	1.000
Lag = 1	1	0.058	0.058	0.023	0.023	—	—	—	—
$T=300$	2	1.000	1.000	1.000	1.000	—	—	—	—
Lag = 2	1	0.201	0.068	0.456	0.155	0.216	0.054	0.841	0.556
$T=300$	2	1.000	1.000	1.000	1.000	1.000	1.000	1.000	1.000
Lag = 3	1	—	—	—	—	0.182	0.057	0.488	0.100
$T=300$	2	—	—	—	—	1.000	1.000	1.000	1.000
Lag = 4	1	—	—	—	—	0.181	0.054	0.223	0.015
$T=300$	2	—	—	—	—	1.000	1.000	1.000	1.000
Lag = AIC	1	0.079	0.064	0.157	0.020	0.188	0.052	0.135	0.023
$T=300$	2	1.000	1.000	1.000	0.996	1.000	1.000	1.000	0.942
Lag = SIC	1	0.058	0.058	0.303	0.060	0.213	0.058	0.256	0.021
$T=300$	2	1.000	1.000	1.000	1.000	1.000	1.000	1.000	1.000

[a]See Footnotes to Table 5. The reported numerical values are the proportion of rejections of the "null hypothesis" of noncausality. All models are compared in pairwise fashion. For example, in this table each entry corresponds to the rejection frequency when two alternative cointegrating models are compared—one with the causal variable of interest included, and one without.

[b]All estimations are based on the assumption that it is known that the data are generated according to a VEC.

TABLE 7 SIC and AIC Approach Rejection Frequencies;[a] Data-Generating Process is (CI-1)[b]

		VAR(1)-(MA1)		VAR(1)-(MA2)		VAR(3)-(MA1)		VAR(3)-(MA2)	
Sample	Lag	AIC	SIC	AIC	SIC	AIC	SIC	AIC	SIC
	1	0.329	0.113	0.410	0.205	—	—	—	—
$T=100$	2	0.262	0.036	0.386	0.082	0.321	0.050	0.595	0.248
	3	—	—	—	—	0.212	0.010	0.370	0.037
	4	—	—	—	—	0.202	0.006	0.261	0.018
	AIC	0.341	0.113	0.306	0.020	0.254	0.017	0.263	0.010
	SIC	0.329	0.113	0.410	0.114	2.326	0.058	0.417	0.092
	1	0.359	0.081	0.420	0.156	—	—	—	—
$T=200$	2	0.272	0.018	0.427	0.063	0.315	0.033	0.759	0.300
	3	—	—	—	—	0.205	0.012	0.432	0.034
	4	—	—	—	—	0.177	0.003	0.256	0.003
	AIC	0.363	0.081	0.212	0.005	0.214	0.011	0.178	0.002
	SIC	0.359	0.081	0.399	0.048	0.307	0.030	0.358	0.019
	1	0.349	0.089	0.458	0.168	—	—	—	—
$T=300$	2	0.264	0.014	0.506	0.079	0.350	0.031	0.825	0.368
	3	—	—	—	—	0.193	0.003	0.496	0.040
	4	—	—	—	—	0.166	0.000	0.268	0.00
	AIC	0.354	0.089	0.198	0.001	0.199	0.003	0.154	0.00
	SIC	0.349	0.089	0.352	0.028	0.276	0.022	0.299	0.004

[a]See footnotes to Table 7.
[b]All estimations are based on the assumption that it is known that the data are generated according to a levels VAR.

models are estimated in levels, the tabulated size of the *ex ante* tests often approaches zero very quickly as T increases. regardless of the true DGP.

7. Difference stationary models perform exceptionally poorly, in all cointegrated cases, suggesting that difference-type Wald tests do not provide a short cut to testing for "short-run" noncausality when the true DGP is a VEC. This is because least squares becomes very confused in the context of this type of mis-specification, and parameter estimates are hence very inaccurate. A final result worth noting is that even though the empirical sizes of many tests (with the exception of SIC and *ex ante* type model selection based tests) are often too high, suggesting that the null of Granger causality may be rejected too often, the use of F versions of the Wald-type tests does not appear to decrease the empirical sizes to any great extent, for the sample sizes that we consider. For this reason, we report rejection frequencies based only on the χ^2 versions of our test statistics.

TABLE 8 SIC and AIC Approach Rejection Frequencies;[a] Data Generating Process is (CI-1)[b]

		VAR(1)-(MA1)		VAR(1)-(MA2)		VAR(3)-(MA1)		VAR(3)-(MA2)	
Sample	Lag	AIC	SIC	AIC	SIC	AIC	SIC	AIC	SIC
	1	—	—	—	—	—	—	—	—
$T=100$	2	1.00	0.999	1.00	1.00	0.721	0.449	0.972	0.916
	3	—	—	—	—	0.949	0.668	0.990	0.905
	4	—	—	—	—	0.946	0.512	0.993	0.835
	AIC	1.00	0.945	0.998	0.922	0.892	0.609	0.977	0.631
	SIC	—	—	1.00	1.00	0.737	0.467	0.986	0.897
	1	—	—	—	—	—	—	—	—
$T=200$	2	1.00	1.00	1.00	1.00	0.914	0.727	0.998	0.995
	3	—	—	—	—	1.00	0.954	1.00	0.998
	4	—	—	—	—	1.00	0.914	1.00	0.995
	AIC	1.00	0.980	1.00	0.988	0.997	0.942	0.998	0.833
	SIC	0.00	0.00	1.00	1.00	0.932	0.776	1.00	0.995
	1	—	—	—	—	—	—	—	—
$T=300$	2	1.00	1.00	1.00	1.00	0.971	0.856	1.00	1.00
	3	—	—	—	—	1.00	0.998	1.00	1.00
	4	—	—	—	—	1.00	0.993	1.00	1.00
	AIC	1.00	1.00	1.00	0.997	1.00	0.996	1.00	0.931
	SIC	0.00	0.00	1.00	1.00	0.987	0.938	1.00	1.00

[a]See footnotes to Table 7.
[b]All estimations are based on the assumption that it is known that the data are generated according to a VAR in differences.

4.1 Cointegration Models

In all of the cointegrated models (please see attached tables) the sequential Wald-type tests, the Toda and Yamamoto Wald-type tests, and the model selection based SIC type causality tests perform quite well. This is most evident when the true DGPs are VAR(1) and VAR(3) without an MA component. In these cases, λ_{LW} type tests have empirical sizes ranging from 0.049 to 0.056 for VAR(1)—$T=200$ or 300, and approximately 0.055 for VAR(3)—$T=300$ (se Table 1). In these cases, the best size is usually (except for VAR(1) – MA1) associated with lag selection based on the AIC. In fact, it turns out that the AIC is also useful for selecting lags in the context of our model selection based tests (see below). Thus, the more parsimonious method for choosing lags (SIC) performs better primarily for our least complex models. One reason for this finding may be that when complex DGPs are specified (as is usually the case when macroeconomic data are analysed), the specification of surplus lags helps standard maximum likelihood estimation techniques to become less confused. This point is particularly important, given the tendency of our simplest TYDL-type tests to perform so well. Put another way, if model complexity and estimation confusion is an issue, then avoiding the estimation of cointegrating vectors, ranks, etc., may be a useful approach—even if the true DGP is cointegrated. This indeed appears to be the case for more complex models, as the TYDL approach appears to perform as well as all of the other standard hypothesis testing approaches, at least in the context of empirical size. Of course, the cost of estimating models in levels, and ignoring cointegrating restrictions, is a reduction in efficiency.

Tests based on difference model estimations perform very poorly all the time. However, the Wald tests based on λ_{ST} have sizes that are usually between 10 and 15%. Tests based on models estimated using differenced data perform very poorly whenever the true DGP is a VEC. However, λ_{ST} tests have sizes that are usually between 10 and 15%, with the better sizes coming at the expense of larger samples, but not necessarily less model complexity. For example, in Table 1, for $T = 300$, the λ_{ST} test has size $= 0.108$ in the VAR(3)—MA1–(CI-1) case when estimation is based on four lags, while the analogous test for the VAR(1) (i.e., two lags used in estimation) has size $= 0.128$. Note that the reason for the relatively good performance of these two models, each of which is estimated with an extra lag, is probably related to the argument used to show that the λ_{LW} type tests are asymptotically valid. That is, even when we use the standard Wald test, as long as an extra lag is included, the size of the test is not too far from the nominal size, even though the asymptotic χ^2 distribution is clearly not valid.

The results for the tests based on the λ_{FM} statistic are somewhat less encouraging than those of the Toda and Yamamoto and sequential Wald-type tests. In all of our experiments, the size of these tests appears to be in the 15–30% range, regardless of sample size, model specification, etc. Although these results are not too bad, relatively, it seems apparent that a number of the other tests perform better. As the λ_{FM} test statistics rely on some optimal choice of bandwidth parameter as well as kernel function, for example, our results should be viewed with caution, as more suitable FM–VAR procedures may be available. To try and avoid this problem somewhat, we used three different lag truncation parameters to define the number of autocovariance terms used to compute the spectrum at zero frequency. These have been used elsewhere, and are: (1) $l = 0$, (2) $l = integer\, 4 \times (T/100)^{0.25}$), and (3) $l = integer\, (12 \times (T/100)^{0.25})$. Further, we used the Parzen kernel function for the calculation of long-run covariance matrix estimates, as well as Andrews' (1991) automatic bandwidth selection procedures. Finally, an AR(1) filter was used to flatten the spectrum of the error around the zero frequency. Our

TABLE 9 MSE, MAD, and MAPE Type *Ex Ante* Predictive Ability;[a] Data-Generating Process is (CI-1)[b]

		VAR(1)-(MA1)			VAR(1)-(MA2)			VAR(3)-(MA1)			VAR(3)-(MA2)		
Lag/sample	CI rank	MSE	MAD	MAPE	MSE	MAD	MAPE	MSE	MAD	MAPE	MSE	MAD	MAPE
Lag = 1	1	0.166	0.158	0.114	0.288	0.243	0.185	—	—	—	—	—	—
$T = 100$	2	0.147	0.144	0.105	0.260	0.228	0.178	—	—	—	—	—	—
Lag = 2	1	0.013	0.016	0.014	0.014	0.019	0.016	0.011	0.017	0.015	0.024	0.025	0.016
$T = 100$	2	0.063	0.082	0.071	0.145	0.152	0.125	0.066	0.058	0.048	0.125	0.113	0.073
Lag = 3	1	—	—	—	—	—	—	0.010	0.015	0.013	0.012	0.012	0.008
$T = 100$	2	—	—	—	—	—	—	0.043	0.047	0.033	0.055	0.060	0.050
Lag = 4	1	—	—	—	—	—	—	0.008	0.010	0.014	0.012	0.010	0.006
$T = 100$	2	—	—	—	—	—	—	0.035	0.036	0.033	0.051	0.048	0.041
Lag = AIC	1	0.164	0.153	0.111	0.005	0.007	0.007	0.010	0.017	0.016	0.007	0.011	0.005
$T = 100$	2	0.144	0.142	0.105	0.080	0.071	0.065	0.048	0.051	0.037	0.045	0.044	0.031
Lag = SIC	1	0.166	0.158	0.114	0.077	0.077	0.058	0.014	0.018	0.016	0.023	0.018	0.010
$T = 100$	2	0.147	0.144	0.105	0.173	0.180	0.138	0.070	0.061	0.049	0.075	0.070	0.058
Lag = 1	1	0.595	0.519	0.373	0.716	0.630	0.441	—	—	—	—	—	—
$T = 200$	2	0.564	0.495	0.357	0.713	0.625	0.431	—	—	—	—	—	—
Lag = 2	1	0.025	0.017	0.011	0.017	0.022	0.014	0.013	0.012	0.009	0.063	0.061	0.047
$T = 200$	2	0.332	0.288	0.199	0.524	0.464	0.303	0.232	0.192	0.123	0.385	0.344	0.236
Lag = 3	1	—	—	—	—	—	—	0.008	0.004	0.009	0.023	0.024	0.014
$T = 200$	2	—	—	—	—	—	—	0.224	0.171	0.117	0.304	0.248	0.157
Lag = 4	1	—	—	—	—	—	—	0.009	0.012	0.014	0.011	0.012	0.006

$T=200$	2	—	—	—	—	—	—	0.146	0.109	0.074	0.221	0.172	0.177
Lag = AIC	1	0.572	0.493	0.352	0.007	0.005	0.008	0.009	0.006	0.010	0.008	0.010	0.003
$T=200$	2	0.550	0.476	0.346	0.238	0.219	0.155	0.220	0.163	0.117	0.138	0.124	0.078
Lag = SIC	1	0.595	0.519	0.373	0.015	0.022	0.016	0.014	0.009	0.009	0.019	0.018	0.010
$T=200$	2	0.564	0.495	0.357	0.462	0.406	0.266	0.233	0.188	0.122	0.252	0.207	0.137
Lag = 1	1	0.832	0.768	0.556	0.909	0.858	0.633	—	—	—	—	—	—
$T=300$	2	0.806	0.750	0.554	0.902	0.851	0.626	—	—	—	—	—	—
Lag = 2	1	0.030	0.033	0.023	0.039	0.040	0.032	0.019	0.015	0.011	0.074	0.072	0.048
$T=300$	2	0.617	0.522	0.368	0.801	0.730	0.509	0.469	0.378	0.246	0.634	0.542	0.340
Lag = 3	1	—	—	—	—	—	—	0.021	0.020	0.015	0.025	0.028	0.015
$T=300$	2	—	—	—	—	—	—	0.417	0.343	0.218	0.477	0.417	0.245
Lag = 4	1	—	—	—	—	—	—	0.020	0.019	0.011	0.005	0.008	0.006
$T=300$	2	—	—	—	—	—	—	0.276	0.235	0.158	0.384	0.318	0.205
Lag = AIC	1	0.788	0.732	0.527	0.007	0.005	0.007	0.019	0.020	0.014	0.005	0.008	0.010
$T=300$	2	0.788	0.735	0.544	0.394	0.312	0.224	0.407	0.332	0.215	0.214	0.180	0.108
Lag = SIC	1	0.832	0.768	0.556	0.017	0.024	0.023	0.024	0.018	0.014	0.008	0.012	0.005
$T=300$	2	0.806	0.750	0.554	0.696	0.590	0.406	0.444	0.352	0.232	0.392	0.326	0.202

[a]The approaches reported in this table are based on the use of *ex ante* MSE, MAD, and MAPE forecasting criteria. The lags used are listed, as is the number of observations in the sample. The DGPs used to generate the data are given across the columns of the of rejections of the "null hypothesis" of noncausality. All models are compared in pairwise fashion. For example, in this table each entry corresponds to the rejection frequency when two alternative cointegrating models are compared—one with the causal variable of interest included, and one without. The choice of which model is "best" is based on an examination of the forecasts errors used in the calculation of the MSE, MAD, and MAPE. In particular, predictive accuracy tests in the spirit of Diebold and Mariano (1995) are carried out (see above discussion).

[b]All estimations are based on the assumption that it is known that the data are generated according to a VEC.

Table 10 MSE, MAD, and MAPE Type *Ex Ante* Predictive Ability;[a] Data-Generating Process is (CI-1)[b]

		VAR(1)-(MA1)			VAR(1)-(MA2)			VAR(3)-(MA1)			VAR(3)-(MA2)		
Sample	Lag	MSE	MAD	MAPE	MSE	MAD	MAPE	MSE	MAD	MAPE	MSE	MAD	MAPE
	1	0.018	0.012	0.024	0.016	0.018	0.021	–	–	–	–	–	–
$T=100$	2	0.006	0.007	0.007	0.016	0.016	0.014	0.013	0.015	0.013	0.021	0.021	0.013
	3	–	–	–	–	–	–	0.015	0.012	0.016	0.008	0.008	0.009
	4	–	–	–	–	–	–	0.009	0.007	0.012	0.004	0.013	0.012
	AIC	0.018	0.014	0.025	0.003	0.005	0.009	0.015	0.013	0.014	0.008	0.011	0.013
	SIC	0.018	0.012	0.024	0.019	0.021	0.017	0.013	0.016	0.014	0.010	0.013	0.009
	1	0.017	0.017	0.025	0.022	0.020	0.028	–	–	–	–	–	–
$T=200$	2	0.007	0.009	0.010	0.012	0.015	0.016	0.013	0.012	0.018	0.036	0.039	0.037
	3	–	–	–	–	–	–	0.004	0.006	0.015	0.012	0.010	0.011
	4	–	–	–	–	–	–	0.006	0.009	0.013	0.009	0.004	0.011
	AIC	0.017	0.017	0.025	0.004	0.004	0.009	0.005	0.007	0.016	0.002	0.004	0.003
	SIC	0.017	0.017	0.025	0.010	0.013	0.013	0.011	0.008	0.015	0.012	0.007	0.013
	1	0.027	0.020	0.023	0.027	0.031	0.039	–	–	–	–	–	–
$T=300$	2	0.009	0.013	0.013	0.010	0.014	0.014	0.010	0.012	0.018	0.047	0.049	0.026
	3	–	–	–	–	–	–	0.006	0.011	0.017	0.013	0.014	0.015
	4	–	–	–	–	–	–	0.007	0.009	0.010	0.003	0.009	0.014
	AIC	0.026	0.020	0.022	0.002	0.012	0.008	0.005	0.011	0.017	0.003	0.004	0.010
	SIC	0.027	0.020	0.023	0.006	0.008	0.012	0.013	0.012	0.018	0.003	0.009	0.013

[a]See footnotes to Table 9.
[b]All estimations are based on the assumption that it is known that the data are generated according to a levels VAR in differences.

TABLE 11 MSE, MAD, and MAPE Type *Ex Ante* Predictive Ability;[a] Data-Generating Process is (CI-1)[b]

		VAR(1)-(MA1)			VAR(1)-(MA2)			VAR(3)-(MA1)			VAR(3)-(MA2)		
Sample	Lag	MSE	MAD	MAPE	MSE	MAD	MAPE	MSE	MAD	MAPE	MSE	MAD	MAPE
	1	—	—	—	—	—	—	—	—	—	—	—	—
$T=100$	2	0.068	0.064	0.041	0.083	0.083	0.073	0.031	0.029	0.025	0.063	0.071	0.042
	3	—	—	—	—	—	—	0.043	0.039	0.039	0.048	0.045	0.028
	4	—	—	—	—	—	—	0.038	0.033	0.032	0.044	0.046	0.033
	AIC	0.055	0.00	0.00	0.069	0.076	0.059	0.034	0.034	0.035	0.036	0.043	0.025
	SIC	—	—	—	0.086	0.094	0.080	0.031	0.028	0.023	0.061	0.059	0.031
	1	0.00	0.00	0.00	0.00	0.00	0.00	—	—	—	—	—	—
$T=200$	2	0.180	0.175	0.128	0.301	0.266	0.186	0.054	0.049	0.035	0.156	0.139	0.084
	3	—	—	—	—	—	—	0.081	0.077	0.047	0.142	0.142	0.077
	4	—	—	—	—	—	—	0.088	0.072	0.060	0.147	0.128	0.081
	AIC	0.180	0.120	0.100	0.182	0.164	0.115	0.078	0.074	0.049	0.101	0.101	0.074
	SIC	0.00	0.00	0.00	0.290	0.256	0.174	0.057	0.052	0.039	0.144	0.132	0.073
	1	—	—	—	0.00	0.00	0.00	—	—	—	—	—	—
$T=300$	2	0.371	0.298	0.190	0.542	0.446	0.304	0.090	0.070	0.059	0.242	0.193	0.106
	3	—	—	—	—	—	—	0.162	0.141	0.092	0.257	0.207	0.122
	4	—	—	—	—	—	—	0.159	0.119	0.088	0.259	0.210	0.126
	AIC	0.412	0.353	0.157	0.349	0.279	0.201	0.160	0.136	0.091	0.176	0.156	0.086
	SIC	—	—	—	0.505	0.390	0.280	0.131	0.110	0.081	0.264	0.218	0.131

[a]See footnotes to Table 9.
[b]All estimations are based on the assumption that it is known that the data are generated according to a VAR in differences.

TABLE 12 Summary of SIC and AIC Criteria Approach to Causal Analysis;[a] Data-Generating Process is (CI-2)

DGP/ sample	Lag	AIC	SIC	Model selected—AIC CI	Levels	Difference	Model selected-SIC CI	Levels	Difference
VAR(1)-(MA1)	1	0.962	0.994	0.961	0.039	—	0.994	0.006	—
$T=100$	2	0.388	0.123	0.858	0.117	0.025	0.911	0.044	0.045
	AIC	0.931	0.950	0.953	0.043	0.004	0.983	0.008	0.009
	SIC	0.962	0.994	0.961	0.039	—	0.994	0.006	—
	1	0.951	0.994	0.949	0.051	—	0.994	0.006	—
$T=200$	2	0.351	0.082	0.863	0.135	0.002	0.952	0.044	0.004
	AIC	0.932	0.958	0.946	0.053	0.001	0.992	0.007	0.001
	SIC	0.951	0.994	0.949	0.051	—	0.994	0.006	—
	1	0.946	0.997	0.943	0.057	—	0.997	0.003	—
$T=300$	2	0.334	0.061	0.876	0.124	0.000	0.958	0.042	0.000
	AIC	0.925	0.946	0.941	0.059	0.000	0.994	0.006	0.000
	SIC	0.946	0.997	0.943	0.057	—	0.997	0.003	—
VAR(1)-(MA2)	1	0.945	0.984	0.913	0.087	—	0.980	0.020	—
$T=100$	2	0.440	0.123	0.909	0.091	0.000	0.986	0.014	0.000
	AIC	0.353	0.046	0.804	0.134	0.062	0.919	0.056	0.025
	SIC	0.436	0.121	0.892	0.101	0.007	0.971	0.019	0.010

	1	0.935	0.986	0.909	0.091	—	0.986	0.014	—
$T=200$	2	0.457	0.082	0.938	0.062	0.000	0.990	0.010	0.000
	AIC	0.233	0.007	0.839	0.132	0.029	0.971	0.027	0.002
	SIC	0.335	0.030	0.916	0.084	0.000	0.989	0.011	0.000
	1	0.932	0.980	0.901	0.099	—	0.979	0.021	—
$T=300$	2	0.465	0.063	0.915	0.085	0.000	0.990	0.010	0.000
	AIC	0.225	0.002	0.835	0.145	0.020	0.959	0.041	0.000
	SIC	0.294	0.011	0.895	0.105	0.000	0.989	0.011	0.000
VAR(3)-(MA1)	2	0.388	0.094	0.885	0.115	0.00	0.955	0.043	0.002
$T=100$	3	0.293	0.035	0.828	0.157	0.015	0.938	0.054	0.008
	4	0.248	0.013	0.798	0.165	0.037	0.941	0.055	0.004
	AIC	0.333	0.049	0.832	0.152	0.016	0.936	0.055	0.009
	SIC	0.402	0.118	0.881	0.116	0.003	0.953	0.044	0.003
$T=200$	2	0.366	0.057	0.895	0.105	0.000	0.964	0.036	0.000
	3	0.287	0.024	0.873	0.126	0.001	0.933	0.064	0.003
	4	0.234	0.008	0.848	0.143	0.009	0.923	0.071	0.006
	AIC	0.294	0.025	0.870	0.125	0.005	0.928	0.068	0.004
	SIC	0.365	0.056	0.893	0.107	0.000	0.955	0.044	0.001
$T=300$	2	0.402	0.041	0.886	0.114	0.000	0.967	0.033	0.000
	3	0.264	0.011	0.859	0.141	0.000	0.940	0.060	0.000
	4	0.215	0.001	0.840	0.159	0.001	0.932	0.068	0.000
	AIC	0.274	0.010	0.865	0.135	0.000	0.940	0.060	0.000
	SIC	0.330	0.026	0.874	0.126	0.000	0.947	0.053	0.000

(*continued*)

Table 12 (Continued)

DGP/				Model selected—AIC			Model selected—SIC		
sample	Lag	AIC	SIC	CI	Levels	Difference	CI	Levels	Difference
VAR(3)-(MA2)	2	0.671	0.381	0.912	0.088	0.000	0.932	0.068	0.000
$T=100$	3	0.404	0.049	0.927	0.072	0.001	0.978	0.021	0.001
	4	0.276	0.017	0.908	0.087	0.005	0.975	0.019	0.006
	AIC	0.293	0.015	0.834	0.130	0.036	0.936	0.055	0.009
	SIC	0.376	0.051	0.921	0.077	0.002	0.977	0.021	0.002
$T=200$	2	0.745	0.383	0.907	0.093	0.000	0.945	0.055	0.000
	3	0.392	0.041	0.924	0.076	0.000	0.988	0.012	0.000
	4	0.259	0.003	0.899	0.100	0.001	0.986	0.012	0.002
	AIC	0.200	0.002	0.821	0.153	0.026	0.963	0.037	0.000
	SIC	0.270	0.008	0.899	0.098	0.003	0.986	0.013	0.001
$T=300$	2	0.783	0.404	0.905	0.095	0.000	0.914	0.086	0.000
	3	0.384	0.025	0.882	0.118	0.000	0.974	0.026	0.000
	4	0.238	0.001	0.868	0.132	0.000	0.981	0.019	0.000
	AIC	0.174	0.000	0.810	0.174	0.016	0.959	0.041	0.000
	SIC	0.224	0.002	0.860	0.139	0.001	0.981	0.018	0.001

[a]See footnotes to Table 6.

TABLE 13 Summary of SIC and AIC Criteria Approach to Causal Analysis;[a] Data-Generating Process is (CI-2)[b]

		VAR(1)-(MA1)		VAR(1)-(MA2)		VAR(3)-(MA1)		VAR(3)-(MA2)	
Lag/sample	CI rank	AIC	SIC	AIC	SIC	AIC	SIC	AIC	SIC
Lag = 1	1	0.064	0.064	0.003	0.003	—	—	—	—
$T = 100$	2	1.00	1.00	1.00	1.00	—	—	—	—
Lag = 2	1	0.218	0.081	0.192	0.049	0.209	0.067	0.297	0.109
$T = 100$	2	0.983	0.757	1.00	1.00	0.996	0.915	1.00	0.999
Lag = 3	1	—	—	—	—	0.193	0.041	0.196	0.022
$T = 100$	2	—	—	—	—	0.913	0.379	0.999	0.970
Lag = 4	1	—	—	—	—	0.195	0.020	0.148	0.006
$T = 100$	2	—	—	—	—	0.714	0.096	0.997	0.773
Lag = AIC	1	0.086	0.074	0.220	0.026	0.218	0.049	0.190	0.010
$T = 100$	2	0.999	0.982	0.983	0.697	0.914	0.497	0.942	0.414
Lag = SIC	1	0.064	0.064	0.197	0.048	0.209	0.064	0.195	0.019
$T = 100$	2	1.00	1.00	1.00	0.991	0.991	0.893	0.998	0.914
Lag = 1	1	0.068	0.068	0.00	0.00	—	—	—	—
$T = 200$	2	1.00	1.00	1.00	1.00	—	—	—	—
Lag = 2	1	0.219	0.087	0.175	0.020	0.196	0.043	0.357	0.108
$T = 200$	2	1.00	0.965	1.00	1.00	1.00	0.999	1.00	1.00
Lag = 3	1	—	—	—	—	0.193	0.058	0.181	0.012
$T = 200$	2	—	—	—	—	0.998	0.745	1.00	1.00

(continued)

TABLE 13 (Continued)

Lag/sample	CI rank	VAR(1)-(MA1) AIC	VAR(1)-(MA1) SIC	VAR(1)-(MA2) AIC	VAR(1)-(MA2) SIC	VAR(3)-(MA1) AIC	VAR(3)-(MA1) SIC	VAR(3)-(MA2) AIC	VAR(3)-(MA2) SIC
Lag = 4	1	—	—	—	—	0.177	0.027	0.141	0.002
$T = 200$	2	—	—	—	—	0.943	0.248	1.00	0.992
Lag = AIC	1	0.085	0.073	0.147	0.013	0.202	0.058	0.142	0.010
$T = 200$	2	1.00	0.992	0.998	0.810	0.995	0.723	0.994	0.590
Lag = SIC	1	0.068	0.068	0.160	0.007	0.205	0.051	0.149	0.005
$T = 200$	2	1.00	1.00	1.00	1.00	1.00	0.947	1.00	0.981
Lag = 1	1	0.069	0.069	0.002	0.002	—	—	—	—
$T = 300$	2	1.00	1.00	1.00	1.00	—	—	—	—
Lag = 2	1	0.194	0.079	0.171	0.021	0.208	0.035	0.398	0.120
$T = 300$	2	1.00	0.998	1.00	1.00	1.00	1.00	1.00	1.00
Lag = 3	1	—	—	—	—	0.172	0.061	0.171	0.008
$T = 300$	2	—	—	—	—	1.00	0.944	1.00	1.00
Lag = 4	1	—	—	—	—	0.165	0.041	0.116	0.00
$T = 300$	2	—	—	—	—	0.989	0.451	1.00	1.00
Lag = AIC	1	0.083	0.069	0.142	0.022	0.180	0.059	0.118	0.011
$T = 300$	2	1.00	0.999	1.00	0.902	1.00	0.918	0.999	0.734
Lag = SIC	1	0.069	0.069	0.140	0.007	0.201	0.058	0.116	0.00
$T = 300$	2	1.00	1.00	1.00	1.00	1.00	0.966	1.00	0.997

[a] See footnotes to Table 5.
[b] All estimations are based on the assumption that it is known that the data are generated according to a VEC.

TABLE 14 Summary of SIC and AIC Criteria Approach to Causal Analysis;[a] Data-Generating Process is (CI-2)[b]

Sample	Lag	VAR(1)-(MA1)		VAR(1)-(MA2)		VAR(3)-(MA1)		VAR(3)-(MA2)	
		AIC	SIC	AIC	SIC	AIC	SIC	AIC	SIC
	1	0.341	0.115	0.554	0.349	—	—	—	—
$T = 100$	2	0.270	0.040	0.358	0.078	0.295	0.043	0.602	0.301
	3	—	—	—	—	0.221	0.014	0.346	0.037
	4	—	—	—	—	0.201	0.008	0.229	0.011
	AIC	0.344	0.113	0.282	0.018	0.253	0.022	0.240	0.005
	SIC	0.341	0.115	0.353	0.072	0.304	0.051	0.320	0.036
	1	0.301	0.080	0.563	0.329	—	—	—	—
$T = 200$	2	0.241	0.022	0.360	0.053	0.283	0.028	0.617	0.288
	3	—	—	—	—	0.202	0.006	0.327	0.031
	4	—	—	—	—	0.186	0.00	0.209	0.001
	AIC	0.308	0.078	0.193	0.002	0.208	0.007	0.160	0.00
	SIC	0.301	0.080	0.270	0.019	0.271	0.026	0.218	0.005
	1	0.310	0.057	0.599	0.342	—	—	—	—
$T = 300$	2	0.224	0.011	0.352	0.043	0.296	0.015	0.687	0.292
	3	—	—	—	—	0.182	0.002	0.312	0.020
	4	—	—	—	—	0.164	0.00	0.187	0.001
	AIC	0.307	0.054	0.165	0.001	0.191	0.002	0.139	0.00
	SIC	0.310	0.057	0.220	0.006	0.230	0.005	0.179	0.001

[a]See footnotes to Table 5
[b]All estimations are based on the assumption that it is known that the data are generated according to a levels VAR.

Table 15 Summary of SIC and AIC Criteria Approach to Causal Analysis; [a]Data-Generating Process is (CI-2)[b]

		VAR(1)-(MA1)		VAR(1)-(MA2)		VAR(3)-(MA1)		VAR(3)-(MA2)	
Sample	Lag	AIC	SIC	AIC	SIC	AIC	SIC	AIC	SIC
	1	—	—	—	—	—	—	—	—
$T=100$	2	0.829	0.567	0.995	0.937	0.385	0.164	0.309	0.109
	3	—	—	—	—	0.401	0.093	0.646	0.214
	4	—	—	—	—	0.364	0.047	0.884	0.388
	AIC	0.828	0.609	0.968	0.702	0.406	0.111	0.870	0.277
	SIC	0.00	0.00	0.994	0.924	0.386	0.164	0.724	0.282
	1	—	—	—	—	—	—	—	—
$T=200$	2	0.978	0.828	1.00	0.999	0.594	0.280	0.332	0.108
	3	—	—	—	—	0.641	0.182	0.861	0.411
	4	—	—	—	—	0.550	0.068	0.996	0.797
	AIC	0.930	0.512	0.997	0.874	0.642	0.186	0.993	0.548
	SIC	0.00	0.00	1.00	0.996	0.601	0.260	0.982	0.762
	1	—	—	—	—	—	—	—	—
$T=300$	2	0.995	0.948	1.00	1.00	0.722	0.382	0.403	0.131
	3	—	—	—	—	0.789	0.271	0.961	0.613
	4	—	—	—	—	0.718	0.111	1.00	0.947
	AIC	0.963	0.870	1.00	0.930	0.787	0.264	0.998	0.747
	SIC	—	—	1.00	1.00	0.761	0.316	1.00	0.944

[a]See Footnotes to Table 6b

[b]All estimations are based on the assumption that it is known that the data are generated according to a VAR in differences.

results did not depend on the specification of l, however, and so in Table 1 we report results only for the case where $l=0$.

Sequential Wald test results are given in Tables 2–4. Single lag cases are not reported for λ_{P3}, as the test is not applicable in this case. In these tables, lag = 1 corresponds to zero lags when the data are differenced, so that the corresponding error-correction models used to calculate λ_{P1}, and λ_{P2} have no short-run dynamics. In general, this has no effect on our results, as the DGPs are designed so as to impact causality both through the short-run dynamics (if there is any) as well as through the long-run dynamics (the error-correction part). In most cases, the best size tests are associated with one lag and with cointegrating rank either (1) selected to be the truth (one) or (2) selected using a Johansen likelihood ratio trace test without a constant. In fact, the very similar sizes for the two "good" cointegrating (CI) rank cases arise because the Johansen method is usually able to "select" the true CI rank, even in the presence of substantial model complexity [i.e., for VAR(3) models with an MA component). However, of note is that when the (CI-1)–VAR(3)–MA2 results are examined (see Table 2), the λ_{P1}–λ_{P3} sizes range from "bests" of 0.194–0.234 ($T=100$); 0.118–0.161 ($T=200$); and 0.099–0.149 ($T=300$), when lags are selected using the AIC, and are always worse when lags are selected using the SIC. Interestingly, for the CI-2 case (see Table 3), SIC lag selection leads to empirical sizes closer to nominal in the VAR(3)–MA(2) case, suggesting that the evidence in favor of using the AIC for more complex models is not always clearcut. Indeed, for corresponding λ_{LW} tests, size based on the SIC lag selection criterion is usually better than that based on the AIC criterion, and is: 0.130 ($T=100$), 0.109 ($T=200$), and 0.090 ($T=300$) for CI-1. In fact, the only DGPs for which AIC is preferred over SIC for lag selection are VAR(3)–MA2 specifications. This again underscored the usefulness of the AIC only in more complex models.

The *ex ante* forecasting based model selection type approaches perform quite well (see Tables 9–11, 13, and 14). In particular, for complex DGPs [e.g., VAR(3)–MA2] the reported sizes of the MSE, MAD, an MAPE tests are very small (from 0.002 to 0.015) for lags selected using either the AIC or the SIC. As is generally the case, though, lag selection based on the SIC is dominated by selection based on the AIC for more complex models. Even given the clear success of the *ex ante* approaches, it should be noted that our experimental design is very simple. In addition, there is much research currently underway in the field of econometrics into the construction and performance of out-of-sample predictive ability tests, so that much remains to be learned.

The in-sample model selection approaches fare very well also (see Tables 5–8 and 12–15). In particular, the SIC seems to provide a very reasonable alternative to standard hypothesis testing, when the null is Granger noncausality. In Tables 5 and 10 note that the number of "wins" (based on a comparison of model selection criteria for the different models) for a given "true" DGP and for each of the estimated cointegrating, difference stationary, and levels stationary models are reported. In this way, we can get an idea as to which type of model is preferred by AIC and SIC. As mentioned above, the way these numbers are reported is as follows. From amongst all lag and CI rank combinations, the AIC and SIC-best models are used in each simulation to "test" for noncausality. If the AIC-best model is a cointegration model, then a "win" is tallied for "cointegration," if the AIC-best model is a difference model, then a "win" is tallied for "differences," and so on. The rest of the tables on in-sample model selection report rejection frequencies. Notice in these tables that the AIC and SIC tend to select the "true" DGP (e.g., VEC models are found to be preferred whenever data are generated according to VECs). Even casual inspection of the results suggests that the SIC approach for checking Granger causality performs better than the AIC. In addition, both model selection criteria are somewhat sensitive to the combination of DGP and estimation method. For example, results are very poor when data are generated according to a VEC, but levels or difference stationary models are estimated, and fixing the CI rank at 2 when it is actually 1 also results in poor performance. Thus, although this sort of model selection appears promising, there do appear to be frailties with this approach (as with all approaches when used with mis-specified models).

4.2 Difference Stationary Models

In order to examine the performance of Wald tests directly using differenced data (λ_{DS}), we also generated a number of difference stationary DGPs, and applied all seven types of causality tests. The results are omitted here for the sake of brevity, but are reported in Swanson et al., and Pisu (SOP: 1996, Tables 17–23). Of note is that the test based on λ_{DS} is very precise for VAR(1) ($T=100$, 200, 300), and for the VAR (3) ($T=300$), when the correct number of lags is used (e.g., the sizes are 0.054, 0.05, 0.052, and 0.055, respectively). However, when an MA component (MA2) is introduced, the standard difference stationary type Wald tests are dominated by the levels type Toda and Yamamoto tests for all lags, as well as all sample sizes for the VAR(1) case. For example, for VAR(1)–MA2 sizes for the λ_{DS} test based on lags selected using the AIC are 0.176, 0.111, and 0.095, for $T=100$, 200, and 300, respectively. These are the "best" sizes available for this case. The corresponding values associated with the λ_{LW} type test are 0.122, 0.099, and 0.080. In addition, it is worth noting that sequential Wald tests yield rather uninspired results when faced with difference stationary data (see Tables 18–21 in SOP). Indeed, in most cases [except for VAR(3)–MA DGPs] the sizes of λ_{P1}–λ_{P3} are in the 1–20% range.

Findings based on the use of in-sample model selection procedures are qualitatively the same as when VEC DGPs are generated. However, the SIC-type causality test seems to perform seven better, when lags are chosen using the AIC. This may be because AIC infrequently picks too few lags for inclusion in the estimated VAR model.

5 CONCLUSIONS

In this chapter we have discussed and compared five different Wald-type tests of the null hypothesis of noncausality, as well as two different tests based on model selection approach to assessing the predictive ability of the variable(s) whose causal effect is of interest. Our main purpose was to gather together and discuss a reasonable number of recent tests, some of which are asymptotically valid under the null hypothesis regardless as to whether or not the data are integrated of order zero, integrated of higher order, or cointegrated of some order. We also examined a number of model selection based approaches to assessing causal ordering, including those based on in-sample complexity penalized likelihood criteria, and those based on the examination of *ex ante* forcasting ability. Finally, we obtained results that have some implications for empirical work. For example, our examination suggests that three types of tests perform reasonably well (in terms of size), regardless as to the integratedness and cointegratedness properties of the variables in the system, and regardless as to whether or not the model is correctly specified. The first is the standard Wald-type test, which is based on levels regressions, and which adds "surplus" lags to account for integratedness and cointegratedness. This test is due to Toda and Yamamoto (1995) and Dolado and Lütkepohl (1996). It should be noted, however, that the local power properties of such tests may indeed be suspect, although concrete results on this matter have yet to be published. The second consists of the model selection based approach of using the Schwarz information criterion to select a "best" model, and then either "reject" or "accept" the noncausality null based on whether the causal variable of interest appears in the "best" model. In is perhaps worth stressing that model selection type approaches focus more on predictive ability, and as such may be particularly useful when the issue at hand is forecastability. In-sample hypothesis testing, on the other hand, may be preferable when the statistical significance of particular coefficients in a model is the issue being examined (e.g., when testing economic theories).

ACKNOWLEDGMENTS

The authors wish to thank Lutz Killian, seminar participants at the 1997 Winter Meetings of the Econometric Society, and seminar participants at the University of Pennsylvania for useful comments on an earlier version of the paper on which this chapter is based. Swanson thanks the Private Enterprise Research Center at Texas A&M University for financial support.

REFERENCES

Akaike, H., 1973. Information theory and an extension of the maximum likelihood principle. In B.N. Petrov and F. Csaki, eds. *Second International Symposium on Information Theory*. Akademiai Kiado, Budapest, pp. 267–281.

Akaike, H., 1974. A new look at the statistical model identification. *IEEE Transactions on Automatic Control* **19**, 716–723.

Andrews, D.W.K. 1991. Heteroskedasticity and autocorrelation consistent covariance matrix estimation. *Econometrica* **59**, 817–858.

Ashley, R., Granger, C.W.J., and Schmalensee, R., 1980. Advertising and aggregate consumption: An analysis of causality. *Econometrica* **48**, 1149–1167.

Chao, J., Corradi, V., and Swanson, N.R., 2000. An out of sample test for Granger causality. *Macroeconomic Dynamics*, **5**, 598–620.

Choi, I., 1993. Asymptotic normality of least squares estimates for higher order autoregressive integrated processes with some applications. *Econometric Theory* **9**, 263–282.

Clark, T., 2000. Finite-sample properties of tests for equal forecast accuracy. *Journal of Forecasting*, **18**, 489–504.

Clark, T. and McCracken, M.W., 1999. Granger causality and tests of equal forecast accuracy and encompassing. Working Paper, Federal Reserve Bank of Kansas City.

Corradi, V. and Swanson, N.R., 2000a. Predictive ability with cointegrated variables. *Journal of Econometrics*, **110**, 353–381.

Corradi, V. and Swanson, N.R., 2000b. A consistent test for nonlinear out of sample predictive accuracy. Working Paper, Rutgers University, New Brunswick, New Jersey.

Den Haan, W. and Levin, A. 1995. Inferences from parametric and non-parametric covariance matrix estimation procedures. NBER Technical Working Paper 195.

Diebold, F.X. and Lopez, J.A., 1995. Forcast evaluation and combination. In G.S. Maddala and C.R. Rao, eds. *Handbook of Statistics*. North Holland, Amsterdam.

Diebold, F.X. and Mariano, R.S., 1995. Comparing predictive accuracy. *Journal of Business and Economic Statistics*, **13**, 253–263.

Dolado, J.J. and Lütkepohl, H., 1996. Making Wald tests work for cointegrated VAR systems. *Econometric Reviews* **15**, 369–386.

Engle, R.F. and Brown, S.J., 1986. Model selection for forecasting. *Applied Mathematics and Computation* **20**, 313–327.

Engle, R.F. and Granger, C.W.J., 1987. Co-integration and error correction: representation, estimation, and testing. *Econometrica* **55**, 251–276.

Fair, R.C. and Shiller, R.J. 1990. Comparing information in forecasts from econometric models. *American Economic Review* **80**, 375–389.

Fomby, T.B. and Guilkey, D.K. 1978. On choosing the optimal level of significance for the Durbin–Watson test and the Bayesian alternative. *Journal of Econometrics*, **8**, 203–213.

Geweke, J., Meese, R., and Dent, W., 1983. Comparing alternative tests of causality in temporal systems. *Journal of Econometrics* **21**, 161–194.

Gonzalo, J. and Lee, T.-H., 1996. Relative power of the t-type tests for stationary and unit root processes. *Journal of Time Series Analysis* **17**, 37–47.

Granger, C.W.J., 1969. Investigating causal relation by econometric models and cross spectral methods. *Econometrica* **37**, 428–438.

Granger, C.W.J., 1980. Testing for causality: A personal viewpoint. *Journal of Economic Dynamics and Control* **2**, 329–352.

Granger, C.W.J. 1988. Some recent developments in a concept of causality. *Journal of Econometrics* **39**, 1199–211.

Granger, C.W.J. and Lin, J.-,. 1995. Causality in the long run. *Econometric Theory*, **11**, 530–536.

Granger, C.W.J., King, M.L., and White, H., 1995. Comments on testing economic theories and the use of model selection criteria. *Journal of Econometrics* **67**, 173–187.

Harvey, D.I., Leybourne, S.J., and Newbold, P., 1997. Tests for forecast encompassing. *Journal of Business and Economic Statistics* **16**, 254–259.

Johansen, S., 1988. Statistical analysis of cointegration vectors. *Journal of Economics and Dynamics Control* **12**, 231–254.

Johansen, S., 1991. Estimation and hypothesis testing of ointegration vectors in Gaussian vector autoregressive models. *Econometrica* **59**, 1551–580.

Johansen, S. and Juselius, K. 1990. Maximum likelihood estimation and inference on cointegration—with applications to the demand for money. *Oxford Bulletin of Economics and Statistics* **52**, 169–210.

Lütkepohl, H., 1991. *Introduction to Multiple Time Series Analysis*. Springer-Verlag, New York.

McCracken, M.W., 1999. Asymptotics for out of sample tests of causality. Working Paper, Louisiana State University.

Phillips, P.C.B., 1995. Fully modified least squares and vector autoregression. *Econometrica* **63**, 1023–1078.

Phillips, P.C.B. and Hansen, B. 1990. Statistical inference in instrumental variables regressions with I(1) processes. *Review of Economic Studies* **53**, 473–496.

Rissanen, J., 1978. Modeling by shortest data description. *Automatica* **14**, 465–471.

Schwarz, G., 1978. Estimating the dimension of a model. *Annals of Statistics* **6**, 461–464.

Sims, C.A., 1972. Money, income, and causality. *American Economic Review* **62**, 540–552.

Sims, C.A., Stock, J.H., and Watson, M.W., 1990. Inference in linear time series models with some unit roots. *Econometrica* **58**, 113–144.

Sin, C.-Y. and White, H., 1995. Information criteria for selecting possibly misspecified parametric models. *Journal of Econometrics* **71**, 207–225.

Swanson, N.R., 1998. Money and output viewed through a rolling window. *Journal of Monetary Economics* **41**, 455–474.

Swanson, N.R. and White, H., 1995. A model selection approach to assessing the information in the term structure using linear models and artificial neural networks. *Journal of Business and Economic Statistics* **13**, 265–275.

Swanson, N.R. and White, H., 1997a. A model selection approach to real-time macroeconimic forecasting using linear models and artificial neural networks. *Review of Economics and Statistics* **79**, 540–550.

Swanson, N.R. and White, H., 1997b. Forecasting economic time series using adaptive versus nonadaptive and linear versus nonlinear econometric models. *International Journal of Forecasting* **13**, 439–461.

Swanson, N.R. Ozyildirium, A., and Pisu, M., 1996. A comparison of alternative causality and predictive ability tests in the presence of integrated and co-integrated variables. Working Paper, Rutgers University, New Brunswick, New Jersey.

Toda, H.Y. and Phillips, P.C.B., 1993. Vector autoregressions and causality. *Econometrica* **61**, 1367–1393.

Toda, H.Y. and Phillips, P.C.B., 1994. Vector autoregression and causality: a theoretical overview and simulation study. *Econometric Reviews* **13**, 259–285.

Toda, H.Y. and Yamamoto, T., 1995. Statistical inference in vector autoregressions with possibly integrated processes. *Journal of Econometrics* **66**, 225–250.

West, K.D., 1996. Asymptotic inference about predictive ability. *Econometrica* **64**, 1067–1084.

West, K. and McCracken, M.W., 1998. Regression based tests of predictive ability. *International Economic Review* **39**, 817–840.

White, H., 2000. A reality check for data snooping. *Econometrica* **68**, 1097–1126.

6

Finite Sample Performance of the Empirical Likelihood Estimator Under Endogeneity

Ron C. Mittelhammer
Washington State University, Pullman, Washington, U.S.A.

George G. Judge
University of California at Berkeley, Berkeley, California, U.S.A.

1 INTRODUCTION

There is a growing body of evidence that traditional asymptotically efficient moment-based estimators for the linear structural model may have large biases for the relatively small sample sizes usually encountered in applied economic research (see, e.g., Newey and Smith, 2000). Furthermore, instrumental variables that are utilized in the context of a set of moment–orthogonality conditions may be only weakly correlated with the endogenous variables in the model and consequently the parameters of the structural model are only poorly or weakly identified. In these situations it is generally recognized that, for both traditional and nontraditional estimators, bias and variability problems may arise and that large sample normal approximations provide a poor basis for finite sample performance (see, e.g., Nelson and Startz, 1990; Maddala and Jeong, 1992; Bound et al., 1995; Stock and Wright, 2000). Given these general problem areas the focus of this chapter is on semiparametric estimation and inference relative to the response parameters for a linear structural statistical model. Using Monte

Carlo sampling procedures and a range of underlying data-sampling processes, we provide finite sample comparisons of the two-stage least squares (2SLS) and the empirical likelihood (EL) estimator for recovering the unknown parameters of the linear structural statistical model.

In finite sample situations where the instrumental variables are either well or weakly correlated with the endogenous variable(s) in question and given that the orthogonal moment conditions hold, we seek answers to the following questions:

1. Do empirical likelihood (EL) estimators exhibit substantial small sample bias and, in terms of bias, is the EL estimator superior to traditional estimators?
2. In terms of precision is the EL estimator superior to traditional estimators?
3. In terms of inference relating to structural model parameters, do traditional testing procedures, relative to EL testing procedures, lead to more accurate coverage but wider confidence intervals?
4. What is the size performance of the traditional and EL inference procedures relative to testing the validity of the moment restrictions?

The format of the chapter is as follows: In Section 2 the linear structural statistical model is defined and the competing estimators are specified. In Section 3 the design of the sampling experiment is presented and the alternative data-sampling processes are defined. In Section 4 the sampling results as they relate to estimation and inference are reported. The implications of the results are summarized in Section 5. Some supplementary information in table form is presented in the Appendix.

2 STATISTICAL MODELS AND ESTIMATORS

Consider a single equation in a system of equations that has the semi-parametric linear statistical response model form $\mathbf{y} = \mathbf{X\beta} + \varepsilon$, where we observe a vector of sample observations $y = (y_1, y_2, \ldots, y_n)$, $\mathbf{X}$ is a $(n \times k)$ matrix of realizations of stochastic variables, ε is an unobservable random noise vector with mean $\mathbf{0}$ and covariance matrix $\sigma^2\mathbf{I}_n$, and $\boldsymbol{\beta} \in \mathbf{B}$ is a $(k \times l)$ vector of unknown parameters. If one or more of the regressors is correlated with the equation noise, then $E[n^{-1}\mathbf{X}'\varepsilon] \neq \mathbf{0}$ or plim $[n^{-1}\mathbf{X}'\varepsilon] \neq \mathbf{0}$ and traditional Gauss–Markov procedures such as the least squares (LS) estimator, or equivalently the method of moments (MOM)–extremum estimator $\hat{\boldsymbol{\beta}}_{\text{mom}} = \arg_{\boldsymbol{\beta}\in\mathbf{B}}[n^{-1}\mathbf{X}'(\mathbf{y} - \mathbf{X\beta}) = \text{o}]$, are biased and inconsistent, with unconditional expectation and probability limit given by $E[\hat{\boldsymbol{\beta}}] \neq \boldsymbol{\beta}$ and plim $[\hat{\boldsymbol{\beta}}] \neq \boldsymbol{\beta}$.

Given a sampling process characterized by nonorthogonality of **X** and ε, to avoid the use of strong underlying distributional assumptions it is conventional to introduce additional information in the form of independent realizations of a $(n \times m), m \geq k$ random matrix **Z** of instrumental variables, whose elements are correlated with **X** but uncorrelated with ε. This information is introduced into the statistical model by using a sample analog to the orthogonality of instruments and model noise as

$$\mathbf{h}(\mathbf{y}, \mathbf{X}, \mathbf{Z}; \beta) = n^{-1}[\mathbf{Z}'(\mathbf{y} - \mathbf{X}\beta)] \xrightarrow{p} \mathbf{o} \tag{1}$$

If $m = k$ the sample moments are solved for the basic instrumental variable (IV) estimator $\hat{b}_{\text{iv}} = (\mathrm{Z'X})^{-1}\mathrm{Z'Y}$, which is consistent and asymptotically normally distributed (Mittelhammer et al., 2000). When $m > k$ the 2SLS estimator, $\boldsymbol{\beta}_{2sls} = (\mathbf{X}'\mathbf{P}_z\mathbf{X})^{-1}\mathbf{X}'\mathbf{P}_z\mathbf{Y}$, where $\mathbf{P}_z = \mathbf{Z}(\mathbf{Z}'\mathbf{Z})^{-1}\mathbf{Z}'$ is the projection matrix based on **Z**, and the GMM estimator, are popular choices with optimal asymptotic properties.

2.1 The MEL Estimator

Alternatively, the maximum empirical likelihood (MEL) approach (Owen, 1988, 1991; Qin and Lawless, 1994; Mittelhammer et al., 2000) offers another way to seek a unique solution for $\boldsymbol{\beta}$ using the instrumental variable-based moment information in Section 2.1. Under the MEL concept, empirical likelihood weights based on multinomial distributions supported on a sample of observed values are used to reduce the infinite dimensional problem of nonparametric likelihood estimation to a finite dimensional one. In an instrumental variable context for the linear structural equation, this leads to the following extremum-type estimator for the parameter vector $\boldsymbol{\beta}$ that defines the MEL estimator $\hat{\boldsymbol{\beta}}_{\text{EL}}$ as

$$\hat{\boldsymbol{\beta}}_{\text{El}} = \arg\max_{\beta}[l_{\text{E}}(\boldsymbol{\beta})] \tag{2}$$

where

$$\ell_{\text{E}}(\boldsymbol{\beta}) \equiv \max_{w}\left\{ n^{-1}\sum_{i=1}^{n}\ln(w_i) \middle| \sum_{i=1}^{n} w_i\mathbf{z}'_i \middle| (y_i - x_i\boldsymbol{\beta}) = 0, \right. \\ \left. \sum_{i=1}^{n} w_i = 1, w_i \geq 0\,\forall i,\ \boldsymbol{\beta} \in \mathbf{B} \right\} \tag{3}$$

can be interpreted as a semiparametric profile empirical likelihood for $\boldsymbol{\beta}$ (Murphy and Van Der Vaart, 2000). If this Owen MEL criterion is used, the estimation objective involves finding the feasible weights $\hat{\mathbf{w}}$ that maximize the empirical probabilities assigned to the observed set of observations.

The Lagrange function associated with the empirical likelihood formulation is

$$L(\mathbf{w}, \boldsymbol{\alpha}, \eta) = n^{-1} \sum_{i=1}^{n} \ln(w_i) - \boldsymbol{\alpha}' \sum_{i=1}^{n} w_i z'_i (y_i - \mathbf{x}_i \boldsymbol{\beta}) - \eta \left(\sum_{i=1}^{n} w_i - 1 \right) \quad (4)$$

and the optimal w_i can be expressed in terms of $\boldsymbol{\beta}$ and $\boldsymbol{\alpha}$ as

$$w_i(\boldsymbol{\alpha}, \boldsymbol{\beta}) = \left[n(\boldsymbol{\alpha}' \mathbf{z}'_i (y_i - \mathbf{x}_i \boldsymbol{\beta}) + 1) \right]^{-1} \quad (5)$$

where $\boldsymbol{\alpha}$ is the $m \times 1$ optimal Lagrange multiplier vector and β solves Eq. (2). The EL estimator $\hat{\beta}_{\mathrm{EL}}$ is consistent, asymptotically normally distributed, and asymptotically efficient. A consistent empirical representation of the asymptotic distribution is given by

$$\hat{\beta}_{\mathrm{EL}} \overset{a}{\sim} N\left(\beta, n^{-1} \left[\left(\sum_{i=1}^{n} \hat{w}_i \mathbf{X}'_i \mathbf{Z}_i \right) \left[\sum_{i=1}^{n} \hat{w}_i \hat{\varepsilon}_i^2 \mathbf{Z}'_i \mathbf{Z}_i \right]^{-1} \left(\sum_{i=1}^{n} \hat{w}_i \mathbf{X}'_i Z_i \right)' \right]^{-1} \right) \quad (6)$$

or equivalently,

$$\hat{\beta}_{\mathrm{EL}} \overset{a}{\sim} N\left(\beta, n^{-1} \left[\mathrm{X}'(\mathrm{Z} \odot \hat{\mathbf{w}}) \times \left[(\mathbf{Z} \odot (\hat{\mathbf{w}}^{1/2} \odot \hat{\varepsilon}))'(\mathbf{Z} \odot (\hat{\mathbf{w}}^{1/2} \odot \hat{\varepsilon})) \right]^{-1} (\mathbf{Z} \odot \hat{\mathbf{w}})' \mathbf{X} \right]^{-1} \right) \quad (7)$$

where $\odot$ denotes the extended Hadamard (elementwise) product, and $\hat{\varepsilon} = \mathbf{Y} - \mathbf{X}\hat{\beta}_{\mathrm{EL}}$. The $\hat{\mathbf{w}}$ vector in Eq. (6) or (7) can be replaced by the vector $n^{-1}\mathbf{1}$ to define an alternative asymptotically valid representation of the covariance matrix of the EL estimator.

Calculating the solution to the MEL estimation problem will generally require that a computer-driven maximization algorithm be employed. When $m = k$, the solution to the EL extremum problems leads back to the standard IV estimator $\hat{\boldsymbol{\beta}}_{\mathrm{iv}}$ with $\hat{w}_i = n^{-1}$. When $m > k$, the estimating equations overdetermine the unknown parameter values to be recovered and a nontrivial EL solution results. Computing solutions to EL-constrained optimization problems can present formidable numerical challenges. This computing problem arises because, in the neighborhood of the solution to such problems, the gradient matrix associated with the moment constraints approaches an ill-conditioned state of being less than full rank. This occurs by design in these types of problems because the fundamental method by which the EL method addresses the overdetermined nature of the moment conditions is by finding EL weights $\hat{w}_i$, for $i = 1, \ldots, n$, that ultimately transform the moment equations into a functionally dependent, lower rank (equal to $k < m$) system of equations capable of being solved uniquely for the parameters. This creates instability in gradient-based constrained optimization algorithms regarding the representation of the feasible spaces and feasible directions for such problems.

2.2 MEL Inference

In terms of inference the EL function $\ell_E(\boldsymbol{\beta})$ bears a strong analogy to the likelihood function used in traditional ML procedures for developing tests and confidence regions that are defined in terms of likelihood ratios. In this context, the empirical likelihood ratio (ELR) for the test of the hypothesis $\boldsymbol{\beta} = \boldsymbol{\beta}_0$, or more generally for testing the linear combinations hypothesis $\mathbf{c}\beta = \mathbf{r}$, is given by [recall Eq. (2)]

$$LR_{EL}(y) = \frac{\max_{\beta}[\ell_E(\boldsymbol{\beta})\, st\, \mathbf{c}\beta = \mathbf{r}]}{\max_{\beta} \ell_E(\boldsymbol{\beta})} \tag{8}$$

and

$$-2\ln(LR_{EL}(\mathbf{Y})) \text{ is distributed as chi square } (k, 0) \tag{9}$$

under H_0 when $m \geq k$. One can also base tests of $\mathrm{c}\beta = \mathbf{r}$ on the Wald criterion in the usual way utilizing the inverse of the asymptotic covariance matrix exhibited in Eq. (6) or (7) as the weight matrix of a quadratic form in the vector $\mathbf{c}\hat{\beta}_{EL} = -\mathbf{r}$, as well as construct tests based on the Lagrange multipliers associated with the constraints $\mathbf{c}\beta = \mathbf{r}$ imposed on the EL optimization problem, Eq. (2) or (3).

Interest may also center on testing the validity of the moment equations, the null hypothesis then being $H_0 : E[\mathbf{h}(\mathbf{y}, \mathbf{X}, \mathbf{Z}\beta)] = \mathbf{o}$. A natural way to make this test is to use an ELR statistic based on a ratio of empirical likelihoods for which the EL weights are either constrained to satisfy the moment conditions (the numerator of the ratio) or else are unconstrained by the moment conditions, in which case the empirical likelihood is based on $w_i = n^{-1}, \forall i$. If the moment constraints are correct, the ratio is distributed asymptotically as a central chi square, with degrees of freedom equal to $m - k$, m being the number of functionally independent moment conditions and k being the number of unknown parameters involved in the specification of the moment conditions. Details of the empirical implementation of tests used in the Monte Carlo experiments contained in this chapter are presented in Section 4. For additional discussion of the ELR and other tests, see Chapters 12 and 13 in Mittelhammer et al. (2000).

2.3 Remark on Performance Evaluations

Although the estimators are asymptotically equivalent, and are also equivalent in finite samples if $m = k$, when $m > k$ the sampling properties of the 2SLS and EL estimators can differ in finite samples. Since each of the estimators involves a bias and a variance component, this raises the question of how to evaluate estimator performance. In some cases a researcher may be willing to accept some degree of bias if it improves precision in estimation and prediction. In situations such as this where bias–variance trade offs are being considered, the squared error loss (SEL) measure $L(\boldsymbol{\beta}, \hat{\boldsymbol{\beta}}) = \|\boldsymbol{\beta} - \hat{\boldsymbol{\beta}}\|^2$ or $L(\mathbf{y}, \mathbf{X}\hat{\boldsymbol{\beta}}) = \|\mathbf{y} - \mathbf{x}\hat{\boldsymbol{\beta}}\|^2$ provides a basis for comparing the performance of the two competing estimators and this is one basis we use for comparing the performance of the estimators in Section 4.

3 DESIGN OF SAMPLING EXPERIMENTS

In terms of the EL-type formulation of Section 2, the solution for the optimal weights cannot be expressed in closed form. Moreover, the finite sample probability distributions of the traditional 2SLS estimator are generally intractable. Consequently, the finite sample properties of the aforementioned estimation procedures cannot be derived from direct evaluation of the estimator's functional form. Therefore, we use a Monte Carlo sampling experiment to identify and compare the repeated sampling performance of the estimators in terms of the expected squared error loss associated with estimating $\boldsymbol{\beta}$ and predicting $\mathbf{y}$. We also provide results relating to the average bias and variances associated with the competing estimators. While these results are specific to the collection of particular Monte Carlo experiments analyzed, the extensive sampling evidence reported does provide an indication of the types of relative performance that can occur over a range of well and weakly identified data-sampling scenarios.

3.1 Experimental Design

Consider a sampling process of the following form:

$$Y_{i1} = Z_{i1}\beta_1 + Y_{i2}\beta_2 + e_i = \mathbf{X}\boldsymbol{\beta} + \varepsilon_i \tag{10}$$

$$Y_{i2} = \sum_{j=1}^{4} \pi_j Z_{ij} + v_i = \mathbf{Z}_i\pi + v_i \tag{11}$$

where $\mathbf{X}_i = (Z_{i1}, Y_{i2})$ and $i = 1, 2, \ldots, n$. The two-dimensional vector of unknown parameters, $\boldsymbol{\beta}$, in Eq. (10) is arbitrarily set equal to the vector $[-1, 2]'$. The outcomes of the (6×1) random vector $[Y_{i2}, \varepsilon_i, Z_{i1}, Z_{i2}, Z_{i3}, Z_{i4}]$ are generated *iid* from a multivariate normal distribution with a zero mean vector and unit variances for all random variables other than ε, which has a variance of 4. Also, there are various other conditions relating to the correlations existent among the six scalar random variables. The values of the π_j's in Eq. (11) are clearly determined by the regression function between Y_{i2} and $[Z_{i1}, Z_{i2}, Z_{i3}, Z_{i4}]$, which is itself a function of the covariance specification relating to the marginal normal distribution associated with the (5×1) random vector $[Y_{i2}, Z_{i1}, Z_{i2}, Z_{i3}, Z_{i4}]$. Thus, the π_j's generally change as the scenario postulated for the correlation matrix of the sampling process changes. In this sampling design, the outcomes of $[Y_{i1}, V_i]$ are then calculated by applying Eqs. (10) and (11) to the outcomes of $[Y_{i2}, Z_{i1}, Z_{i2}, Z_{i3}, Z_{i4}]$.

3.2 Monte Carlo Scenario Characteristics

Regarding the details of the sampling scenarios simulated for this set of Monte Carlo (MC) experiments, the focus was on sample sizes of $n = 25$, 50, 100, and 250. The outcomes of ε_i were generated independently of the vector $[Z_{i1}, Z_{i2}, Z_{i3}, Z_{i4}]$ so that the correlations between ε_i and the Z_{ij}'s were zeros, thus fulfilling a fundamental condition for $[Z_{i1}, Z_{i2}, Z_{i3}, Z_{i4}]$ to be considered a set of valid instrumental variables for estimating the unknown parameters in Eq. (10). Regarding the degree of nonorthogonality–identifiability in Eq. (10), a correlation of 0.75 between the random variables Y_{i2} and ε_i was utilized to simulate a relatively strong correlation–nonorthogonality relationship between the explanatory variable Y_{i2} and the equation noise ε_i.

For each sample size, alternative scenarios were examined relating to both the degree of correlation existing between each of the instruments in the matrix $\mathbf{Z}$ and the $\mathbf{Y}_2$ variable, and the levels of collinearity existing among the instrumental variables themselves. By varying the degrees of intercorrelation among the variables, the overall correlation of the instrumental variables with $\mathbf{Y}_2$ is effected, and contributes to determining the overall effectiveness of the set of instruments in predicting values of the endogenous $\mathbf{Y}_2$.

The major characteristics of each of the scenarios are delineated in Table 1. In general, the scenarios range from very weak but independent instruments to stronger but highly collinear instruments. All told, there were 10 different MC experimental designs in combination with the four different samples, resulting in 40 different sampling scenarios in which to observe estimator and inference behavior. For a useful summary of experimental designs used by others to represent weakly determined situations, see Zivot (2001).

The reported sampling properties relating to estimation objectives are based on 2500 MC repetitions, and include estimates of the empirical risks,

TABLE 1 Monte Carlo Experiment Characteristics

Scenario	$\rho_{Y_2,e}$	ρ_{Y_2,z_1}	ρ_{Y_2,z_j}	ρ_{z_1,z_j}	ρ_{z_j,z_k}	$\rho_{Y_2,\hat{Y}_2}$	$\rho^2_{Y_1,\hat{Y}_1}$
1	0.75	0.1	0.4	0.25	0.1	0.65	0.86
2	0.75	0.3	0.3	0	0	0.60	0.88
3	0.75	0.4	0.4	0.3	0.3	0.58	0.90
4	0.75	0.5	0.5	0.6	0.6	0.60	0.92
5	0.75	0.1	0.1	0	0	0.20	0.86
6	0.75	0.1	0.1	0.5	0.5	0.13	0.86
7	0.75	0.05	0.05	0.75	0.75	0.06	0.86
8	0.75	0.02	0.02	0.9	0.9	0.02	0.85
9	0.75	0.6	0.6	0.9	0.9	0.62	0.96
10	0.75	0.6	0.25	0.5	0.5	0.60	0.96

$\beta = [-1,2]'$ and $\sigma^2_{e_i} = 4$ for scenarios 1–10, and $\sigma^2_{Y_{2i}} = \sigma^2_{Z_{ij}} = 1$, $\forall i$ and j. $\rho_{Y_2,e}$ denotes the correlation between Y_2 and e, and measures the degree of nonorthogonality; ρ_{Y_2,z_1} denotes the correlation between Y_2 and Z_1; ρ_{Y_2,z_j} denotes the common correlation between Y_2 and each of the three instrumental variables not included in the first equation; ρ_{z_1,z_j} denotes the common correlation between Z_1 and the remaining three Z_j's; ρ_{z_j,z_k} denotes the common correlation between the three instrumental variables not included in the first equation; $\rho_{Y_2,\hat{Y}_2}$ denotes the population correlation between $\mathbf{Y}_2$ and $\hat{\mathbf{Y}}_2 = \mathbf{Z}\pi$; and $R^2_{Y_1,\hat{Y}_1}$ denotes the population squared correlation between $\mathbf{Y}_1$ and $\hat{\mathbf{Y}}_1 = \mathbf{X}\hat{\beta}$.

based on a SEL measure, associated with estimating $\boldsymbol{\beta}$ with $\hat{\boldsymbol{\beta}}$ (parameter estimation risk) and estimating $\mathbf{y}$ with $\hat{\mathbf{y}}$ (predictive risk). We also report on the average estimated bias in the estimates, bias $(\hat{\boldsymbol{\beta}}) = E[\hat{\boldsymbol{\beta}}] - \boldsymbol{\beta}$, and the average estimated variances of the estimates, var$(\hat{\boldsymbol{\beta}}_i)$.

Overall, three different estimators were examined including the 2SLS estimator, the least squares (LS) estimator, and the EL estimator. Estimator outcomes were obtained using GAUSS software, and in particular, the EL-type solutions were calculated using the GAUSS-constrained optimization application module provided by Aptech Systems, Maple Valley, WA, U.S.A.

4 SAMPLING RESULTS

The sampling designs specified in Section 3 range from well-behaved data-sampling processes, involving instrumental variables that have the recommended correlation structure for which the traditional 2SLS estimator would be expected to perform reasonably well, to other extremes involving low correlation of the IVs with the included endogenous variable and weak identification, high multicollinearity, and poorer fitting models. In reporting the sampling results we discuss the empirical evidence regarding the performance of the 2SLS and EL estimators for the well- and ill-behaved sampling models for samples sizes of $n = 25, 50, 100$, and 250. In each case we report estimation sampling performance relating to empirical mean parameter squared error loss (SEL), mean squared prediction error (MSPE), bias, and variance. In terms of inference, we report sampling performance relating to test sizes for tests of moment validity, as well as coverage and power observations for tests on structural model parameters. To conserve space and facilitate discussion, some of the sampling results are reported in graphical form, and we generally defer observations on LS estimator performance to the results tables. Appendix tables contain detailed information underlying the graphs and contain sampling results for alternative sample sizes and the bias and variance sampling results for β_1 and β_2.

4.1 Discussion

4.1.1 Empirical Parameter SEL Performance

The well-behaved data-sampling processes that appear strongly identified and that involve low correlations among the instruments, moderate to high correlations between Y_2 and the Z_j's, and low or moderate correlation between Y_2 and Z_1 are represented by data-sampling processes (DSPs) defined by scenarios 1, 2, 3, and 4. In Figure 4.1 the empirical parameter

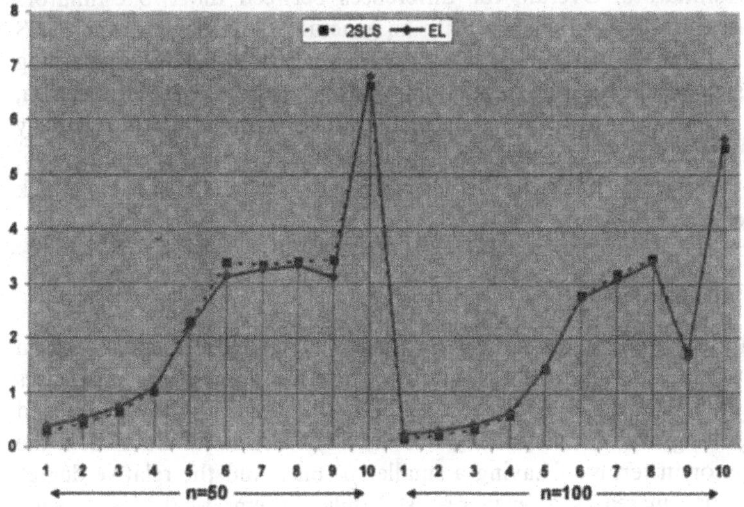

FIGURE 1 Parameter SEL for selected DSPs.

estimation risks for these DSPs and the 2SLS and EL estimators are presented for samples sizes of $n = 50$ and 100. In Fig. 1, for sample sizes $n = 50$ and 100, the 2SLS estimator is seen to be superior in empirical mean parameter SEL for all four scenarios, but the empirical risk differences are small. For $n = 250$, the 2SLS estimator continues to exhibit empirical risk dominance, but at this sample size the nominal levels of risk are small for either estimator, with the empirical risk differences between the two estimators also being relatively small. For $n = 25$ the 2SLS remains risk superior in the first three scenarios, but EL is superior in the fourth, the risk differences between the estimation procedures again being relatively small.

The ill-behaved, weakly identified DSPs are defined by scenarios 5–10. These sampling processes depart from the a priori IV ideal type of correlation structure and are consistent with sampling processes found in practice where the instruments have low correlation with the individual endogenous variable, and the parameters are more poorly identified. The empirical parameter risks of the 2SLS and EL estimators for these sampling designs are depicted in Fig. 1 for sample sizes $n = 50$ and 100. In the set of graphed comparisons the EL estimator is superior in three-quarters of the scenarios, where only in scenario 10 were there any effective risk gains from using 2SLS in place of EL. The results in Table A.1 indicate that the same relative risk behavior occurs when $n = 5$, while as n increases to 250 the performance of the 2SLS estimator improves relative to the EL estimator, as does the overall performance of both estimators. Overall, the differences between the two estimators in empirical mean parameter SEL is generally small, with the 2SLS estimator being a marginally better choice in well-behaved sampling processes characterized by moderate-to-strong parameter identification, and the EL estimator being marginally superior in cases of relatively poorly behaved sampling processes in which there is higher collinearity, weaker correlation between instruments and Y_2, and relatively weak parameter identification.

4.1.2 Empirical Bias

The empirical bias in estimating β_2 for $n = 50$ and 100 is presented in Fig. 2, and the biases for both β_1 and β_2, as well as results for sample sizes $n = 25$ and 250, are given in Table A.2. In the case of the well-behaved sampling designs 1–4, and for $n \geq 50$, the 2SLS estimator dominates the EL estimator in terms of having a smaller β_2 bias, and the relative degree of dominance increases as n increases. While the empirical bias of each estimator decreases as n increases, the bias of the EL estimator is more

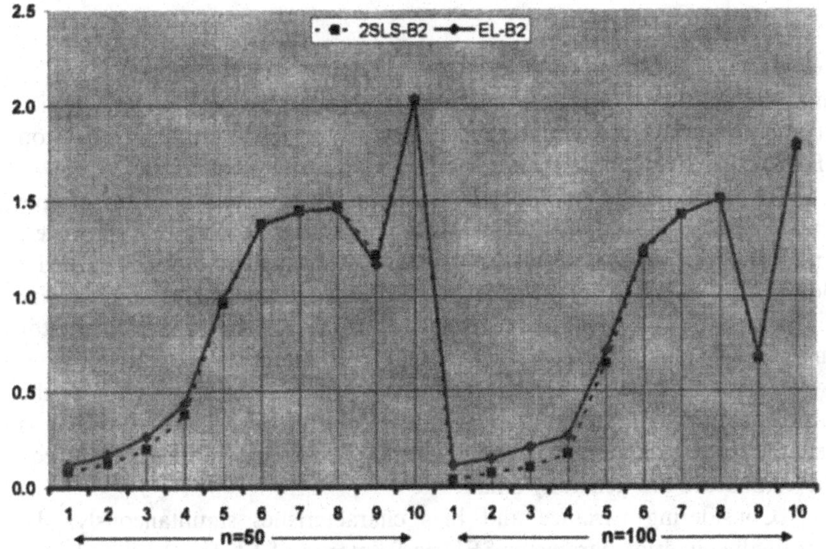

FIGURE 2 Empirical estimation biases for selected DSPs.

persistent and was still 5 to 8% of the true β_2 value for $n = 250$. As expected, the LS estimator exhibited a large relative bias and this bias did not decrease as n increased. Interestingly, for the smallest sample size of $n = 25$, the EL estimator dominates the 2SLS estimator in terms of β_2 bias across all four scenarios.

In terms of the empirical bias in estimating β_2 for the ill-behaved sampling scenarios 5–10, both estimators have substantial bias. For $n = 25$ the EL estimator has slightly better β_2-bias performance than the 2SLS estimator. As sample sizes increases, the biases of both estimators decrease, but the 2SLS estimator tends to be slightly better than the EL estimator in the majority of cases. For the severely ill-behaved scenarios 7 and 8 in which the instruments are extremely poor in predicting outcomes of $\mathbf{Y}_2$, there is no perceptible decrease in estimator bias as the sample size increases. Overall, except in the case of both very small sample sizes and ill-behaved data-sampling processes, the 2SLS estimator is generally marginally better than the EL estimator in terms of β_2 bias.

The comparison between 2SLS and EL is not as clear cut when comparing β_1-bias results, where EL is superior in more cases, and EL superiority is not isolated to the smallest sample size but rather occurs in at least half of the cases for each of the four sample sizes. However, the large majority of the bias levels for theβ_1 parameter are nominally quite small, and the bias differences are of generally little practical consequence.

4.1.3 Empirical Variance

For scenarios 1–4, the empirical variance of the 2SLS estimator for β_2 is smaller than the EL estimator in a majority of the cases. However, the size of the variance differences is small. As n increases the sampling precision of both of the estimators increases (see Fig. 3 and Table A.3).

In terms of the empirical β_2 variance for scenarios 5–10, as sample size increases the empirical variances decrease (see Fig. 3). However, similar to the case of bias, the rate of decline in variance is markedly slower for the severely ill-behaved scenarios 7 and 8. As sample size increases, the EL estimator tends to dominate the 2SLS estimator empirically in terms of achieving a smaller estimator β_2 variance.

In the case of β_1-variance comparisons, the 2SLS is superior in the large majority of cases. However, the β_1 variance of each estimator is generally much smaller than for β_2, and the variance differences between the estimators are generally small.

Considering variance and bias characteristics simultaneously, the marginally superior parameter SEL performance of EL for the ill-behaved DSPs can generally be attributed to reductions in the variances of the estimators of the structural parameters, and generally *not* to superior bias characteristics of the EL estimator.

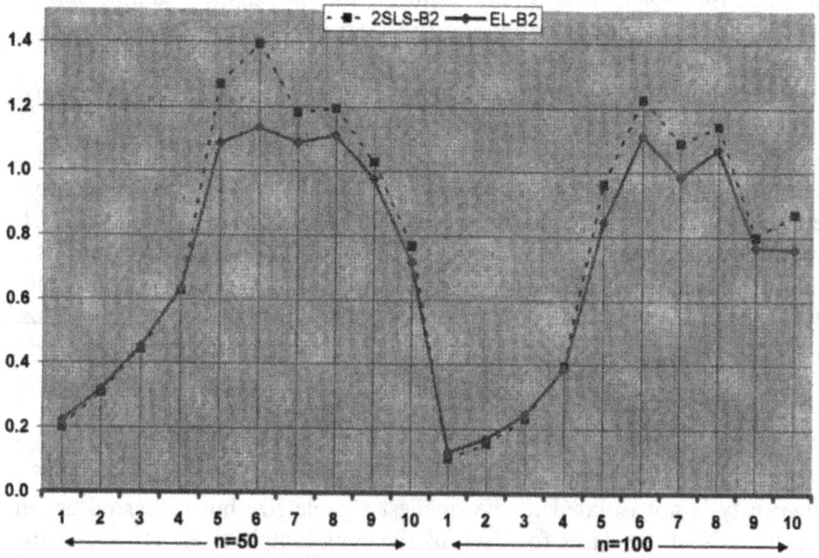

FIGURE 3 Empirical estimator variances for selected DSPs.

4.1.4 Empirical Prediction Errors

In terms of mean squared prediction error, the EL estimator is superior to the 2SLS estimator in the large majority of cases, and the frequency and level of superiority increases as n increases (see Table A.1). The EL estimator dominates the 2SLS estimator in average squared prediction risk when $n \geq 100$. The only cases in which 2SLS exhibited any effective gain in prediction risk was for the smallest sample size and for well-behaved data-sampling processes, although even in these cases the gain was far from dramatic. Of course, in terms of this performance metric, the least squares estimator dominates over all sample sizes and scenarios.

4.2 Inference

We now focus on presenting empirical evidence regarding the sizes of tests of moment information validity and we also present results relating to coverage and power relating to tests of the values of structural parameters.

4.2.1 Testing Moment Equation Validity

Four different tests of moment information validity are presented in Fig. 4 and all four of the tests share the same asymptotic Chi square $(m-k, 0)$

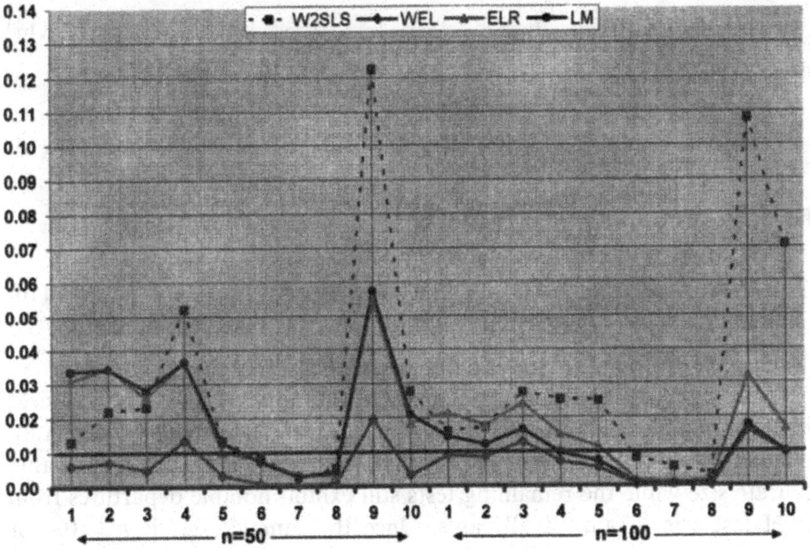

FIGURE 4 Sizes of test of moment validity-target size = 0.01.

distribution under H_0. These tests include two tests based on the usual GMM-motivated Wald-type quadratic form and use either the 2SLS or EL estimates (the W2SLS and WEL tests, respectively), as

$$W = \left(\sum_{i=1}^{n} \mathbf{h}(\mathbf{Y}_i, \mathbf{X}_i, \mathbf{Z}_i, \hat{\beta})\right)' \left[\sum_{i=1}^{n} \mathbf{h}(\mathbf{Y}_i, \mathbf{X}_i, \mathbf{Z}_i, \hat{\beta})\mathbf{h}(\mathbf{Y}_i, \mathbf{X}_i, \mathbf{Z}_i, \hat{\beta})'\right]^{-1} \times \left(\sum_{i=1}^{n} \mathbf{h}(\mathbf{Y}_i, \mathbf{X}_i, \mathbf{Z}_i, \hat{\beta})\right) \tag{12}$$

where $\mathbf{h}(\mathbf{y}_i, \mathbf{x}_i, \mathbf{z}_i, \hat{\beta}) \equiv \mathbf{z}'_i(y_i - \mathbf{x}_i\hat{\beta})$, and $\hat{\beta}$ is either the 2SLS or EL estimator of β.

The empirical likelihood ratio test (ELR) is based on the ratio of constrained (by the moment equations) and unconstrained empirical likelihood functions, as defined in Section 2.2, and is calculated as

$$ELR = -2\left(\sum_{i=1}^{n} \ln(\hat{w}_i) + n\ln(n)\right) \tag{13}$$

The Lagrange multiplier test (the LM test) is based on the asymptotic normal distribution of the Lagrange multipliers in the EL optimization problem, includes a robust estimator of the asymptotic covariance matrix of this distribution (see Imbens et al., 1998), and is defined by

$$LM = n\hat{\alpha}'[\widehat{\mathbf{cov}}(\hat{\alpha})]^{-1}\hat{\alpha} \tag{14}$$

where $\widehat{\mathbf{cov}}(\hat{\alpha}) = \mathbf{R}\mathbf{Q}^{-1}\mathbf{R}'$,

$$\mathbf{R} \equiv \sum_{i=1}^{n} \hat{w}_i \mathbf{h}(\mathbf{Y}_i, \mathbf{X}_i, \mathbf{Z}_i, \hat{\beta}_{\mathrm{EL}})\mathbf{h}(\mathbf{Y}_i, \mathbf{X}_i, \mathbf{Z}_i, \hat{\beta}_{EL})' \tag{15}$$

and

$$\mathbf{Q} \equiv \sum_{i=1}^{n} \hat{w}_i^2 \mathbf{h}(\mathbf{Y}_i, \mathbf{X}_i, \mathbf{Z}_i, \hat{\beta}_{\mathrm{EL}})\mathbf{h}(\mathbf{Y}_i, \mathbf{X}_i, \mathbf{Z}_i, \hat{\beta}_{\mathrm{EL}})' \tag{16}$$

The target size for all of these tests in the MC experiments was 0.01.

For the well-behaved data-sampling scenarios 1–4, all of the test sizes are notably different from the target test size when $n = 25$ (see Table A.4). At $n = 50$ (see Fig. 4 and Table A.4), the WEL test has reasonably accurate size while the remaining tests still exhibit notable departures from target test size in almost all cases. Once the sample sizes is $n = 100$ or larger, all of the tests exhibit test sizes in the neighborhood of the target 0.01 size, with 2SLS tending to be the least accurate on average.

Regarding the tests of moment equation validity for the ill-behaved sampling scenarios, the performance of the 2SLS Wald-based approach is seen to be relatively erratic across the data-sampling processes. However, the majority of cases are such that the realized test size is within 0.02 of the target size of 0.01 (see Fig. 4 and Table A.4). When $n \geq 50$, all three test procedures based on the EL estimator perform reasonably well in terms of approximating the target test size of 0.01 in all but one scenario, and the results are substantially more stable than for the 2SLS estimator.

4.2.2 Coverage and Power

Empirical confidence interval coverage for β_2 is presented in Fig. 5 for $n = 50$ and $n = 100$, and results for all sample sizes are presented in Table A.5. The confidence intervals were calculated in the usual way by centering the intervals at either the 2SLS or EL estimate, and then forming interval bounds at $\pm t_{1-\tau/2}$ standard deviations from the estimate. The value $t_{1-\tau/2}$ is the $100(\tau/2)\%$ quantile of the T-distribution, with degrees of freedom equal to $n-k$, and the standard deviation used in the definition of the confidence interval is estimated based on the asymptotic covariance matrix of either the 2SLS or EL estimators. This corresponds via duality to a test of the hypothesis $H_o : \beta_2 = 2$ (truth).

The target coverage is 0.99 in these experiments, and for the well-behaved data sampling scenarios 1–4, both 2SLS and EL provide reasonably accurate coverage when $n \geq 100$. On average, EL confidence

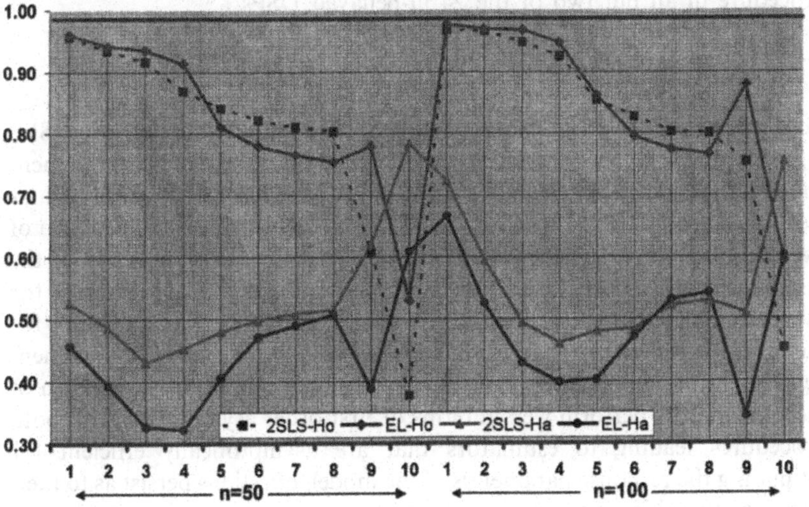

FIGURE 5 Covergae $\beta_2 = 2(H_o)$ (target = 0.99) and power $\beta_2 = 1(H_a)$.

intervals provide the more accurate coverage, with the two approaches becoming effectively indistinguishable once a sample size of $n = 250$ is achieved. In terms of power relative to the incorrect hypothesis $Ha: \beta_2 = 1$, the confidence intervals based on 2SLS are on average more powerful than the EL confidence intervals in excluding the false β_2 value, although again once $n = 250$ is achieved, the two procedures produce results that are virtually the same. Confidence interval width is, of course, one-to-one with the variance characteristics exhibited in Fig. 3 and Table A.3, and thus in all but one of the 16 well-behaved DSPs, the 2SLS procedure produces confidence intervals with slightly smaller width.

For the poorly behaved sampling scenarios 5–10 (see Fig. 5 and Table A.5) neither 2SLS nor EL provide reasonably accurate coverage for any level of n except perhaps in two cases when $n = 250$. Unlike the well-behaved DSPs, EL and 2SLS confidence intervals vacillate in terms of coverage superiority for all sample sizes, although the difference in performance between the two procedures diminishes as n increases. In terms of power relative to the incorrect hypothesis $Ha: \beta_2 = 1$, the confidence intervals based on 2SLS are in more instances more powerful than the EL confidence intervals, although the EL-based procedure is superior in one-third of the ill-behaved scenarios, and differences in power between the two procedures dissipates as n increases. Again noting that the width of confidence intervals is one-to-one with the variance characteristics exhibited in Fig. 3 and Table A.3, confidence interval width produced by the 2SLS procedure is inferior to confidence intervals generated by the EL procedure in all but two of the 24 ill-behaved DSPs.

5 OVERVIEW

In statistical models consisting of linear structural equations, the 2SLS estimator has long been the estimator of choice when the number of moment conditions–IV variables exceeded the number of unknown response parameters in the equation in question. The 2SLS estimator solves the problem of overidentification by taking a particular rank-k linear combination of the instruments. In contrast, the nontraditional EL estimator transforms the overdetermined moments problem into a set of equations that is solvable for the model parameters by imposing a functional dependence on the moment equations through choice of sample observation weights. Although both of these estimators perform well in terms of first-order asymptotics, with both procedures leading to estimators that are asymptotically efficient for estimating the response parameters of the model, questions persist as to their small sample bias and variance performance in estimation, and their coverage, interval width, and power characteristics in terms of inference.

Furthermore, in line with sampling processes often found in practice, there are questions concerning the possible estimation and inference impacts of IVs that are only weakly correlated with right-hand side endogenous variables so that response parameters are only weakly identified or determined.

Given these questions and corresponding conjectures that appear in the literature, in this chapter we have attempted to provide some empirical evidence concerning 2SLS and EL estimator performance by simulating a range of sampling processes and observing empirical repeated sampling behavior of the estimation procedures. While MC sampling results are never definitive, we feel that the base results presented in this chapter provide important insights into the relative sampling performance of these two general moment-based estimators for a range of DSPs. From this study the following finite sample performance results are suggested:

1. In well-behaved data-sampling situations both the 2SLS and EL estimators perform much like their parametric counterparts and, in terms of both estimation and inference, provide a good basis for avoiding likelihood specifications.
2. In weak instrument situations both 2SLS and EL estimators lead to sampling results exhibiting large biases and variances for estimated model parameters.
3. In weak instrument situations neither estimator performs well in terms of coverage or size relating to hypotheses about the values of model parameters.
4. When the sampling process is not well behaved and the IVs exhibit ill-posed sampling characteristics often found in practice, neither estimator dominates performance wise. In these cases the EL estimator exhibits frequent empirical superiority in many of the performance categories such as estimator precision, and parameter and prediction SEL.
5. The EL estimator faces computational problems in the weakly determined sampling designs and in some extreme cases it is difficult to obtain convergence.
6. Seemingly well-determined and viable IV sampling processes do not necessarily lead to satisfactory estimator and inference performance in small samples, as illustrated by data-sampling design 10.
7. Estimator variability and inference performance is notably effected by moderate to high correlation between the RHS (y, z) variables in the structural equation.
8. In applied work there is usually some significant level of correlation between the *estimated* y_2 variable(s) and the error process. This results in inflated estimation biases and the bias for

both estimators is in the directions of the LS estimator. This is notably the case for scenario 10.

9. The small sample performance of the EL estimator is encouraging and raises the question as to the possible normative performance of other EL-like (measures of divergence) estimators.
10. The strong showing of the LS estimator in several performance categories should perhaps reinvigorate thinking not only about how to combine moment conditions but also how best to combine estimators.

Overall, the results in this chapter underscore that the asymptotic optimality of the 2SLS estimator does not in any way guarantee performance superiority in sample sizes often found in empirical econometric practice. Furthermore, new nontraditional methods of estimation, such as the EL method, represent credible alternatives that deserve further attention from both applied and theoretical econometricians.

REFERENCES

Altonji, J. and Segal, L., 1996. Small sample bias in GMM estimation of covariance structures.*Journal of Business and Economic Statistics* **14**, 353–366.

Bound, J., Jaeger, D., and Baker, R., 1995. Problems with instrumental variable estimation when the correlation between the instruments and the endogenous variables is weak. *Journal of the American Statistical Association* **90**, 443–450.

DiCiccio, T., Hall, P., and Romano, J., 1991. Empirical likelihood is Bartlett-correctable. *Annals of Statistics* **19**, 1053–1061.

Imbens, G.W., Spady, R.H., and Johnson, P., 1998. Information theoretic approaches to inference in moment condition models. *Econometrica*, **66**, 333–357.

Judge, G., Hill, R., Griffiths, W., Lutkepohl, H., Lee, and Lee, T., 1985. *The Theory and Practice of Econometrics.* Wiley, New York.

Maddala, G.S. and Jeong, J., 1992. On the exact small sample distribution of the instrumental variable estimator. *Econometrica* **60**, 181–183.

Mittelhammer, R., Judge, G., and Miller, D., 2000. *Econometric Foundations.* Cambridge University Press, New York.

Murphy, A. and Van Der Vaart, A., 2000. On profile likelihood. *Journal of the American Statistical Association* **95**, 449–485.

Nelson, C.R., Startz, R., and Zivot, E., 1998. Valid confidence intervals and inference in the presence of weak instruments. *International Economic Review* **39**, 1119–1144.

Nelson, C.R. and Startz, R., 1990. Some further results on the exact small sample properties of the instrumental variable estimator. *Econometrica* **58**, 967–976.

Newey, W.K. and Smith, R.J., 2000. Asymptotic bias and equivalence of GMM and GEL estimators. Working paper, MIT.

Owen, A., 1988. Empirical likelihood ratio confidence intervals for a single functional. *Biometrika* **75**, 237–249.

Owen, A., 1991. Empirical likelihood for linear models. *Annals of Statistics* **19**, 1725–1747.

Owen, A., 2000. *Empirical Likelihood.* Chapman and Hall, New York.

Qin, J. and Lawless, J., 1994. Empirical likelihood and general estimating equations. *Annals of Statistics* **22**, 300–325.

Staiger, D. and Stock, J.H., 1997. Instrumental variables regression with weak instruments. *Econometrica* **65**, 557–586.

Stock, J. H. and Wright, J. H., 2000. GMM with weak identification. *Econometrica* **68**, 1055–1096.

Tauchen, G., 1986. Statistical properties of generalized method-of-moments estimators of structural parameters obtained from financial market data. *Journal of Business and Economic Statistics* **4**, 397–425.

Wald, A., 1943. Tests of statistical hypotheses concerning several parameters when the number of observations is large. *Transactions of the American Mathematical Society* **54**, 426–482.

Zivot, E., 2001. Monte Carlo designs for IV regressions with weak instruments. Working paper, University of Washington.

APPENDIX

TABLE A.1 Empirical Parameter and Prediction SEL

		Empirical parameter SEL			Empirical prediction SEL		
N	MC no.	2SLS	EL	OLS	2SLS	EL	OLS
25	1	0.623	0.797	2.456	84.756	91.852	39.900
25	2	0.877	1.004	3.068	79.536	86.863	34.740
25	3	1.469	1.563	3.846	79.186	82.492	30.550
25	4	2.233	2.067	5.113	64.555	68.164	23.041
25	5	3.111	3.028	2.494	68.020	68.454	39.637
25	6	3.706	3.621	2.515	69.012	68.764	39.818
25	7	3.701	3.683	2.415	68.402	69.953	40.189
25	8	3.954	3.911	2.423	74.140	74.846	40.223
25	9	5.515	5.098	7.558	31.142	35.579	11.104
25	10	7.521	7.668	7.582	21.582	22.934	11.223
50	1	0.294	0.392	2.389	186.108	186.720	82.972
50	2	0.436	0.536	3.051	184.688	183.488	73.420
50	3	0.647	0.746	3.772	179.075	175.114	63.439
50	4	1.029	1.100	5.042	162.575	159.391	47.684
50	5	2.298	2.217	2.393	153.343	146.956	82.628
50	6	3.390	3.144	2.361	146.821	137.962	82.888

(continued)

TABLE A.1 (Continued)

		Empirical parameter SEL			Empirical prediction SEL		
N	MC no.	2SLS	EL	OLS	2SLS	EL	OLS
50	7	3.349	3.270	2.310	138.509	135.744	84.073
50	8	3.408	3.340	2.340	139.301	137.138	84.103
50	9	3.441	3.120	7.463	90.272	94.751	23.302
50	10	6.632	6.804	7.495	48.633	47.963	23.393
100	1	0.151	0.224	2.362	388.208	373.070	169.364
100	2	0.212	0.288	3.004	382.803	369.176	149.499
100	3	0.323	0.406	3.729	377.965	356.553	129.597
100	4	0.571	0.626	5.020	368.642	348.021	97.687
100	5	1.427	1.436	2.346	331.090	313.998	169.349
100	6	2.780	2.732	2.338	290.640	281.056	168.684
100	7	3.164	3.069	2.302	276.616	268.038	171.378
100	8	3.453	3.388	2.281	283.292	277.510	172.662
100	9	1.718	1.650	7.486	271.411	268.673	47.652
100	10	5.489	5.642	7.462	120.502	114.088	47.657
250	1	0.056	0.101	2.339	988.491	942.095	428.353
250	2	0.080	0.130	2.971	981.058	928.068	377.937
250	3	0.126	0.181	3.705	981.470	916.850	328.024
250	4	0.203	0.258	4.991	960.103	899.439	248.126
250	5	0.540	0.623	2.331	903.073	855.621	427.718
250	6	1.847	1.863	2.326	739.784	712.408	427.574
250	7	3.035	2.954	2.290	666.916	647.323	432.441
250	8	3.193	3.166	2.271	663.558	656.334	434.720
250	9	0.735	0.762	7.480	885.993	857.236	120.403
250	10	2.809	3.104	7.450	404.469	386.364	120.326

TABLE A.2 Empirical Biases

		Bias: B[2]			Bias: B[1]			Sum of Biases		
N	MC no.	2SLS	EL	OLS	2SLS	EL	OLS	2SLS	EL	OLS
25	1	0.172	0.128	1.506	−0.019	0.028	−0.157	0.153	0.155	1.350
25	2	0.304	0.246	1.635	−0.101	−0.094	−0.491	0.203	0.152	1.144
25	3	0.462	0.457	1.787	−0.177	−0.111	−0.711	0.285	0.346	1.076
25	4	0.761	0.727	2.001	−0.377	−0.260	−0.995	0.385	0.467	1.006
25	5	1.246	1.231	1.519	−0.130	−0.123	−0.156	1.117	1.107	1.363

(*continued*)

TABLE A.2 (Continued)

		Bias: B[2]			Bias: B[1]			Sum of Biases		
N	MC no.	2SLS	EL	OLS	2SLS	EL	OLS	2SLS	EL	OLS
25	6	1.459	1.446	1.527	−0.144	−0.121	−0.153	1.315	1.325	1.374
25	7	1.492	1.475	1.499	−0.081	−0.072	−0.079	1.411	1.403	1.420
25	8	1.496	1.478	1.504	−0.022	−0.027	−0.020	1.474	1.452	1.484
25	9	1.698	1.602	2.347	−1.019	−0.919	−1.407	0.679	0.683	0.940
25	10	2.185	2.167	2.350	−1.313	−1.363	−1.411	0.872	0.804	0.939
50	1	0.080	0.112	1.514	0.000	0.026	−0.149	0.080	0.137	1.364
50	2	0.121	0.167	1.654	−0.039	−0.051	−0.495	0.082	0.116	1.158
50	3	0.195	0.261	1.787	−0.081	−0.032	−0.717	0.114	0.228	1.069
50	4	0.378	0.437	1.998	−0.185	−0.119	−0.997	0.192	0.317	1.000
50	5	0.963	0.987	1.515	−0.094	−0.092	−0.148	0.869	0.895	1.367
50	6	1.379	1.372	1.506	−0.144	−0.106	−0.152	1.235	1.266	1.354
50	7	1.448	1.442	1.493	−0.075	−0.057	−0.078	1.373	1.385	1.415
50	8	1.464	1.459	1.506	−0.024	−0.021	−0.027	1.440	1.438	1.479
50	9	1.217	1.167	2.338	−0.724	−0.618	−1.402	0.494	0.549	0.936
50	10	2.023	2.032	2.342	−1.213	−1.287	−1.406	0.810	0.744	0.936
100	1	0.033	0.111	1.518	−0.008	0.010	−0.154	0.026	0.122	1.363
100	2	0.070	0.143	1.651	−0.023	−0.037	−0.495	0.047	0.106	1.156
100	3	0.099	0.203	1.785	−0.036	−0.025	−0.715	0.062	0.178	1.070
100	4	0.170	0.264	1.998	−0.087	−0.046	−1.000	0.083	0.219	0.998
100	5	0.646	0.711	1.513	−0.058	−0.061	−0.149	0.588	0.649	1.364
100	6	1.223	1.248	1.510	−0.119	−0.062	−0.150	1.105	1.186	1.360
100	7	1.426	1.425	1.503	−0.076	−0.055	−0.078	1.350	1.370	1.425
100	8	1.510	1.509	1.498	−0.025	−0.026	−0.026	1.485	1.483	1.472
100	9	0.669	0.689	2.343	−0.409	−0.342	−1.408	0.261	0.347	0.936
100	10	1.779	1.799	2.341	−1.067	−1.161	−1.403	0.712	0.638	0.938
250	1	0.015	0.092	1.517	−0.001	0.000	−0.152	0.014	0.092	1.366
250	2	0.026	0.113	1.647	−0.010	−0.032	−0.494	0.015	0.081	1.154
250	3	0.032	0.137	1.784	−0.013	−0.016	−0.714	0.019	0.121	1.070
250	4	0.071	0.171	1.996	−0.036	−0.011	−0.998	0.035	0.160	0.997
250	5	0.284	0.376	1.515	−0.028	−0.030	−0.152	0.256	0.345	1.362
250	6	0.931	0.977	1.513	−0.091	−0.024	−0.149	0.840	0.953	1.364
250	7	1.437	1.439	1.507	−0.076	−0.048	−0.078	1.362	1.391	1.429
250	8	1.500	1.501	1.502	−0.029	−0.022	−0.029	1.471	1.479	1.473
250	9	0.246	0.297	2.344	−0.145	−0.111	−1.407	0.101	0.186	0.937
250	10	1.226	1.278	2.340	−0.739	−0.853	−1.404	0.488	0.425	0.936

TABLE A.3 Empirical Variances

		Variance: B[2]			Variance: B[1]			Sum of variances		
N	MC no.	2SLS	EL	OLS	2SLS	EL	OLS	2SLS	EL	OLS
25	1	0.417	0.474	0.080	0.176	0.306	0.083	0.593	0.780	0.163
25	2	0.574	0.601	0.074	0.200	0.334	0.078	0.775	0.934	0.153
25	3	0.923	0.926	0.072	0.301	0.416	0.074	1.224	1.342	0.146
25	4	1.130	1.037	0.058	0.381	0.434	0.060	1.511	1.471	0.118
25	5	1.383	1.239	0.080	0.158	0.259	0.083	1.541	1.498	0.163
25	6	1.409	1.305	0.079	0.147	0.211	0.081	1.555	1.516	0.160
25	7	1.331	1.288	0.079	0.138	0.215	0.081	1.469	1.503	0.160
25	8	1.564	1.506	0.079	0.151	0.219	0.082	1.716	1.724	0.161
25	9	1.123	1.159	0.034	0.469	0.529	0.035	1.593	1.688	0.069
25	10	0.737	0.780	0.034	0.288	0.336	0.034	1.024	1.116	0.068
50	1	0.201	0.221	0.037	0.087	0.158	0.039	0.288	0.379	0.076
50	2	0.309	0.317	0.035	0.110	0.188	0.036	0.419	0.505	0.071
50	3	0.444	0.452	0.032	0.158	0.226	0.034	0.602	0.677	0.066
50	4	0.629	0.632	0.029	0.224	0.262	0.028	0.852	0.895	0.057
50	5	1.270	1.086	0.038	0.093	0.149	0.038	1.363	1.235	0.075
50	6	1.395	1.137	0.036	0.073	0.113	0.035	1.468	1.250	0.071
50	7	1.184	1.089	0.037	0.063	0.098	0.037	1.246	1.188	0.074
50	8	1.198	1.112	0.036	0.065	0.099	0.036	1.264	1.210	0.072
50	9	1.028	0.976	0.016	0.408	0.401	0.016	1.436	1.377	0.032
50	10	0.771	0.718	0.016	0.298	0.300	0.017	1.069	1.018	0.033
100	1	0.107	0.126	0.018	0.043	0.086	0.018	0.150	0.212	0.036
100	2	0.153	0.168	0.017	0.053	0.098	0.018	0.206	0.266	0.035
100	3	0.232	0.244	0.017	0.080	0.120	0.016	0.312	0.364	0.032
100	4	0.396	0.386	0.014	0.138	0.168	0.014	0.534	0.554	0.028
100	5	0.963	0.846	0.018	0.043	0.081	0.018	1.006	0.927	0.036
100	6	1.228	1.112	0.018	0.042	0.060	0.018	1.269	1.172	0.036
100	7	1.091	0.984	0.018	0.034	0.051	0.018	1.125	1.035	0.037
100	8	1.143	1.069	0.018	0.028	0.041	0.018	1.171	1.111	0.036
100	9	0.801	0.768	0.008	0.302	0.291	0.007	1.103	1.058	0.015
100	10	0.871	0.764	0.008	0.314	0.295	0.008	1.185	1.059	0.015
250	1	0.039	0.054	0.007	0.016	0.038	0.007	0.056	0.092	0.014
250	2	0.058	0.069	0.007	0.022	0.047	0.007	0.079	0.116	0.013
250	3	0.094	0.104	0.006	0.032	0.058	0.007	0.125	0.162	0.013
250	4	0.144	0.153	0.005	0.053	0.076	0.006	0.197	0.229	0.011
250	5	0.439	0.438	0.007	0.019	0.042	0.007	0.458	0.481	0.014
250	6	0.950	0.872	0.007	0.022	0.035	0.007	0.973	0.907	0.014
250	7	0.951	0.859	0.007	0.013	0.023	0.007	0.964	0.882	0.014
250	8	0.931	0.894	0.007	0.011	0.018	0.007	0.943	0.912	0.014
250	9	0.470	0.470	0.003	0.184	0.193	0.003	0.654	0.662	0.006
250	10	0.559	0.537	0.003	0.201	0.206	0.003	0.760	0.743	0.006

TABLE A.4 Empirical Test Sizes for Nominal Size 0.01 Chi-Square Moment Validity Test

		Test size for moment validity test			
N	MC no.	W2SLS	WEL	ELR	LM
25	1	0.015	0.003	0.054	0.121
25	2	0.022	0.005	0.072	0.134
25	3	0.026	0.002	0.072	0.131
25	4	0.037	0.008	0.086	0.142
25	5	0.008	0.002	0.026	0.062
25	6	0.004	0.000	0.016	0.043
25	7	0.004	0.001	0.016	0.044
25	8	0.004	0.001	0.013	0.036
25	9	0.046	0.016	0.101	0.160
25	10	0.008	0.001	0.029	0.064
50	1	0.013	0.006	0.031	0.034
50	2	0.022	0.007	0.035	0.034
50	3	0.023	0.005	0.026	0.028
50	4	0.052	0.014	0.036	0.036
50	5	0.013	0.003	0.010	0.013
50	6	0.008	0.001	0.008	0.007
50	7	0.003	0.000	0.003	0.003
50	8	0.005	0.001	0.003	0.004
50	9	0.122	0.020	0.055	0.057
50	10	0.028	0.003	0.018	0.021
100	1	0.016	0.009	0.021	0.014
100	2	0.017	0.008	0.018	0.012
100	3	0.027	0.013	0.024	0.016
100	4	0.025	0.007	0.015	0.010
100	5	0.025	0.005	0.011	0.007
100	6	0.008	0.000	0.002	0.001
100	7	0.006	0.001	0.001	0.000
100	8	0.003	0.000	0.002	0.001
100	9	0.108	0.016	0.032	0.018
100	10	0.071	0.010	0.017	0.010
250	1	0.011	0.009	0.012	0.005
250	2	0.018	0.013	0.018	0.007
250	3	0.019	0.012	0.016	0.008
250	4	0.014	0.009	0.012	0.005
250	5	0.030	0.010	0.014	0.008
250	6	0.014	0.006	0.006	0.003
250	7	0.010	0.004	0.005	0.001

(*continued*)

Table A.4 (Continued)

		Test size for moment validity test			
N	MC no.	W2SLS	WEL	ELR	LM
250	8	0.006	0.001	0.002	0.000
250	9	0.060	0.016	0.021	0.010
250	10	0.150	0.023	0.031	0.014

W2SLS and WEL are Wald-type tests, ELR is based on the empirical likelihood ratio, and LM utilizes the Lagrange multipliers of the EL-constrained optimization problem.

Table A.5 Confidence Interval Coverage (Target = 0.99) for Containing True $\beta_2 = 2$ and Power in Excluding Incorrect $\beta_2 = 1$

		Coverage		Power	
N	MC no.	2SLS	EL	2SLS	EL
25	1	0.934	0.927	0.446	0.387
25	2	0.902	0.899	0.444	0.353
25	3	0.855	0.866	0.456	0.350
25	4	0.754	0.800	0.524	0.384
25	5	0.823	0.747	0.497	0.472
25	6	0.798	0.712	0.502	0.511
25	7	0.818	0.717	0.478	0.501
25	8	0.820	0.731	0.482	0.500
25	9	0.472	0.610	0.705	0.522
25	10	0.361	0.461	0.788	0.669
50	1	0.956	0.960	0.525	0.456
50	2	0.934	0.942	0.486	0.393
50	3	0.917	0.935	0.429	0.327
50	4	0.869	0.914	0.452	0.323
50	5	0.841	0.811	0.480	0.404
50	6	0.821	0.780	0.498	0.470
50	7	0.810	0.766	0.509	0.489
50	8	0.804	0.754	0.514	0.506
50	9	0.608	0.780	0.620	0.388
50	10	0.377	0.530	0.784	0.608
100	1	0.968	0.976	0.723	0.666
100	2	0.966	0.970	0.595	0.525
100	3	0.948	0.968	0.494	0.429
100	4	0.925	0.947	0.460	0.398

(*continued*)

TABLE A.5 (Continued)

		Coverage		Power	
N	MC no.	2SLS	EL	2SLS	EL
100	5	0.853	0.861	0.479	0.402
100	6	0.826	0.795	0.484	0.471
100	7	0.802	0.775	0.521	0.530
100	8	0.800	0.767	0.528	0.541
100	9	0.754	0.878	0.509	0.344
100	10	0.452	0.593	0.754	0.600
250	1	0.979	0.982	0.951	0.938
250	2	0.977	0.978	0.862	0.840
250	3	0.966	0.971	0.746	0.730
250	4	0.964	0.967	0.599	0.595
250	5	0.923	0.930	0.452	0.404
250	6	0.841	0.810	0.498	0.518
250	7	0.799	0.756	0.553	0.584
250	8	0.784	0.755	0.548	0.584
250	9	0.896	0.926	0.462	0.421
250	10	0.579	0.690	0.686	0.635

7

Testing for Unit Roots in Semiannual Data

Sandra G. Feltham
Ministry of Human Resources, Government of British Columbia, Victoria, British Columbia, Canada

David E. A. Giles
University of Victoria, Victoria, British Columbia, Canada

1 INTRODUCTION

There is now a well-established literature relating to the problem of testing for nonstationarity in seasonal economic time-series data. For example, some of the earlier literature on this topic can be found in Hylleberg (1992), and a more recent overview is given by Franses (1997). This issue is of considerable importance as the distinction needs to be drawn between unit roots at the zero frequency, and unit roots at some or all of the seasonal frequencies in the case of nonannual data. An incorrect identification of the nature of such unit roots would lead to inappropriate filtering of the series prior to its use in regression analysis, say, as well as inadequate testing for possible cointegration between one such series and another.

A seasonal series is one that has a spectrum with distinct peaks at the seasonal frequencies, $\theta=(2\pi j/s)$, $j=1, 2, \ldots, s-1$, where s is the number of seasons in the year. The frequencies of interest with quarterly data, for example, are at 0, 1/4, 1/2, and 3/4 cycles, or $\theta=0$, $\pi/2$, π, and $3\pi/2$. An integrated series is one that has an infinite mass in its spectrum at some frequency, θ. In contrast, the spectrum of a stationary series is finite, but nonzero, at all frequencies. If a seasonal series has a unit root at *all* of its frequencies, it is said to be "seasonally integrated", or SI(s), and it needs to be s-differenced to make it stationary. Early contributions (e.g., Dickey et al., 1984) to the problem of testing for nonstationarity in seasonal time series considered only a null of SI(s), and an alternative hypothesis of stationarity. That is, they did not consider the possibility of unit roots at only the seasonal (nonzero) frequencies, and neither did they allow for unit roots at a subset of the seasonal frequencies. The more recent test proposed by Kunst (1997) also falls into this category. In the case of quarterly data, this shortcoming was rectified partially by Osborne et al. (1988), and more fully with the tests proposed by Hylleberg et al. (hereafter HEGY) (1990) and Ghysels et al. (1994). Franses (1991) and Beaulieu and Miron (1993) provided a similar testing framework for monthly data.*

Economic time-series data recorded only on a semiannual basis are quite common. For example, many companies report their financial statements on this basis, at least on a provisional basis. The purpose of this chapter is to investigate the properties of a HEGY testing framework in the case of semiannual data, and to compare it with conventional Dickey–Fuller (DF) (1981) testing for unit roots at the zero frequency in such time series. In Section 2 we describe the testing procedure, and Section 3 discusses the asymptotic distributions of our test statistics. Section 4 deals with the finite-sample percentiles for these tests, and reports some small-sample power results. The application of the semiannual HEGY tests is illustrated in Section 5, and some concluding remarks and recommendations appear in Section 6.

2 A TESTING FRAMEWORK

2.1 Background

In this section the framework for applying HEGY-type unit root tests is outlined for the case of semiannual data. Recall the autoregressive model for a stochastic seasonal process, x_t:

$$\Psi(B)x_t = \varepsilon_t \tag{1}$$

where $\varepsilon_t \sim \text{iid}(0, \sigma^2)$, $\Psi(B)$ is a polynomial in the backshift operator, B, and the roots of $\Psi(B) = 0$ determine whether or not the series is stationary. For the purposes of the following derivation the data-generating

*See also Franses and Hobijn (1997). Smith and Taylor (1998) provide a recent elegant discussion of the HEGY tests for quarterly data.

process of x_t is assumed to be free of any deterministic components. Following HEGY (1990) and Beaulieu and Miron (1993), $\Psi(B)$ can be expressed in terms of elementary polynomials and a remainder:

$$\Psi(B) = \sum_{k=1}^{2} \lambda_k \Delta(B) \frac{1 - \delta_k(B)}{\delta_k(B)} + \Delta(B)\Psi^*(B) \tag{2}$$

where $\delta_k(B) = 1 - (1/\theta_k)B$, k = 1, 2; $\Delta(B) = \prod_{k=1}^{p} \delta_k(B)$; $\lambda_k = \Psi(\theta_k)/\prod_{i=k} \times \delta_j(\theta_k)$; and $\Psi^*(B)$ is a remainder with roots outside the unit circle.

The θ_k's are the unit roots. In the case of semiannual data we have $\theta_1 = 1$ and $\theta_2 = -1$, and so $\delta_1(B) = (1 - B)$, $\delta_2(B) = (1 + B)$, and $\Delta(B) = (1 - B^2)$. Substituting these into Eq. (2) yields

$$\Psi(B) = \lambda_1(B)(1 + B) + \lambda_2(-B)(1 - B) + (1 - B^2)\Psi^*(B) \tag{3}$$

Next, let $\pi_1 = -\lambda_1$ and $\pi_2 = -\lambda_2$, and substitute the right-hand side of Eq. (3) into the autoregression equation $\Psi(B)x_t = \varepsilon_t$, so we have

$$-\pi_1(B)(1 + B)x_t - \pi_2(-B)(1 - B)x_t + (1 - B^2)\Psi^*(B)x_t = \varepsilon_t$$

This can be rewritten in the form of a regression equation on which the tests can be based:

$$\Psi^*(B)z_{3t} = \pi_1 z_{1t-1} + \pi_2 z_{2t-1} + \varepsilon_t \tag{4}$$

where

$$\begin{aligned} z_{1t} &= (1 + B)x_t = x_t + x_{t-1} \\ z_{2t} &= -(1 - B)x_t = -(x_t - x_{t-1}) \\ z_{3t} &= (1 - B^2)x_t = x_t - x_{t-2} \\ &(t = 1, 2, 3, \ldots, n) \end{aligned} \tag{5}$$

2.2 Implementation of Tests

To apply the unit root tests with semiannual data, Eq. (4) is estimated by ordinary least squares (OLS). Using a "t-test," the null hypothesis of a unit root at the zero frequency ($\pi_1 = 0$) is tested against the one-sided alternative hypothesis $\pi < 0$, which is equivalent to testing $\Psi(1) = 0$ versus the alternative of stationarity, $\Psi(1) > 0$. Similarly, the hypothesis of a unit root at the π frequency ($\pi_2 = 0$) is tested against the alternative $\pi_2 < 0$. In addition, the "F-statistic" for $\pi_1 = \pi_2 = 0$ may be used to test if unit roots exist at both frequencies simultaneously: i.e., if the series is "seasonally integrated." The asymptotic and finite-sample percentiles of these nonstandard "t" and "F" statistics are considered below, and various critical values are given in Table 2.

These tests are evaluated under the assumption that $\Psi^*(B)=1$. However, as Beaulieu and Miron (1993) point out, if $\Psi(B)$ is of order greater than S, then $\Psi^*(B)\neq 1$ and the fitted model must be "augmented" with lagged values of the dependent variable as extra regressors in order to whiten the errors. This follows from the findings of Said and Dickey (1984) in the context of the familiar (augmented) Dickey–Fuller (ADF) tests for unit roots at the zero frequency. These augmentations will introduce a finite-sample distortions in the null distributions of the "t" and "F" tests that will die out asymptotically, provided that the correct number of lags is added. In particular, the number of such augmentation terms, q, must be allowed to increase as the sample size increases. Beaulieu and Miron (1990) investigated these finite-sample distortions in the case of monthly data, and found that the true sizes (significance levels) of the individual unit root "t" tests were no higher than those implied by the asymptotic critical values. They therefore suggest that this distortion may cause underrejection of the unit root hypotheses in small samples with such data.

3 ASYMPTOTIC NULL DISTRIBUTIONS

3.1 HEGY-Type Tests for Semiannual Data

This section presents the asymptotic null distributions for the half-yearly data tests under consideration in this paper. Fuller (1976) proves that, when $\rho=-1$, the limiting null distributions of $\hat{\rho}$ and the associated test statistics in a simple random walk model, a random walk with drift, and a random walk with drift and trend, are simply the mirror image of the limiting distributions when $\rho=1$ (Dickey and Fuller, 1979). Moreover, HEGY (1990) show that the analysis of Chan and Wei (1988) can be used to extract the asymptotic distribution theory for the HEGY "t" tests from the results of Dickey and Fuller (1979) and Fuller (1976).

The asymptotic null distributions for our two "t" statistics follow directly from the analyses* of HEGY (1990) and of Beaulieu and Miron (1993). The latter provide the derivation of the asymptotic distributions under the assumption that ε_t is a martingale difference† sequence with constant variance. A sequence of random scalars, $\{\varepsilon_t\}_{t=1}^{\infty}$, is a martingale difference sequence if $E(\varepsilon_t)=0$ for all t and $E(\varepsilon_t|\varepsilon_{t-1},\ldots,\varepsilon_1)=0$ for $t=2, 3,\ldots,n$ (Hamilton, 1994, p. 189).

*See also Phillips (1987) and Stock (1988).

†The assumption of a martingale difference sequence is stronger than that of serially uncorrelated errors, but weaker than independence (Hamilton, 1994, p.190).

When the unit root tests are applied in the context of a regression model without any deterministic terms the statistic t_1 has the same limiting null distribution as the familiar DF "t-test" statistic:

$$t_1 \xrightarrow{L} \frac{\left(\int_0^1 [W_1(r)]^2 \mathrm{d}W_1\right)}{\left(\int_0^1 [W_1(r)]^2 \mathrm{d}r\right)^{1/2}} = \frac{\frac{1}{2}\{[W(1)]^2 - 1\}}{\left\{\int_0^1 [W(r)]^2 \mathrm{d}r\right\}^{1/2}} \tag{6}$$

where $W(r)$ denotes a standard Brownian motion and $\int_0^1$ is a stochastic integral (Ghysels et al., 1994; Hamilton, 1994). Further, the asymptotic null distribution of t_2 is the mirror image of the usual DF distribution; i.e., the negative of t_2 has asymptotic distribution in Eq. (6). Essentially, these results follow from the fact that z_{1t} and z_{2t} in Eq. (5) are asymptotically uncorrelated.

Ghysels et al. (1994) show that this relationship with the usual DF "t-tests" also holds in finite samples. By the same arguments, this will also be true for semiannual data. Further, from the results of Engle et al., (1993), they prove that, for quarterly seasonal data the asymptotic null distributions for "F" statistics for testing either (1) that the series is seasonally integrated or (2) that there are unit roots at *all* seasonal frequencies, are the same as the limiting distributions of the sum of the appropriate squared "t-statistics." Similarly, the asymptotic null distribution of our F_{12} statistic, for seasonally integrated half-yearly data, follows directly from the limiting distributions of the sum of the squares of t_1 and t_2. So, in the case of no drift, no trend, and no seasonal dummy variable in the fitted regression we have

$$F_{12} \xrightarrow{L} \frac{1}{2}\left\{\frac{\left(\int_0^1 [W_1(r)]^2 \mathrm{d}W_1\right)^2}{\left(\int_0^1 [W_1(r)]^2 \mathrm{d}r\right)} + \frac{\left(\int_0^1 [W_2(r)]^2 \mathrm{d}W_2\right)^2}{\left(\int_0^1 [W_2(r)]^2 \mathrm{d}r\right)}\right\} \tag{7}$$

where $W_i(r)$ for $i=1, 2$ denote independent standard Brownian motions (Ghysels et al., 1994).

Next, consider the asymptotic null distributions of the tests when deterministic terms are added to the fitted regression equation [Eq. (4)]. The general effects of the various deterministic terms are discussed and then the actual distributions are presented following the notation of Beaulieu and Miron (1993). The addition of drift or trend terms affects the asymptotic null distribution of t_1 because these components have their spectral mass at the zero frequency (HEGY, 1990). However, the asymptotic distribution of t_2 is independent of the drift and trend terms as z_{2t} is asymptotically orthogonal to terms that are not periodic (Beaulieu and Miron, 1993). Recall that the asymptotic distribution of the F_{12}

statistic is related to those of the squared t_1 and squared t_2 statistics. Thus, the asymptotic null distribution of the F_{12} statistic will also change with the addition of deterministic terms, as the distribution of t_1 changes when a drift and/or trend is added to the fitted regression

In the case of the "t" statistics, the effect of adding a seasonal dummy variable into the "HEGY regression" is the reverse of that of adding constant and trend components. Once a constant term is included in the regression, the addition of a seasonal dummy variable does not affect the asymptotic null distribution of t_1 any further. The subsequent addition of a seasonal dummy variable does, however, change the asymptotic null distribution of t_2, and so the asymptotic properties of F_{12} also change in this case. Again, this all follows for reasons analogous to those in the case of quarterly data. Some further insights into these, and other relationships between the tests, in the case of quarterly data are given by Smith and Taylor (1998).

The forms of the various asymptotic distributions can now be summarized. Let the distributions of t_1 and t_2, when deterministic terms are excluded from the regression equation, be denoted as

$$t_1 \xrightarrow{L} \frac{N_1^0}{D_1^0}; \qquad t_2 \xrightarrow{L} \frac{N_2^0}{D_2^0} \tag{8}$$

Following the notation of Beaulieu and Miron (1993) N_1^x and N_2^x represent the part of the numerator that differs from that in the above t_1 and t_2 distributions, respectively. Similarly D_1^x and D_2^x represent the part of the denominator that differs. The four variations of deterministic terms in the fitted regression model are represented by $x = \mu$, τ, ξ, and $\xi\tau$, corresponding to regressions with a constant, a constant and trend, a constant and seasonal dummy, and finally a constant, trend, and seasonal dummy. Thus, the general form of the limiting null distributions for the "t-statistics" can be written as

$$\begin{aligned} t_1 &\xrightarrow{L} \frac{\left(\int_0^1 [W_1(r)]^2 \, dW_1\right)^2 + N_1^X}{\left(\int_0^1 [W_1(r)]^2 \, dr + D_1^X\right)^{1/2}} \\ t_2 &\xrightarrow{L} \frac{\left(\int_0^1 [W_2(r)]^2 \, dW_2\right)^2 + N_2^X}{\left(\int_0^1 [W_2(r)]^2 \, dr + D_2^X\right)^{1/2}} \end{aligned} \tag{9}$$

The five possible asymptotic null distributions for each of the test statistics are summarized in Table 1. Only the parts of the numerator and denominator that differ from the no-drift, no-trend model, are given for the "t" statistics. This is done to provide visual representation of the distributions that are the same. The full asymptotic null distribution of the F_{12} statistic is given in terms of these components.

3.2 Dickey-Fuller Tests and Semiannual Data

Several properties of the DF "t-test" have been analyzed in the context of seasonal data. For example, HEGY (1990) showed it to have asymptotic equivalence to their t_1 statistic in the "no drift/no trend" case. Ghysels et al. (1994) demonstrated that when unit roots exist at seasonal frequencies the DF "t-test" is still a valid test for unit roots at the zero frequency, provided that the usual DF regression is appropriately "augmented" with $l \geq s-1$ lagged values of the differenced series. They use Monte Carlo simulations to see how the existence of seasonal unit roots influences the finite-sample size and power of the test. Their results show that, in the case of quarterly data, when $l < s-1$ the size of the DF test is severely distorted. When $l \geq s-1$ the size is close to its nominal size, provided that the data-generating process is free of negative moving average components.

Although we have not explicitly explored here these characteristics of the DF "t-test" in the context of semiannual data, intuitively it is clear that corresponding results to those above will hold (with $s=2$).

3.3 Moving-Average Errors

Although the HEGY test provides for identification of roots at different frequencies, it has been found to suffer from size distortions from negative MA components in the data-generating process. In particular, for the data-generating process:

$$Y_t = \alpha_d Y_{t-d} + u_t \tag{10}$$

where $u_t = \varepsilon_t + \theta_1 \varepsilon_{t-1} + \theta_4 \varepsilon_{t-4}$, and $\theta_1 = -0.9$, the near cancellation of (1−B) in the autoregressive component of u_t makes the unit root at the zero frequency difficult to detect. This also explains why the DF test suffers from size distortions with negative MA components, because it has the same distribution as the HEGY t_1 test when $l \geq s-1$. This is not a trivial problem as such situations arise frequently in practice.

A large bias in size or very low power was also found to occur when seasonal dummies are in the data-generating process, but are not included in the regression. This provides justification for including deterministic terms or additional lags even though they may be irrelevant and thus may reduce the power of the test.

TABLE 1 Effects of Deterministic Terms on Asymptotic Distributions of Semiannual Unit Root Tests

Model	t_1	t_2	F_{12}
ND, NT, NS	$\frac{N_1^0}{D_1^0}$	$\frac{N_2^0}{D_2^0}$	$\frac{N_{12}^0}{D_{12}^0} = \left(\frac{N_1^0}{D_1^0}\right)^2 + \left(\frac{N_2^0}{D_2^0}\right)^2$
D, NT, NS	$N_1^{\mu} = -W_1(1)\int_0^1 W_1(1)\mathrm{d}r$	$N_2^{\mu} = 0$	$\left(\frac{\left(\int_0^1 [W_1(r)]^2 \mathrm{d}W_1\right)^2 + N_1^{\mu}}{\left(\int_0^1 [W_1(r)]^2 \mathrm{d}r + D_1^{\mu}\right)^{1/2}}\right)^2 + \left(\frac{N_2^0}{D_2^0}\right)^2$
	$D_1^{\mu} = -\left(\int_0^1 W_1(1)\mathrm{d}r\right)^2$	$D_2^{\mu} = 0$	
D, T, NS	$N_1^{\tau} = -4W_1(1)\int_0^1 W_1(r)\mathrm{d}r$ $+6\int_0^1 W_1(r)\mathrm{d}r \int_0^1 r\mathrm{d}W_1(r)$ $-12\int_0^1 rW_1(r)\mathrm{d}r \int_0^1 r\mathrm{d}W_1(r)$ $+6W_1(1)\int_0^1 rW_1(r)\mathrm{d}r$	$N_2^{\tau} = 0$	$\left(\frac{\left(\int_0^1 [W_1(r)]^2 \mathrm{d}W_1\right)^2 + N_1^{\tau}}{\left(\int_0^1 [W_1(r)]^2 \mathrm{d}r + D_1^{\tau}\right)^{1/2}}\right)^2 + \left(\frac{N_2^0}{D_2^0}\right)^2$
	$D_1^{\tau} = -4\left(\int_0^1 W_1(r)\mathrm{d}r\right)^2 - 12\left(\int_0^1 rW_1(r)\mathrm{d}r\right)^2$ $+12\int_0^1 W_1(r)\mathrm{d}r \int_0^1 rW_1(r)\mathrm{d}r$	$D_2^{\tau} = 0$	

D, NT, S	$N_1^{\xi} = N_1^{\mu}$	$N_2^{\xi} = -W_2(1)\int_0^1 W_2(1)\,dr$	$\left(\frac{\left(\int_0^1 [W_1(r)]^2 dW_1\right)^2 + N_1^{\mu}}{\left(\int_0^1 [W_1(r)]^2 dr + D_1^{\mu}\right)^{1/2}}\right)^2 +$
	$D_1^{\xi} = D_1^{\mu}$	$D_2^{\xi} = -(\int_0^1 W_2(1)\,dr)^2$	$\left(\frac{\left(\int_0^1 [W_2(r)]^2 dW_2\right)^2 + N_2^{\xi}}{\left(\int_0^1 [W_2(r)]^2 dr + D_2^{\xi}\right)^{1/2}}\right)^2$
D, T, S	$N_1^{\xi\tau} = N_1^{\tau}$	$N_2^{\xi\tau} = N_2^{\xi}$	$\left(\frac{\left(\int_0^1 [W_1(r)]^2 dW_1\right)^2 + N_1^{\tau}}{\left(\int_0^1 [W_1(r)]^2 dr + D_1^{\tau}\right)^{1/2}}\right)^2 +$
	$D_1^{\xi\tau} = D_1^{\tau}$	$D_2^{\xi\tau} = D_2^{\xi}$	$\left(\frac{\left(\int_0^1 [W_2(r)]^2 dW_2\right)^2 + N_2^{\xi}}{\left(\int_0^1 [W_2(r)]^2 dr + D_2^{\xi}\right)^{1/2}}\right)^2$

4 FINITE-SAMPLE RESULTS

4.1 Critical Values

We have investigated the small-sample distributions of our three semiannual unit root tests under both the null and alternative hypotheses using Monte Carlo simulations. All of these simulations were implemented using SHAZAM (1997) on a DEC Alpha 3000 workstation. The data-generating process used was $x_t = x_{t-2} + \varepsilon_t$, where the ε_t values were generated by the normal random number generator in SHAZAM. Five different variations of the basic HEGY auxiliary regression were fitted to the generated data: drift, seasonal dummy variable, and trend (D, S, T); drift, seasonal dummy, and no trend (D, S, NT); drift, no seasonal dummy variable, and trend (D, NS, T); drift, no seasonal dummy variable, and no trend (D, NS, NT); and no drift, no seasonal dummy variable, and no trend (ND, NS, NT)

The initial value for y_t was set to zero and then the first 200 observations were dropped to "wash out" the effect of this initial value. All simulations were based on 20,000 repetitions. Critical values associated with the 1st, 5th, and 10th percentiles of the underlying distributions were generated for the "t_1 test" of H_0: $\pi_1 = 0$ and for the "t_2 test" of H_0: $\pi_2 = 0$. Corresponding critical values were calculated for the "F_{12} test" of H_0: $\pi_1 = \pi_2 = 0$ by determining the 90th, 95th, and 99th percentiles of the empirical distributions. These results are presented in Table 2.

There is no discernable change in the t_2 critical values when a drift and/or trend are added to the fitted regressions, as was indicated by the asymptotic distribution theory in Section 3. In contrast, the critical values for t_1 change when either a drift or trend is added to the fitted model on which the tests are based. The *subsequent* addition of a seasonal dummy variable does not then change the finite sample distribution of t_1, but it does change that of t_2. As predicted by the asymptotic distribution theory, the "F_{12} statistic" critical values change with each variation to the deterministic components in the fitted regression, as its distribution is the same as the distribution of the sum of the squares of t_1 and t_2.

It is also worth noting the very close accordance between the no-drift, no-trend values for t_1 in Table 2, the corresponding DF critical values of MacKinnon (1991), and the HEGY (1990) π_1 values. For example, with a 10% significance level and $n = 200$, the critical values are −1.6163 in Table 2, −1.6165 from MacKinnon, and −1.62 from HEGY. The critical values for t_2 (i.e., for testing H_0: $\pi_2 = 0$) in Table 2 can be compared with the HEGY (1990) critical values for the fitted model with no deterministic terms. For the 10% level and $n = 200$ the two statistics are −1.6112 and −1.61, respectively.

The results in Table 2 are based on a data-generating process (d.g.p.) of $x_t = x_{t-2} + \varepsilon_t$. That is, unit roots are present at both the zero and π frequencies—the semi-annual series is "seasonally integrated." An alternative approach is to consider a d.g.p. that has a unit root at only one of these frequencies, but then still base the unit roots tests on the HEGY-type integrating regression as in Eq. (4). Critical values were also generated for these tests. In the test for a unit root only at the zero frequency, a simple random walk was used as the d.g.p.; in the second case the d.g.p. had a unit root at the π frequency, but not at the zero frequency. For completeness, these results appear in Table 3.

4.2 Powers of the Tests

The Monte Carlo experiment was also used to simulate the size-adjusted powers of the test statistics developed in this chapter, for the following three fitted regressions: drift, seasonal dummy variable, and trend (D, S, T); drift, seasonal dummy variable, and no trend (D, S, NT); and drift, no seasonal dummy, with trend (D, NS, T).

Using the d.g.p. of Table 2, namely, $x_t = \rho x_{t-2} + \varepsilon_t$, the size-adjusted powers of the tests of $\pi_1 = 0$, $\pi_2 = 0$, and $\pi_1 = \pi_2 = 0$ were simulated for $\rho = [0.0(0.1)1.0]$ and various sample sizes (n). Each power curve was generated using 20,000 repetitions. The results of these empirical power simulations for the t_1, t_2, and F_{12} tests are given in Tables 4–6, for the case where the d.g.p. is as in Table 2; and in Tables 7 and 8 for the case where the d.g.p. is as in Table 3. Illustrative plots appear in Figs. 1–3 and Figs. 4 and 5, respectively.

These results show that the power curves for large sample sizes have the basic expected shape. That is, they move towards 100% as ρ moves away from 1. There is reasonable power when $n = 100$, but very low size-adjusted power when the sample size is small. This is no surprise given other results for similar such unit root tests. The patterns in the results as the sample size is increased reflect the consistency of the tests. Changing the significance level of the test yields the usual trade off between the size and power. As expected, the tests based on auxiliary regressions without a drift or trend have more power than their counterparts with deterministic terms, because the d.g.p. does not include deterministic terms.

5 SOME APPLICATIONS

The application of the testing procedures discussed in this chapter is illustrated here for four quite different semiannual economic time series. In each case, we compare the results obtained when the possibility of seasonal

TABLE 2 Critical Values for the HEGY-Type Unit Root Tests; d.g.p.: $(1-B^2)x_t=\varepsilon_t$

Model	n	$\pi_1=0$			$\pi_2=0$			$\pi_1=\pi_2=0$		
		1%	5%	10%	1%	5%	10%	99%	95%	90%
D, S, T										
	20	−4.3597	−3.5654	−3.1873	−3.7030	−2.9599	−2.5935	14.9449	9.8711	7.9402
	50	−4.1381	−3.4776	−3.1532	−3.5644	−2.8922	−2.5810	11.6604	8.6479	7.3220
	100	−4.0264	−3.4291	−3.1358	−3.4967	−2.8749	−2.5732	11.2051	8.3778	7.1243
	200	−4.0109	−3.4256	−3.1286	−3.4846	−2.9005	−2.5823	10.8262	8.2567	7.0817
	5000	−4.0210	−3.4664	−3.1760	−3.4046	−2.8280	−2.5373	10.4656	8.1290	7.0575
	10000	−4.0463	−3.4625	−3.1837	−3.3761	−2.8242	−2.5255	10.6561	8.2164	7.0964
D, S, NT										
	20	−3.6986	−2.9087	−2.5561	−3.7253	−2.9347	−2.5807	11.5547	7.7163	6.1039
	50	−3.5143	−2.8779	−2.5713	−3.5388	−2.8887	−2.5741	9.5589	6.9001	5.7132
	100	−3.4461	−2.8687	−2.5585	−3.4909	−2.8738	−2.5714	9.2908	6.7591	5.6672
	200	−3.4391	−2.8578	−2.5643	−3.4787	−2.9034	−2.5850	9.0919	6.7064	5.5854
	5000	−3.4903	−2.9109	−2.5926	−3.4038	−2.8276	−2.5375	8.6661	6.5033	5.5156
	10000	−3.4773	−2.8959	−2.6009	−3.3763	−2.8243	−2.5256	8.7691	6.5342	5.5520
D, NS, T										
	20	−4.2762	−3.4796	−3.1092	−2.5974	−1.7879	−1.4403	11.5195	7.5930	6.0760
	50	−4.1008	−3.4568	−3.1371	−2.5615	−1.8881	−1.5436	9.6565	6.8921	5.7726
	100	−4.0231	−3.4295	−3.1281	−2.5731	−1.9098	−1.5720	9.2573	6.8333	5.7206
	200	−3.9998	−3.4220	−3.1246	−2.5131	−1.9334	−1.6107	8.8741	6.7787	5.6860

	5000	−4.0211	−3.4659	−3.1765	−2.4668	−1.8768	−1.5657	8.8779	6.8213	5.8105
	10000	−4.0463	−3.4628	−3.1838	−2.5147	−1.8792	−1.5568	9.0445	6.8076	5.8415
D, NS, NT										
	20	−3.6524	−2.8802	−2.5381	−2.6349	−1.8600	−1.4787	8.7567	5.6446	4.4119
	50	−3.5091	−2.8748	−2.5667	−2.5841	−1.9209	−1.5578	7.3945	5.2131	4.2230
	100	−3.4477	−2.8775	−2.5592	−2.5794	−1.9143	−1.5798	7.3461	5.1094	4.1850
	200	−3.4490	−2.8577	−2.5624	−2.5216	−1.9385	−1.6111	7.2181	5.1364	4.1707
	5000	−3.4906	−2.9113	−2.5927	−2.4656	−1.8770	−1.5657	6.9546	5.0897	4.1968
	10000	−3.4772	−2.8956	−2.6007	−2.5147	−1.8791	−1.5567	7.0410	5.0871	4.1873
ND, NS, NT										
	20	−2.7309	−1.9283	−1.5399	−2.7428	−1.9265	−1.5310	6.0730	3.6772	2.7541
	50	−2.6158	−1.9105	−1.5636	−2.6117	−1.9351	−1.5661	5.3603	3.3859	2.5819
	100	−2.6042	−1.9164	−1.5943	−2.5846	−1.9196	−1.5866	5.0504	3.2839	2.5656
	200	−2.5915	−1.9499	−1.6163	−2.5291	−1.9404	−1.6112	6.0312	3.6146	2.7172
	5000	−2.6200	−1.9774	−1.6621	−2.4661	−1.8767	−1.5657	4.9073	3.2119	2.5234
	10000	−2.6273	−1.9975	−1.6690	−2.5145	−1.8792	−1.5567	4.8699	3.2301	2.5243

TABLE 3 Critical Values for the Individual Seasonal Unit Root Tests

		d.g.p.: $(1-B)x_t=\varepsilon_t$			d.g.p.: $(1+B)x_t=\varepsilon_t$		
		$\pi_1=0$			$\pi_2=0$		
Model	N	1%	5%	10%	1%	5%	10%
D, S, T							
	20	−4.5573	−3.6889	−3.2840	−3.7952	−2.9836	−2.6175
	50	−4.1575	−3.5107	−3.1967	−3.5812	−2.8933	−2.5852
	100	−4.0394	−3.4492	−3.1518	−3.5009	−2.8949	−2.5741
	200	−4.0077	−3.4346	−3.1314	−3.5119	−2.9053	−2.5844
	5000	−4.0074	−3.4647	−3.1744	−3.4046	−2.8304	−2.5388
	10,000	−4.0408	−3.4647	−3.1822	−3.3689	−2.8246	−2.5250
D, S, NT							
	20	−3.7920	−2.9720	−2.6209	−3.7809	−3.0182	−2.6364
	50	−3.5273	−2.9010	−2.5920	−3.5849	−2.9001	−2.5903
	100	−3.4967	−2.8749	−2.5696	−3.5092	−2.8979	−2.5790
	200	−3.4686	−2.8608	−2.5611	−3.5123	−2.9051	−2.5885
	5000	−3.4745	−2.9079	−2.5920	−3.4036	−2.8308	−2.5390
	10,000	−3.4758	−2.8964	−2.6018	−3.3691	−2.8247	−2.5252
D, NS, T							
	20	−4.5285	−3.6806	−3.2892	−2.6241	−1.8378	−1.4749
	50	−4.1618	−3.5155	−3.2006	−2.5660	−1.9044	−1.5534
	100	−4.0525	−3.4448	−3.1542	−2.5878	−1.9164	−1.5787
	200	−4.0103	−3.4370	−3.1307	−2.5088	−1.9405	−1.6174
	5000	−4.0074	−3.4642	−3.1746	−2.4541	−1.8796	−1.5657
	10,000	−4.0411	−3.4643	−3.1821	−2.5170	−1.8804	−1.5566
D, NS, NT							
	20	−3.7839	−2.9961	−2.6377	−2.7150	−1.8909	−1.5147
	50	−3.5449	−2.9164	−2.6002	−2.5933	−1.9269	−1.5708
	100	−3.4896	−2.8807	−2.5719	−2.6030	−1.9253	−1.5859
	200	−3.4728	−2.8623	−2.5638	−2.5151	−1.9440	−1.6178
	5000	−3.4745	−2.9074	−2.5918	−2.4543	−1.8797	−1.5656
	10,000	−3.4757	−2.8967	−2.6014	−2.5171	−1.8805	−1.5566
ND, NS, NT							
	20	−2.7678	−1.9597	−1.5652	−2.7994	−1.9621	−1.5620
	50	−2.6501	−1.9284	−1.5774	−2.6260	−1.9524	−1.5868
	100	−2.6190	−1.9283	−1.6007	−2.6156	−1.9335	−1.5918
	200	−2.6011	−1.9519	−1.6247	−2.5143	−1.9463	−1.6194
	5000	−2.6332	−1.9771	−1.6620	−2.4545	−1.8796	−1.5656
	10,000	−2.6241	−1.9972	−1.6697	−2.5169	−1.8806	−1.5567

Table 4 Power of the t_1 Test; d.g.p.: $(1-B^2)x_t=\varepsilon_t$

	ρ										
	1.00	0.90	0.80	0.70	0.60	0.50	0.40	0.30	0.20	0.10	0.00
D, S, T											
1%											
$n=20$	1.00	1.44	1.46	1.86	2.43	2.84	3.99	5.80	8.28	11.71	18.17
$n=50$	1.00	1.28	2.05	3.61	7.09	12.93	24.56	41.94	61.44	79.63	92.04
$n=100$	1.00	1.96	5.88	17.41	41.18	71.77	91.67	99.07	99.91	100.00	100.00
5%											
$n=20$	5.00	5.83	6.63	7.42	9.31	11.73	14.72	19.55	26.65	33.98	44.89
$n=50$	5.00	6.26	8.92	15.43	26.00	40.87	59.31	77.72	90.12	96.65	99.27
$n=100$	5.00	9.06	22.27	48.32	78.19	94.92	99.51	99.99	100.00	100.00	100.00
10%											
$n=20$	10.00	11.31	12.44	14.01	17.14	21.09	25.60	32.84	41.72	50.93	61.79
$n=50$	10.00	12.38	17.38	27.48	42.57	59.78	76.74	90.27	96.61	99.15	99.89
$n=100$	10.00	16.83	36.61	66.88	90.83	98.68	99.90	100.00	100.00	100.00	100.00
D, S, NT											
1%											
$n=20$	1.00	1.45	1.69	2.07	3.23	4.36	6.38	10.15	15.33	21.65	32.44
$n=50$	1.00	1.71	3.32	7.00	15.20	28.10	47.50	68.91	85.55	94.78	98.79
$n=100$	1.00	3.13	12.40	35.15	68.80	91.97	99.03	99.98	100.00	100.00	100.00
5%											
$n=20$	5.00	6.52	7.85	10.09	14.02	18.75	24.53	34.09	44.27	55.76	68.28
$n=50$	5.00	8.13	14.63	26.70	44.85	65.44	82.71	94.22	98.39	99.61	99.95
$n=100$	5.00	13.15	37.73	72.53	94.80	99.41	99.97	100.00	100.00	100.00	100.00

(continued)

TABLE 4 (Continued)

	ρ										
	1.00	0.90	0.80	0.70	0.60	0.50	0.40	0.30	0.20	0.10	0.00
10%											
$n=20$	10.00	12.45	15.14	18.54	24.78	31.69	40.04	51.44	62.87	73.09	82.97
$n=50$	10.00	15.39	25.85	42.71	63.87	82.32	93.21	98.36	99.60	99.93	99.99
$n=100$	10.01	24.78	58.10	88.19	98.89	99.91	100.00	100.00	100.00	100.00	100.00
D, NS, T											
1%											
$n=20$	1.00	1.44	1.62	1.88	2.61	3.28	4.24	6.57	9.07	13.06	19.97
$n=50$	1.00	1.34	2.25	3.87	7.52	13.80	26.04	43.96	63.38	80.99	92.87
$n=100$	1.00	1.93	5.97	17.46	41.32	71.93	91.84	99.12	99.91	100.00	100.00
5%											
$n=20$	5.01	6.50	7.53	8.20	10.29	13.13	16.31	21.92	29.45	37.68	48.91
$n=50$	5.00	6.57	9.42	15.81	26.92	41.85	60.21	78.69	90.78	97.00	99.34
$n=100$	5.00	8.91	22.19	48.19	78.16	94.89	99.50	99.99	100.00	100.00	100.00
10%											
$n=20$	10.00	12.41	13.81	15.52	19.00	23.60	28.20	36.01	45.03	54.95	65.92
$n=50$	10.00	12.85	18.01	28.12	43.38	60.57	77.43	90.85	96.80	99.22	99.90
$n=100$	10.00	16.95	36.92	67.22	91.00	98.73	99.93	100.00	100.00	100.00	100.00

Table 5 Power of the t_2 Test; d.g.p.: $(1-B^2)x_t=\varepsilon_t$

	ρ										
	1.00	0.90	0.80	0.70	0.60	0.50	0.40	0.30	0.20	0.10	0.00
D, S, T											
1%											
$n=20$	1.00	1.22	1.62	2.28	3.15	4.44	6.81	10.04	15.06	22.33	32.01
$n=50$	1.00	1.48	2.81	6.67	13.39	25.50	44.65	66.16	83.26	93.75	98.45
$n=100$	1.00	3.06	10.89	32.64	65.30	90.60	98.60	99.95	100.00	100.00	100.00
5%											
$n=20$	5.00	5.89	7.43	9.48	12.87	17.46	23.38	31.23	41.57	52.30	64.86
$n=50$	5.00	7.85	13.94	25.95	43.72	64.57	82.26	93.80	98.32	99.68	99.96
$n=100$	5.00	14.04	38.52	70.91	93.85	99.45	99.97	100.00	100.00	100.00	100.00
10%											
$n=20$	10.00	11.50	14.57	18.13	23.20	30.16	38.55	48.95	60.08	70.59	80.85
$n=50$	10.00	15.08	25.43	42.34	63.43	81.61	93.11	98.33	99.65	99.96	100.00
$n=100$	10.00	24.91	58.06	87.90	98.70	99.94	100.00	100.00	100.00	100.00	100.00
D, S, NT											
1%											
$n=20$	1.00	1.20	1.56	2.08	2.93	4.12	6.13	9.39	14.33	21.43	31.31
$n=50$	1.00	1.53	3.08	7.02	14.08	26.59	46.06	67.71	84.40	94.36	98.64
$n=100$	1.00	3.07	10.91	32.18	64.86	90.33	98.64	99.95	100.00	100.00	100.00
5%											
$n=20$	5.00	6.13	7.54	9.83	13.34	18.00	23.94	32.32	43.15	54.39	66.91
$n=50$	5.01	7.91	13.94	26.05	43.74	64.70	82.48	93.87	98.32	99.68	99.96
$n=100$	5.00	14.08	38.57	70.66	93.80	99.43	99.98	100.00	100.00	100.00	100.00

(continued)

TABLE 5 (Continued)

	ρ										
	1.00	0.90	0.80	0.70	0.60	0.50	0.40	0.30	0.20	0.10	0.00
10%											
$n=20$	10.00	11.87	14.57	18.31	23.90	30.98	39.56	50.11	61.53	71.88	82.17
$n=50$	9.99	15.15	25.45	42.58	63.89	81.94	93.35	98.42	99.67	99.97	100.00
$n=100$	10.00	25.01	58.28	87.59	98.65	99.91	100.00	100.00	100.00	100.00	100.00
D, NS, T											
1%											
$n=20$	1.00	2.45	4.14	6.76	11.06	17.20	25.07	36.50	48.87	62.21	74.77
$n=50$	1.00	4.77	12.31	27.43	51.28	74.27	90.17	97.39	99.46	99.92	100.00
$n=100$	1.00	11.21	42.15	80.00	97.39	99.87	100.00	100.00	100.00	100.00	100.00
5%											
$n=20$	5.00	13.57	20.93	30.92	43.56	56.70	69.29	80.78	89.25	94.13	97.39
$n=50$	5.00	20.52	44.15	70.51	89.79	97.65	99.56	99.95	100.00	100.00	100.00
$n=100$	5.00	40.84	85.60	98.85	99.96	100.00	100.00	100.00	100.00	100.00	100.00
10%											
$n=20$	10.00	25.45	37.51	50.40	65.41	78.03	86.74	93.30	96.60	98.50	99.37
$n=50$	10.00	37.39	67.40	89.11	97.87	99.67	99.95	100.00	100.00	100.00	100.00
$n=100$	10.00	64.16	96.48	99.88	100.00	100.00	100.00	100.00	100.00	100.00	100.00

Table 6 Power of the F_{12} Test; d.g.p.: $(1+B^2)x_t=\varepsilon_t$

	ρ										
	1.00	0.90	0.80	0.70	0.60	0.50	0.40	0.30	0.20	0.10	0.00
D, S, T											
1%											
$n=20$	1.00	1.02	1.21	1.70	2.44	3.61	5.40	8.16	12.72	19.63	28.75
$n=50$	1.00	1.63	3.94	8.77	19.48	38.81	62.66	82.36	94.00	98.67	99.73
$n=100$	1.00	3.39	16.29	50.37	85.82	98.49	99.94	100.00	100.00	100.00	100.00
5%											
$n=20$	5.01	5.64	6.94	8.43	11.49	15.71	21.19	28.96	38.73	50.80	62.72
$n=50$	5.01	8.19	14.97	29.53	49.65	72.64	89.47	97.06	99.36	99.92	99.99
$n=100$	5.00	14.56	46.47	84.52	98.55	99.92	100.00	100.00	100.00	100.00	100.00
10%											
$n=20$	10.01	11.49	13.43	16.81	21.46	27.39	35.68	45.41	56.94	68.39	78.11
$n=50$	10.01	15.43	26.17	45.91	67.68	86.15	96.03	99.11	99.91	100.00	100.00
$n=100$	10.00	26.16	65.41	94.11	99.73	100.00	100.00	100.00	100.00	100.00	100.00
D, S, NT											
1%											
$n=20$	1.01	1.18	1.69	2.48	3.83	5.92	9.38	14.36	22.05	33.15	45.66
$n=50$	1.01	2.19	5.49	13.84	30.43	54.96	78.79	92.67	98.16	99.72	99.97
$n=100$	1.00	4.76	25.44	67.57	94.70	99.70	100.00	100.00	100.00	100.00	100.00
5%											
$n=20$	5.00	6.13	7.85	10.81	15.15	21.43	29.93	40.56	53.30	66.27	77.14
$n=50$	5.00	9.64	20.79	41.22	65.62	86.09	96.32	99.26	99.92	100.00	100.00
$n=100$	5.01	19.44	61.33	93.79	99.78	100.00	100.00	100.00	100.00	100.00	100.00

(*continued*)

Table 6 (Continued)

	ρ										
	1.00	0.90	0.80	0.70	0.60	0.50	0.40	0.30	0.20	0.10	0.00
10%											
$n=20$	10.00	12.16	15.45	20.72	27.39	36.39	47.89	59.82	72.00	81.90	89.15
$n=50$	10.00	18.35	35.10	59.53	81.81	94.72	99.02	99.81	99.98	100.00	100.00
$n=100$	10.01	33.09	78.42	98.36	99.96	100.00	100.00	100.00	100.00	100.00	100.00
D, NS, T											
1%											
$n=20$	1.01	1.37	1.82	2.77	4.04	6.16	9.90	14.79	22.98	33.89	45.70
$n=50$	1.00	2.27	5.71	13.97	30.22	54.30	78.31	92.33	98.02	99.71	99.94
$n=100$	1.00	5.33	26.62	68.92	94.98	99.72	100.00	100.00	100.00	100.00	100.00
5%											
$n=20$	5.01	6.63	8.88	12.03	16.54	23.05	31.73	42.99	55.22	68.18	78.25
$n=50$	5.01	10.59	22.10	42.16	66.56	86.34	96.44	99.25	99.91	100.00	100.00
$n=100$	5.01	20.11	61.40	93.68	99.70	100.00	100.00	100.00	100.00	100.00	100.00
10%											
$n=20$	10.00	13.12	16.67	22.01	28.39	38.30	48.80	60.92	72.59	82.47	89.20
$n=50$	10.01	19.14	35.90	59.88	81.56	94.57	98.91	99.81	99.99	100.00	100.00
$n=100$	10.00	34.16	78.64	98.25	99.95	100.00	100.00	100.00	100.00	100.00	100.00

Table 7 Power of the t_1 Test; d.g.p.: $(1-B)x_t=\varepsilon_t$

	ρ										
	1.00	0.90	0.80	0.70	0.60	0.50	0.40	0.30	0.20	0.10	0.00
D, S, T											
1%											
$n=20$	1.00	1.13	1.38	2.03	2.58	3.82	5.19	6.49	8.65	11.15	13.99
$n=50$	1.00	1.91	4.05	10.49	20.56	34.37	49.95	65.13	75.84	85.13	91.46
$n=100$	1.00	4.69	23.67	58.87	86.32	97.21	99.51	99.94	100.00	100.00	100.00
5%											
$n=20$	5.00	5.60	6.83	9.01	11.56	15.53	19.58	24.24	28.59	34.57	39.58
$n=50$	5.00	8.22	16.78	33.01	51.80	69.10	82.38	90.67	94.97	97.96	99.01
$n=100$	5.00	18.46	56.69	88.89	98.34	99.81	99.99	100.00	100.00	100.00	100.00
10%											
$n=20$	10.00	11.10	13.44	17.14	21.75	27.02	33.49	38.91	44.95	52.21	57.85
$n=50$	10.00	15.48	29.41	50.08	69.36	83.11	92.20	96.58	98.35	99.47	99.78
$n=100$	10.01	31.40	75.05	96.06	99.65	99.97	100.00	100.00	100.00	100.00	100.00
D, S, NT											
1%											
$n=20$	1.00	1.58	2.03	3.51	5.19	7.92	10.86	14.88	18.34	23.80	28.44
$n=50$	1.00	3.16	8.77	22.26	41.03	60.13	76.25	86.96	93.07	97.08	98.63
$n=100$	1.00	8.69	42.70	81.94	96.97	99.64	99.96	100.00	100.00	100.00	100.00

(continued)

Table 7 (Continued)

	ρ										
	1.00	0.90	0.80	0.70	0.60	0.50	0.40	0.30	0.20	0.10	0.00
5%											
$n=20$	5.00	7.39	10.32	15.72	21.38	28.45	35.98	44.13	50.76	58.72	65.34
$n=50$	5.00	12.92	30.76	55.31	76.85	89.11	95.74	98.53	99.35	99.80	99.95
$n=100$	5.00	31.44	81.52	98.14	99.91	99.99	100.00	100.00	100.00	100.00	100.00
10%											
$n=20$	10.00	14.12	19.21	27.02	35.45	44.44	53.27	61.87	68.62	75.14	80.36
$n=50$	10.00	23.15	48.71	73.44	89.41	96.17	98.83	99.67	99.85	99.98	100.00
$n=100$	10.00	49.72	92.97	99.65	99.97	100.00	100.00	100.00	100.00	100.00	100.00
D, NS, T											
1%											
$n=20$	1.00	1.11	1.37	2.08	2.62	3.71	5.00	6.73	8.88	11.49	14.22
$n=50$	1.00	1.82	3.98	10.36	20.40	34.16	49.88	64.93	75.91	85.14	91.43
$n=100$	1.00	4.53	23.14	58.13	85.80	97.05	99.48	99.93	100.00	100.00	100.00
5%											
$n=20$	5.00	5.67	6.88	8.98	11.72	15.64	19.60	24.31	29.05	34.75	39.66
$n=50$	5.00	8.18	16.78	32.80	51.80	68.99	82.35	90.70	94.97	97.93	99.03
$n=100$	5.00	18.69	57.03	89.08	98.43	99.80	99.99	100.00	100.00	100.00	100.00
10%											
$n=20$	10.00	11.15	13.32	16.91	21.67	26.73	33.05	39.03	45.08	52.21	58.04
$n=50$	10.00	15.46	29.16	49.85	69.23	83.01	92.34	96.61	98.35	99.44	99.79
$n=100$	9.99	31.37	74.83	96.07	99.63	99.97	100.00	100.00	100.00	100.00	100.00

TABLE 8 Power of the t_2 Test; d.g.p.: $(1+B)x_t=\varepsilon_t$

	ρ										
	1.00	0.90	0.80	0.70	0.60	0.50	0.40	0.30	0.20	0.10	0.00
D, S, T											
1%											
$n=20$	1.00	1.39	2.25	3.60	5.17	7.83	11.18	14.44	18.40	23.65	28.53
$n=50$	1.00	2.56	8.09	20.29	38.09	56.83	73.85	85.13	91.66	96.24	98.12
$n=100$	1.00	9.20	43.94	82.53	97.13	99.65	99.96	100.00	100.00	100.00	100.00
5%											
$n=20$	5.00	7.25	10.43	15.15	20.46	27.74	35.75	42.92	49.79	57.83	63.68
$n=50$	5.00	12.79	31.65	55.87	77.19	89.49	96.01	98.46	99.44	99.90	99.97
$n=100$	4.99	31.76	80.57	98.04	99.88	99.99	100.00	100.00	100.00	100.00	100.00
10%											
$n=20$	10.00	13.78	19.59	26.87	34.56	44.31	53.33	61.31	68.04	75.08	79.90
$n=50$	10.00	23.48	48.87	74.27	89.88	96.25	98.94	99.56	99.90	99.99	99.99
$n=100$	10.00	50.86	92.92	99.71	100.00	100.00	100.00	100.00	100.00	100.00	100.00
D, S, NT											
1%											
$n=20$	1.00	1.49	2.24	3.55	5.29	7.77	11.15	14.57	18.71	24.07	29.04
$n=50$	1.00	2.54	8.08	20.08	37.72	56.65	73.77	85.09	91.65	96.34	98.14
$n=100$	1.00	8.94	43.40	82.03	97.07	99.63	99.96	100.00	100.00	100.00	100.00
5%											
$n=20$	4.99	6.91	9.70	14.46	19.69	26.45	34.44	41.39	48.43	56.47	62.64
$n=50$	5.00	12.59	31.30	55.61	76.85	89.34	95.97	98.45	99.44	99.88	99.97
$n=100$	4.99	31.51	80.40	98.03	99.88	99.99	100.00	100.00	100.00	100.00	100.00

(continued)

TABLE 8 (Continued)

	ρ										
	1.00	0.90	0.80	0.70	0.60	0.50	0.40	0.30	0.20	0.10	0.00
10%											
$n=20$	10.00	13.60	19.13	26.41	34.19	43.78	52.74	60.94	67.69	74.99	79.71
$n=50$	10.00	23.42	48.55	73.96	89.77	96.17	98.94	99.54	99.90	99.99	99.99
$n=100$	10.00	50.48	92.79	99.68	100.00	100.00	100.00	100.00	100.00	100.00	100.00
D, NS, T											
1%											
$n=20$	1.00	3.88	8.28	14.50	22.22	31.02	41.14	50.60	59.17	67.26	73.66
$n=50$	1.00	10.96	35.38	64.81	85.81	94.66	98.47	99.48	99.83	99.99	99.99
$n=100$	1.00	34.25	88.16	99.40	99.99	100.00	100.00	100.00	100.00	100.00	100.00
5%											
$n=20$	5.00	18.69	33.27	48.71	62.80	73.75	82.46	88.24	91.70	95.19	96.64
$n=50$	5.00	38.34	76.65	94.59	98.88	99.77	99.95	99.99	100.00	100.00	100.00
$n=100$	5.00	78.08	99.43	99.98	100.00	100.00	100.00	100.00	100.00	100.00	100.00
10%											
$n=20$	10.00	34.42	54.40	70.50	82.34	89.37	93.99	96.44	97.75	98.83	99.28
$n=50$	10.00	61.55	92.01	98.79	99.80	99.96	100.00	100.00	100.00	100.00	100.00
$n=100$	10.00	93.04	99.94	100.00	100.00	100.00	100.00	100.00	100.00	100.00	100.00

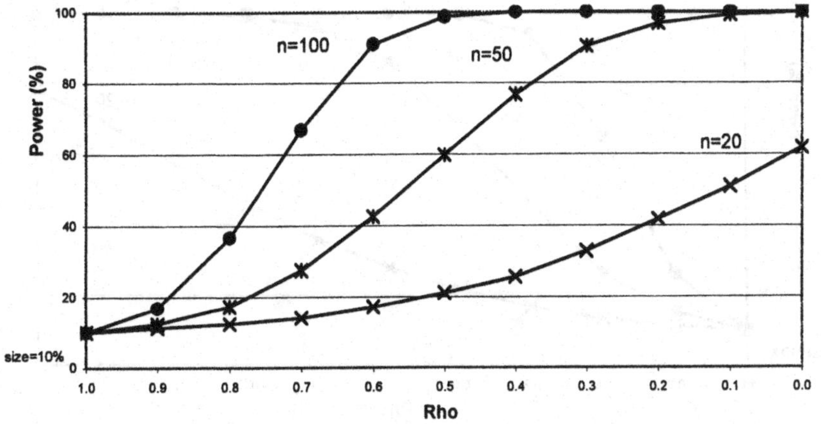

FIGURE 1 Power of the t_1 test. Drift, seasonal dummy, and trend in regression; 10% size. [d.g.p.: $(1-B^2)x_t=\varepsilon_t$.]

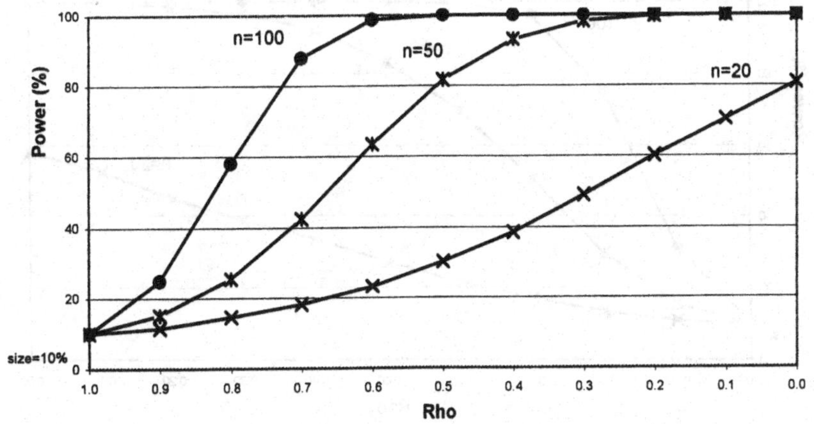

FIGURE 2 Power of the t_2 test. Drift, seasonal dummy, and trend in regression; 10% size. [d.g.p.: $(1-B^2)x_t=\varepsilon_t$.]

unit roots is entertained, with those obtained when the ADF tests are used to test for unit roots only at the zero frequency. The series are published only in semiannual form, and are:

1. New Zealand total knitted fabric sales (tonnes), June 1965 to December 1997, series SEPH.SATTD (Statistics New Zealand, 1998).

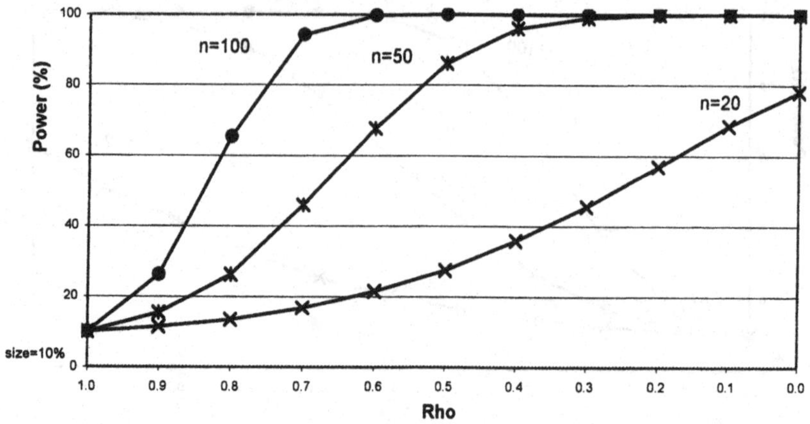

FIGURE 3 Power of the F_{12} test. Drift, seasonal dummy, and trend in regression; 10% size. [d.g.p.: $(1-B^2)x_t=\varepsilon_t$.]

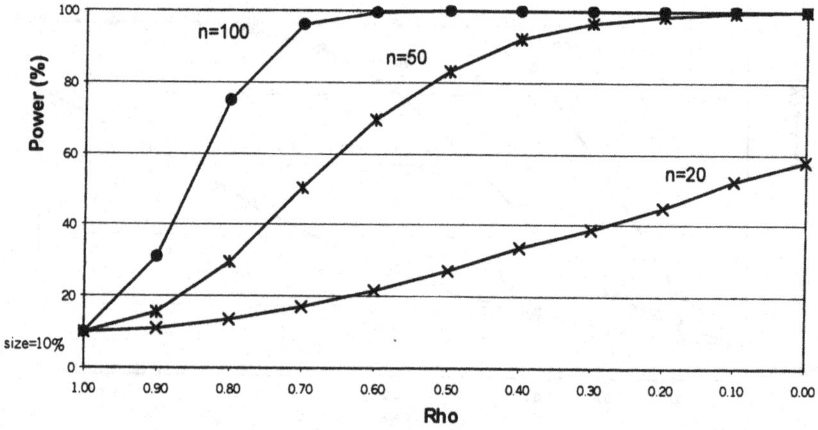

FIGURE 4 Power of the t_1 test. Drift, seasonal dummy, and trend in regression; 10% size. [d.g.p.: $(1-B)x_t=\varepsilon_t$.]

2. Canadian precast concrete price index (1981 = 100), 1977–1992, matrix 421 (Statistics Canada, 1998).
3. Canadian production of marketable gas (millions of cubic meters, oil equivalent), June 1979–December 1989 (Petroleum Monitoring Agency Canada, various years).
4. U.S.A. six-monthly *increase* in number of cellular phone subscribers, December 1984–December 1995 (Waterman, 1998).

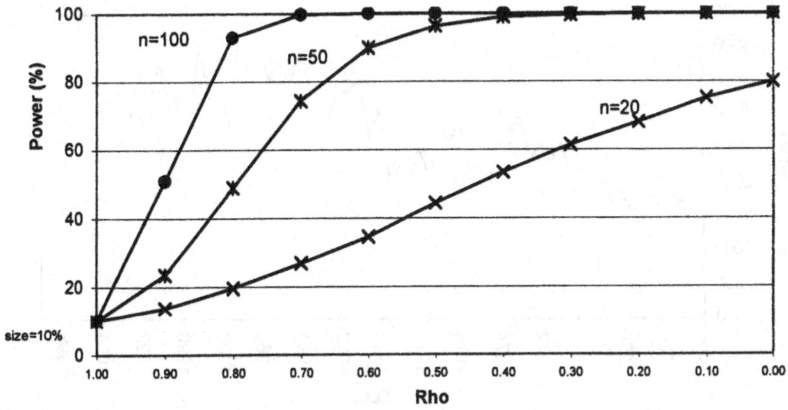

FIGURE 5 Power of the t_2 test. Drift, seasonal dummy, and trend in regression; 10% size. [d.g.p.: $(1+B)x_t=\varepsilon_t$.]

These series are shown* in Fig. 6, and Table 9 presents the results of testing for a unit root only at the zero frequency, using the ADF tests. We allowed for drift and/or trend in the ADF regressions, and used the strategy of Dolado et al. (1990) to determine their final inclusion.† The "augmentation level," q, for the regressions was chosen as the minimum required to obtain "clean" autocorrelation and partial autocorrelation functions for the associated residuals‡; and a seasonal dummy variable was included with any drift term.§ As can be seen, the ADF tests suggest in each case that the series are $I(1)$. Accordingly, one would infer that the series can be made stationary via simple first-differencing.

*In the case of series (4) the plotted series is for the number of cellphones, rather than the six-monthly increase in this number. The latter series is the one that is analyzed.

†In Table 9, t_{dt} denotes the ADF unit root "t-test" with drift and trend terms included in the fitted regression; F_{ut} is the corresponding ADF "F-test" for a unit root and zero trend; t_{d} is the unit root "t-test" with a drift but no trend in the fitted regression; F_{ud} is the corresponding "F-test" for a unit root and a zero drift; and t is the ADF unit root test when the fitted regression has no drift or trend term included. Finite-sample critical values for our "t-tests" and "F-tests" come from MacKinnon (1991) and from Dickey and Fuller (1979, 1981), respectively.

‡The simulation results of Dods and Giles (1995) favor this approach, in terms of low pre-test size distortion, for sample sizes such as ours. In fact the results are not sensitive to the method used to choose this augmentation level.

§The need to include this dummy variable follows from the results of Ghysels et al. (1994). There was no evidence that any of the series other than series (4) may be $I(2)$.

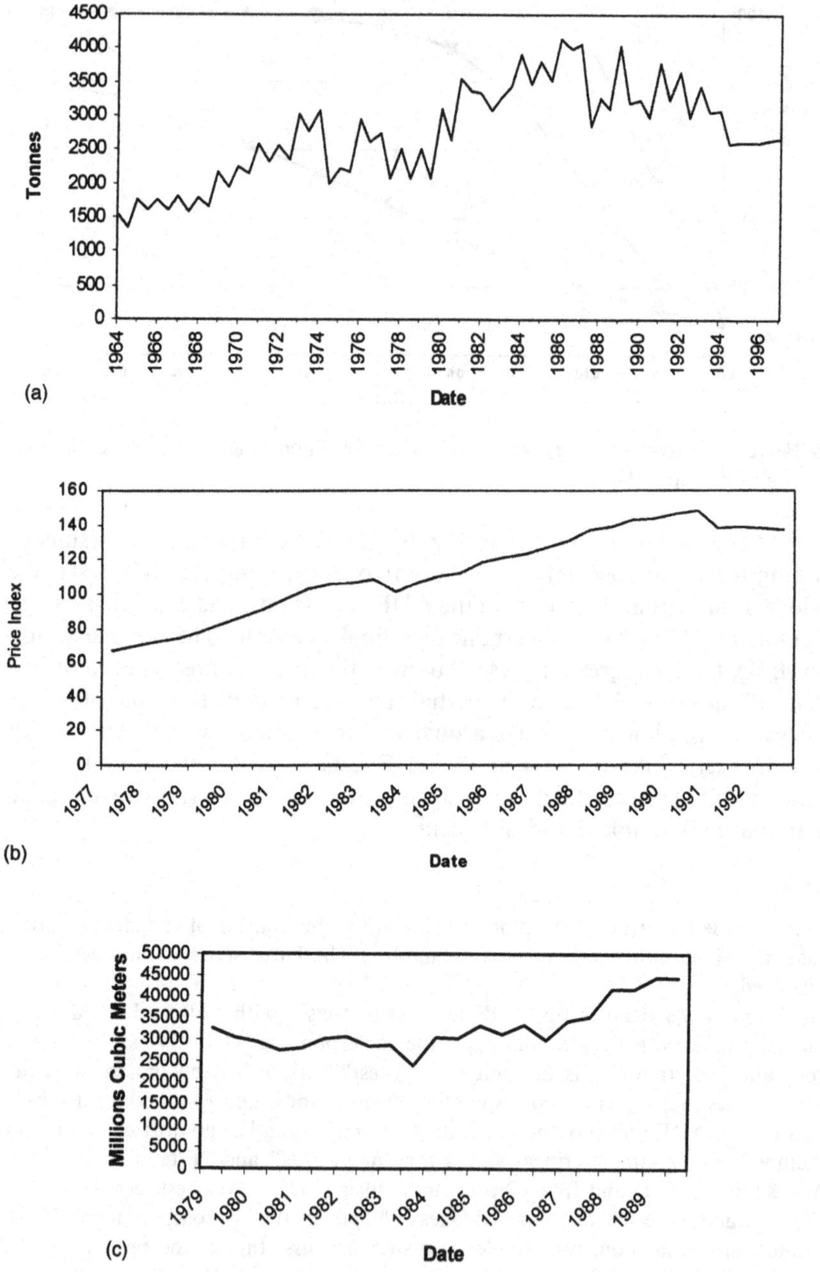

FIGURE 6 (a) New Zealand knitted fabric sales; (b) Canadian precast concrete price index (1981 = 100); (c) Canadian gas production; (d) Number of U.S. cellphones.

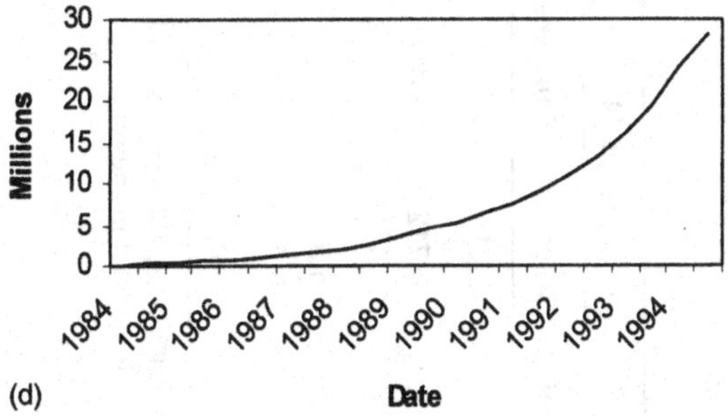

FIGURE 6 continued.

In Table 10 we show the results of applying the semiannual HEGY tests outlined in this chapter. In the cases 1, 3, and 4 we find that in fact the series are SI(1)—i.e., they each have unit roots at both the zero and π frequencies. So, in fact *two-period differencing* is the appropriate filter needed to make these series stationary. The earlier conclusion that series 2 has a unit root (only) at the zero frequency is upheld in Table 10, but these examples certainly illustrate the need for care when applying standard ADF tests to semiannual data.

6 CONCLUSIONS

In this chapter we have discussed a framework for testing for unit roots in time-series data that are reported at a maximum frequency of twice a year. In such cases, the possibility of unit roots at the zero and/or π frequencies arises, and the testing strategies developed by Hylleberg et al. (1990) and Beaulieu and Miron (1993) provide a natural framework to exploit. The asymptotic null distributions of the tests we propose are derived, discussed, and summarized, and in certain cases they can be linked to existing results associated with the well-known tests of Dickey and Fuller (1979, 1981). Simulated percentiles for the finite-sample null distributions of the test statistics are tabulated for various sample sizes and combinations of deterministic terms in the fitted regressions. The powers of the tests are simulated under a similar range of situations, and their application is illustrated with several actual semiannual time series.

The analysis in this chapter may be extended in several directions. The obvious one is to investigate the properties of corresponding tests for cointegration. Other topics of interest include the size robustness of the tests to autocorrelation in the data-generating process, to the method used to select an appropriate "augmentation level," and to structural breaks in the data. These remain matters for future research.

TABLE 9 Results of ADF Tests

Fitted model[a]	Test	Series (1)		Series (2)		Series (3)		Series (4)	
		Statistic	Outcome	Statistic	Outcome	Statistic	Outcome	Statistic	Outcome
$n(q)$		66 (1)		32 (3)		22 (3)		22(1)	
D, S, T	t_{dt}	−0.473		−0.439		−0.930		1.712	I(1)
	F_{ut}	1.824		2.281		2.009		10.194	
D, S, NT	t_d	−1.946		−1.900	I(1)	0.989		n.a.	
	F_{ud}	3.226		4.571		1.971		n.a.	
ND, NS, NT	t	1.837	I(1)	n.a.		1.848	I(1)	n.a.	

[a] n is the sample size; q is the "augmentation level" for the tests; D, S, T denotes "drift, seasonal dummy, trend"; D, S, NT denotes "drift, seasonal dummy, no trend"; ND, NS, NT denotes "no drift, no seasonal dummy, no trend." The other notation is defined in the text.

TABLE 10 Results of Seasonal Unit Root Tests

Fitted model	Test	Series (1)		Series (2)		Series (3)		Series (4)	
		Statistic	Outcome	Statistic	Outcome	Statistic	Outcome	Statistic	Outcome
D, S, T									
(q)		(2)		(2)		(1)		(1)	
	t_1	−0.439	Zero root	−0.473	Zero root	−1.075	Zero root	−0.382	Zero root
	t_2	−2.369	π root	−5.072		−2.303	π root	0.054	π root
	F_{12}	2.887	SI(1)	13.113		4.561	SI(1)	0.120	SI(1)
D, S, NT									
(q)		(2)		(2)		(1)		(1)	
	t_1	−1.900	Zero root	−1.946	Zero root	0.219	Zero root	0.963	Zero root
	t_2	−2.419	π root	−5.376		−2.569	π root	−0.355	π root
	F_{12}	4.896	SI(1)	20.866		3.405	SI(1)	0.626	SI(1)
D, NS, T									
(q)		(2)		(0)		(1)		(1)	
	t_1	−0.405	Zero root	−0.464	Zero root	−1.661	Zero root	−0.516	Zero root
	t_2	−1.103	π root	−5.152		−1.253	π root	0.494	π root
	F_{12}	0.685	SI(1)	13.530		2.434	SI(1)	0.149	SI(1)

(*continued*)

TABLE 10 (Continued)

		Series (1)		Series (2)		Series (3)		Series (4)	
Fitted model	Test	Statistic	Outcome	Statistic	Outcome	Statistic	Outcome	Statistic	Outcome
D, NS, NT									
(q)		(2)		(0)		(2)		(1)	
	t_1	−1.896	Zero root	−2.003	Zero root	1.420	Zero root	0.837	Zero root
	t_2	−1.145	π root	−5.458		−0.510	π root	0.055	π root
	F_{12}	2.518	SI(1)	21.551		1.332	SI(1)	0.967	SI(1)
ND, NS, NT									
(q)		(2)		(0)		(2)		(1)	
	t_1	0.434	Zero root	1.837	Zero root	1.752	Zero root	1.276	Zero root
	t_2	−1.181	π root	−6.440		−0.748	π root	−0.130	π root
	F_{12}	0.799	SI(1)	38.202		1.935	SI(1)	1.718	SI(1)

ACKNOWLEDGMENTS

The authors are grateful to Robert Draeseke, Judith Giles, Nilanjana Roy, Gugsa Werkneh, and participants in the University of Victoria Econometrics Colloquium, for their helpful comments on an earlier version of this work.

REFERENCES

Beaulieu, J.J. and Miron, J.A., 1993. Seasonal unit roots in aggregate U.S. data. *Journal of Econometrics* **55**, 305–328.

Chan, N.H. and Wei, C.Z., 1988. Limiting distributions of least squares estimates of unstable autoregressive processes. *Annals of Statistics* **16**, 367–401.

Dickey, D.A. and Fuller, W.A., 1979. Distribution of the estimators for autoregressive time series with a unit root. *Journal of the American Statistical Association* **74**, 427–431.

Dickey, D.A. and Fuller, W.A., 1981. Likelihood ratio statistics for autoregressive time series with a unit root. *Econometrica* **49**, 1057–1072.

Dickey, D.A., Hasza, D.P., and Fuller, W.A., 1984. Testing for unit roots in seasonal time series. *Journal of the American Statistical Association* **79**, 355–367.

Dods, J.L. and Giles, D.E.A., 1995. Alternative strategies for 'augmenting' the Dickey–Fuller test: Size-robustness in the face of pre-testing. *Journal of Statistical Computation and Simulation* **53**, 243–258.

Dolado, J.J., Jenkinson, T., and Sosvilla-Rivero, S., 1990. Cointegration and unit roots. *Journal of Economic Surveys* **4**, 249–273.

Franses, P.H., 1991. Seasonality, nonstationarity and the forecasting of monthly time series. *International Journal of Forecasting* **7**, 199–208.

Franses, P.H., 1998. *Time Series Models for Business and Economic Forecasting*. Cambridge University Press, Cambridge, U.K.

Franses, P.H. and Hobijn, B., 1997. Critical values for unit root tests in seasonal time series. *Journal of Applied Statistics* **24**, 25–47.

Fuller, W.A., 1976. *Introduction to Statistical Time Series*. Wiley, New York.

Ghysels, E., Lee, H.S., and Noh, J., 1994. Testing for unit roots in seasonal time series: some theoretical extensions and a Monte Carlo investigation. *Journal of Econometrics* **62**, 415–442.

Hylleberg, S., 1992. *Modelling Seasonality*. Oxford University Press, Oxford, U. K.

Hylleberg, S., Engle, R.F., Granger, C.W.J., and Yoo, B.S., 1990. Seasonal integration and cointegration. *Journal of Econometrics* **44**, 215–238.

Kunst, R.M., 1997. Testing for cyclical non-stationarity in autoregressive processes. *Journal of Time Series Analysis* **18**, 325–330.

MacKinnon, J.G., 1991. Critical values for cointegration tests. In: R.F. Engle and C.W.J. Granger (eds.), *Long-Run Economic Relationships*. Oxford University Press, Oxford, pp. 267–276.

Osborn, D.R. and Rodrigues, P.M.M., 1998. The asymptotic distributions of seasonal unit root tests: a unifying approach. Discussion Paper 9811, School of Economic Studies, University of Manchester, U.K.

Osborn, D.R., Chui, A.P.L., Smith, J.P., and Birchenhall, C.R., 1988. Seasonality and the order of integration of consumption. *Oxford Bulletin of Economics and Statistics* **50**, 361–377.

Petroleum Monitoring Agency Canada, various years. Canadian Petroleum Industry Monitoring Reports.

Phillips, P.C.B., 1987. Time series regression with a unit root. *Econometrica* **55**, 277–302.

Said, S.E., Dickey, D.A., 1984. Testing for unit roots in autoregressive–moving average models of unknown order. *Biometrika* **71**, 599–607.

SHAZAM, 1997. SHAZAM Econometrics computer program, *User's Reference Manual*, Version 8.0. McGraw-Hill, New York.

Smith, R.J. and Taylor, A.M.R., 1998. Additional critical values and asymptotic representations for seasonal unit root tests. *Journal of Econometrics* **85**, 269–288.

Statistics Canada, 1998. Cansim Main Index.

Statistics New Zealand, 1998. INFOS Databank.

Stock, J.H., 1988. A class of tests for integration and cointegration. Mimeograph Kennedy School of Government, Harvard University, Cambridge, MA.

Waterman, R., 1998. <www-stat.wharton.upenn.edu/~waterman/fsw/datasets/txt/Cellular.txt>.

8

Using Simulation Methods for Bayesian Econometric Models

John Geweke

University of Iowa, Iowa City, Iowa, U.S.A.

William McCausland

Université de Montréal, Montréal, Québec, Canada

John Stevens

Board of Governors of the Federal Reserve System,
Washington, D.C., U.S.A.

1 INTRODUCTION

Subjective uncertainty is a central concept in economic theory and applied economics. In economic theory, it characterizes the beliefs of economic agents about the state of their environment. In applied economics, subjective uncertainty describes the situation of investigators who assess competing models based on their implications for what might be observed, and the circumstances of decision makers who must act, given limited information. With the application of the expected utility paradigm in increasingly richer environments, explicit distributional assumptions have become common, but closed-form analytical expressions for the distribution of observables are typically unobtainable. In this environment, simulation methods—the representation of probability distributions by related finite samples—have become important tools in economic theory.

In applied economics the possibility of proceeding strictly analytically is also remote. Even in the simplest typical situation the investigator or decision maker must proceed knowing the observables which are random variables in models of behavior, but not knowing the specification of tastes and technology that the theorist takes as fixed. Bayesian inference formalizes the applied economics problem in exactly this way: given a distribution over competing models and the prediction of each model for observables, the distribution of competing models *conditional* on the observables is well defined. However, the technical tasks in moving from even such well-specified models and data to the conditional distribution over models are more daunting than those found in economic theory. In the past decade very substantial progress has been made in the development of simulation methods suited to this task. Section 2 of this chapter reviews the conditional distributions of interest to the investigator or decision maker. Section 3 describes how modern simulation methods permit access to these distributions, using some simple examples to illustrate the methods. Section 4 sets forth recently developed numerical procedures for the explicit comparison of econometric models. The last section of the chapter takes up simple, effective ways of simultaneously realizing the rich promise of explicitly Bayesian methods and dealing with the desire of decision makers to change investigators' assumptions at low cost.

Throughout, the chapter demonstrates the practical implementation of all these methods in the Bayesian analysis, computation, and communication (BACC) software. This embeds Bayesian tools in mathematical software such as Matlab and Gauss. It provides a seamless integration of commands for Bayesian inference with built-in general-purpose commands for computation, graphics, and flow control that are already familiar to many users. The BACC web site, http://www.cirano.gc.ca/~bacc, provides full details and the software itself.

2 BAYESIAN INFERENCE

This section provides a brief overview of Bayesian inference with reference to its application in economics. The purpose is to set the contribution of simulation methods in an explicit context of concepts and notation. Every attempt has been made to distill a large literature in statistics to what is essential to Bayesian inference as it is usually applied to economic problems. If this endeavor has been successful then this section also provides a sufficient introduction for econometricians with little or no grounding in Bayesian methods to appreciate some of the contributions, both realized and potential, of simulation methods to economic science.

Most of the material here is standard, reflecting much more comprehensive treatments including Jeffreys (1939, 1961), Zellner (1971), Berger (1985), Bernardo and Smith (1994), and Poirier (1995). At two junctures the exposition departs from the usual development. The first is the concept of a complete model (Section 2.1), which is the specification of a proper predictive distribution over an explicit class of events. This concept can be a clarifying analytical device.

The other deviation from the standard treatment is the decomposition of the marginal likelihood in terms of predictive densities (Section 2.3). This development was first provided explicitly by Geisel (1977), but has largely been ignored in the subsequent literature. The decomposition is the quantitative expression of the fact that predictive power is the scientifically relevant test of the validity of a hypothesis (Jeffreys, 1939; Friedman, 1953).

This review concentrates entirely on exact, finite sample methods. As is the case in non-Bayesian statistics, given suitable regularity conditions there exist useful asymptotic approximations to the exact, finite sample results. Bernardo and Smith (1994, Sect. 5.3) provide an accessible introduction to these results. Asymptotic methods are complementary to, rather than a prerequisite for, the posterior simulation methods taken up subsequently in Section 3.

2.1 Basic Concepts and Notation

Bayesian inference takes place in the context of one or more parametric econometric models. Let $\mathbf{y}_t$ denote a $p \times 1$ vector of observable random vectors over a sequence of discrete time units $t = 1, 2, \ldots$. The history of the sequence $\{\mathbf{y}_t\}$ at time t is given by $\mathbf{Y}_t = \{\mathbf{y}_s\}_{s=1}^{t} \in \Psi_t$; $\mathbf{Y}_0 = \{\emptyset\}$. A *model*, A, specifies a corresponding sequence of probability density functions $\mathrm{p}(\mathbf{y}_t|\mathbf{Y}_{t-1}, \theta, A)$ in which θ is a $k \times 1$ vector of unknown parameters, $\theta \in \Theta \subseteq \Re^k$, and A denotes the model.* In this section we shall condition on a single model, but subsequently, in Section 2.3, several models will be entertained simultaneously.

The probability density function (p.d.f.) of $\mathbf{Y}_T$, conditional on the model A and parameter vector θ, is $\mathrm{p}(\mathbf{Y}_T|\theta, A) = \prod_{t=1}^{T} \mathrm{p}(\mathbf{y}_t|\mathbf{Y}_{t-1}, \theta, A)$. Conditional on observed $\mathbf{Y}_T$, the *likelihood function* is any function

*Throughout, $\mathrm{p}(\cdot)$ denotes a generic probability density function with respect to a measure $d\nu(\cdot)$ and $\mathrm{P}(\cdot)$ a generic cumulative distribution function. The conditioning set makes clear the specific distribution or density intended. The measure $d\nu(\cdot)$ permits continuous, discrete, and mixed random variables.

$L(\theta; \mathbf{Y}_T, A) \propto p(\mathbf{Y}_T|\theta, A)$. If the model specifies that the $\mathbf{y}_t$ are independent and identically distributed then $p(\mathbf{y}_t|\mathbf{Y}_{t-1}, \theta, A) = p(\mathbf{y}_t|\theta, A)$ and $p(\mathbf{Y}_T|\theta, A) = \prod_{t=1}^{T} p(\mathbf{y}_t|\theta, A)$. More generally, the index t may pertain to cross-sections, to time series, or both, but time-series models and language are used here for specificity.

If, in addition, the model A also provides the distribution of θ, then it also provides the joint distribution of θ and $\mathbf{Y}_T$. In particular, if $p(\theta|A)$ denotes the *prior density*, then

$$p(\mathbf{Y}_T, \theta|A) = p(\theta|A) \prod_{t=1}^{T} p(\mathbf{y}_t|\mathbf{Y}_{t-1}, \theta, A) = p(\theta|A)p(\mathbf{Y}_T|\theta, A) \tag{1}$$

but we also have

$$p(\mathbf{Y}_T, \theta|A) = p(\theta|\mathbf{Y}_T, A)p(\mathbf{Y}_T|A) \tag{2}$$

in which

$$p(\mathbf{Y}_T|A) = \int_\Theta p(\mathbf{Y}_T|\theta, A)p(\theta|A)\, d\nu(\theta) \tag{3}$$

is the *marginal likelihood** of model A. Therefore,

$$p(\theta|\mathbf{Y}_T, A) = p(\mathbf{Y}_T|\theta, A)p(\theta|A)/p(\mathbf{Y}_T|A) \propto p(\mathbf{Y}_T|\theta, A)p(\theta|A)$$

is the *posterior density* of θ in model A, so long as

$$\int_\Theta p(\mathbf{Y}_T|\theta, A)p(\theta|A)\, d\nu(\theta) \tag{4}$$

is absolutely convergent. This last condition is typically, but not necessarily, satisfied and easy to verify. For example, boundedness of the likelihood function $p(\mathbf{Y}_T|\theta, A)$ in θ is sufficient, since $\int_\Theta p(\theta|A)\, d\nu(\theta) = 1$. However, if the likelihood function is unbounded, it is vital to confirm the absolute convergence of expression (4). Equations (1) and (2) are central, either explicitly or implicitly, to scientific learning. The former is used to express the reduction of reality to θ inherent in the model A, and the latter is used to learn about reality from the perspective of this particular simplification. This section outlines the basic principles of the explicit, or Bayesian, approach to learning.

In addition to the data density $p(\mathbf{Y}_T|\theta, A)$ and the prior density $p(\theta|A)$, a model also specifies a density $p(\omega|\mathbf{Y}_T, \theta, A)$ for a vector of interest $\omega \in \Omega \subseteq R^\ell$. This vector represents entities that the model is intended to describe. Whereas θ is specific to A, ω remains the same across

---

*This terminology dates at least to Raiffa and Schlaifer (1961, Sect. 2.1), which also treats these topics.

models. For example, suppose that a model specifies a first-order autoregressive process for y_t, $(y_t - \theta_1) = \theta_2(y_{t-1} - \theta_1) + \varepsilon_t$ with $\varepsilon_t \overset{IID}{\sim} \mathrm{N}(0, \theta_3)$. If $\omega' = (y_{T+1}, y_{T+2})$, the first two postsample observations, then $\mathrm{p}(\omega|\mathbf{Y}_T, \theta, A)$ is a bivariate normal density with mean and variance:

$$\begin{bmatrix} \theta_1 + \theta_2(y_T - \theta_1) \\ \theta_1 + \theta_2^2(y_T - \theta_1) \end{bmatrix} \quad \text{and} \quad \theta_3 \begin{bmatrix} 1 & \theta_2 \\ \theta_2 & 1 + \theta_2^2 \end{bmatrix}$$

respectively. If a second model specifies a second-order autoregressive process for y_t, $(y_t - \theta_1) = \theta_2(y_{t-1} - \theta_1) + \theta_3(y_{t-2} - \theta_1) + \varepsilon_t$ with $\varepsilon_t \overset{IID}{\sim} \mathrm{N}(0, \theta_4)$, then $\mathrm{p}(\omega|\mathbf{Y}_T, \theta, A)$ is again a bivariate normal density, but with mean and variance:

$$\begin{bmatrix} \theta_1 + \theta_2(y_T - \theta_1) + \theta_3(y_{T-1} - \theta_1) \\ \theta_1 + (\theta_2^2 + \theta_3)(y_T - \theta_1)(y_T - \theta_1) + \theta_2\theta_3(y_{T-1} - \theta_1) \end{bmatrix}$$

$$\text{and} \quad \theta_4 \begin{bmatrix} 1 & \theta_2 \\ \theta_2 & 1 + \theta_2^2 \end{bmatrix}$$

respectively. Since $\mathrm{p}(\omega|\mathbf{Y}_T, \theta, A)$ implies marginal distributions for subvectors of ω, one need not explicitly elaborate all of ω. Indeed, much scientific discourse can be interpreted as specification of ω. A *complete model* consists of three components: $\mathrm{p}(\mathbf{Y}_T|\theta, A)$, $\mathrm{p}(\theta|A)$, and $\mathrm{p}(\omega|\mathbf{Y}_T, \theta, A)$.

Without loss of generality, let the objective of inference when there is one model be

$$\mathrm{E}[\mathrm{h}(\omega)|\mathbf{Y}_T, A] \tag{5}$$

for suitably chosen $\mathrm{h}(\cdot)$. This formulation includes several special cases of interest. If a hypothesis restricts θ to a set Θ_0, then, by taking $\mathrm{h}(\omega) = \chi_{\Theta_0}(\theta)$, we have $\mathrm{E}[\mathrm{h}(\omega)|\mathbf{Y}_T, A] = P(\theta \in \Theta_0|\mathbf{Y}_T, A)$, the posterior probability that the hypothesis is true.* To illustrate, consider the hypothesis that $\{y_t\}$ is stationary in the example just presented. In the first-order autoregression, take $\Theta = \{\theta_2\text{: } -1 < \theta_2 < 1\}$. In the second-order autoregression, take $\Theta = \{(\theta_2, \theta_3)\text{: } 1 - \theta_2 z - \theta_3 z^2 = 0 \Rightarrow |z| > 1\}$. Here, and in general, nuisance parameters pose no particular difficulties.

Another important class of cases arises from prediction problems, $\omega' = (\mathbf{y}_{T+1}, \ldots, \mathbf{y}_{T+f})$. Through the appropriate choice of $\mathrm{h}(\omega)$ this category includes expected values, turning point probabilities, and

*Here, and throughout, $\chi_S(z)$ is the indicator function $\chi_S(z) = 1$ if $z \in S$ and $\chi_S(z) = 0$ if $z \notin S$.

predictive intervals. In the example just set forth, suppose that $y_{T-2} < y_{T-1} < y_T$. If a turning point at time t is said to occur if $y_{t-2} < y_{t-1} < y_t > y_{t+1} > y_{t+2}$, then a turning point at time T is the set of events $\Omega^* = \{\omega\colon\ \omega_2 < \omega_1 < y_T\}$. Hence, for $\mathrm{h}(\omega) = \chi_{\Omega^*}(\omega)$, $\mathrm{E}[\mathrm{h}(\omega)|\mathbf{Y}_T, A]$ is the probability of a turning point at time T, where T is the end of the sample.

Yet another useful class of functions is $\mathrm{h}(\omega) = \mathrm{L}(a_1, \omega) - \mathrm{L}(a_2, \omega)$, in which $\mathrm{L}(a, \omega)$ denotes the loss incurred if action a is taken and then the realization of the vector of interest is ω. To examine a specific case, suppose that in our example y_t is the logarithm of tax revenue in period t. A policy maker must either commit (a_1) or not commit (a_2) to a program that utilizes tax revenues in periods $T+1$ and $T+2$. Then, the policy maker's loss function $\mathrm{L}(a, \omega)$ might be monotone decreasing in $\omega_1 + \omega_2$ for $a = a_1$, and monotone increasing in $\omega_1 + \omega_2$ for $a = a_2$, and consequently $\mathrm{h}(\omega)$ is monotone decreasing in $\omega_1 + \omega_2$. The solution of the decision problem is to commit to the project if $\mathrm{E}[\mathrm{h}(\omega)|\mathbf{Y}_T, A] < 0$ and not commit if $\mathrm{E}[\mathrm{h}(\omega)|\mathbf{Y}_T, A] > 0$.

The posterior moment, Eq. (5), can be expressed:

$$
\begin{aligned}
\mathrm{E}[\mathrm{h}(\omega)|\mathbf{Y}_T, A] &= \int_\Theta \int_\Omega \mathrm{h}(\omega)\mathrm{p}(\omega|\theta, \mathbf{Y}_T, A)\mathrm{p}(\theta|\mathbf{Y}_T, A)\,\mathrm{d}\nu(\omega)\,\mathrm{d}\nu(\theta) \\
&= \int_\Theta \int_\Omega \mathrm{h}(\omega)\mathrm{p}(\omega|\theta, \mathbf{Y}_T, A)\mathrm{p}^*(\theta|\mathbf{Y}_T, A)\,\mathrm{d}\nu(\omega)\,\mathrm{d}\nu(\theta) \Big/ \\
&\qquad \int_\Theta \mathrm{p}^*(\theta|\mathbf{Y}_T, A)\,\mathrm{d}\nu(\theta) \qquad (6)
\end{aligned}
$$

where $\mathrm{p}^*(\theta|\mathbf{Y}_T, A) \propto \mathrm{p}(\theta|\mathbf{Y}_T, A) \propto \mathrm{p}(\theta|A)\mathrm{p}(\mathbf{Y}_T|\theta, A)$ is any *posterior density kernel* for θ.* It clearly matters not which posterior kernel is used. However, the problem of evaluating integrals—one in the numerator, the other in the denominator—remains paramount.

The importance of verifying the absolute convergence of the integral in the denominator of the right-hand side of Eq. (6) has already been noted. It is, of course, equally important to verify the absolute convergence of the numerator of Eq. (6). Together, both conditions are equivalent to the existence of the posterior moment Eq. (5). It is straightforward to verify these convergence conditions in the examples discussed above.

*More generally, any non-negative function proportional to a probability density is a density kernel.

Many of these ideas can be illustrated in the standard linear model. For an observable $T \times 1$ vector of dependent variables $\mathbf{y}$ and $T \times k$ matrix of fixed covariates $\mathbf{X}$*

$$\mathbf{y} = \mathbf{X}\beta + \varepsilon; \qquad \varepsilon|\mathbf{X} \sim \mathrm{N}(\mathbf{0}, h^{-1}\mathbf{I}_T); \qquad \mathrm{rank}(\mathbf{X}) = k \tag{7}$$

The parameter h is the *precision* of the i.i.d. disturbances, $\varepsilon_1, \ldots, \varepsilon_T$; it is the inverse of $\mathrm{var}(\varepsilon_t) = \sigma^2$.† Consider the independent prior distributions for β and h:

$$\beta \sim \mathrm{N}(\underline{\beta}, \underline{\mathbf{H}}^{-1}) \tag{8}$$

$$\underline{s}^2 h \sim \chi^2(\underline{\nu}) \tag{9}$$

where $\underline{\mathbf{H}}$ is a fixed precision matrix, $\underline{\beta}$ is a fixed mean vector, and $\underline{s}^2$ and $\underline{\nu}$ are fixed scalars. In any given application Eqs. (8) and (9) are not necessarily adequate expressions of prior beliefs.‡ However, the specification in Eqs. (8) and (9) has attractive analytical properties that will become clear in due course. Moreover, in many cases it is straightforward to modify the posterior distribution implied by the prior distributions Eqs. (8) and (9), to express the posterior distributions corresponding to Eq. (7) and alternative prior distributions, using simple numerical methods described in Section 5.

From Eq. (8):

$$\mathrm{p}(\beta) = (2\pi)^{-k/2} |\underline{\mathbf{H}}|^{1/2} \exp[-0.5(\beta - \underline{\beta})' \underline{\mathbf{H}} (\beta - \underline{\beta})] \tag{10}$$

and from Eq. (9):

$$\mathrm{p}(h) = [2^{\underline{\nu}/2}\Gamma(\underline{\nu}/2)]^{-1} (\underline{s}^2)^{\underline{\nu}/2} h^{(\underline{\nu}-2)/2} \exp(-\underline{s}^2 h/2) \tag{11}$$

Since Eq. (7) is equivalent to the conditional data density:

$$\mathrm{p}(\mathbf{y}|\mathbf{X}, \beta, h) = (2\pi)^{-T/2} h^{T/2} \exp[-0.5h(\mathbf{y} - \mathbf{X}\beta)'(\mathbf{y} - \mathbf{X}\beta)] \tag{12}$$

---

*If instead $\mathbf{X}$ is random with p.d.f. $\mathrm{p}(\mathbf{X}|\eta)$, $\mathrm{p}(\beta, h, \eta|A) = \mathrm{p}(\beta, h|A)\mathrm{p}(\eta|A)$ and $\mathrm{p}(\omega|\mathbf{y}, \mathbf{X}, \beta, h, \eta) = \mathrm{p}(\omega|\mathbf{y}, \mathbf{X}, \beta, h)$, then $\mathbf{X}$ is ancillary and the analysis that follows still pertains. For further discussion of ancillarity see Bernardo and Smith (1994, Sect. 5.1.4). The condition of weak exogeneity in the econometrics literature (Engle et al., 1983; Steel and Richard, 1991) is closely related.

†More generally, the precision of any random variable is the inverse of its variance.

‡Nor is Eq. (7), necessarily. We return to this important question in greater depth in Sections 2.3 and 2.4.

a posterior density kernel is the product of Eqs. (10)–(12):

$$(2\pi)^{-(T+k)/2}[2^{\underline{\nu}/2}\Gamma(\underline{\nu}/2)]^{-1} \tag{13a}$$

$$\times |\underline{\mathbf{H}}|^{1/2}(\underline{s}^2)^{\underline{\nu}/2} \tag{13b}$$

$$\times h^{(T+\underline{\nu}-2)/2}\exp(-\underline{s}^2 h/2) \tag{13c}$$

$$\times \exp\{-0.5[(\beta-\underline{\beta})'\underline{\mathbf{H}}(\beta-\underline{\beta}) + h(\mathbf{y}-\mathbf{X}\beta)'(\mathbf{y}-\mathbf{X}\beta)]\} \tag{13d}$$

To simplify this expression, complete the square in β of the term in brackets in Eq. (13d), yielding

$$(\beta-\underline{\beta})'\underline{\mathbf{H}}(\beta-\underline{\beta}) + h(\mathbf{y}-\mathbf{X}\beta)'(\mathbf{y}-\mathbf{X}\beta) = (\beta-\bar{\beta})'\overline{\mathbf{H}}(\beta-\bar{\beta}) + Q$$

where

$$\overline{\mathbf{H}} = \underline{\mathbf{H}} + h\mathbf{X}'\mathbf{X} \tag{14}$$

$$\bar{\beta} = \overline{\mathbf{H}}^{-1}(\underline{\mathbf{H}}\underline{\beta} + h\mathbf{X}'\mathbf{y}) = \overline{\mathbf{H}}^{-1}(\underline{\mathbf{H}}\underline{\beta} + h\mathbf{X}'\mathbf{X}\mathbf{b}) \tag{15}$$

and

$$\begin{aligned} Q &= h\mathbf{y}'\mathbf{y} + \underline{\beta}'\overline{\mathbf{H}}\underline{\beta} - \bar{\beta}'\mathbf{H}\bar{\beta} \\ &= h\nu s^2 + (\mathbf{b}-\bar{\beta})'h\mathbf{X}'\mathbf{X}(\mathbf{b}-\bar{\beta}) + (\underline{\beta}-\bar{\beta})'\underline{\mathbf{H}}(\underline{\beta}-\bar{\beta}) \end{aligned} \tag{16}$$

with $\mathbf{b}$ denoting the coefficients in the ordinary least squares fit of $\mathbf{y}$ to $\mathbf{X}$, $\mathbf{b} = (\mathbf{X}'\mathbf{X})^{-1}\mathbf{X}'\mathbf{y}$; $s^2 = (\mathbf{y}-\mathbf{X}\mathbf{b})'(\mathbf{y}-\mathbf{X}\mathbf{b})/\nu$ and $\nu = T-k$. If Eq. (13) is interpreted as a function of β only, that function must be a posterior density kernel for β conditional on h, and our square completion shows that $p(\beta|h,\mathbf{y},\mathbf{X}) \propto \exp\{-0.5(\beta-\bar{\beta})'\overline{\mathbf{H}}(\beta-\bar{\beta})\}$. Consequently,

$$\beta|(h,\mathbf{y},\mathbf{X}) \sim \mathrm{N}(\bar{\beta},\ \overline{\mathbf{H}}^{-1}) \tag{17}$$

Interpreting Eq. (13) as a function of h alone:

$$p(h|\beta,\mathbf{y},\mathbf{X}) \propto h^{(T+\underline{\nu}-2)/2}\exp\{-0.5[\underline{s}^2 + (\mathbf{y}-\mathbf{X}\beta)'(\mathbf{y}-\mathbf{X}\beta)]h\}$$

and consequently,

$$[\underline{s}^2 + (\mathbf{y}-\mathbf{X}\beta)'(\mathbf{y}-\mathbf{X}\beta)]h|(\beta,\mathbf{y},\mathbf{X}) \sim \chi^2(T+\underline{\nu}) \tag{18}$$

The distributions in Eqs. (8) and (9) are special cases of conditionally conjugate priors (to be defined shortly). They are attractive because they lead to the tractable results [Eqs. (17) and (18)]. Yet these results are not directly useful, for they do not provide distributions conditional only on the data and prior information. However, they form the basis of an attractive simulation method discussed in Section 3.3.

2.2 Conjugate Prior Distributions

The prior distribution $p(\theta|A)$ is a representation of belief in the context of model A. In selecting a prior or data distribution, the richer the class of functional forms from which to choose the more adequate the representation of prior beliefs possible. On the other hand, the choice is constrained by the tractability of the posterior density $p(\theta|\mathbf{Y}_T, A) \propto p(\theta|A)p(\mathbf{Y}_T|\theta, A)$, which is jointly determined by the choice of functional forms for the data density and prior density. The search for rich tractable classes of prior distributions may be formalized by considering classes of prior densities, $p(\theta|A) = p(\theta|\gamma, A)$, where γ is a parameter vector that indexes prior beliefs. We have used this approach in the linear model: e.g., the prior distribution $\beta \sim N(\underline{\beta}, \underline{\mathbf{H}}^{-1})$ is indexed by $\underline{\beta}$ and $\underline{\mathbf{H}}$.

Suppose the model $p(\mathbf{Y}_T|\theta, A)$ has sufficient statistic $\mathbf{s}_T = ((s_T)_1, \ldots, (s_T)_r)' = s_T(\mathbf{Y}_T)$, when r is fixed as T varies, and $(s_T)_1 = T$. Then, the *conjugate family of prior densities with respect to* $p(\mathbf{Y}_T|\theta, A)$ is $\{p(\theta|\gamma, A), \gamma \in \Gamma\}$, where

$$p(\theta|\gamma, A) \propto p[(s_{\gamma_1})_j = \gamma_j \; (j = 2, \ldots, r)|\theta, A]$$

and

$$\Gamma = \left\{\gamma : \int_{\Theta} p[(s_{\gamma_1})_j = \gamma_j \; (j = 2, \ldots, r)|\theta, A] \, d\theta < \infty\right\}$$

The kernel of any conjugate prior density may be interpreted as a likelihood function corresponding to a data set $\mathbf{Z}_{\gamma_1}$ with sufficient statistic $\mathbf{s}'_{\gamma_1} = (\gamma_2, \ldots, \gamma_r)$. To the extent one can represent prior beliefs arising from notional data with the same probability density functional form as the actual data, a conjugate prior distribution will provide a good representation of belief. By construction $p(\mathbf{Y}_T|\theta, A) \propto p^*(\mathbf{s}_T|\theta, A)$ and $p(\theta|A) \propto p^*(\gamma|\theta, A)$, where the proportionality is in θ, and $p^*(\mathbf{s}_T|\theta, A)$ and $p^*(\gamma|\theta, A)$ have exactly the same functional form in θ. Hence, $p(\theta|\mathbf{Y}_T, A) \propto p^*(\mathbf{s}_T|\theta, A)p^*(\gamma|\theta, A)$. It is often the case that the functional form of $p(\theta|\mathbf{Y}_T, A)$ is the same as that of $p^*(\mathbf{s}_T|\theta, A)$, and it is this feature that makes the posterior density tractable.*

To extend this idea, let $\theta' = (\theta'_1, \theta'_2)$ and fix $\theta_2 = \theta_2^0$. Suppose the model $p(\mathbf{Y}_T|\theta_1, \theta_2 = \theta_2^0, A)$ has sufficient statistic $\underset{r^* \times 1}{\mathbf{s}_T^*} = s_T^*(\mathbf{Y}_T)$, r^* is fixed, and $(s_T^*)_1 = T$. Then, the *conditionally conjugate family of prior densities with respect to* $p(\mathbf{Y}_T|\theta_1, \theta_2 = \theta_2^0, A)$ is $\{p(\theta_1|\gamma^*, A), \gamma^* \in \Gamma^*\}$, with $\Gamma^* = \{\gamma^*:$

---

*Indeed, one can begin with this property as the definition of conjugate; see Berger (1985, Sect. 4.2.2) and Poirier (1995, Sect. 6.7). The definition here is that used by Bernardo and Smith (1994, Sect. 5.2.1) and Zellner (1971, Sect. 2.3). For the exponential family of distributions (which includes the standard linear model) the two are equivalent [see Bernardo and Smith (1994), Proposition 5.4].

$\int_{\Theta_1} \text{p}[(s^*_{\gamma_1})_j = \gamma^*_j(j = 2, \ldots, r^*)|\theta_1, \theta_2 = \theta^0_2, A]\, d\theta_1 < \infty\}$ and $\text{p}(\theta_1|\gamma^*, A) \propto \text{p}[(s^*_{\gamma_1})_j = \gamma^*_j(j = 2, \ldots, r^*)|\theta_1, \theta_2 = \theta^0_2, A]$.

The prior distributions Eqs. (8) and (9), are conditionally conjugate, but not conjugate, in the linear model, Eq. (7). In this example, the prior density for $\theta' = (\beta', h)$ is indexed by $\gamma = \{\underline{\beta}, \underline{\mathbf{H}}, \underline{s}^2, \underline{\nu}\}$. In the linear model, because

$$\text{p}(\mathbf{y}|\mathbf{X}, \beta, h) \propto h^{T/2} \exp\{-0.5h[\nu s^2 + (\beta - \mathbf{b})'\mathbf{X}'\mathbf{X}(\beta - \mathbf{b})]\} \tag{19}$$

the vector $\mathbf{s}_T = [T, \mathbf{b}, s^2, \mathbf{X}'\mathbf{X}]$ is a sufficient statistic.* Conditioning on $h = h_0$, $\text{p}(\mathbf{y}|\mathbf{X}, \beta) \propto \exp[-0.5(\beta - \mathbf{b})' h_0 \mathbf{X}'\mathbf{X}(\beta - \mathbf{b})]$. Since $\text{p}(\beta) \propto \exp[-0.5 (\beta - \underline{\beta})' \, \underline{\mathbf{H}}(\beta - \underline{\beta})]$, the prior density, Eq. (10), is conditionally conjugate. Likewise, conditioning on $\beta = \beta_0$, $\text{p}(\mathbf{y}|\mathbf{X}, h) \propto h^{T/2} \exp(-\bar{s}^2 h/2)$ where $\bar{s}^2 = \nu s^2 + (\beta_0 - \mathbf{b})'\mathbf{X}'\mathbf{X}(\beta_0 - \mathbf{b})$. Hence, the prior density, Eq. (11), $\text{p}(h) \propto h^{(\underline{\nu}-2)/2} \exp(-\underline{s}^2 h/2)$, is conditionally conjugate.

2.3 Model Comparison and Combination

Often one has under consideration several complete models, say $A_1, \ldots, A_J$:

$$\text{p}(\theta_j|A_j)\ (\theta_j \in \Theta_j),\ \text{p}(\mathbf{Y}_T|\theta_j, A_j),\quad \text{p}(\omega|\mathbf{Y}_T, \theta_j, A_j)\quad (j = 1, \ldots, J)$$

The numbers of parameters in the models need not be the same, and various models may or may not nest one another. If we assign prior probabilities $\text{P}(A_j)$ $(j = 1, \ldots, J)$ to the respective models, with $\sum_{j=1}^J \text{P}(A_j) = 1$, then there is a complete probability structure for $\{A_j, \theta_j\}_{j=1}^J$, $\mathbf{Y}_T$, and ω. There is no essential conceptual distinction between model and prior, since one could just as well regard the entire collection as the model, with $\{\text{P}(A_j), \text{p}_j(\theta_j|A_j)\}_{j=1}^J$ as the characterization of the prior distribution. At an operational level the distinction is usually clear and useful in that one may undertake the essential computations one model at a time.

Suppose that the posterior moment, Eq. (5), is ultimately of interest. The formal solution is

$$\text{E}[\text{h}(\omega)|\mathbf{Y}_T] = \sum_{j=1}^{J} \text{E}[\text{h}(\omega)|\mathbf{Y}_T, A_j]\text{P}(A_j|\mathbf{Y}_T) \tag{20}$$

*This follows from the Neyman factorization criterion [Bernardo and Smith, 1994, Sect.4.5.2]. Less formally, from Eq. (19) it is clear that we only need to know $\mathbf{s}_T$ to write the likelihood function for β and h.

which is known as *model averaging*. Clearly, $E[h(\omega)|\mathbf{Y}_T, A_j]$ is given by Eq. (6) with $A = A_j$. There is nothing new in this part of Eq. (20). From Bayes' rule:

$$\begin{aligned}
P(A_j|\mathbf{Y}_T) &= P(A_j)p(\mathbf{Y}_T|A_j)\Big/\sum_{j=1}^{J} P(A_j)p(\mathbf{Y}_T|A_j) \\
&= P(A_j)\int_{\Theta_j} p(\mathbf{Y}_T|\theta_j, A_j)p(\theta_j|A_j)\,d\nu(\theta_j)\Big/\sum_{j=1}^{J} P(A_j)p(\mathbf{Y}_T|A_j) \\
&\propto P(A_j)\int_{\Theta_j} p(\mathbf{Y}_T|\theta_j, A_j)p(\theta_j|A_j)\,d\nu(\theta_j) = P(A_j)p(\mathbf{Y}_T|A_j) \qquad (21)
\end{aligned}$$

where $p(\mathbf{Y}_T|A_j) = \int_{\Theta_j} p(\mathbf{Y}_T|\theta_j, A_j)p(\theta_j|A_j)\,d\nu(\theta_j)$ is the marginal likelihood of model j, consistent with the definition in Eq. (3). Notice it is important that the properly normalized prior and properly normalized data density, and not arbitrary kernels of these densities, be used in forming the marginal likelihood.

Model averaging thus involves three steps. First, obtain the posterior moments, Eq. (6), corresponding to each model. Second, obtain the relative values of $P(A_j|\mathbf{Y}_T)$ from Eq. (21). Finally, obtain the posterior moment using Eq. (20), which now only involves simple arithmetic, recognizing that $\sum_{j=1}^{J} P(A_j|\mathbf{Y}_T) = 1$. Variation of the prior model probabilities $P(A_j)$ is a trivial step, as is the revision of the posterior moment following the introduction of a new model or deletion of an old one from the conditioning set of models.

From Eq. (21), for any pair of models A_j and A_k:

$$P(A_j|\mathbf{Y}_T)/P(A_k|\mathbf{Y}_T) = [P(A_j)/P(A_k)][p(\mathbf{Y}_T|A_j)/p(\mathbf{Y}_T|A_k)] \qquad (22)$$

This ratio of probabilities is the *posterior odds ratio* in favor of model j versus model k. It is invariant with respect to the addition and deletion of models from the set $\{A_j\}_{j=1}^{J}$ under consideration, so long as the prior probabilities $\{P(A_j)\}_{j=1}^{J}$ are changed in a logically consistent fashion—i.e., ratios $P(A_j)/P(A_k)$ remain unchanged for all included models.* The posterior odds ratio is expressed in Eq. (22) as the product of the *prior odds ratio* in favor of model j versus model k, $P(A_j)/P(A_k)$, and the *Bayes factor* in favor of model j versus model k, $p(\mathbf{Y}_T|A_j)/p(\mathbf{Y}_T|A_k)$.

*This property is analogous to the independence of irrelevant alternatives in the qualitative choice literature; see Poirier (1997).

In the case of the standard linear model it is straightforward to work out the marginal likelihood and Bayes factors if h is fixed. The product of the properly normalized prior and data densities is

$$\mathrm{p}(\beta)\mathrm{p}(\mathbf{y}|\mathbf{X},\beta,h) = (2\pi)^{-(T+k)/2} h^{T/2} |\underline{\mathbf{H}}|^{1/2} \exp\{-0.5 \times [h(\mathbf{y}-\mathbf{X}\beta)'(\mathbf{y}-\mathbf{X}\beta) + (\beta-\underline{\beta})'\underline{\mathbf{H}}(\beta-\underline{\beta})]\} \tag{23}$$

The term in brackets may be expressed as

$$(\beta-\bar{\beta})'\overline{\mathbf{H}}(\beta-\bar{\beta}) + Q \tag{24}$$

with $\bar{\beta}$, $\overline{\mathbf{H}}$ and Q as defined in Eqs. (14)–(16). Substituting Eq. (24) into (23), the marginal likelihood is

$$\begin{aligned}
&\int_{\Re^k} \mathrm{p}(\beta)\mathrm{p}(\mathbf{y}|\mathbf{X},\beta,h)\,\mathrm{d}\beta \\
&= (2\pi)^{-(T+k)/2} h^{T/2} |\underline{\mathbf{H}}|^{1/2} \int_{\Re^k} \exp\{-0.5[(\beta-\bar{\beta})'\overline{\mathbf{H}}(\beta-\bar{\beta}) + Q]\}\,\mathrm{d}\beta \\
&= (2\pi)^{-T/2} h^{T/2} |\underline{\mathbf{H}}|^{1/2} |\overline{\mathbf{H}}|^{-1/2} \exp(-Q/2) \\
&= (2\pi)^{-T/2} h^{T/2} (|\underline{\mathbf{H}}|/|\overline{\mathbf{H}}|)^{1/2} \exp\{-0.5[hvs^2 \\
&\quad + (\mathbf{b}-\bar{\beta})'h\mathbf{X}'\mathbf{X}(\mathbf{b}-\bar{\beta}) + (\underline{\beta}-\bar{\beta})'\underline{\mathbf{H}}(\underline{\beta}-\bar{\beta})]\}
\end{aligned} \tag{25}$$

From the last expression it is apparent that the marginal likelihood of a linear model depends on more than the least squares fit of $\mathbf{y}$ to $\mathbf{X}$, which is measured by the sum of squared residuals vs^2. It also depends on the squared Euclidean distance of the least squares fit $\mathbf{b}$ from the posterior mean $\bar{\beta}$, using the data-based norm $h\mathbf{X}'\mathbf{X}$; the squared distance of the prior mean $\underline{\beta}$ from the posterior mean $\bar{\beta}$, using the prior-based norm $\underline{\mathbf{H}}$; and the fraction of posterior precision accounted for by prior precision, as measured by $|\underline{\mathbf{H}}|/|\overline{\mathbf{H}}|$.*

2.4 Hierarchical Priors and Latent Variables

A *hierarchical prior distribution* expresses the prior in two or more steps. The two-step case specifies a model:

$$\mathrm{p}^{(1)}(\mathbf{Y}_T|\theta,\lambda,A) \tag{26}$$

---

*Equation (21) shows that the marginal likelihood of model j, $\mathrm{p}(\mathbf{Y}_T|A_j)$, is the measure of how well model A_j predicted the observed data $\mathbf{Y}_T$ that is relevant for the comparison of model j with any other models. For a formal development of this idea, see Geweke (2001).

with a prior density for $\theta \in \Theta$ conditional on a vector of *hyperparameters* $\phi \in \Phi$:

$$\mathrm{p}^{(2)}(\theta|\phi, A) \tag{27}$$

and a prior density for ϕ and $\lambda \in \Lambda$:

$$\mathrm{p}^{(3)}(\phi, \lambda|A) \tag{28}$$

it being understood in Eq. (26) that $\mathrm{p}^{(1)}(\mathbf{Y}_T|\theta, \lambda, A) = \mathrm{p}(\mathbf{Y}_T|\theta, \lambda, \phi, A)$.

The full prior density for all parameters and hyperparameters is

$$\mathrm{p}(\theta, \phi, \lambda|A) = \mathrm{p}^{(3)}(\phi, \lambda|A)\mathrm{p}^{(2)}(\theta|\phi, A) \tag{29}$$

There is no fundamental difference between this prior density and the one described in Section 2.1, since

$$\mathrm{p}(\theta, \lambda) = \int_{\Phi} \mathrm{p}^{(2)}(\theta|\phi, A)\mathrm{p}^{(3)}(\phi, \lambda|A)\, d\nu(\phi)$$

However, the hierarchical formulation is often so convenient as to render fairly simple the analysis of posterior densities that would otherwise be quite difficult. Given a hierarchical prior, one may express the full posterior density:

$$\mathrm{p}(\theta, \lambda, \phi|\mathbf{Y}_T, A) \propto \mathrm{p}^{(1)}(\mathbf{Y}_T|\theta, \lambda, A)\mathrm{p}^{(2)}(\theta|\phi, A)\mathrm{p}^{(3)}(\phi, \lambda|A) \tag{30}$$

A *latent variable model* expresses the likelihood function in two or more steps. In the two-step case the likelihood function may be written:

$$\mathrm{p}^{(1)}(\mathbf{Y}_T|\mathbf{Z}_T^*, \lambda, A) \tag{31}$$

where $\mathbf{Z}_T^* \in \tilde{\mathbf{Z}}_T$ is a matrix of latent variables and $\lambda \in \Lambda$. The model for $\mathbf{Z}_T^*$ is

$$\mathrm{p}^{(2)}(\mathbf{Z}_T^*|\phi, A) \tag{32}$$

and the prior density for $\phi \in \Phi$ and λ is

$$\mathrm{p}^{(3)}(\phi, \lambda|A) \tag{33}$$

The full prior density for all parameters and unobservable variables is

$$\mathrm{p}(\mathbf{Z}_T^*, \lambda, \phi|A) = \mathrm{p}^{(3)}(\phi, \lambda|A)\mathrm{p}^{(2)}(\mathbf{Z}_T^*|\phi, A) \tag{34}$$

and the full posterior density is

$$\mathrm{p}(\mathbf{Z}_T^*, \lambda, \phi|\mathbf{Y}_T, A) \propto \mathrm{p}^{(1)}(\mathbf{Y}_T|\mathbf{Z}_T^*, \lambda, A)\mathrm{p}^{(2)}(\mathbf{Z}_T^*|\phi, A)\mathrm{p}^{(3)}(\phi, \lambda|A) \tag{35}$$

Comparing Eqs. (26)–(30) with (31)–(35), it is apparent that the latent variable model is formally identical to a model with a two-stage hierarchical prior, the latent variables corresponding to the intermediate level of the hierarchy. With appropriate marginalization of Eq. (35) one may obtain $p(\mathbf{Z}_T^*|\mathbf{Y}_T, A)$, which fully reflects uncertainty about the parameters. If one is interested only in λ and ϕ, these distributions may also be obtained by marginalization of Eq. (35). Marginalization requires integration over $\mathbf{Z}_T^*$, which is possible analytically only in special cases. If the problem is approached using the simulation methods described beginning in the Section 3, then this integration simply amounts to discarding simulated values of $\mathbf{Z}_T^*$.

A simple example of a latent variable model is provided by the textbook probit model:

$$\mathbf{y}^* = \mathbf{X}\beta + \varepsilon, \quad \varepsilon|\mathbf{X} \sim \mathrm{N}(\mathbf{0}, \mathbf{I}_T), \quad \mathrm{rank}(\mathbf{X}) = k, \quad d_t = \chi_{[0,\infty)}(y_t^*) \tag{36}$$

in which the $T \times k$ matrix of covariates $\mathbf{X} = [\mathbf{x}'_1, \ldots, \mathbf{x}'_T]'$ and decision vector $\mathbf{d}' = (d_1, \ldots, d_T)$ are observed, but $\mathbf{y}'^* = (y_1^*, \ldots, y_T^*)$ is latent. To complete the model take

$$\beta \sim \mathrm{N}(\underline{\beta}, \underline{\mathbf{H}}^{-1}) \tag{37}$$

In the equivalent formulation of this model using a hierarchical prior the parameter vector is $(\mathbf{y}^*, \beta)$. The first level of the hierarchical prior is $\beta \sim \mathrm{N}(\underline{\beta}, \underline{\mathbf{H}}^{-1})$, corresponding to $p^{(3)}$ with $\phi = \beta$. The second level is $\mathbf{y}^*|(\beta, \mathbf{X}) \sim \mathrm{N}(\mathbf{X}\beta, \mathbf{I})$, with $\theta = \mathbf{y}^*$ in the hierarchical prior interpretation and $\mathbf{Z}_T^* = \mathbf{y}^*$ in the latent variable interpretation. (There is no analog of λ in this example.) The data distribution is

$$p(\mathbf{d}|\mathbf{y}^*) = \prod_{t=1}^{T} [\chi_{[0,\infty)}(y_t^*)\, d_t + \chi_{(-\infty,0)}(y_t^*)(1 - d_t)]$$

Either formulation leads to the same joint distribution for β, $\mathbf{y}^*$, and $\mathbf{d}$:

$$\begin{aligned} p(\beta, \mathbf{y}^*, \mathbf{d}|\mathbf{X}) = {} & (2\pi)^{-(T+k)/2}|\underline{\mathbf{H}}|^{1/2} \exp[-0.5(\beta - \underline{\beta})'\underline{\mathbf{H}}(\beta - \underline{\beta})] \\ & \times \prod_{t=1}^{T} \exp[-0.5(y_t^* - \beta'\mathbf{x}_t)^2][\chi_{([0,\infty)}(y_t^*)\, d_t \\ & + \chi_{(-\infty,0)}(y_t^*)(1 - d_t)] \end{aligned} \tag{38}$$

The main conceptual point is that since Bayesian inference conditions on the observables $(\mathbf{d}, \mathbf{X})$, parameters and latent variables have the same standing as unknown entities whose joint distribution with the observables is given by the model. As we shall see in in Section 3.3, this formulation provides a basis for computations as well.

3 POSTERIOR SIMULATION METHODS

The objective of inference in a single model:

$$E[h(\omega)|\mathbf{Y}_T, A] = \int_\Theta \int_\Omega h(\omega)p(\omega|\theta, \mathbf{Y}_T, A)p(\theta|\mathbf{Y}_T, A)\, d\nu(\omega)\, d\nu(\theta),$$

can be evaluated analytically only in a few specific simple cases. This section describes simulation methods for obtaining a sequence of strongly consistent approximations to $E[h(\omega)|\mathbf{Y}_T, A]$, and Section 4 will take up the process of model averaging. In most applications, it is generally straightforward to find a function $g(\mathbf{Y}_T, \theta)$, possibly random, with the property:

$$\begin{aligned} E[g(\mathbf{Y}_T, \theta)|\mathbf{Y}_T, \theta, A] &= E[h(\omega)|\mathbf{Y}_T, \theta, A] \\ &= \int_\Omega h(\omega)p(\omega|\mathbf{Y}_T, \theta, A)\, d\omega = \bar{g}(y_T, \theta) \end{aligned} \tag{39}$$

Finding this function is trivial if $h(\omega)|(\mathbf{Y}_T, \theta, A)$ is deterministic.* If $h(\omega)$ is random, then it is often straightforward to take $\omega \sim p(\omega|\mathbf{Y}_T, \theta, A)$ and then $g(\mathbf{Y}_T, \theta) = h(\omega)$. In either case,

$$E[h(\omega)|\mathbf{Y}_T, A] = \int_\Theta \bar{g}(\mathbf{Y}_T, \theta)p(\theta|\mathbf{Y}_T, A)\, d\nu(\theta) = \bar{g}$$

More generally, one may be able to find a function satisfying Eq. (39), but for which

$$\operatorname{var}[g(\mathbf{Y}_T, \theta)|\mathbf{Y}_T, \theta, A] < \operatorname{var}[h(\omega)|\mathbf{Y}_T, \theta, A] \tag{40}$$

The turning point example of Section 2.1 provides an illustration. Recall that in this example the objective was to evaluate $P(y_{T+2} < y_{T+1} < y_T|\mathbf{Y}_T)$ and to this end we took $\omega' = (y_{T+1}, y_{T+2})$. One could draw $\omega|(\mathbf{Y}_T, \theta)$ and use the random function $g(\mathbf{Y}_T, \theta) = h(\omega) = \chi_{\Omega^*}(\omega)$. Alternatively, one could draw only $\omega_1|(\mathbf{Y}_T, \theta)$ and use the random function $g(\mathbf{Y}_T, \theta) = P(\omega_2 < \omega_1|\mathbf{Y}_T, \theta)$, which requires only the ability to evaluate the univariate standard normal c.d.f. Yet a third alternative is to employ the deterministic function $g(\mathbf{Y}_T, \theta) = P(\omega_2 < \omega_1 < y_T|\mathbf{Y}_T, \theta)$ using bivariate quad-

*The *evaluation* of $g(\mathbf{Y}_T, \theta)$ may not be trivial at all. For example, Bajari (1997) has functions of interest whose evaluation requires the solution of a system of nonlinear differential equations.

rature. In each case $E[g(\mathbf{Y}_T, \theta)|\mathbf{Y}_T, \theta] = P(y_{T+2} < y_{T+1} < y_T|\mathbf{Y}_T, \theta)$ but $\mathrm{var}[g(\mathbf{Y}_T, \theta)|\mathbf{Y}_T, \theta]$ is greatest in the first alternative, less in the second, and zero in the third.*

Throughout we shall make use of the notation $g(\mathbf{Y}_T, \theta)$, it always being implicit that Eq. (39) is satisfied.

If one could also make a sequence of independent draws $\{\theta^{(m)}\}$ from the posterior distribution, then by choosing $\omega^{(m)} \sim p(\omega|\mathbf{Y}_T, \theta^{(m)}, A)$ one could guarantee $M^{-1} \sum_{m=1}^{M} h(\omega^{(m)}) \xrightarrow{a.s.} E[h(\omega)|\mathbf{Y}_T, A]$. However, direct simulation from the posterior distribution is rarely possible. This section describes methods for obtaining a sequence $\{\theta^{(m)}\}_{m=1}^{\infty}$, and an associated weighting function $w(\theta)$, with the property that if $E[h(\omega^{(m)})|\theta^{(m)}] = E[h(\omega)|\mathbf{Y}_T, \theta^{(m)}, A]$ for a corresponding sequence $\{\omega^{(m)}\}_{m=1}^{\infty}$, then

$$\sum_{m=1}^{M} w(\theta^{(m)}) h(\omega^{(m)}) \Big/ \sum_{m=1}^{M} w(\theta^{(m)}) \xrightarrow{a.s.} E[h(\omega)|\mathbf{Y}_T, A]$$

The ability to generate such sequences has improved greatly in the past 10 years, due in large part to the development of Markov chain Monte Carlo (MCMC) methods and the dramatic decrease in the cost of computing. We begin by reviewing two more established methods, acceptance and importance sampling, and then move on to the Gibbs sampler and the Hastings–Metropolis algorithm as examples of MCMC. This is followed by a more abstract development of MCMC theory, a description of some of the hybrid procedures that make MCMC a powerful tool for posterior simulation, and a discussion of the evaluation of approximation error. The section concludes with a description of some public domain software for posterior simulation and two simple examples. The emphasis here is on concepts and practicality; only references to proofs of theorems are given. A more general and extensive introduction is provided by Gelman et al. (1995). A concise presentation of the relevant continuous state space Markov chain theory that underlies MCMC procedures is given in Tierney (1994).

3.1 Acceptance Sampling

Acceptance sampling is the algorithm that underlies the generation of random variables from most familiar univariate distributions like the normal and the gamma (Press et al., 1992). The idea behind acceptance

*In some cases the left-hand side of Eq. (40) can be made quite small indeed, and asymptotically it may be made to approach zero (Geweke, 1988).

sampling is to generate a random vector* from a distribution that is similar, in an appropriate sense, to the posterior distribution, and then to accept that drawing with a probability that depends on the drawn value of the vector. If this acceptance probability function is chosen correctly, then the accepted values will have the desired distribution.

Theorem 1

Suppose that $p^*(\theta|\mathbf{Y}_T, A)$ is any kernel of the posterior density $p(\theta|\mathbf{Y}_T, A)$. Let $s^*(\theta)$ be a source density kernel with respect to the same measure $d\nu(\theta)$ as $p(\theta|A)$, with support S and the property:

$$0 \leq p^*(\theta|\mathbf{Y}_T, A)/s^*(\theta) \leq a < \infty \, \forall \theta \in \Theta \tag{41}$$

Suppose that at each iteration a candidate $\tilde{\theta}$ is drawn from the source density, and $\theta^{(m)}$ is set to $\tilde{\theta}$ with probability $p^*(\tilde{\theta}|\mathbf{Y}_T, A)/[as^*(\tilde{\theta})]$. This process is repeated until $\theta^{(m)}$ is set to a candidate $\tilde{\theta}$. Then $\theta^{(m)} \overset{IID}{\sim} p(\theta|\mathbf{Y}_T, A)$.

Proof

See Geweke (1999, Theorem 3.1.1).

A successful application of acceptance sampling has three requirements. First, there must be a source density corresponding to a distribution from which it is efficient and convenient to make i.i.d. draws. Second, there must be a known upper bound on the ratio of the posterior density to the source density. Finally, the frequency with which $\theta^{(m)}$ set to $\tilde{\theta}$ must not be so small that the whole algorithm is impractical. The upper bound must be established analytically, whereas efficiency can be evaluated through experimentation. Notice that draws from the source density may (and usually do) involve acceptance sampling: e.g., if the source density is a normal or gamma density, the software used to draw from this density very likely employs acceptance sampling, a fact typically transparent to the software user.

Acceptance sampling produces an i.i.d. sequence $\{\theta^{(m)}\}$. It follows from the strong law of large numbers that $\bar{g}_M = M^{-1} \times \sum_{m=1}^{M} g(\mathbf{Y}_T, \theta^{(m)}) \xrightarrow{a.s.} \bar{g}$. If in addition $\sigma^2 = var[g(\mathbf{Y}_T, \theta)|\mathbf{Y}_T, A]$ exists then,

*We ignore the distinction between the mathematical properties of a sequence of random variables and the properties of (what is properly called) a pseudo-random variable sequence created using a computer. For a discussion of these issues see Geweke (1996) and references therein.

from the Lindberg–Levy central limit theorem, $M^{1/2}[\bar{g}_m - \bar{g}] \xrightarrow{d} \mathrm{N}(0, \sigma^2)$, and a second application of the strong law of large numbers yields $\hat{\sigma}^2 = M^{-1}\sum_{m=1}^{M}[\mathrm{g}(\mathbf{Y}_T, \theta^{(m)}) - \bar{g}_M] \xrightarrow{a.s.} \sigma^2$. Thus, if the posterior variance of the function of interest exists, a central limit theorem may be used in the usual way to assess the numerical accuracy of the approximation of $\mathrm{E}[\mathrm{h}(\omega)|\mathbf{Y}_T, A]$ by $M^{-1}\sum_{m=1}^{M}\mathrm{g}(\mathbf{Y}_T, \theta^{(m)})$.

3.2 Importance Sampling

Rather than accept only a fraction of the draws from the source density, it is possible to retain all of them, and consistently approximate the posterior moment by appropriately weighting the draws. The probability density function of the source distribution is then called the *importance sampling density*, a term due to Hammersly and Handscomb (1964), who were among the first to propose the method. It appears to have been introduced to the econometrics literature by Kloek and van Dijk (1978). To help distinguish between acceptance and importance sampling we shall indicate the importance sampling distribution by its density $j(\theta)$ with respect to the same measure $d\nu(\theta)$ as the prior density $\mathrm{p}(\theta|A)$. Let $j^*(\theta)$ be any kernel of $j(\theta)$, and let $p^*(\theta|\mathbf{Y}_T, A)$ be any kernel of $\mathrm{p}(\theta|\mathbf{Y}_T, A)$.

Theorem 2

Suppose $\mathrm{E}[\mathrm{g}(\mathbf{Y}_T, \theta)|\mathbf{Y}_T, A]$ exists, and the support of $j(\theta)$ includes Θ. Then

$$\bar{g}_M = \sum_{m=1}^{M}\mathrm{g}(\mathbf{Y}_T, \theta^{(m)})\mathrm{w}(\theta^{(m)}) \Big/ \sum_{m=1}^{M}\mathrm{w}(\theta^{(m)}) \xrightarrow{a.s.} \mathrm{E}[\mathrm{g}(\mathbf{Y}_T, \theta)|\mathbf{Y}_T, A] = \bar{g}$$

where $\mathrm{w}(\theta) = \mathrm{p}^*(\theta|\mathbf{Y}_T, A)/j^*(\theta)$ is the corresponding *weighting function*. If in addition both $\mathrm{E}[\mathrm{w}(\theta)|\mathbf{Y}_T, A] = \int_\Theta \mathrm{w}(\theta)\mathrm{p}(\theta|\mathbf{Y}_T, A)\,d\nu(\theta)$ and $\mathrm{var}[\mathrm{g}(\mathbf{Y}_T, \theta)|\mathbf{Y}_T, A]$ exist, then

$$M^{1/2}(\bar{g}_M - \bar{g}) \xrightarrow{d} \mathrm{N}(0, \sigma^2) \tag{42}$$

and

$$\hat{\sigma}_M^2 = M\sum_{m=1}^{M}[\mathrm{g}(\mathbf{Y}_T, \theta^{(m)}) - \bar{g}_M]^2\mathrm{w}(\theta^{(m)})^2 \Big/ \left[\sum_{m=1}^{M}\mathrm{w}(\theta^{(m)})\right]^2 \xrightarrow{a.s.} \sigma^2 \tag{43}$$

Proof

See Geweke (1989b, Theorems 1 and 2).

This result provides a practical way to assess approximation error and also indicates conditions in which the method of importance sampling will work well. Small variance in $w(\theta)$, perhaps reflecting close upper and lower bounds on $w(\theta)$, will lead to small values of σ^2 relative to $\text{var}[g(\mathbf{Y}_T, \theta)|\mathbf{Y}_T, A]$. Of course, the existence of $E[w(\theta)|\mathbf{Y}_T, A]$ and $\text{var}[g(\mathbf{Y}_T, \theta)|\mathbf{Y}_T, A]$ must be verified analytically. The following implication of Theorem 2 is often useful in the latter undertaking.

Corollary 1

If $\text{var}[g(\mathbf{Y}_T, \theta)|\mathbf{Y}_T, A]$ exists and the weighting function $w(\theta) = p^*(\theta|\mathbf{Y}_T, A)/j^*(\theta)$ is bounded, then Eqs. (42) and (43) are true.

The hypothetical special case $j(\theta) \propto p(\theta|\mathbf{Y}_T, A)$ corresponds to i.i.d. sampling from the posterior distribution, since the weighting function is then constant. In this case, $\sigma^2 = \text{var}[g(\mathbf{Y}_T, \theta)|\mathbf{Y}_T, A]$, which can serve as a benchmark in evaluating the adequacy of $j(\theta)$ in all other cases. The ratio $\text{var}[g(\theta, \mathbf{Y}_T)|\mathbf{Y}_T, A]/\sigma^2$ has been termed the *relative numerical efficiency* of the importance sampling approximation to $E[g(\mathbf{Y}_T, \theta)|\mathbf{Y}_T, A]$ (Geweke, 1989b): it indicates the ratio of iterations using $p(\theta|\mathbf{Y}_T, A)$ itself as the importance sampling density, to the number using $j(\theta)$, required to achieve the same accuracy of approximation of $\bar{g}$. Since both the numerator and denominator of the ratio $\text{var}[g(\theta, \mathbf{Y}_T)|\mathbf{Y}_T, A]/\sigma^2$ can be approximated consistently as the number of draws M increases, this is a practical indication of the computational efficiency of importance sampling. Relative numerical efficiency much less than 1.0 (less than 0.1, certainly less than 0.01) indicates poor imitation of $p(\theta|\mathbf{Y}_T, A)$ by $j(\theta)$, possibly the existence of a better importance sampling distribution or the failure of the underlying convergence conditions for Eq. (43).

Importance sampling is an important useful tool in modifying prior distributions. Suppose that models A_1 and A_2 are distinguished only by their prior densities $p(\theta|A_j)$, $j = 1, 2$. Suppose that one has available an i.i.d. sample from the posterior density $p(\theta|\mathbf{Y}_T, A_1) \propto p(\theta|A_1)p(\mathbf{Y}_T|\theta, A_1)$. If $p(\theta|A_2)/p(\theta|A_1)$ is bounded above, then $p(\theta|\mathbf{Y}_T, A_1)$ is an importance sampling density for $p(\theta|\mathbf{Y}_T, A_2)$ that satisfies the conditions of Corollary 1. The weighting function is $w(\theta) = p(\theta|A_2)/p(\theta|A_1)$. Thus, one may change the prior distribution without reworking the entire problem. The ability to do so makes conditionally conjugate prior distributions, of the kind discussed in Section 2 in conjunction with the standard linear model, attractive as reporting devices because an investigator's results, produced with such priors, may be modified by a client with different priors. This idea will be developed more fully in Section 5.

3.3 The Gibbs Sampler

The Gibbs sampler is an algorithm that has been used with noted success in many econometric models. It is one example of a wider class of procedures known as *Markov chain Monte Carlo* (MCMC). In these procedures the idea is to construct a Markov chain with state space Θ, and unique invariant distribution $p(\theta|\mathbf{Y}_T, A)$. Following an initial transient or *burn-in* phase, simulated values from the chain are used to approximate $E[g(\mathbf{Y}_T, \theta)|\mathbf{Y}_T, A]$. The Gibbs sampler in the form described here was first fully developed by Gelfand and Smith (1990).

The Gibbs sampler begins with a partition, or *blocking*, of θ, $\theta' = (\theta'_{(1)}, \ldots, \theta'_{(B)})$. In applications, the blocking is chosen so that it is possible to draw from each of the conditional p.d.f.'s, $p(\theta_{(b)}|\mathbf{Y}_T, \theta_{(a)}\ (a < b), \theta_{(a)}\ (a > b), A)$. This blocking can arise naturally, if the prior distributions for the $\theta_{(b)}$ are independent and each is conditionally conjugate. To motivate the key idea underlying the Gibbs sampler, suppose—contrary to fact—that there existed a single drawing $\theta^{(0)}$, $\theta'^{(0)} = (\theta'_{(1)}(0), \ldots, \theta'_{(B)}(0))$, from $p(\theta|\mathbf{Y}_T, A)$. Successively make drawings from the conditional distributions as follows:

$$
\begin{aligned}
&\theta^{(1)}_{(1)} \sim p\left(\cdot|\mathbf{Y}_T, \theta^{(0)}_{(2)}, \ldots, \theta^{(0)}_{(B)}, A\right)\\
&\theta^{(1)}_{(2)} \sim p\left(\cdot|\mathbf{Y}_T, \theta^{(1)}_{(1)}, \theta^{(0)}_{(3)}, \ldots, \theta^{(0)}_{(B)}, A\right)\\
&\quad\vdots\\
&\theta^{(1)}_{(b)} \sim p\left(\cdot|\mathbf{Y}_T, \theta^{(1)}_{(1)}, \ldots, \theta^{(1)}_{(b-1)}, \theta^{(0)}_{(b+1)}, \ldots, \theta^{(0)}_{(B)}, A\right) \qquad (44)\\
&\quad\vdots\\
&\theta^{(1)}_{(B)} \sim p\left(\cdot|\mathbf{Y}_T, \theta^{(1)}_{(1)}, \ldots, \theta^{(1)}_{(B-1)}, A\right).
\end{aligned}
$$

This defines a transition process from $\theta'^{(0)}$ to $\theta'^{(1)} = \left(\theta'^{(1)}_{(1)}, \ldots, \theta'^{(1)}_{(B)}\right)$. Since $\theta^{(0)} \sim p(\theta|\mathbf{Y}_T, A)$:

$$\left(\theta^{(1)}_{(1)}, \ldots, \theta^{(1)}_{(b-1)}, \theta^{(1)}_{(b)}, \theta^{(0)}_{(b+1)}, \ldots, \theta^{(0)}_{(B)}\right) \sim p(\theta|\mathbf{Y}_T, A)$$

at each step in Eq. (44) by definition of the conditional density. In particular, $\theta^{(1)} \sim p(\theta|\mathbf{Y}_T, A)$.

Iteration of this algorithm produces a sequence $\theta^{(0)}, \theta^{(1)}, \ldots, \theta^{(m)}, \ldots,$ which is a realization of a Markov chain with probability density function kernel for the transition from point $\theta^{(m)}$ to point $\theta^{(m+1)}$ given by

$$\mathrm{K}_G(\theta^{(m)}, \theta^{(m+1)}) = \prod_{b=1}^{B} \mathrm{p}\left[\theta_{(b)}^{(m+1)} | \mathbf{Y}_T, \theta_{(a)}^{(m)}(a > b), \theta_{(a)}^{(m+1)}(a < b), A\right] \quad (45)$$

Any single iterate $\theta^{(m)}$ retains the property that it is drawn from the posterior distribution. For the Gibbs sampler to be practical, it is essential that the blocking be chosen in such a way that one can make the drawings in an efficient manner. In econometrics the blocking is often natural and the conditional distributions familiar. In making the drawings, Eq. (44), acceptance sampling is often useful.

The appeal of the Gibbs sampler is easy to illustrate with the standard linear model, Eqs. (7)–(9): The results, Eqs. (17) and (18), indicate that the blocking $\theta_{(1)} = \beta$, $\theta_{(2)} = h$ meets the criterion that drawings can be made in an efficient manner. The probit model introduced in Section 2.4 is a further example, as noted in Albert and Chib (1993). From Eqs. (36) and (37) it is evident that conditional on the vector of latent variables $\mathbf{y}^*$ the distribution of β is given by Eq. (17) if we use $\mathbf{y}^*$ in place of $\mathbf{y}$ and set $h = 1$. Examination of the kernel of Eq. (38) in $\mathbf{y}^*$ shows that, given β and $\mathbf{X}$, the y_t^* are conditionally independent, with $y_t^* \sim \mathrm{N}(\beta'\mathbf{x}_t, 1)$ truncated to $[0, \infty)$ if $d_t = 1$ and truncated to $(-\infty, 0)$ if $d_t = 0$. An efficient algorithm for drawing from truncated normal distributions is given in Geweke (1991). In both cases, given drawings for the parameters it is straightforward to produce numerical approximations to $\mathrm{E}[\mathrm{h}(\omega)|\mathbf{Y}_T, A]$, as indicated at the start of this section, and, as discussed in Section 2.1, the evaluation of $\mathrm{E}[\mathrm{h}(\omega)|\mathbf{Y}_T, A]$ subsumes most of the uses to which these models are put.

Of course, if it really were possible to make an initial draw from the posterior distribution, then independence Monte Carlo would also be possible. An important remaining task is to elucidate conditions for the distribution of $\theta^{(m)}$ to converge to the posterior for any $\theta^{(0)} \in \Theta$. This is not trivial, because even if $\theta^{(0)}$ were drawn from $\mathrm{p}(\theta|\mathbf{Y}_T, A)$, the argument just given establishes only that any single $\theta^{(m)}$ is also drawn from the posterior distribution. It does not establish that a single sequence $\{\theta^{(m)}\}_{m=1}^{\infty}$ is representative of the posterior distribution. For example, if Θ consists of two disjoint subsets Θ_1 and Θ_2 with $\theta_1 > \theta_2 \; \forall \; \theta_j \in \Theta_j$, then a Gibbs sampler that begins in Θ_1 and has individual elements of Θ as blocks will never visit Θ_2 and vice versa. This situation clearly does not arise in the Gibbs samplers for the standard linear and probit models just described, but evidently a careful development of conditions under which $\{\theta^{(m)}\}$

converges in distribution to the posterior distribution is needed. We outline these developments in Section 3.5.

3.4 The Hastings–Metropolis Algorithm

The Hastings–Metropolis algorithm begins with an arbitrary transition probability density function $q(\mathbf{x}, \mathbf{y})$ indexed by $\mathbf{x} \in \Theta$ and with density argument $\mathbf{y} \in \Theta$, and with an arbitrary starting value $\theta^{(0)} \in \Theta$. The random vector θ^* generated from $q(\theta^{(m)}, \theta^*)$ is a candidate value for $\theta^{(m+1)}$. The algorithm actually sets $\theta^{(m+1)} = \theta^*$ with probability:

$$\begin{aligned}\alpha(\theta^{(m)}, \theta^*) &= \min\left\{\frac{p(\theta^*|\mathbf{Y}_T, A)q(\theta^*, \theta^{(m)})}{p(\theta^{(m)}|\mathbf{Y}_T, A)q(\theta^{(m)}, \theta^*)}, 1\right\} \\ &= \min\left\{\frac{p(\theta^*|\mathbf{Y}_T, A)/q(\theta^{(m)}, \theta^*)}{p(\theta^{(m)}|\mathbf{Y}_T, A)/q(\theta^*, \theta^{(m)})}, 1\right\}\end{aligned} \tag{46}$$

otherwise, the algorithm sets $\theta^{(m+1)} = \theta^{(m)}$. This defines a Markov chain with a generally mixed continuous–discrete transition probability from $\theta^{(m)}$ to $\theta^{(m+1)}$ given by

$$K_H(\theta^{(m)}, \theta^{(m+1)}) = \begin{cases} q(\theta^{(m)}, \theta^{(m+1)})\alpha(\theta^{(m)}, \theta^{(m+1)}) & \text{if } \theta^{(m+1)} \neq \theta^{(m)} \\ 1 - \int_\Theta q(\theta^{(m)}, \theta)\alpha(\theta^{(m)}, \theta)\, d\nu(\theta) & \text{if } \theta^{(m+1)} = \theta^{(m)} \end{cases}$$

This form of the algorithm is due to Hastings (1970). The Metropolis et al. (1953) form takes $q(\theta^{(m)}, \theta^*) = q(\theta^*, \theta^{(m)})$. A simple variant that is often useful is the *independence chain* (Tierney, 1994), whereby $q(\theta^{(m)}, \theta^*) = k(\theta^*)$. Then

$$\alpha(\theta^{(m)}, \theta^*) = \min\left\{\frac{p(\theta^*|\mathbf{Y}_T, A)k(\theta^{(m)})}{p(\theta^{(m)}|\mathbf{Y}_T, A)k(\theta^*)}, 1\right\} = \min\left\{\frac{w(\theta^*)}{w(\theta^{(m)})}, 1\right\}$$

where $w(\theta) = p(\theta|\mathbf{Y}_T, A)/k(\theta)$. The independence chain is closely related to acceptance sampling and importance sampling. In acceptance sampling, if the posterior density is low (high) relative to the source density the probability of acceptance is low (high). In importance sampling, if the posterior density is low (high) relative to the importance sampling the weight assigned to the draw is low (high). In the independence chain, to the extent the posterior density is lower (higher) relative to the proposal than was the case in the previously accepted draw, the probability of accepting the proposed vector is lower (one).

There is a simple two-step argument that motivates the convergence of the sequence $\{\theta^{(m)}\}$ generated by the Hastings–Metropolis algorithm to the posterior. (This approach is due to Chib and Greenberg, 1995.) First, observe that if the transition probability function $p(\theta^{(m)}, \theta^{(m+1)})$ satisfies the *reversibility condition:*

$$p(\theta^{(m)})p(\theta^{(m)}, \theta^{(m+1)}) = p(\theta^{(m+1)})p(\theta^{(m+1)}, \theta^{(m)}) \qquad (47)$$

for stated $p(\cdot)$, then it has $p(\cdot)$ as an invariant distribution. To see this, note that if Eq. (47) holds then

$$\begin{aligned}\int_{\Theta} p(\theta^{(m)})p(\theta^{(m)}, \theta^{(m+1)})\, d\nu(\theta^{(m)}) &= \int_{\Theta} p(\theta^{(m+1)})p(\theta^{(m+1)}, \theta^{(m)})\, d\nu(\theta^{(m)}) \\ &= p(\theta^{(m+1)}) \int_{\Theta} p(\theta^{(m+1)}, \theta^{(m)})\, d\nu(\theta^{(m)}) \\ &= p(\theta^{(m+1)})\end{aligned}$$

For $\theta^{(m+1)} = \theta^{(m)}$, Eq. (47) is satisfied trivially. For $\theta^{(m+1)} \neq \theta^{(m)}$, suppose without loss of generality that $p(\theta^{(m+1)})/q(\theta^{(m)}, \theta^{(m+1)}) > p(\theta^{(m)})/q(\theta^{(m+1)}, \theta^{(m)})$. Then

$$p(\theta^{(m)}, \theta^{(m+1)}) = q(\theta^{(m)}, \theta^{(m+1)})$$

and

$$\begin{aligned}p(\theta^{(m+1)}, \theta^{(m)}) &= q(\theta^{(m+1)}, \theta^{(m)}) \cdot \frac{p(\theta^{(m)})/q(\theta^{(m+1)}, \theta^{(m)})}{p(\theta^{(m+1)})/q(\theta^{(m)}, \theta^{(m+1)})} \\ &= p(\theta^{(m)})q(\theta^{(m)}, \theta^{(m+1)})/p(\theta^{(m+1)})\end{aligned}$$

whence Eq. (47) is satisfied.

In implementing the Hastings–Metropolis algorithm the transition probability density function must share two important properties. First, it must be possible to generate θ^* efficiently from $q(\theta^{(m)}, \theta^*)$. A second key characteristic of a satisfactory transition process is that the unconditional acceptance rate not be so low that the time required to generate a sufficient number of distinct $\theta^{(m)}$ is too great.

In the case of the independence chain the Hastings–Metropolis algorithm will be efficient under essentially the same conditions that the corresponding importance sampling algorithm with the same $j(\theta)$ will be efficient. If there are values of θ for which $p^*(\theta|\mathbf{Y}_T, A)/j(\theta)$ is very much greater than at other values, then the importance sampling algorithm will place very high weights on these values, which are drawn infrequently relative to $p^*(\theta|\mathbf{Y}_T, A)$. The Hastings–Metropolis independence chain will tend to remain at such values for many successive iterations. In either case relative numerical efficiency will, as a consequence, be low.

The Hastings–Metropolis algorithm and the Gibbs sampling algorithm can be combined in a way that has proved quite useful in econometrics. In a two-block Gibbs sampler, suppose that it is straightforward to sample from $p(\theta_{(1)}|\mathbf{Y}_T, \theta_{(2)}, A)$, but the distribution corresponding to $p(\theta_{(2)}|\mathbf{Y}_T, \theta_{(1)}, A)$ is intractable. The Hastings–Metropolis algorithm can be used in these circumstances, and it often provides an efficient solution to the problem. In what has become known as the Metropolis-within-Gibbs procedure, at the $(m+1)$th iteration first draw $\theta^*_{(2)}$ from a proposal density $q(\theta^{(m)}_{(2)}, \theta^*_{(2)}|\theta^{(m+1)}_{(1)})$. Accept this draw with probability

$$\min\left\{\frac{p(\theta^{(m+1)}_{(1)}, \theta^*_{(2)}|\mathbf{Y}_T, A)/q(\theta^{(m)}_{(2)}, \theta^*_{(2)}|\theta^{(m+1)}_{(1)})}{p(\theta^{(m+1)}_{(1)}, \theta^{(m)}_{(2)}|\mathbf{Y}_T, A)/q(\theta^*_{(2)}, \theta^{(m)}_{(2)}|\theta^{(m+1)}_{(1)})}, 1\right\}$$

If $\theta^*_{(2)}$ is accepted then $\theta^{(m+1)}_{(2)} = \theta^*_{(2)}$, and if not then $\theta^{(m+1)}_{(2)} = \theta^{(m)}_{(2)}$. The extension of this procedure to multi-block Gibbs samplers, with a Hastings–Metropolis algorithm used at some (or even all) of the blocks is clear. This procedure is generally known as the Metropolis-within-Gibbs algorithm. For further discussion see Zeger and Karim (1991) and Chib and Greenberg (1995), and for a proof that the posterior distribution is an invariant state of this Markov chain see Chib and Greenberg (1996).

3.5 Some MCMC Theory

Much of the treatment here draws heavily on the work of Tierney (1994), who first used the theory of general state space Markov chains to demonstrate convergence, and Roberts and Smith (1994), who elucidated sufficient conditions for convergence that turn out to be applicable in a wide variety of problems in econometrics.

Let $\{\theta^{(m)}\}_{m=0}^{\infty}$ be a Markov chain defined on $\Theta \subseteq \Re^k$ with transition density $K: \Theta \times \Theta \to \Re^+$ such that, for all ν-measurable $\Theta_0 \subseteq \Theta$,

$$P(\theta^{(m)} \in \Theta_0|\theta^{(m-1)}) = \int_{\Theta_0} K(\theta^{(m-1)}, \theta)\, d\nu(\theta) + r(\theta^{(m-1)})\chi_{\Theta_0}(\theta^{(m-1)}),$$

where $r(\theta^{(m-1)}) = 1 - \int_{\Theta} K(\theta^{(m-1)}, \theta)\, d\nu(\theta)$.

The transition density K is substochastic: it defines only the distribution of accepted candidates. Assume that K has no absorbing states, so that $r(\theta) < 1\ \forall \theta \in \Theta$. The corresponding substochastic kernel over m steps is then defined iteratively,

$$K^{(m)}(\theta^{(0)}, \theta^{(m)}) = \int_{\Theta} K^{(m-1)}(\theta^{(0)}, \theta)K(\theta, \theta^{(m)})\, d\nu(\theta)$$
$$+ K^{(m-1)}(\theta^{(0)}, \theta^{(m)})r(\theta^{(m)}) + [r(\theta^{(0)})]^{m-1}K(\theta^{(0)}, \theta^{(m)}).$$

This describes all m-step transitions that involve at least one accepted move. As a function of $\theta^{(m)}$ it is the p.d.f. with respect to ν of $\theta^{(m)}$, excluding realizations with $\theta^{(n)} = \theta^{(0)} \,\forall\, n = 1, \ldots, m$. For any ν-measurable Θ_0 let $\mathrm{P}^{(m)}(\theta^{(0)}, \Theta_0)$ denote the mth iterate of P:

$$\mathrm{P}^{(m)}(\theta^{(0)}, \Theta_0) = \int_{\Theta_0} \mathrm{K}^{(m)}(\theta^{(0)}, \theta)\, \mathrm{d}\nu(\theta) + (\mathrm{r}(\theta^{(0)}))^m \chi_{\Theta_0}(\theta^{(0)})$$

An invariant distribution of the transition density K is a function $\mathrm{p}(\theta)$ that satisfies

$$\begin{aligned}\mathrm{P}(\Theta_0) &= \int_{\Theta_0} \mathrm{p}(\theta)\, \mathrm{d}\nu(\theta) = \int_{\Theta} \mathrm{P}\big(\theta^{(m)} \in \Theta_0 | \theta^{(m-1)} = \theta\big) \mathrm{p}(\theta)\, \mathrm{d}\nu(\theta) \\ &= \int_{\Theta} \left\{ \int_{\Theta_0} \mathrm{K}(\theta, \theta^*)\, \mathrm{d}\nu(\theta^*) + \mathrm{r}(\theta) \chi_{\Theta_0}(\theta^*) \right\} \mathrm{p}(\theta)\, \mathrm{d}\nu(\theta)\end{aligned}$$

for all ν-measurable Θ_0. Let $\Theta^* = \{\theta \in \Theta : \mathrm{p}(\theta) > 0\}$. The density K is p-*irreducible* if for all $\theta^{(0)} \in \Theta^*$, $\mathrm{P}(\Theta_0) > 0$ implies that $\mathrm{P}^{(m)}(\theta^{(0)}, \Theta_0) > 0$ for some $m \geq 1$.

The transition density K is *aperiodic* if there exists no ν-measurable partition $\Theta = \bigcup_{s=0}^{r-1} \tilde{\Theta}_s$ $(r \geq 2)$ such that

$$\mathrm{P}(\theta^{(m)} \in \tilde{\Theta}_{m \,\mathrm{mod}(r)} | \theta^{(0)} \in \tilde{\Theta}_0) = 1 \;\forall\, m$$

It is *Harris recurrent* if $P[\theta^{(m)} \in \Theta_0 \text{ i.o. } | \theta^{(0)}] = 1$ for all ν-measurable Θ_0 with $\int_{\Theta_0} \mathrm{p}(\theta)\, \mathrm{d}\nu(\theta) > 0$ and all $\theta^{(0)} \in \Theta$.* It follows directly that if a kernel is Harris recurrent, then it is p-irreducible. A kernel whose invariant distribution is proper, and that is both aperiodic and Harris recurrent, is *ergodic* by definition (Tierney, 1994, pp. 1712–1713).

A useful metric in what follows is the total variation norm for signed and bounded measures μ defined over the field of all ν-measurable sets S_ν on $\Theta : \|\mu\| = \sup_{\Theta_0 \in S_\nu} \mu(\Theta_0) - \inf_{\Theta_0 \in S_\nu} \mu(\Theta_0)$.

Theorem 3

Convergence of Continuous State Markov Chains

Suppose $\mathrm{p}(\theta | \mathbf{Y}_T, A)$ is an invariant distribution of the transition density $\mathrm{K}(\theta, \theta^*)$.

(A) If K is $\mathrm{p}(\theta | \mathbf{Y}_T, A)$-irreducible, then $\mathrm{p}(\theta | \mathbf{Y}_T, A)$ is the unique invariant distribution.

---

*The expression "i.o." in $\mathrm{P}[\theta^{(m)} \in \tilde{\Theta}_0 \text{i.o.} | \theta^{(0)}] = 1$ means "infinitely often." The condition is that $\lim_{M \to \infty} \mathrm{P}[\sum_{m=1}^{M} \chi_{\tilde{\Theta}}(\theta^{(m)}) \leq L] = 0 \;\forall\, L$.

(B) If K is $p(\theta|\mathbf{Y}_T, A)$- irreducible and aperiodic, then except possibly for $\theta^{(0)}$ in a set of posterior probability 0, $\|P^{(m)}(\theta^{(0)}, \cdot) - P(\cdot|\mathbf{Y}_T, A)\| \rightarrow 0$. If K is ergodic (i.e., it is also Harris recurrent) then this occurs for all $\theta^{(0)}$.

(C) If K is ergodic with invariant distribution $p(\theta|\mathbf{Y}_T, A)$, then for all $g(\mathbf{Y}_T, \theta)$ absolutely integrable with respect to $p(\theta|\mathbf{Y}_T, A)$ and for all $\theta^{(0)} \in \Theta$,

$$M^{-1}\sum_{m=1}^{M} g(\mathbf{Y}_T, \theta^{(m)}) \xrightarrow{a.s.} \int_{\Theta} g(\mathbf{Y}_T, \theta) p(\theta|\mathbf{Y}_T, \theta)\, d\nu(\theta).$$

Proof

(A) and (B) follow immediately from Theorem 1 and (C) from Theorem 3 in Tierney (1994).

For the Gibbs sampling algorithm we argued informally in Section 3.3 that $p(\theta|\mathbf{Y}_T, A)$ is an invariant distribution. More formally, from Eq. (45) we have for the blocking $\theta' = (\theta'_{(1)}, \theta'_{(2)})$:

$$\begin{aligned}
&\int_{\Theta} K_G(\theta, \theta^*) p(\theta|\mathbf{Y}_T, A)\, d\nu(\theta) \\
&= \int_{\Theta} p\big(\theta^*_{(1)}|\mathbf{Y}_T, \theta_{(2)}, A\big) p\big(\theta^*_{(2)}|\mathbf{Y}_T, \theta^*_{(1)}, A\big) p(\theta|\mathbf{Y}_T, A)\, d\nu(\theta) \\
&= p\big(\theta^*_{(2)}|\mathbf{Y}_T, \theta^*_{(1)}, A\big) \int_{\Theta} p\big(\theta^*_{(1)}|\mathbf{Y}_T, \theta_{(2)}, A\big) p(\theta|\mathbf{Y}_T, A)\, d\nu(\theta) \\
&= p\big(\theta^*_{(2)}|\mathbf{Y}_T, \theta^*_{(1)}, A\big) p\big(\theta^*_{(1)}|\mathbf{Y}_T, A\big) = p(\theta^*|\mathbf{Y}_T, A)
\end{aligned}$$

The general result for more than two blocks follows by induction. Thus, it is the uniqueness of the invariant state that is at issue in establishing convergence of the Gibbs sampler. The following result is immediate and is often easy to apply.

Corollary 2

A Sufficient Condition for Convergence of the Gibbs Sampler

Suppose that for every point $\theta^* \in \Theta$ and every $\Theta_0 \subseteq \Theta$ with the property $P(\theta \in \Theta_0|\mathbf{Y}_T, A) > 0$, it is the case that $P_G(\theta^{(m+1)} \in \Theta_0|\mathbf{Y}_T, \theta^{(m)} = \theta^*, A) > 0$, where $P_G(\cdot)$ is the probability measure induced by the Gibbs sampler. Then the Gibbs transition kernel is ergodic.

Proof

The conditions ensure that P_G is aperiodic and absolutely continuous with respect to $p(\theta|\mathbf{Y}_T, A)$. The result follows from Corollary 1 of Tierney (1994).

Tierney (1994) and Roberts and Smith (1994) show that the convergence properties of the Hastings–Metropolis algorithm are inherited from those of $q(\theta, \theta^*)$: if the Markov chain with transition probability q is aperiodic and $p(\theta|\mathbf{Y}_T, A)$-irreducible, then so is the Markov chain simulated by the Hastings–Metropolis algorithm. This feature leads to a sufficient condition for convergence analogous to Corollary 2.

Theorem 4

A Sufficient Condition for Convergence of the Hastings–Metropolis Algorithm

Suppose that for every point $\theta^* \in \Theta$ and every $\Theta_0 \subseteq \Theta$ with the property $P(\theta \in \Theta_0|\mathbf{Y}_T, A) > 0$, it is the case that $\int_{\Theta_0} q(\theta, \theta^*)\alpha(\theta, \theta^*)\, d\nu(\theta^*) + r(\theta)\chi_{\Theta_0}(\theta) > 0$. Then, the Hastings–Metropolis density $K(\theta, \theta^*) = q(\theta, \theta^*) \times \alpha(\theta, \theta^*)$ is ergodic.

Proof

The conditions ensure that the transition kernel is aperiodic and that $p(\theta^*|\mathbf{Y}_T, A)$ irreducible. Thus, by Corollary 2 of Tierney (1994), the Hastings–Metropolis density is Harris recurrent. Since the kernel is both aperiodic and Harris recurrent, it is ergodic.

The conditions in Corollary 2 and Theorem 4 are sufficient but not necessary. However, they are easier to apply directly than is Theorem 3. For some alternative sufficient conditions, see Geweke (1999).

3.6 Assessing Numerical Accuracy in Markov Chain Monte Carlo

In any practical application one is concerned with the discrepancy $\bar{g}_M - \bar{g}$. A leading analytical tool for assessing this discrepancy is a central limit theorem, if one can be obtained. This was accomplished in Section 3.1 for i.i.d. sampling from the posterior distribution, and in Section 3.2 for importance sampling. The assumption of independence, key to those results, does not apply in MCMC. The weaker assumption of uniform ergodicity yields a central limit theorem, however. Let $P^{(m)}(\theta^{(0)}, \Theta_0)$ denote $P(\theta^{(m)} \in \Theta_0|\theta^{(0)})$ for any $\theta^{(0)} \in \Theta$ and for any $\Theta_0 \subseteq \Theta$ for which $P(\theta \in \Theta_0|\mathbf{Y}_T, A)$ is defined. The Markov chain is *uniformly ergodic* if $\sup_{\theta\in\Theta} \|P^{(m)}(\theta, \cdot) - P(\cdot|\mathbf{Y}_T, A)\| \leq Mr^m$ for some $M > 0$ and some positive $r < 1$.

Tierney (1994, p. 1714) demonstrates two results that are quite useful in establishing uniform ergodicity. First, an independence Metropolis kernel $j(\theta)$ with bounded weighting function $w(\theta) = p(\theta|\mathbf{Y}_T)/j(\theta)$ is uniformly ergodic. (Recalling the similarity between the independence Metropolis kernel and importance sampling and Corollary 1 this result is not surprising.) Second, if one kernel in a mixture of kernels is uniformly ergodic, then the mixture kernel itself is uniformly ergodic.

The interest in uniform ergodicity stems from the following central limit theorem. Note how close this result is to Corollary 1.

Theorem 5

A Central Limit Theorem for Markov Chain Monte Carlo

Suppose $\{\theta^{(m)}\}$ is uniformly ergodic with equilibrium distribution $p(\theta|\mathbf{Y}_T, A)$. Suppose further that $E[g(\mathbf{Y}_T, \theta)|\mathbf{Y}_T, A] = \bar{g}$ and $\text{var}[g(\mathbf{Y}_T, \theta)|\mathbf{Y}_T, A]$ exist and are finite, and let $\bar{g}_M = M^{-1}\sum_{m=1}^{M} g(\mathbf{Y}_T, \theta^{(m)})$. Then there exists finite σ^2 such that

$$M^{1/2}(\bar{g}_M - \bar{g}) \xrightarrow{d} N(0, \sigma^2) \tag{48}$$

Proof

Tierney (1994, Theorem 5), attributed to Cogburn [1972, Corollary 4.2(ii)].

Thus, for any Markov chain $\{\theta^{(m)}\}$ with invariant distribution $p(\theta|\mathbf{Y}_T, A)$, one can guarantee Eq. (48) by mixing the chain with an independence Metropolis kernel and a bounded weighting function, so long as the posterior mean and variance are known to exist. If the likelihood function is bounded, then the prior distribution itself will provide such an independence transition kernel.

Nevertheless, some practical concerns remain. One difficulty is that useful conditions sufficient for approximation of the unknown constant σ^2 have not yet been developed. That is, there is no $\hat{\sigma}_M^2$ for which $\hat{\sigma}_M^2 \to \sigma^2$ as there is for independence and importance sampling. A second difficulty is assessing the sensitivity of $\theta^{(m)}$ to the initial condition $\theta^{(0)}$. For example, in the case of a multimodal distribution, transitions between the neighborhoods of all the modes could be quite slow.

There is an extensive literature on this problem. A good introduction is provided by the papers of Gelman and Rubin (1992) and Geyer (1992) and their discussants. Geweke (1992) developed a consistent estimator of σ^2 in Eq. (48), under the strong condition that conventional time series mixing conditions (e.g., Hannan, 1970, pp. 207–210) apply to $\{\theta^{(m)}\}$. There is no analytical foundation for this assumption, but these methods are now widely used and have proven reliable in the sense that they predict well the behavior of the Markov chain when it is restarted with a new initial condition, in econometric models.

In practice, some robustness to initial conditions is achieved by discarding initial iterations: 10 to 20% is common. By drawing $\theta^{(0)}$ from the prior distribution, using a random number generator with a fresh seed each time, several runs may provide some indication of whether the results are sensitive to initial conditions as they might be, e.g., given near-reducibility of the kind that may arise from severe multimodality. A formal test for sensitivity to initial conditions was developed by Gelman and Rubin (1992). For other tests for sensitivity to initial conditions see Geweke (1992) and Zellner and Min (1995).

3.7 Software

The BACC software provides its users with tools for Bayesian analysis, computation, and communication. Dynamically linked libraries (DLLs) implement these tools as extensions to popular mathematical applications such as Matlab, Splus, and Gauss. The DLLs and full source code are available for free from the website www.cirano.gc.ca/~bacc.* The BACC user has available, within the mathematical application, commands for various Bayesian tools. The **minst** command allows a user to set up a model instance, by identifying one of the available models, and specifying the known quantities of the model, such as data and fixed hyperparameters. The **postsim** command generates a posterior simulation structure for a model instance, from which the user can extract the simulated parameter vectors $\theta^{(m)}$, the evaluations $\log(p(\theta^{(m)})|A)$ of the log prior, and the evaluations $\log(p(Y_T|\theta^{(m)}, A))$ of the log likelihood. The **expect1** and **expectN** commands calculate estimates of moments and their standard errors. These commands are described in detail below. The **mlike** command computes estimates of the marginal likelihood for a model instance. This command is described in detail in Section 4.5.

The BACC user also has available all of the mathematical application's built-in commands for matrix computation, graphics, program flow control, and I/O. The utility of this flexibility is illustrated in Section 5.2, where standard Matlab commands are used to calculate estimates of posterior moments not envisioned by the investigator, and to

*Complete documentation for all software is provided at the website. Since this software will continue to be developed and improved, some details provided in this chapter will become outdated. Users should rely on the website documentation for actual use rather than the descriptions in this chapter, which are intended to provide concrete examples of how Bayesian inference, development, and communication can proceed.

compute estimates of posterior moments for a client whose prior differs from that of the investigator.

The **expect1** command takes as input any (possibly weighted) sample, and calculates estimates of the mean and standard deviation. Users can also request estimates of the numerical standard error of the mean, for as many values of a taper parameter L (discussed below) as they care to provide.

We describe the computations performed by the **expect1** command for the special case where the sample is the sequence $\{g(\mathbf{Y}_T, \theta^{(m)})\}$ of evaluations of a function of interest at the draws $\theta^{(m)}$ of the parameter vector, and the weighting sequence is $\{w^{(m)}\}$. The draws $\theta^{(m)}$ and weights $w^{(m)}$ represent the posterior distribution of the parameter vector.

Suppose there are Monte Carlo interactions, of which the first r are discarded as burn-in iterations. Then numerical approximation of the posterior mean of the function of interest is

$$\tilde{g} = \sum_{m=r+1}^{M} w(\theta^{(m)}) g(\mathbf{Y}_T, \theta^{(m)}) \Big/ \sum_{m=r+1}^{M} w(\theta^{(m)})$$

The numerical approximation of the posterior standard deviation of the function of interest is

$$\left\{ \sum_{m=r+1}^{M} w(\theta^{(m)}) \left[g(\mathbf{Y}_T, \theta^{(m)}) - \tilde{g} \right]^2 \Big/ \sum_{m=r+1}^{M} w(\theta^{(m)}) \right\}^{1/2}$$

Different variants of a numerical standard error for the accuracy of the approximation of the true posterior moment $E[g(\mathbf{Y}_T, \theta) | \mathbf{Y}_T, A]$ by the numerical approximation $\tilde{g}$ are provided in BACC. To describe these, let $n_M = (M-r)^{-1} \times \sum_{m=r+1}^{M} w(\theta^m) g(\theta^{(m)}, \mathbf{Y}_T)$ denote the numerator of $\tilde{g}$ and let $d_M = (M-r)^{-1} \sum_{m=r+1}^{M} w(\theta^m)$ denote the denominator of $\tilde{g}$. Using the conventional asymptotic expansion ("delta method"):

$$\operatorname{var}(n_M/d_M) \approx [\, d_M^{-1} \;\; -n_M d_M^{-2} \,] \begin{bmatrix} \operatorname{var}(n_M) & \operatorname{cov}(n_M, d_M) \\ \operatorname{cov}(d_M, n_M) & \operatorname{var}(d_M) \end{bmatrix} \begin{bmatrix} d_M^{-1} \\ -n_M d_M^{-2} \end{bmatrix}$$

The variants of the numerical standard error, $\text{s.e.}(n_M/d_M) = [\operatorname{var}(n_M/d_M)]^{1/2}$, are based on different approximations of $\operatorname{var}(n_M)$, $\operatorname{cov}(n_M, d_M)$ and $\operatorname{var}(d_M)$.

The first method assumes no serial correlation in $\{\theta^{(m)}\}$, and is appropriate for independence or importance sampling. Following Geweke (1989b), and consistent with Eq. (43), this leads to

$$\text{var}(n_M/d_M) \approx \sum_{m=r+1}^{M} \text{w}(\theta^{(m)})^2[\text{g}(\theta^{(m)}, \mathbf{Y}_T) - \bar{g}]^2 \Bigg/ \left[\sum_{m=r+1}^{M} \text{w}(\theta^{(m)})\right]^2$$

(The square root of the value on the right-hand side is reported by **expect1**.)

In the remaining methods, $\text{var}(n_M)$, $\text{cov}(n_M, d_M)$ and $\text{var}(d_M)$ are approximated using conventional time-series methods for a wide sense stationary process similar to those described in Geweke (1992). In the case of $\text{var}(d_M)$:

$$\text{var}(d_M) \approx (M - r)^{-1} \sum_{s=-L+1}^{L-1} [(L - |s|)/L]c(s) \tag{49}$$

where for $s \geq 0$, $c(s) = c(-s) = (M - r)^{-1} \sum_{m=r+s+1}^{M} [\text{w}(\theta^{(m)}) - d_M][\text{w}(\theta^{(m-s)}) - d_M]$. For $\text{var}(n_M)$, the approximation is the right hand side of Eq. (49) but with

$$c(s) = c(-s) = (M - r)^{-1} \sum_{m=r+s+1}^{M} \left[\text{g}(\theta^{(m)})\text{w}(\theta^{(m)}) - n_M\right] \times \left[\text{g}(\theta^{(m-s)})\text{w}(\theta^{(m-s)}) - n_M\right]$$

where $s \geq 0$. For $\text{cov}(n_M, d_M)$ the approximation is the right-hand side of Eq. (49) but with

$$c(s) = (M - r)^{-1} \sum_{m=\max(r+1, r+s+1)}^{\min(M, M-s)} \left[\text{g}(\theta^{(m)})\text{w}(\theta^{(m)}) - n_M\right]\left[\text{w}(\theta^{(m-s)}) - d_M\right]$$

The remaining variants differ in the value of L chosen. One value of the numerical standard error is reported for each value of L chosen.*

The **expectN** command takes as input several weighted samples. It provides estimates of the numerical standard error for the combined sample, and conventional test statistics for the equality of the moments of the component samples, under the assumption that the samples are independent from one another.

---

*Small values of L assume a more rapid rate of decay in the autocovariance function of $\{\text{g}(\mathbf{Y}_T, \theta^{(m)})\}$. In practice results are usually about the same for the three values. Substantial differences indicate that serial correlation may persist across a substantial fraction of the iterations, and a longer simulation may be warranted.

It computes the same variants of the numerical standard error as **expect1**. If there are J samples, the combined posterior moment approximation is $\tilde{g} = \sum_{j=1}^{J} v_j \tilde{g}_j / \sum_{j=1}^{J} v_j$, where $\tilde{g}_j$ is the mean of the jth sample and v_j is the inverse square of its numerical standard error. The **expectN** command provides the conventional chi-square test of $\tilde{g}_1 = \cdots = \tilde{g}_J$ (in all variants), and the marginal significance level of this test statistic. If the J samples are the same posterior functions of interest, created using J independent initial conditions for the same posterior simulator, then this test is essentially the convergence test proposed by Gelman and Rubin (1992).

Evidence of different values of the moments from different posterior samples is an indication that there may be sensitivity to starting values, or—almost equivalently—that an insufficient number of burn-in iterations were taken in approximating the moments.

3.8 Examples

Two examples illustrate the use of these methods, and will be used in subsequent portions of this chapter as well.* The first example is based on the hedonic model of residential real estate prices discussed by Anglin and Gencay (1996). Their baseline model is a linear regression of the logarithm of sales prices on an intercept and 11 attributes. The attributes are indicated in the first column of Table 1. All variables beginning with "#" are positive integers, log(Lot size) is continuous, and all other variables are dichotomous (1 if present and 0 if not). The data consist of 546 transactions during July, August, and September 1987 in metropolitan Windsor, Ontario. The least squares estimates match those reported in Anglin and Gencay (1996).

The form of the prior distribution is the one discussed in Section 2.1 for the normal linear model. The normal distribution for the coefficient vector has mean $\mathbf{0}$, the precision matrix is diagonal, and standard deviations are chosen to allow reasonable values of the coefficients. The prior distribution of the precision parameter is $0.12h \sim \chi^2(3)$ so that h has prior mean 25 and standard deviation about 20. The posterior simulator is the Gibbs sampling algorithm described in Section 3.3, based on the conditional distributions Eqs. (17) and (18). The posterior simulation structure was created using the BACC commands **minst** and **postsim**. Creation of 10,000 records in the posterior simulation structure required

---

*Data for both examples are available at: http://www.biz.uiowa.edu/faculty/jgeweke/papers.html Data for the first example are also available at: http://qed.econ.queens.ca:80/jae/1996-v11.6/anglin-gencay.

Table 1 Regression Model: Posterior Moments

	Ordinary least squares		Prior		Posterior			
Coefficient (RNE)	Estimate	s.e.	Mean	s.d.	Mean	s.d.	NSE	RNE
Intercept	7.745	0.216	0	11.0	7.725	0.219	0.0021	1.19
Driveway	0.110	0.028	0	0.1	0.104	0.027	0.0003	1.06
Recreation room	0.058	0.026	0	0.1	0.058	0.025	0.0002	1.31
Finished basement	0.104	0.021	0	0.1	0.103	0.021	0.0002	1.10
Gas hot water	0.179	0.043	0	0.1	0.149	0.040	0.0004	0.90
Central air	0.166	0.021	0	0.1	0.159	0.021	0.0002	0.98
#Garage stalls	0.048	0.011	0	0.1	0.049	0.011	0.0001	1.65
Good nbhd	0.132	0.023	0	0.1	0.127	0.020	0.0002	1.67
log(Lot size)	0.303	0.027	0	0.3	0.307	0.027	0.0003	1.11
#Bedrooms	0.034	0.014	0	0.1	0.036	0.014	0.0001	1.38
#Full bathrooms	0.166	0.020	0	0.1	0.161	0.020	0.0002	1.82
#Stories	0.092	0.013	0	0.1	0.093	0.012	0.0001	1.05

roughly 8 sec.* The last four columns of Table 1 report results from **expect1** using this posterior simulation structure, discarding the first 1000 iterations. Initial values here, and in all other examples discussed in this chapter, were drawn from the prior distribution, and in every case the first 1000 draws were discarded. The **expect1** computations took a quarter of a second. Posterior means and standard deviations are close to the least squares values, reflecting the lack of information in the prior relative to the data set. The numerical standard error (NSE) of each posterior mean is given for $L = 0.08(M - r)$ as described in the previous section 3.7.† The NSEs imply accuracy of more than two figures past the decimal in the posterior means, and the relative numerical efficiency (RNE) indicates that numerical accuracy is comparable to what would have been achieved with i.i.d. drawings directly from the posterior distribution.

The second example is a probit model of women's labor force participation, based on the one presented in Geweke and Keane (1999). The data consist of 1555 observations of women in the 1987 Panel Survey of Income Dynamics. The choice variable is 1 if a woman reports positive hours of work for 1987. The covariates are indicated in the first

---

*All execution times are given for a Pentium II processor with 64MB of RAM.
†This value of L is also used to report NSE and RNE in all other examples in this chapter's. Results using other values of L are similar.

column of Table 2. "Black," "Married," and "Kids" are dichotomous. Age is measured in years and interacted with "Married" and its negation. "Education" is years of completed schooling. "Spouse$" is husband's income and "Family$" is unearned household income, in dollars for the year 1987. Work experience is measured in cumulative hours since the woman became a household head or spouse. For each unmarried woman with children, AFDC is the monthly cash support she would receive if she did not work. "Food$" is the monthly food-stamp allotment to which a woman's household would be entitled if she did not work. The last two variables differ according to the state of residence. The prior distributions of the coefficients are independent normal, each with mean zero and standard deviation chosen to permit large but reasonable values to be within two standard deviations of zero. Details of sample screening, variable descriptions, and prior construction are given in Geweke and Keane (1999).

Conventional maximum likelihood estimates, the posterior mode, and approximate posterior standard deviations based in the usual way on the Hessian of the log-posterior at the mode* are presented in Table 2. It

TABLE 2 Probit Model: Likelihood Mode, Prior, and Posterior Mode

	Maximum likelihood		Prior		Posterior	
Coefficient	Mode	Asymptotic s.e.	Mean	Stand. dev.	Mode	Approx. (s.d.)
Intercept	1.22	0.520	0	4.00	1.21	0.177
Black	0.109	0.105	0	0.125	0.0151	0.0773
Age-Single	−0.0611	0.0132	0	0.00417	−0.0102	0.00381
Age-Married	−0.0874	0.0128	0	0.03333	−0.0279	0.00652
Education	0.113	0.0274	0	0.00417	0.00228	0.00411
Married	0.682	0.522	0	0.125	0.180	0.118
Kids	−0.488	0.171	0	0.250	−0.365	0.131
#Kids	−0.0505	0.0552	0	0.125	−0.151	0.0441
Spouse$	-1.78×10^{-5}	3.46×10^{-6}	0	3.57×10^{-6}	-7.18×10^{-6}	2.28×10^{-6}
Family$	1.86×10^{-6}	7.25×10^{-6}	0	3.57×10^{-6}	-9.51×10^{-7}	2.92×10^{-6}
AFDC	-5.02×10^{-4}	3.85×10^{-4}	0	6.25×10^{-4}	-5.68×10^{-4}	3.05×10^{-4}
Food$	−0.00211	6.94×10^{-4}	0	6.25×10^{-4}	−0.00121	4.35×10^{-4}
WorkExp	1.36×10^{-4}	9.50×10^{-6}	0	6.25×10^{-5}	1.16×10^{-4}	8.28×10^{-6}

*For details and the asymptotic justification for this approximation see Bernardo and Smith (1994, Sect. 5.3) and the references cited therein.

is clear from Table 2 that the prior distribution is informative, relative to the data: prior standard deviations of 6 of the 13 coefficients are less than the corresponding maximum likelihood asymptotic standard errors. Posterior standard deviations are less than asymptotic standard errors in every case, and for several coefficients the posterior mode is closer to the prior mode than to the likelihood mode.

The Gibbs sampler for the probit model, developed in Albert and Chib (1993) and described in Section 3.3, is the posterior simulation algorithm used to construct the posterior simulation structure. As with the first example, this structure is created using the BACC commands minst and postsim. Creation of a posterior simulation structure with 10,000 records required 410 sec. Posterior means, standard deviations, and measures of numerical accuracy, based on the last 9000 records, are presented in Table 3. The posterior moments are quite close to the approximation at the posterior mode, very likely reflecting a posterior distribution that is close to multivariate normal. The average relative numerical efficiency for the coefficients indicates that the same accuracy could have been achieved with about half the number of iterations, had an i.i.d. sample been drawn directly from the posterior distribution.

4 MODEL COMPARISON

Section 2.3 demonstrated that given prior probabilities over models, and prior probability distributions for parameter vectors within models, there is a complete theory of model combination and model comparison. The

TABLE 3 Probit Model: Posterior Moments

Coefficient	Mean	Stand. dev.	NSE	RNE
Intercept	1.2135	0.176	0.00376	0.245
Black	0.01538	0.0772	0.00173	0.222
Age-Single	−0.010258	0.00381	4.65×10^{-5}	0.745
Age-Married	−0.028012	0.00650	1.08×10^{-4}	0.406
Education	0.0022905	0.00411	4.62×10^{-5}	0.876
Married	0.17867	0.118	0.00122	1.038
Kids	−0.36476	0.132	0.00262	0.279
#Kids	−0.15209	0.0438	7.30×10^{-4}	0.400
Spouse$	-7.2318×10^{-6}	2.25×10^{-6}	3.62×10^{-8}	0.432
Family$	-1.1449×10^{-6}	2.91×10^{-6}	3.81×10^{-8}	0.648
AFDC	-5.7610×10^{-4}	3.03×10^{-4}	4.25×10^{-6}	0.565
Food$	−0.0012176	4.33×10^{-4}	5.91×10^{-6}	0.597
WorkExp	1.1672×10^{-4}	8.46×10^{-6}	4.24×10^{-7}	0.044

central technical task in implementing the theory is calculation of the marginal likelihood $p(\mathbf{Y}_T|A) = \int_\Theta p(\mathbf{Y}_T|\theta, A)p(\theta|A)\,d\nu(\theta)$. The marginal likelihood cannot, in general, be cast in the form of a posterior moment, Eq. (5), and therefore the posterior simulation methods of Section 3, which have proven useful in obtaining posterior moments in a single model, are not directly applicable to this problem. A decade ago, there were essentially no methods developed for the numerical approximation of marginal likelihoods or Bayes factors and results were limited to a handful of cases for which there were analytical results or asymptotic approximations. Now, it is possible to attain good and generic approximations to marginal likelihoods in most cases; however, some models with large numbers of latent variables remain troublesome.

This section provides several approaches to the approximation of marginal likelihoods, with an emphasis on generic methods that are consistent as the number of simulations increases. Generic methods exclude those that are ingenious but specific to particular situations as well as methods that rely on asymptotic approximations rather than simulation. Many of these methods are discussed in a comprehensive review article by Kass and Raftery (1996). We discuss here a method that works well with importance sampling and the Hastings–Metropolis algorithm; a method specific to the Gibbs sampler; and a generic method that works well with most posterior simulators regardless of the algorithm employed. For the last method, we describe publicly available software designed to work with the most commonly used computing platforms in econometrics. Some examples illustrate the numerical accuracy that can be attained, and provide comparisons of some of the different methods of approximating marginal likelihoods.

4.1 Importance Sampling

Suppose that $j(\theta)$, with support Θ, is the probability density function (not just a kernel) with respect to the measure $d\nu(\theta)$ of an importance sampling distribution for the posterior density $p(\theta|\mathbf{Y}_T, A) \propto p(\theta|A)p(\mathbf{Y}_T|\theta, A)$, where $p(\theta|A)$ is the properly normalized prior density and $p(\mathbf{Y}_T|\theta, A)$ is the properly normalized data density. Define the weighting function $w(\theta) = p(\theta|A)\ p(\mathbf{Y}_T|\theta, A)/j(\theta)$.

Corollary 3

Let $j(\theta)$ be the importance sampling density in an importance sampling algorithm. Suppose the support of $j(\theta)$ includes Θ. Then

$$\bar{w}_M = M^{-1}\sum_{m=1}^{M} w(\theta^{(m)}) \xrightarrow{a.s.} \int_{\Theta} w(\theta)j(\theta)\,d\nu(\theta)$$
$$= \int_{\Theta} p(\theta|A)p(\mathbf{Y}_T|\theta, A)\,d\nu(\theta) = p(\mathbf{Y}_T|A)$$

If $w(\theta)$ is bounded above then

$$M^{1/2}[\bar{w}_M - p(\mathbf{Y}_T|A)] \xrightarrow{d} N(0, \sigma^2) \text{ and}$$
$$M^{-1}\sum_{m=1}^{M}\left[w(\theta^{(m)}) - \bar{w}_M\right]^2 \xrightarrow{a.s.} \sigma^2$$

Proof

Immediate from Theorem 2 and Corollary 1.

The first application of this idea is that of Geweke (1989a); see also Gelfand and Dey (1994) and Raftery (1995). Since $w(\theta^{(m)})$ must be computed from each iteration of the importance sampling algorithm in any event and the normalizing constant for $j(\theta)$ is usually known, this simulation-consistent approximation of $p(\mathbf{Y}_T|A)$ may be obtained at essentially no additional cost.

4.2 The Gibbs Sampler

In the case of the Gibbs sampler there is a different procedure due to Chib (1995) that provides accurate evaluations of the marginal likelihood, at the cost of additional simulations. Suppose that the output from the blocking $\theta' = (\theta'_{(1)}, \ldots, \theta'_{(B)})$ is available, and that the conditional p.d.f.'s $p(\theta_{(j)}|\theta_{(i)}(i \neq j), \mathbf{Y}_T, A)$ can be evaluated in closed form for all j. (This latter requirement is generally satisfied.)

From Eqs. (1) and (2):

$$p(\mathbf{Y}_T|A) = p(\tilde{\theta}|A)p(\mathbf{Y}_T|\tilde{\theta}, A)/p(\tilde{\theta}|\mathbf{Y}_T, A) \tag{50}$$

for any $\tilde{\theta} \in \Theta$. Typically $p(\mathbf{Y}_T|\tilde{\theta}, A)$ and $p(\tilde{\theta}|A)$ can be evaluated in closed form, but $p(\tilde{\theta}|\mathbf{Y}_T, A)$ cannot. A marginal/conditional decomposition of $p(\tilde{\theta}|\mathbf{Y}_T, A)$ is

$$p(\tilde{\theta}|\mathbf{Y}_T, A) = p(\tilde{\theta}_{(1)}|\mathbf{Y}_T, A)p(\tilde{\theta}_{(2)}|\mathbf{Y}_T, \tilde{\theta}_{(1)}, A)\cdots$$
$$\times\, p(\tilde{\theta}_{(B)}|\mathbf{Y}_T, \tilde{\theta}_{(1)}, \ldots, \tilde{\theta}_{(B-1)}, A) \tag{51}$$

The first term in the product of B terms can be approximated from the output of the posterior simulator because

$$M^{-1}\sum_{m=1}^{M}\mathrm{p}\left(\tilde{\theta}_{(1)}|\mathbf{Y}_T,\theta_{(2)}^{(m)},\ldots,\theta_{(B)}^{(m)},A\right)\xrightarrow{a.s}\mathrm{p}(\tilde{\theta}_{(1)}|\mathbf{Y}_T,A)$$

To approximate $\mathrm{p}(\tilde{\theta}_{(b)}|\mathbf{Y}_T,\tilde{\theta}_{(1)},\ldots,\tilde{\theta}_{(b-1)},A)$, first execute the Gibbs sampler with the parameters in the first $b-1$ blocks fixed at the indicated values, thus producing a sequence $\{\theta_{(b),(b+1)}^{(m)},\ldots,\theta_{(b),(B)}^{(m)}\}$ from the conditional posterior. Then

$$M^{-1}\sum_{m=1}^{M}\mathrm{p}\left(\tilde{\theta}_b|\mathbf{Y}_T,\tilde{\theta}_{(1)},\ldots,\tilde{\theta}_{(b-1)},\theta_{(b),(b+1)}^{(m)},\ldots,\theta_{(b),(B)}^{(m)},A\right)$$

$$\xrightarrow{a.s.}\mathrm{p}\left(\tilde{\theta}_{(b)}|\mathbf{Y}_T,\tilde{\theta}_{(1)},\ldots,\tilde{\theta}_{(b-1)},A\right)$$

These approximations are then used in Eqs. (50) and (51) to obtain the approximation to the marginal likelihood. In general, this method is more efficient the greater is $\mathrm{p}(\tilde{\theta}|\mathbf{Y}_T,A)$, so in many applications it is natural to choose $\tilde{\theta}$ near the posterior mode. It is straightforward to apply the methods of Section 3.7 to evaluate the numerical accuracy of the final approximation to the marginal likelihood, using standard delta methods. See Chib (1995) on these and other important practical details.

4.3 Modified Harmonic Mean

Gelfand and Dey (1994) observe that for any p.d.f. $f(\theta)$, whose support is contained in Θ,

$$\mathrm{E}\left[\frac{f(\theta)}{\mathrm{p}(\theta|A)\mathrm{p}(\mathbf{Y}_T|\theta,A)}\bigg|\mathbf{Y}_T,A\right]$$

$$=\int_{\Theta}\frac{f(\theta)}{\mathrm{p}(\theta|A)\mathrm{p}(\mathbf{Y}_T|\theta,A)}\mathrm{p}(\theta|\mathbf{Y}_T,A)\,d\nu(\theta)$$

$$=\int_{\Theta}\frac{f(\theta)}{\mathrm{p}(\theta|A)\mathrm{p}(\mathbf{Y}_T|\theta,A)}\cdot\frac{\mathrm{p}(\theta|A)\mathrm{p}(\mathbf{Y}_T|\theta,A)}{\int_{\Theta}\mathrm{p}(\theta|A)\mathrm{p}(\mathbf{Y}_T|\theta,A)d\nu(\theta)}d\nu(\theta)$$

$$=\frac{\int_{\Theta}f(\theta)\,d\nu(\theta)}{\int_{\Theta}\mathrm{p}(\theta|A)\mathrm{p}(\mathbf{Y}_T|\theta,A)\,d\nu(\theta)}=\mathrm{p}(\mathbf{Y}_T|A)^{-1}\tag{52}$$

Thus, the posterior mean of the function of interest $f(\theta)/\mathrm{p}(\theta|A)\mathrm{p}(\mathbf{Y}_T|\theta,A)$ is $\mathrm{p}(\mathbf{Y}_T|A)^{-1}$. It is, therefore, a candidate for approximation by a posterior simulator. If $f(\theta)/p(\theta|A)p(\mathbf{Y}_T|\theta, A)$ is bounded above, then the approximation is simulation consistent and the rate of convergence is likely to be practical.

It is not difficult to guarantee the boundedness condition in Eq. (52). Consider first the case in which $\Theta = \Re^k$. From the output of the posterior simulator define*

$$\hat{\theta}_M = \sum_{m=1}^{M} w\left(\theta^{(m)}\right)\theta^{(m)} \Big/ \sum_{m=1}^{M} w\left(\theta^{(m)}\right)$$

and

$$\hat{\Sigma}_M = \sum_{m=1}^{M} w\left(\theta^{(m)}\right)\left(\theta^{(m)} - \hat{\theta}_M\right)\left(\theta^{(m)} - \hat{\theta}_M\right)' \Big/ \sum_{m=1}^{M} w(\theta^{(m)})$$

(It is not essential that the posterior mean and variance of θ exist.) Then, for some $p \in (0, 1)$, define $\hat{\Theta}_M = \{\theta : (\theta - \hat{\theta}_M)'\hat{\Sigma}_M^{-1}(\theta - \hat{\theta}_M) \leq \chi^2_{1-p}(k)\}$ and take

$$f(\theta) = p^{-1}(2\pi)^{-k/2}|\hat{\Sigma}_M|^{-1/2}\exp[-0.5(\theta - \hat{\theta}_M)'\hat{\Sigma}_M^{-1}(\theta - \hat{\theta}_M)]\chi_{\hat{\Theta}_M}(\theta) \qquad (53)$$

If the posterior density is uniformly bounded away from 0 on every compact subset of Θ, then the function $f(\theta)/p(\theta|A)p(\theta|\mathbf{Y}_T, A)$ possesses posterior moments of all orders. For a wide range of regular problems, this function will be approximately constant on $\hat{\Theta}_M$, which is nearly ideal. In most situations smaller values of p will result in better behavior of $f(\theta)/p(\theta|A)\ p(\mathbf{Y}_T|\theta, A)$ over the domain $\hat{\Theta}_M$, but greater simulation error due to a smaller number of $\theta^{(m)} \in \hat{\Theta}_M$; there is almost no incremental cost in carrying out the computations for several values of p rather than a single value of p.

So long as $\hat{\Theta}_M \subseteq \Theta$, $\int_{\hat{\Theta}_M} f(\theta)\,d\nu(\theta) = 1$. If not, the domain of integration must be redefined to be $\hat{\Theta}_M \cap \Theta$. In this case a new normalizing constant for $f(\theta)$ can be well approximated by taking a sequence of i.i.d. draws $\{\theta^{(\ell)}\}$ from the original distribution Eq. (53) with domain $\hat{\Theta}_M$, and then averaging $\chi_\Theta(\theta^{(\ell)})$.

Frequently the behavior of $f(\theta)/p(\theta|A)p(\mathbf{Y}_T|\theta, A)$ can be improved by reparameterization of θ to $\zeta = h(\theta)$, where h is a one-to-one function. Of course, the prior density must then be adjusted by the Jacobian of transformation. If the support Z of $p(\zeta|A)$ is $\Re^k$, then $\int_Z f(\zeta)\,d\nu(\zeta) = 1$ for f constructed as indicated in Eq. (53). For example, if this method is used to

---

*The weighting function $w(\theta)$ is defined in Theorem 2 in the case of importance sampling. For MCMC algorithms $w(\theta) = 1$.

approximate the marginal likelihood in the standard linear model, Eqs. (7)–(9), transformation of h to $\log(h)$ guarantees the support condition, and generally results in more accurate approximation of $p(\mathbf{Y}_T|A)$.

The numerical accuracy of the approximation can be evaluated using the methods of Section 3.7, as detailed in Section 4.5.

4.4 Improving Numerical Approximations

In many instances a portion of the marginal likelihood $P(\mathbf{Y}_T|A) = \int_\Theta p(\theta|A)p(\mathbf{Y}_T|\theta, A)\,d\nu(\theta)$ may be evaluated analytically. Suppose that

$$\int_\Theta p(\theta|A)p(\mathbf{Y}_T|\theta, A)\,d\nu(\theta)$$
$$= \int_{\Theta_1}\int_{\Theta_2} p(\theta_1, \theta_2|A)p(\mathbf{Y}_T|\theta_1, \theta_2, A)\,d\nu(\theta_2)\,d\nu(\theta_1) = \int_{\Theta_1} r(\theta_1)\,d\nu(\theta_1)$$

where $r(\theta_1) = \int_{\Theta_2} p(\theta_1, \theta_2|A)p(\mathbf{Y}_T|\theta_1, \theta_2, A)\,d\nu(\theta_2)$ can be evaluated analytically. The modified harmonic mean method can be applied directly to the simulated values $\theta_1^{(m)}$, using $r(\theta_1^{(m)})$ in lieu of $p(\theta^{(m)}|A)p(\mathbf{Y}_T|\theta^{(m)}, A)$ and tailoring $f(\theta)$ to $r(\theta)$ rather than to $p(\theta|A)p(\mathbf{Y}_T|\theta, A)$. Similar adjustments can be made for importance sampling. Because the dimension of integration is lower, the resulting approximation will typically be more accurate. In the case of the method employing the Gibbs sampler described in Section 4.2 this preliminary evaluation will eliminate at least one of the blocks for which the auxiliary simulations must be undertaken.

An example of this procedure is provided by earlier results for the standard linear model. The entire posterior kernel in standard form is Eq. (13), but in Section 2.3 we found the marginal likelihood conditional on h, Eq. (25). The latter expression is a function of a single unknown parameter, whereas the former is a function of $k+1$ unknown parameters.

The probit model described in Section 2.4 provides a second example. In this case there are $T+k$ unknown parameters (T latent variables and k coefficients). The modified harmonic mean method in this case is completely unwieldy, since it would require storing a very large amount of posterior simulator output, and generation of the requisite $T+k$ truncated normal random variables would require the factorization of a matrix of the same order. In this case, integration of the T latent variables is straightforward, and leads to the product of the prior density for the coefficients and the likelihood function as typically written:

$$(2\pi)^{-k/2}|\underline{\mathbf{H}}_{\beta}|^{1/2}\exp[-.5(\beta-\underline{\beta})'\underline{\mathbf{H}}_{\beta}(\beta-\underline{\beta})]$$
$$\times\prod_{t=1}^{T}\{d_t\Phi(\beta'\mathbf{x}_t)+(1-d_t)[1-\Phi(\beta'\mathbf{x}_t)]\}$$

More generally, in models with latent variables, accurate evaluation of the marginal likelihood requires that it be possible to perform the integration over the space of such variables analytically.

4.5 Software

The **mlike** command computes estimates of the log marginal likelihood using the modified harmonic mean posterior simulation method, given a model instance for which the user has generated a posterior simulation structure using the **postsim** command. The user optionally supplies a vector of values of the p parameter of Section 4.3. Otherwise the single default value of 0.5 applies. For each value of p, the mlike command computes

$$\left[\sum_{m=1}^{M}\mathrm{w}(\theta^{(m)})\right]^{-1}\left[\sum_{m=1}^{M}\mathrm{w}(\theta^{(m)})\mathrm{f}(\theta^{(m)})/\mathrm{p}(\theta^{(m)}|A)\mathrm{p}(\mathbf{Y}_T|\theta^{(m)},A)\right] \qquad (54)$$

and reports minus the logarithm of this value. Also, for each value of p, it optionally provides numerical standard errors for each value of L (described in Section 3.7) supplied by the user. The numerical standard error for the log marginal likelihood is obtained as the standard error of the quantity in Eq. (54) divided by that quantity. This latter standard error is computed as described in Section 3.7.

Parameter transformations to improve the behavior of $\mathrm{f}(\theta)/\mathrm{p}(\theta|A)\mathrm{p}(\mathbf{Y}_T|\theta)$, and simulations to handle cases where $\hat{\Theta}_M \not\subset \Theta$, discussed in Section 4.3, are performed automatically by the **mlike** command. The parameter transformations and the indicator function χ_{Θ} form part of the definition of a model, supplied by the model developer. They are invisible to the user.

4.6 Examples

For the regression model described in Section 3.8, the marginal likelihood was approximated using the BACC command **mlike**. The software incorporates a reparameterization of the precision from h to $\log h$. After this reparameterization $\Theta_A=\Re^k$, the **mlike** execution time is under half a second for this case.

The top panel of Table 4.1 provides results using $p=0.9$, 0.5, and 0.1 in Eq. (53). Computation with $p=0.9$ provides the most accurate assessment. In view of the good approximation of the posterior by a multivariate normal distribution, it is not surprising that a more inclusive $f(\theta)$ yields more accurate results. Differences in approximations for different values of p are consistent with the numerical standard errors (NSEs).

To illustrate the use of the approximated marginal likelihood in the construction of Bayes factors, two variants on the model set forth in Section 3.8 were constructed by making two changes in the prior distribution. In the first change the mean of the prior distributions of all slope coefficients is shifted to the value of the standard deviation which in turn is unchanged: i.e., the prior distribution is changed from $N(0, 0.1^2)$ to $N(.1, 0.1^2)$ for all covariates except log (Lot size), for which the prior distribution is shifted from $N(0, 0.3^2)$ to $N(0.3, 0.3^2)$. Log marginal likelihood approximations for this model are given in the middle panel of Table 4. Finally, the standard deviations in these priors are reduced by

TABLE 4 Regression Model: Marginal Likelihoods

	Log marginal likelihood	NSE
First prior (zero center)		
$p=0.9$	46.081	0.004
$p=0.5$	46.082	0.011
$p=0.1$	46.093	0.030
Second prior (nonzero center)		
$p=0.9$	52.148	0.004
$p=0.5$	52.151	0.011
$p=0.1$	52.162	0.031
Third prior (nonzero center, higher precision)		
$p=0.9$	56.365	0.004
$p=0.5$	56.369	0.011
$p=0.1$	56.383	0.031

TABLE 5 Probit Model: Marginal Likelihoods

	Log marginal likelihood	NSE
$p=0.9$	−564.71	0.0033
$p=0.5$	−564.70	0.0142
$p=0.1$	−564.69	0.0335

half: now all priors are $N(0.1, 0.05^2)$ except for log (Lot size), which is $N(0.3, 0.15^2)$. Log marginal likelihoods for this model are shown in the lower panel of Table 4.

Using the approximation based on $p = 0.9$, the log Bayes factor in favor of the last model, versus the first, is approximately 10.285 and the associated NSE is 0.006. Thus, the Bayes factor is almost certainly (based on NSE×3) between 28,767 and 29,822.

In the probit model example, the marginal likelihood can be approximated using the modified harmonic mean method implemented in mlike. The evaluation of the likelihood function for the probit model as discussed at the end of Section 4.4 is used. The results are shown in Table 5. For the same reasons as in the regression model the approximation is quite accurate and is better for larger values of p.

5 BAYESIAN COMMUNICATION

For a subjective Bayesian decision maker the computation of the posterior moments $E[h(\omega)|\mathbf{Y}_T, A]$ for suitable models, priors, and functions of interest is typically the final objective of inference. For an investigator reporting results for other potential decision makers,. however, the situation is different. In the language of Hildreth (1963) these decision makers are *remote clients,* whose priors and functions of interest are not known to the investigator.

What should the investigator report? Traditionally, published papers report a few posterior moments and more rarely some indication of sensitivity to prior distributions, and alternative data densities may be given. Such information is generally much too limited. At the other extreme, the investigator may simply report some likelihood functions, but this leaves most of the work to the client. Investigators almost never report marginal likelihoods, thereby leaving unrealized the promise inherent in model averaging.

5.1 Posterior Reweighting

An investigator will have carried through formal inference for a set of models $A_1, \ldots, A_J$. This collection will reflect the process of model development, and a public report of the investigator's work should at least summarize this process. In the ideal situation described by Poirier (1988), clients have agreed to disagree in terms of the prior. Since the set of models that exists in any meaningful sense is the set publicly reported, collectively investigators will have provided the grand model in which variation of the prior is the basis of formal discourse in normal science.*

Corresponding to each model $A_1, \ldots, A_J$ included in an investigation, there is a posterior simulation structure of the form described in Section 3.7. It is a simple matter to save these simulation structures and make them available at an ftp or web site, and for any client to obtain them for the purpose of the manipulations described here.

Given the posterior simulation structure a client can immediately compute numerical approximations to posterior moments not reported or even considered by the investigator. Specifically, suppose a client wishes to know $\mathrm{E}[\mathrm{h}(\omega)|\mathbf{Y}_T, A]$ where $\omega \sim \mathrm{p}(\omega|\mathbf{Y}_T, A)$ is specified by the client and $\mathrm{p}(\mathbf{Y}_T|\theta, A)$ and $\mathrm{p}(\theta|A)$ have been specified by the investigator. Corresponding to each $\theta^{(m)}$ reported in the posterior simulation structure the client forms $\mathrm{g}(\mathbf{Y}_T, \theta^{(m)})$ with the property $\mathrm{E}[\mathrm{g}(\mathbf{Y}_T, \theta)|\mathbf{Y}_T, \theta, A] = \mathrm{E}[\mathrm{h}(\omega)|\mathbf{Y}_T, \theta, A]$, and then computes $\bar{g}_M = \sum_{m=1}^{M} \mathrm{w}(\theta^{(m)})\mathrm{g}(\theta^{(m)}) / \sum_{m=1}^{M} \mathrm{w}(\theta^{(m)})$. If the investigator's posterior simulator is ergodic then $\bar{g}_M \xrightarrow{a.s.} \bar{g} = \mathrm{E}[\mathrm{h}(\omega)|\mathbf{Y}_T, A]$ and if it is uniformly ergodic then $M^{1/2}(\bar{g}_M - \bar{g}) \xrightarrow{d} \mathrm{N}(0, \sigma^2)$. For simple functions $\mathrm{h}(\omega)$ computation of $\bar{g}_M$ amounts to spreadsheet arithmetic. More elaborate functions of interest may involve simulations, but in all cases these computations are precisely those which economists undertake as a routine matter when investigating the implications of a model.

For example, a client reading a research report might be skeptical that the investigator's model, prior, and data set provide much information about the effects of an interesting change in a policy variable on the outcome in question. If the simulator output matrix is available electronically the client can obtain the exact (up to numerical approximation error, which can also be evaluated) answer to his query, without arising from his office chair, in considerably less time than required to read the research report.

The social contribution of the investigator in this context is clear. She enables clients to incorporate the effects of uncertainty about parameters in a specified model consisting of $\mathrm{p}(\mathbf{Y}_T|\theta, A)$ and $\mathrm{p}(\theta|A)$, in reaching conclusions or decisions of the client's choosing, which can be addressed by the model. This contribution extends in an obvious way to uncertainty about models, so long as a posterior simulation structure has been provided by an investigator for each model considered.

With a small amount of additional effort the client can modify many of the investigator's assumptions. Suppose the client wishes to evaluate $\mathrm{E}[\mathrm{h}(\omega)|\mathbf{Y}_T, A^*]$, where the model A^* differs from the model A only in the

*The term normal science is used here as in Kuhn (1970).

specification of the prior distribution $p(\theta|A^*) \neq p(\theta|A)$; i.e., $p(\mathbf{Y}_T|\theta, A) = p(\mathbf{Y}_T|\theta, A^*)$. Suppose further that the support of the investigator's prior distribution includes the support of the client's prior. The investigator's posterior density may be then regarded as an importance sampling density for the client's posterior density. The client reweights the investigator's $\{\theta^{(m)}\}_{m=1}^M$ using the function:

$$w(\theta;\, A^*) = \frac{p(\theta|\mathbf{Y}_T, A^*)}{p(\theta|\mathbf{Y}_T, A)} \propto \frac{p(\theta|A^*)p(\mathbf{Y}_T|\theta, A^*)}{p(\theta|A)p(\mathbf{Y}_T|\theta, A)} = \frac{p(\theta|A^*)}{p(\theta|A)}$$

The client then approximates his posterior moment $E[g(\theta)|\mathbf{Y}_T, A^*]$ by

$$\bar{g}_M^* = \frac{\sum_{m=1}^M w(\theta^{(m)};\, A^*)w(\theta^{(m)})g(\mathbf{Y}_T\theta^{(m)})}{\sum_{m=1}^M w(\theta^{(m)};\, A^*)w(\theta^{(m)})}$$

Geweke (2000) shows that if $\{\theta^{(m)}\}$ is ergodic with invariant distribution $p(\theta|\mathbf{Y}_T, A)\,d\nu(\theta)$, $E[g(\mathbf{Y}_T, \theta)|\mathbf{Y}_T, A^*]$ exists and is finite, and the support of $p(\theta|A)$ includes the support of $p(\theta|A^*)$, then $\bar{g}_M^* \xrightarrow{a.s.} E[g(\mathbf{Y}_T, \theta)|\mathbf{Y}_T, A^*] = \bar{g}^*$; and if in addition $\{\theta^{(m)}\}$ is uniformly ergodic, $\text{var}[g(\mathbf{Y}_T, \theta)|\mathbf{Y}_T, A]$ exists and is finite, and $p(\theta|A^*)/p(\theta|A)$ is bounded above, then there exists $\sigma^2 > 0$ such that $M^{1/2}(\bar{g}_M^* - \bar{g}^*) \xrightarrow{d} N(0, \sigma^2)$.

Efficiency of the reweighting scheme requires some similarity of $p(\theta|A^*)$ and $p(\theta|A)$, as illustrated subsequently in Section 5.3. In particular, both reasonable convergence rates and the use of a central limit theorem to assess numerical accuracy essentially require that $p(\theta|A^*)/p(\theta|A)$ be bounded. Across a set of diverse clients this condition is more likely to be satisfied the more diffuse is $p(\theta|A)$, and is trivially satisfied for the (possibly improper) prior $p(\theta|A) \propto \chi_\Theta(\theta|A)$ if the client's prior is bounded. In the latter case the reweighting scheme will be efficient so long as the client's prior is uninformative relative to the likelihood function. This condition is stated precisely in Theorem 2 of Geweke (1989b). Relative numerical efficiency will indicate situations in which the reweighting scheme is inefficient. If the investigator chooses to use an improper prior for reporting, it is of course incumbent on her to verify the existence of the posterior distribution and convergence of her posterior simulator.

Including $p(\theta^{(m)}|A)$ in the standard posterior simulation structure avoids the need for every client who wishes to impose his own priors to re-evaluate the investigator's prior. Of course, $p(\theta|A^*)$ need not be the client's subjective prior: it may simply be a device by which the client, functioning as another investigator, explores the robustness of results with respect to alternative reasonable priors.

The reweighting scheme permits some updating of the investigator's results at relatively low cost. If observations $T+1, \ldots, T+f$ beyond the T originally used have become available, then

$$\begin{aligned} p(\theta|\mathbf{Y}_{T+f}, A) &\propto p(\theta|A)p(\mathbf{Y}_{T+f}|\theta, A) \\ &= p(\theta|A)p(\mathbf{Y}_T|\theta, A)\prod_{s=T+1}^{T+f} p(\mathbf{y}_s|\mathbf{Y}_{s-1}, \theta, A) \\ &\propto p(\theta|\mathbf{Y}_T, A)\prod_{s=T+1}^{T+f} p(\mathbf{y}_s|\mathbf{Y}_{s-1}, \theta, A) \end{aligned}$$

The client therefore forms the approximation to the updated posterior moment $E[g(\mathbf{Y}_{T+f}, \theta)|\mathbf{Y}_{T+f}, A]$:

$$\bar{g}_M^* = \frac{\sum_{m=1}^{M} w(\theta^{(m)}; \mathbf{Y}_{T+f})w(\theta^{(m)})g(\mathbf{Y}_{T+f}, \theta^{(m)})}{\sum_{m=1}^{M} w(\theta^{(m)}; \mathbf{Y}_{T+f})w(\theta^{(m)})} \xrightarrow{a.s.} E[g|\mathbf{Y}_{T+f}, \theta)|\mathbf{Y}_{T+f}, A] = \bar{g}^*$$

with $w(\theta; \mathbf{Y}_{T+f}) = \prod_{s=T+1}^{T+f} p(\mathbf{y}_s|\mathbf{Y}_{s-1}, \theta, A)$. If f is small relative to T, and there is no major change in the data generating process between T and $T+f$, the new approximation will be efficient. However, as f grows, efficiency diminishes and at some point the approximation $\bar{g}^*$ becomes too inaccurate to be useful. This process also requires evaluation of the likelihood function, which usually involves more technical difficulties than evaluation of priors or functions of interest.

5.2 Software

No special BACC commands are required for reweighting or computing new functions of interest. The general-purpose commands of the mathematical application itself are sufficient. The user simply calculates new functions of interest and/or computes new weights and passes them as input to the **expect1** command.

5.3 Examples

Let us return to the regression model of hedonic pricing introduced in Section 3.8. Suppose that an investigator wishes to provide a posterior simulation structure with the intention that clients will impose their own priors by reweighting the output. To this end the investigator should use a prior distribution that is uninformative relative to the data. Illustrating what such an investigator might do, prior distributions $N(0, 1^2)$ for all slope coefficients, except $N(0, 3^2)$ for log(Lot size), and $0.04h \sim \chi^2(1)$ for precision, were chosen and the corresponding posterior simulation structure with 10,000 replications was created. To mimic a client, the tightest of the three prior distributions described in Section 4.6 was chosen: $N(0.1, .05^2)$ for all slope coefficients except $N(0.3, 0.15^2)$ for lot size, and $0.12h \sim \chi^2(3)$ for precision. A new vector of log weights was created by taking the log of the client's prior minus the log of the investigator's prior. Then **expect1** was used with this new vector of log weights to obtain posterior moments for the coefficients; this required a quarter of a second.

The results of this exercise are displayed in Table 6. The left panel provides the results that would have been obtained had the client directly executed the posterior simulator corresponding to his prior. The accuracy of the numerical approximation of the posterior moments is similar to that exhibited for the less informative prior in Table 1. The right panel displays the results the client obtains by reweighting the investigator's simulator output. The chi-squared statistics from **expectN** (last column) comparing the posterior means approximated in these two different ways indicate no difficulties with the assessment of numerical accuracy through the numerical standard errors. Relative numerical efficiency for the coefficient posterior means ranges from 0.23 to 0.29: overall, the client obtains the same numerical accuracy he would have achieved executing the simulator directly with about 2500 iterations. This would have required 2.5 sec and the use of the **minst** and **postsim** commands, whereas the reweighting took about 3.5 sec and the simpler **expect1** command. Reweighting of the simulator output holds little advantage in this example due to the simplicity of simulating the regression model.

A similar exercise was conducted for the probit example introduced in Section 3.8. To mimic what an investigator might do, a posterior simulation structure for the posterior distribution, with a prior distribution, in which all standard deviations were 10 times larger than those indicated in Table 2, was created. This simulator output was then reweighted to reflect the prior distribution in Table 2. This attempt failed completely—two draws received more than 99.99% of the total weight. To appreciate the reason for this failure recall that the prior distribution used in Section 3.8 is highly informative relative to the sample. This is evident from inspection of Table 2.

To complete this example, the investigator was recast in the role of a client—i.e., the client now has the priors indicated in Table 2 except that the standard deviations are 10 times larger than shown there. Imagine an investigator who uses a prior distribution with standard deviations 10 times larger yet—i.e., the standard deviations are 100 times those shown for the prior distribution in Table 2. This client's reweighting of this investigator's simulator output yields the results displayed in the right half of Table 7. The chi-squared statistic (again

TABLE 6 Regression Model: Comparison of Direct MCMC with Client's Reweighting

	Posterior (tightest prior, direct Gibbs)				Posterior (reweighting of MCMC from more diffuse prior)				
Coefficient	Mean	Stand. dev.	NSE	RNE	Mean	Stand. dev.	NSE	RNE	χ^2
Intercept	7.7283	0.2144	0.0021	1.186	7.7317	0.2157	0.0056	0.165	0.569
Driveway	0.10808	0.02455	0.00026	1.022	0.10792	0.02469	0.00042	0.379	0.749
Rec room	0.068476	0.02307	0.00021	1.328	0.068839	0.023219	0.00049	0.249	0.496
Fin basement	0.10304	0.01950	0.00020	1.090	0.10320	0.01956	0.00037	0.306	0.711
Gas hot water	0.14254	0.03316	0.00036	0.928	0.14145	0.03351	0.00062	0.320	0.131
Central air	0.15408	0.01961	0.00021	0.970	0.15376	0.01947	0.00042	0.239	0.487
#Garage stalls	0.051947	0.01119	0.00009	1.640	0.052116	0.011217	0.00018	0.410	0.412
Good nbhd	0.12561	0.02043	0.00017	1.665	0.12538	0.020962	0.00041	0.288	0.609
log(Lot size)	0.30469	0.02625	0.00026	1.097	0.30433	0.026346	0.00069	0.163	0.623
#Bedrooms	0.040706	0.013615	0.00012	1.368	0.040502	0.013479	0.00024	0.340	0.455
#Full baths	0.15519	0.018767	0.00015	1.830	0.15477	0.018699	0.00040	0.238	0.324
#Stories	0.093422	0.012002	0.00012	1.055	0.093754	0.012062	0.00027	0.230	0.256

TABLE 7 Probit Model: Comparison of Direct MCMC with Client's Reweighting

	Posterior (Gibbs, direct MCMC)				Posterior (reweighting of MCMC from more diffuse prior)				
Coefficient	Mean	Stand. dev.	NSE	RNE	Mean	Stand. dev.	NSE	RNE	χ^2
Intercept	1.545	0.4618	0.0147	0.110	1.528	0.4502	0.0137	0.121	0.405
Black	0.1072	0.1048	0.0020	0.310	0.1064	0.1029	0.0026	0.168	0.809
Age-Single	−0.05702	0.01208	0.00033	0.153	−0.05611	0.01255	0.00035	0.145	0.056
Age-Married	−0.08414	0.01239	0.00024	0.308	−0.08357	0.01208	0.00046	0.076	0.273
Education	0.07848	0.02247	0.00060	0.157	0.07862	0.02261	0.00078	0.094	0.890
Married	0.6779	0.4744	0.0136	0.134	0.6776	0.4636	0.0106	0.213	0.987
Kids	−0.5035	0.1746	0.0058	0.102	−0.5020	0.1633	0.0080	0.046	0.879
#Kids	−0.05377	0.05461	0.00090	0.413	−0.05201	0.05384	0.00154	0.136	0.324
Spouse$	-1.702×10^{-5}	3.436×10^{-6}	8.29×10^{-8}	0.191	-1.700×10^{-5}	3.378×10^{-6}	1.22×10^{-7}	0.085	0.883
Family$	1.043×10^{-6}	6.397×10^{-6}	1.51×10^{-7}	0.200	5.819×10^{-7}	6.269×10^{-6}	2.84×10^{-7}	0.054	0.152
AFDC	-5.342×10^{-4}	3.782×10^{-4}	6.07×10^{-6}	0.431	-5.529×10^{-4}	3.744×10^{-4}	1.02×10^{-5}	0.148	0.116
Food$	−0.002225	6.946×10^{-4}	1.41×10^{-5}	0.269	−0.002266	6.880×10^{-4}	2.45×10^{-5}	0.088	0.143
WorkExp	1.372×10^{-4}	9.392×10^{-6}	2.48×10^{-7}	0.159	1.367×10^{-4}	9.503×10^{-6}	4.69×10^{-7}	0.046	0.435

from expectN) for comparison of the posterior means again indicates no problem with numerical standard errors (NSEs). The relative numerical efficiencies (RNEs) show that the investigator's simulator output with 10,000 records provides about the same information the client would have obtained with 1000 records directly from simulator output for his posterior. This would have required the minst and postsim commands and about 40 sec of execution time, whereas the reweighting required less than 4 sec and the simpler expect1 command. (Notice that the posterior means, in Table 7, are much closer to the maximum likelihood statistics in Table 2 than are the posterior means in the same table, which correspond to the previous more informative prior.)

These examples underscore both the potential efficiency of Bayesian communication through posterior reweighting, and its limitations. The efficiency comes about because reweighting software is simple and generic, whereas posterior simulators are model specific and impose greater computational demands. (This advantage increases dramatically in more complex models.) The limitations arise from the need for some similarity of the investigator's and client's posterior distributions. As argued and illustrated here, the investigator should use a prior distribution that is uninformative relative to the sample. In this situation a client's reweighting will be successful for priors that are moderately, but not greatly, informative relative to the sample.

ACKNOWLEDGMENT

This chapter is a revision of the article "Using Simulation Methods for Bayesian Econometric Models: Inference, Development, and Communication" that appeared, together with comments and reply, as issue no. 1 in volume 18 of *Econometric Reviews*, early in 1999. The illustrations in this chapter use a more recent version of BACC than did the earlier article, and the article treats aspects of model development that are not included in this chapter. Financial support from NSF grants SBR-9600040, SBR-9731037, and SBR-0214303 is gratefully acknowledged.

REFERENCES

Albert, J. and Chib, S., 1993. Bayesian analysis of binary and polychotomous response data. *Journal of the American Statistical Association* **88**, 669–679.

Anglin, P.M. and Gencay, R., 1996. Semiparametric estimation of a hedonic price function. *Journal of Applied Econometrics* **11**, 633–648.

Bajari, P., 1997. The first price sealed bid auction with asymmetric bidders: Theory and applications. Unpublished Ph.D. dissertation, University of Minnesota, Minneapolis, Minnesota.

Berger, J. O., 1985. *Statistical Decision Theory and Bayesian Analysis*, 2nd ed. Springer–Verlag, New York.

Bernardo, J.M. and Smith, A.F.M., 1994. *Bayesian Theory*. Wiley, New York.

Chib, S., 1995. Marginal likelihood from the Gibbs output. *Journal of the American Statistical Association* **90**, 1313–1321.

Chib, S. Greenberg, and E., 1995. Understanding the Metropolis–Hastings algorithm. *American Statistician* **49**, 327–335.

Chib, S. and Greenberg, E., 1996. Markov chain Monte Carlo simulation methods in econometrics. *Econometric Theory* **12**, 409–431.

Cogburn, R., 1972. The central limit theorem for Markov processes. In *Proceedings of the Sixth Berkeley Symposium on Mathematical Statistics and Probability 2*, University of California Press, Berkeley, CA, pp. 485–512.

Engle, R.F., Hendry, D.F., and Richard, J.F., 1983. Exogeneity. *Econometrica* **51**, 277–304.

Friedman, M., 1953. The methodology of positive economics. In: M. Friedman, ed. *Essays in Positive Economics*. University of Chicago Press, Chicago, pp. 3–43.

Geisel, M.S., 1977. Bayesian comparisons of simple macroeconomic models. In: S.E. Fienberg and A. Zellner, eds. *Studies in Bayesian Econometrics and Statistics in Honor of Leonard J. Savage*. North–Holland, Amsterdam, pp. 227–256.

Gelfand, A.E. and Dey, D.K., 1994. Bayesian model choice: Asymptotics and exact calculations. *Journal of the Royal Statistical Society Series B* **56**, 501–514.

Gelfand, A.E. and Smith, A.F.M., 1990. Sampling based approaches to calculating marginal densities. *Journal of the American Statistical Association* **85**, 398–409.

Gelman, A. and Rubin, D.B. 1992. Inference from iterative simulation using multiple sequences. *Statistical Science* **7**, 457–472.

Gelman, A., Carlin, J.B., Stern, H.S., and Rubin, D.B., 1995. *Bayesian Data Analysis*. Chapman and Hall, London.

Geweke, J., 1988. Antithetic acceleration of Monte Carlo integration in Bayesian Inference. *Journal of Econometrics* **38**, 73–90.

Geweke, J., 1989a. Exact predictive densities in linear models with ARCH disturbances. *Journal of Econometrics* **40**, 63–86.

Geweke, J., 1989b. Bayesian inference in econometric models using Monte Carlo integration. *Econometrica* **57**, 1317–1340.

Geweke, J., 1991. Efficient simulation from the multivariate normal and Student-t distributions subject to linear constraints. In: E. M. Keramidas, ed., *Computing Science and Statistics: Proceedings of the Twenty–Third Symposium on the Interface*. Interface Foundation of North America, Inc., Fairfax, pp. 571–578.

Geweke, J., 1992. Evaluating the accuracy of sampling–based approaches to the calculation of posterior moments. In: J.O. Berger, J.M. Bernardo, A.P. Dawid, and A.F.M. Smith eds. *Proceedings of the Fourth Valencia International Meeting on Bayesian Statistics.* Oxford University Press, Oxford, pp. 169–194.

Geweke, J., 1999. Using simulation methods for Bayesian econometric models: Inference, development, and communication. (With comments and rejoinder). *Econometric Reviews* **18**, 1–73.

Geweke, J., 2000. Bayesian communication software: The BACC system. American Statistical Association, Section 2000, *Proceedings of the Section on Bayesian Statistical Sciences*, 40–49.

Geweke, J., 2001. Bayesian econometrics and forecasting. *Journal of Econometrics*, **100**, 11–15.

Geweke, J. and Keane, M., 1999. Mixture of normals probit models. In: C. Hsiao, K. Lahiri, L-F Lee, and M. H. Pesaran, eds. *Analysis of Panels and Limited Dependent Variables: In Honor of G. S. Maddala.* Cambridge University Press, Cambridge, pp. 49–78.

Geyer, C.J., 1992. Practical Markov chain Monte Carlo. *Statistical Science* **7**, 473–481.

Hammersly, J.M. and Handscomb, D.C., 1964. *Monte Carlo Methods.* Methuen, London.

Hannan, E.J., 1970. *Multiple Time Series.* Wiley, New York.

Hastings, W.K., 1970. Monte Carlo sampling methods using Markov chains and their applications. *Biometrika* **57**, 97–109.

Hildreth, C., 1963. Bayesian statisticians and remote clients. *Econometrica* **31**, 422–438.

Jeffreys, H., 1939. *Theory of Probability*, 1st ed. Oxford University Press, Oxford.

Jeffreys, H., 1961. *Theory of Probability*, 3rd ed. Oxford University Press, Oxford.

Kass, R.E. and Raftery, A.E. 1995. Bayes factors. *Journal of the American Statistical Association* **90**, 773–795.

Kloek, T. and van Dijk, H.K., 1978. Bayesian estimates of equation system parameters: An application of integration by Monte Carlo. *Econometrica* **46**, 1–20.

Kuhn, T.S., 1970. *The Structure of Scientific Revolutions*, 2nd ed. University of Chicago Press, Chicago.

Metropolis, N., Rosenbluth, A.W., Rosenbluth, M.N., Teller, A.H., and Teller, E., 1953. Equation of state calculations by fast computing machines. *Journal of Chemical Physics* **21**, 1087–1092.

Poirier, D.J., 1988. Frequentist and subjectivist perspectives on the problems of model building in economics. (With discussion and rejoinder.) *Journal of Economic Perspectives* **2**, 121–170.

Poirier, D.J., 1995. *Intermediate Statistics and Econometrics: A Comparative Approach.* MIT Press, Cambridge, MA.

Poirier, D.J., 1997. Comparing and choosing between two models with a third model in the background. *Journal of Econometrics* **78**, 139–152.

Press, W.H., Flannery, B.P., Teukolsky, S.A., and Vetterling, W.T., 1992. *Numerical Recipes: The Art of Scientific Computing*, 2nd ed. Cambridge University Press, Cambridge.

Raftery, A.E., 1996. Hypothesis testing and model selection via posterior simulation, in Gilks, W.R. et al., eds., *Markov Chain Monte Carlo in Practice*. Chapman and Hall, London..

Raiffa, H. and Schlaifer, R., 1961. *Applied Statistical Decision Theory*. Harvard University Press, Cambridge, MA.

Roberts, G.O. and Smith, A.F.M., 1994. Simple conditions for the convergence of the Gibbs sampler and Metropolis–Hastings algorithms. *Stochastic Processes and Their Applications* **49**, 207–216.

Steel, M. and Richard, J.F., 1991. Bayesian multivariate exogeneity analysis: An application to a U.K. money demand equation. *Journal of Econometrics* **49**, 239–274.

Tierney, L., 1994. Markov chains for exploring posterior distributions. (With discussion and rejoinder.) *Annals of Statistics* **22**, 1701–1762.

Zeger, S.L. and Karim, M.R., 1991. Generalized linear models with random effects: A Gibbs sampling approach. *Journal of the American Statistical Association* **86**, 79–86.

Zellner, A., 1971. *An Introduction to Bayesian Inference in Econometrics*. Wiley, New York.

Zellner, A. and Min, C., 1995. Gibbs sampler convergence criteria. *Journal of the American Statistical Association* **90**, 921–927.

9

Bayesian Inference in the Seemingly Unrelated Regressions Model

William E. Griffiths
University of Melbourne, Melbourne, Australia

1 INTRODUCTION

Zellner's idea of combining several equations into one model to improve estimation efficiency (Zellner, 1962) ranks as one of the most successful and lasting innovations in the history of econometrics. The resulting seemingly unrelated regressions (SUR) model has generated a wealth of both theoretical and empirical contributions. Reviews of work on or involving the SUR model can be found in Srivastava and Dwivedi (1979), Judge et al. (1985), Srivastava and Giles (1987), and Fiebig (2001). It was also Zellner (1971) who popularized Bayesian inference in econometrics generally and described the SUR model within the context of Bayesian inference. However, at that time, convenient methods for deriving or estimating marginal posterior density functions and moments for individual SUR coefficients were not generally available. Subsequently, analytical results were derived for some special cases (Drèze and Morales, 1976; Richard and Tompa, 1980; Richard and Steel, 1988; Steel 1992) and importance sampling was suggested as a means for estimating marginal posterior density functions and their moments (Kloek and van Dijk, 1978). More recently, the application of Markov-Chain Monte Carlo (MCMC) methodology to Bayesian inference has made available a new range of numerical methods that make Bayesian estimation of the SUR model more convenient and accessible. The literature of MCMC is extensive; for a general appreciation of its scope and purpose, see Tierney (1994), Albert and Chib (1996), Chen et al. (2000), Chib and Greenberg (1996), Gilks et al. (1996), Tanner (1996), and the chapter by Geweke et al. (this volume). For application of MCMC to the SUR model, see, e.g., Percy (1992, 1996), Chib and Greenberg (1995), Griffiths and Chotikapanich (1997), and Griffiths et al. (2000).

The objective of this chapter is to provide a practical guide to computer-aided Bayesian inference for a variety of problems that arise in applications of the SUR model. We describe examples of problems, models, and algorithms that have been placed within a general framework in the chapter by Geweke et al. (this volume); our chapter can be viewed as complementary to that one. The model is described in Section 2; the joint, conditional, and marginal posterior density functions that result from a noninformative prior are derived. In Section 3 we describe how to use sample draws of parameters from their posterior desnsities to estimate posterior quantities of interest; two Gibbs sampling algorithms and a Metropolis–Hastings algorithm are given. Modifications necessary for nonlinear equations, equality restrictions and inequality restrictions are presented in Sections 4, 5, and 6, respectively. Three applications are described in Section 7. Section 8 contains methodology for forecasting. Some extensions are briefly mentioned in Section 9 and a few concluding remarks are given in Section 10.

2 MODEL SPECIFICATION AND POSTERIORS FROM A NONINFORMATIVE PRIOR

Consider M equations written as

$$y_i = X_i\beta_i + e_i \quad i = 1, 2, \ldots, M \tag{1}$$

where y_i is a T-dimensional vector of observations on a dependent variable, X_i is a $(T \times K_i)$ matrix of observations on K_i nonstochastic explanatory variables, possibly including a constant term, β_i is a K_i-dimensional vector of unknown coefficients that we wish to estimate, and e_i is a T-dimensional unobserved random vector. The M equations can be combined into one large model written as

$$\begin{bmatrix} y_1 \\ y_2 \\ \vdots \\ y_M \end{bmatrix} = \begin{bmatrix} X_1 & & & \\ & X_2 & & \\ & & \ddots & \\ & & & X_M \end{bmatrix} \begin{bmatrix} \beta_1 \\ \beta_2 \\ \vdots \\ \beta_M \end{bmatrix} + \begin{bmatrix} e_1 \\ e_2 \\ \vdots \\ e_M \end{bmatrix} \tag{2}$$

which we then write compactly as

$$y = X\beta + e \tag{3}$$

where y is of dimension $(TM \times 1)$, X is of dimension $(TM \times K)$, with $K = \sum_{i=1}^{M} K_i$, β is $(K \times 1)$ and e is $(TM \times 1)$. We assume that the distribution for e is given by

$$e \sim N(0, \Sigma \otimes I_T) \tag{4}$$

Thus, the errors in each equation are homoskedastic and not autocorrelated. There is, however, contemporaneous correlation between corresponding errors in different equations. The variance in the error of the ith equation we denote by σ_{ii}, the ith diagonal element of Σ. The covariance between two corresponding errors in different equations (say i and j), we write as σ_{ij}, an off-diagonal element of Σ.

Using $f(\cdot)$ as generic notation for a probability density function (p.d.f.), the likelihood function for β and Σ can be written as

$$f(y|\beta, \Sigma) = (2\pi)^{-MT/2}|\Sigma|^{-T/2} \exp\{-\tfrac{1}{2}(y - X\beta)'(\Sigma^{-1} \otimes I_T)(y - X\beta)\} \tag{5}$$

This p.d.f. can also written as

$$f(y|\beta, \Sigma) = (2\pi)^{-MT/2}|\Sigma|^{-T/2} \exp\{-\tfrac{1}{2}\mathrm{tr}(A\Sigma^{-1})\} \tag{6}$$

where A is an $(M \times M)$ matrix with (i,j)th element given by

$$[A]_{ij} = (y_i - X_i\beta_i)'(y_j - X_j\beta_j) \tag{7}$$

Note that A can also be written as

$$A = (Y - X^*B)'(Y - X^*B) \tag{8}$$

where Y is the $(T \times M)$ matrix $Y = (y_1, y_2, \ldots, y_M)$, X^* is the $(T \times K)$ matrix $X^* = (X_1, X_2, \ldots, X_M)$, and B is the $(K \times M)$ matrix:

$$B = \begin{bmatrix} \beta_1 & & & \\ & \beta_2 & & \\ & & \ddots & \\ & & & \beta_M \end{bmatrix} \tag{9}$$

Result (9) on page 42 of Lütkepohl (1996) can be used to establish the equivalence of Eqs. (5) and (6). Specically,

$$\begin{aligned} \mathrm{tr}[(Y - X^*B)'(Y - X^*B)\Sigma^{-1}] &= [\mathrm{vec}(Y - X^*B)]'[\Sigma^{-1} \otimes I_T]\mathrm{vec}(Y - X^*B) \\ &= (y - X\beta)'(\Sigma^{-1} \otimes I_T)(y - X\beta) \end{aligned} \tag{10}$$

Two prior p.d.f.s. will be considered in this chapter; they are the conventional noninformative prior (see, e.g., Zellner 1971, ch. 8):

$$f(\beta, \Sigma) = f(\beta)\, f(\Sigma) \propto |\Sigma|^{-(M+1)/2} \tag{11}$$

and another prior that imposes inequality restrictions on β, but is otherwise noninformative. The inequality prior and its consequences will be considered later in the chapter. The noninformative prior in Eq. (11) is chosen to provide objectivity in reporting, not because we believe total ignorance is prevalent. Geweke et al. (Chap. 8 in this volume) discuss how to modify results to accommodate the prior of a specific client.

2.1 Joint Posterior p.d.f. for (β, Σ)

Applying Bayes' theorem to the prior p.d.f. in Eq. (11) and the likelihood function in Eqs. (5) and (6) yields the joint posterior p.d.f. for β and Σ:

$$\begin{aligned} f(\beta, \Sigma \mid y) &\propto f(y \mid \beta, \Sigma) f(\beta, \Sigma) \\ &\propto |\Sigma|^{-(T+M+1)/2} \exp\{-\tfrac{1}{2}(y - X\beta)'(\Sigma^{-1} \otimes I_T)(y - X\beta)\} \\ &= |\Sigma|^{-(T+M+1)/2} \exp\{-\tfrac{1}{2}\mathrm{tr}(A\Sigma^{-1})\} \end{aligned} \tag{12}$$

In the remainder of this section we describe a number of marginal and conditional posterior p.d.f.'s that are derived from Eq. (12). These p.d.f.'s will prove useful in later sections when we discuss methods for estimating quantities of interest. We will assume that interest centers on individual coefficients, say the kth coefficient in the ith equation β_{ik}, and, more generally, on some functions of the coefficients, say $g(\beta)$. Forecasting future values y^* will also be considered. The relevant p.d.f.'s that express our uncertain postsample knowledge about these quantities are the marginal p.d.f.'s $f(\beta_{ik} \mid y)$, $f(g(\beta) \mid y)$, and $f(y^* \mid y)$, respectively. Typically, we report results by graphing these p.d.f.'s, and tabulating their means, standard deviations, and probabilities for regions of interest. Describing the tools for doing so is the major focus of this chapter.

2.2 Conditional Posterior p.d.f. for ($\beta | \Sigma$)

The term in the exponent of Eq. (12) can be written as

$$\begin{aligned} (y - X\beta)'(\Sigma^{-1} \otimes I_T)(y - X\beta) &= (y - X\hat{\beta})'(\Sigma^{-1} \otimes I_T)(y - X\hat{\beta}) \\ &\quad + (\beta - \hat{\beta})'X'(\Sigma^{-1} \otimes I_T)X(\beta - \hat{\beta}) \end{aligned} \tag{13}$$

where $\hat{\beta} = [X'(\Sigma^{-1} \otimes I_T)X]^{-1}X'(\Sigma^{-1} \otimes I_T)y$. It follows that the conditional posterior p.d.f. for β given Σ is the multivariate normal p.d.f.:

$$f(\beta \mid \Sigma, y) \propto \exp\{-\tfrac{1}{2}(\beta - \hat{\beta})'X'(\Sigma^{-1} \otimes I_T)X(\beta - \hat{\beta})\} \quad (14)$$

with posterior mean equal to the generalized least squares estimator:

$$E(\beta \mid y, \Sigma) = \hat{\beta} = [X'(\Sigma^{-1} \otimes I_T)X]^{-1}X'(\Sigma^{-1} \otimes I_T)y \quad (15)$$

and posterior covariance matrix equal to

$$V(\beta \mid y, \Sigma) = [X'(\Sigma^{-1} \otimes I_T)X]^{-1} \quad (16)$$

The last two expressions are familiar ones in sampling theory inference for the SUR model. They show that the traditional SUR estimator, written as

$$\hat{\hat{\beta}} = [X'(\hat{\Sigma}^{-1} \otimes I_T)X]^{-1}X(\hat{\Sigma}^{-1} \otimes I_T)y \quad (17)$$

where $\hat{\Sigma}$ is a two-step estimator or a maximum likelihood estimator, can be viewed as the mean of the conditional posterior p.d.f. for β given $\hat{\Sigma}$. The traditional covariance matrix estimator $[X'(\hat{\Sigma}^{-1} \otimes I_T)X]^{-1}$ can be viewed as the conditional covariance matrix from the same p.d.f. Since this p.d.f. does not take into account uncertainty from not knowing Σ (the fact that $\hat{\Sigma}$ is an estimate is not recognized), it overstates the reliability of our information about β. This dilemma was noted by Fiebig and Kim (2000) in the context of an increasing number of equations.

2.3 Marginal Posterior p.d.f. for β

The more appropriate representation of our uncertainty about β is the marginal posterior p.d.f. $f(\beta \mid y)$. It can be shown that this p.d.f. is given by

$$f(\beta \mid y) = \int f(\beta, \Sigma \mid y)\, d\Sigma \propto |A|^{-T/2} \quad (18)$$

The integral in Eq. (18) is performed by using properties of the inverted Wishart distribution (see, e.g., Zellner 1971, p. 395). For $f(\beta \mid y) \propto |A|^{-T/2}$ to be proper, we require $T \geq M + \text{rank}(X^*)$ (Griffiths et al., 2002). Also, this p.d.f. is not of a standard recognizable form. Except for special cases, analytical expressions for its normalizing constant and moments are not available. Estimating these moments, and marginal p.d.f.'s for individual coefficients β_{ik}, is considered in Section 3; first, we describe some more p.d.f.'s that will prove to be useful.

2.4 Conditional Posterior p.d.f. for $(\beta_1 \mid \beta_2, \ldots, \beta_M)$

It is possible to show that the posterior p.d.f. for the coefficient vector from one equation, conditional on those from other equations, is a multivariate t-distribution.

To derive this result, we will consider the posterior p.d.f. for β_1, conditional on $(\beta_2, \beta_3, \ldots, \beta_M)$. We write a partition of $(Y - X^*B)$ into its first and remaining $(M-1)$ columns as

$$Y - X^*B = (y_1 - X_1\beta_1 \quad E_{(1)})$$

The corresponding partition of A is

$$A = \begin{bmatrix} (y_1 - X_1\beta_1)'(y_1 - X_1\beta_1) & (y_1 - X_1\beta_1)'E_{(1)} \\ E_{(1)}'(y_1 - X_1\beta_1) & E_{(1)}'E_{(1)} \end{bmatrix}$$

Using a result on the determinant of a partitioned matrix, we have

$$|A| = |E_{(1)}E_{(1)}'|((y_1 - X_1\beta_1)'(y_1 - X_1\beta_1) - (y_1 - X_1\beta_1)'E_{(1)}(E_{(1)}'E_{(1)})^{-1}E_{(1)}'(y_1 - X_1\beta_1))$$

Defining $Q_{(1)} = I_T - E_{(1)}(E_{(1)}'E_{(1)})^{-1}E_{(1)}'$ and $\tilde{\beta}_1 = (X_1'Q_{(1)}X_1)^{-1}X_1'Q_{(1)}y_1$, the second term in the above equation can be written as

$$(y_1 - X_1\beta_1)'Q_{(1)}(y_1 - X_1\beta_1) = (y_1 - X_1\tilde{\beta}_1)'Q_{(1)}(y_1 - X_1\tilde{\beta}_1) + (\beta_1 - \tilde{\beta}_1)'X_1'Q_{(1)}X_1(\beta_1 - \tilde{\beta}_1)$$

Collecting all these results, substituting into Eq. (18), and letting $|E_{(1)}'E_{(1)}|$ be absorbed into the proportionality constant, we can write

$$f(\beta_1 \mid y, \beta_2, \beta_3, \ldots, \beta_M) \propto \left[\nu_1 + \frac{(\beta_1 - \tilde{\beta}_1)'X_1'Q_{(1)}X_1(\beta_1 - \tilde{\beta}_1)}{\tilde{s}_1^2} \right]^{-(K_1+\nu_1)/2} \tag{19}$$

where $\nu_1 = T - K_1$ and $\tilde{s}_1^2 = (y_1 - X_1\tilde{\beta}_1)'Q_{(1)}(y_1 - X_1\tilde{\beta}_1)/\nu_1$. Equation (19) is in the form of a multivariate t-distribution with degrees of freedom ν_1, mean $\tilde{\beta}_1$, and covariance matrix $(\nu_1/(\nu_1 - 2))\tilde{s}_1^2(X_1'Q_{(1)}X_1)^{-1}$. See, e.g., Zellner (1971, p. 383). The conditional posterior p.d.f.'s for other β_i are similarly defined.

2.5 Conditional Posterior p.d.f. for $(\Sigma|\beta)$

Viewing the joint posterior p.d.f. in Eq. (12) as a function of only Σ yields the conditional posterior p.d.f. for Σ given β. It is the inverted Wishart p.d.f. (see, e.g., Zellner, 1971, p. 395):

$$f(\Sigma|\beta, y) \propto |\Sigma|^{-(T+M+1)/2} \exp\{-\tfrac{1}{2}\text{tr}(A\Sigma^{-1})\} \tag{20}$$

It has T degrees of freedom, and parameter matrix A.

2.6 Marginal Posterior pdf for Σ

The marginal p.d.f. for Σ, obtained by using the result in Eq. (13), and then using properties of the multivariate normal p.d.f. to integrate out β, is given by

$$\begin{aligned} f(\Sigma\,|\,y) &= \int f(\beta, \Sigma y\,|\,d\beta \\ &\propto \left|X'(\Sigma^{-1}\otimes I_T)X\right|^{-1/2}|\Sigma|^{-(T+M+1)/2} \\ &\quad\times \exp\{-\tfrac{1}{2}(y - X\hat{\beta})'(\Sigma^{-1}\otimes I_T)(y - X\hat{\beta})\} \\ &= \left|X'(\Sigma^{-1}\otimes I_T)X\right|^{-1/2}|\Sigma|^{-(T+M+1)/2}\exp\{-\tfrac{1}{2}\text{tr}(\hat{A}\Sigma^{-1})\} \end{aligned} \tag{21}$$

where $\hat{A}$ is an $(M \times M)$ matrix with the (i,j)th element given by

$$[\hat{A}]_{ij} = (y_i - X_i\hat{\beta}_i)'(y_j = X_j\hat{\beta}_j)$$

The posterior p.d.f. in Eq. (21) is not an analytically tractable one whose moments are known. However, as we will see, we can draw observations from it using the Gibb's sampler.

3 ESTIMATING POSTERIOR QUANTITIES

Given the intractability of the posterior p.d.f. $f(\beta\,|\,y) \propto A|^{-T/2}$, methods for estimating marginal posterior p.d.f's for individual coefficients β_{ik}, their moments, and probabilities of interest, are required. Suppose that we have draws $\beta^{(1)}, \beta^{(2)}, \ldots, \beta^{(N)}$ taken from $f(\beta\,|\,y)$ and, possibly, draws $\Sigma^{(1)}, \Sigma^{(2)}, \ldots, \Sigma^{(N)}$ taken from $f(\Sigma\,|\,y)$. We will describe a number of ways one can proceed to estimate the desired quantities; then, we discuss how the required posterior draws can be obtained.

3.1 Estimating Posterior p.d.f.s

A simple way to estimate the marginal posterior p.d.f. of β_{ik}, say, is to construct a histogram of draws of that parameter. Joining the midpoints of the histogram classes provides a continuous representation of the p.d.f., but typically, it will be jagged one unless some kind of smoothing procedure is employed. Alternatively, one can obtain a smooth p.d.f., and a more efficient estimate, by averaging conditional posterior p.d.f's for the quantity of interest. In this case, for conditional posterior p.d.f's one can use the t-distributions defined by Eq. (19), or, if draws on both β and Σ are available, the normal distributions defined by Eq. (14).

Considering the t-distribution first, an estimate of $f(\beta_{ik} \mid y)$ is given by

$$\begin{aligned}\hat{f}(\beta_{ik}\mid y) &= \frac{1}{N}\sum_{\ell=1}^{N} f(\beta_{ik}\mid y, \beta_1^{(\ell)}, \ldots, \beta_{i-1}^{(\ell)}, \beta_{i+1}^{(\ell)}, \ldots, \beta_M^{(\ell)}) \\ &= c\frac{1}{N}\sum_{\ell=1}^{N} \frac{1}{\sqrt{\tilde{s}_i^{2(\ell)} q_{(i)kk}^{(\ell)}}}\left[\nu_i + \frac{(\beta_{ik} - \tilde{\beta}_{ik}^{(\ell)})^2}{\tilde{s}_i^{2(\ell)} q_{(i)kk}^{(\ell)}}\right]^{-(\nu_i+1)/2}\end{aligned} \tag{22}$$

The univariate t-distribution that is being averaged in Eq. (22) is the conditional p.d.f. for a single coefficient from β_i, obtained from the multivariate t-distribution in Eq. (19), after generalizing from β_1 to β_i. The previously undefined terms in Eq. (22) are the constant:

$$c = \frac{\Gamma[(\nu+1)/2]\nu^{\nu/2}}{\Gamma(\nu/2)\sqrt{\pi}}$$

where $\Gamma(\cdot)$ is the gamma function, the conditional posterior mean $\tilde{\beta}_{ik}$, which is the k-element in $\tilde{\beta}_i$, and $q_{(i)kk}$, which is the kth diagonal element of $(X_i' Q_{(i)} X_i)^{-1}$. To plot the p.d.f. in Eq. (22), we choose a grid of values for β_{ik} (50–100 is usually adequate), and for each value of β_{ik}, we compute the average in Eq. (22). These averages are plotted against the β_{ik}.

Alternatively, the conditional normal distributions in Eq. (14) can be averaged over Σ. In this case an estimate of the marginal posterior p.d.f. for β_{ik} is given by

$$\begin{aligned}\hat{f}(\beta_{ik}\mid y) &= \frac{1}{N}\sum_{\ell=1}^{N} f(\beta_{ik}\mid y, \Sigma^{(\ell)}) \\ &= \frac{1}{\sqrt{2\pi}}\frac{1}{N}\sum_{\ell=1}^{N}\frac{1}{\sqrt{h_{(i)kk}^{(\ell)}}}\exp\left\{-\frac{1}{2h_{(i)kk}^{(\ell)}}(\beta_{ik} - \hat{\beta}_{ik}^{(\ell)})^2\right\}\end{aligned} \tag{23}$$

where $\hat{\beta}_{ik}$ is the k-element in the ith vector component of $\hat{\beta}$ [see Eq. (15)] and $h_{(i)kk}$ is the k-diagonal element in the ith diagonal block of $[X'(\Sigma^{-1} \otimes I_T)X]^{-1}$ [see Eq. (16)]. As in Eq. (22), the average in Eq. (23) is computed for, and plotted against, a grid of values for β_{ik}.

3.2 Estimating Posterior Means and Standard Deviations

Corresponding to the three ways given for estimating posterior p.d.f.'s, there are three ways of estimating their posterior means and variances. The first way is to use the sample mean and covariance matrix of the draws. That is,

$$\hat{E}(\beta \mid y) = \frac{1}{N}\sum_{\ell=1}^{N} \beta^{(\ell)} = \bar{\beta} \tag{24}$$

and

$$\hat{V}(\beta \mid y) = \frac{1}{N-1}\sum_{\ell=1}^{N} (\beta^{(\ell)} - \bar{\beta})(\beta^{(\ell)} - \bar{\beta})' \tag{25}$$

The second and third approaches use the results: (1) an unconditional mean is equal to the mean of the conditional means, and (2) the unconditional variance is equal to the mean of the conditional variances plus the variance of the conditional means. Applying these two results to the conditional posterior p.d.f. in Eq. (19) yields:

$$\hat{E}(\beta_i \mid y) = \frac{1}{N}\sum_{\ell=1}^{N} \tilde{\beta}_i^{(\ell)} = \frac{1}{N}\sum_{\ell=1}^{N} (X_i' Q_{(i)}^{(\ell)} X_i)^{-1} X_i' Q_{(i)}^{(\ell)} y_i = \bar{\tilde{\beta}}_i \tag{26}$$

$$\begin{aligned} \hat{V}(\beta_i \mid y) = {} & \left(\frac{\nu_i}{\nu_i - 2}\right)\frac{1}{N}\sum_{\ell=1}^{N} \tilde{s}_i^{2(\ell)} (X_i' Q_{(i)}^{(\ell)} X_i)^{-1} \\ & + \frac{1}{N-1}\sum_{\ell=1}^{N} \left(\tilde{\beta}_i^{(\ell)} - \bar{\tilde{\beta}}_i\right)\left(\tilde{\beta}_i^{(\ell)} - \bar{\tilde{\beta}}_i\right)' \end{aligned} \tag{27}$$

Applying the two results to the normal conditional posterior p.d.f.'s in Eq. (14) yields:

$$\hat{E}(\beta \mid y) = \frac{1}{N}\sum_{\ell=1}^{N} \hat{\beta}^{(\ell)} = \frac{1}{N}\sum_{\ell=1}^{N} [X'(\Sigma^{-1(\ell)} \otimes I_T)X]^{-1} X'(\Sigma^{-1(\ell)} \otimes I_T)y = \bar{\hat{\beta}} \tag{28}$$

and

$$\begin{aligned} \hat{V}(\beta \mid y) = {} & \frac{1}{N}\sum_{\ell=1}^{N} [X'(\Sigma^{-1(\ell)} \otimes I_T)X]^{-1} \\ & + \frac{1}{N-1}\sum_{\ell=1}^{N} (\hat{\beta}^{(\ell)} - \bar{\hat{\beta}})(\hat{\beta}^{(\ell)} - \bar{\hat{\beta}})' \end{aligned} \tag{29}$$

Clearly, using the sample means and standard deviations from Eqs. (24) and (25) is much easier than using the conditional quantities in Eqs. (26)–(29). However, averaging conditonal moments generally leads to more efficient estimates.

3.3 Estimating Probabilities

Often, we are interested in reporting the probability that β_{ik} lies with a particular interval or finding an interval with a prespecified probability content. In sampling theory, inference intervals with 95% probability content are popular. An estimate of the probability that β_{ik} lies in a particular interval is given by the proportion of draws that lie within that interval. Alternatively, one can find conditional probabilities and average them, along the lines that the conditional means are averaged in Eq. (26) and (28). Using the conditional normal distribution as an example, we can estimate the probability that β_{ik} lies in the interval (a, b) as

$$\hat{P}(a < \beta_{ik} < b) = \frac{1}{N}\sum_{\ell=1}^{N} P(a < (\beta_{ik} \mid \Sigma^{(\ell)}) < b) \tag{30}$$

Order statistics can be used to obtain an interval with a prespecified probability content. For example, for a 95% probability interval for β_{ik}, we can take the 0.025 and 0.975 empirical quantiles of the draws of the β_{ik}.

3.4 Functions of β

How do we proceed if we are interested in some functions of β, say $g(\beta)$? Examples of such functions considered later in this chapter are monotonicity and curvature conditions in a cost system, and the relative magnitudes of equaivalence scales in household expenditure functions. Examples outside the context of SUR models are the evaluation of consumer surplus (Griffiths, 1999) and the stationary region in a time-series model (Geweke, 1988).

If it is possible to derive, analytically, conditional distributions of the form $f(g(\beta_1) \mid y, \beta_2, \ldots, \beta_M)$ or $f(g(\beta) \mid y, \Sigma)$, then one can work with these conditional distributions along the lines described above. However, the ability to proceed analytically is rare, given that $g(\beta)$ is frequently nonlinear and of lower dimension than β. Instead, we can compute values $g(\beta^{(\ell)})$, $\ell = 1, 2, \ldots, N$, from the draws of β. These values can be placed in a histogram to estimate the p.d.f. of $g(\beta)$. Their sample mean and variance can be used to estimate the corresponding posterior mean and variance. Probabilities can be estimated using the proportion of values in a given region and order statistics can be used to find an interval with a given probability content.

3.5 Gibbs Sampling with β and Σ

We now turn to the question of how to obtain draws β and Σ from their respective marginal posterior p.d.f.'s. One possible way is to use an MCMC algorithm known as Gibbs sampling. In this procedure draws are made iteratively from the conditional posterior p.d.f's. Specifically, given a particular starting value for Σ, say $\Sigma^{(0)}$, the ℓth draw from the Gibbs sampler ($\beta^{(l)}$, $\Sigma^{(l)}$) is obtained using following two steps:

1. Draw $\beta^{(\ell)}$ from $f(\beta|\Sigma^{(\ell-1)}, y)$
2. Draw $\Sigma^{(\ell)}$ from $f(\Sigma|\beta^{(\ell)}, y)$

Making these draws is straightforward, given that the two conditional posterior p.d.f.'s are normal and inverted Wishart, respectively. (See Appendix for details.) MCMC theory suggests that, after a sufficiently large number of draws, the Markov Chain created by the draws will converge. After convergence, the subsequent draws can be viewed as draws from the marginal posterior p.d.f's $f(\beta\,|\,y)$ and $f(\Sigma\,|\,y)$. It is these draws that can be used to present results in the desired fashion. Draws taken prior to the point at which convergence is assumed to have taken place are sometimes called the "burn in"; they are discarded. A large number of diagnostics have been suggested for assessing whether convergence has taken place. See, e.g. Cowles and Carlin (1996). Assessing whether convergence has taken place is similar to assessing whether a time series is stationary. Thus, visual inspection of a graph of the sequence of draws, testing whether the mean and variance are the same at the beginning of the chain as at the end of the chain, and testing whether two or more separately run chains have the same mean and variance, are ways of checking for convergence.

Since we are using a sample of draws of β and Σ to estimate posterior means and standard deviations, and other relevant population quantities, the accuracy of the estimates is a concern. Estimation accuracy is assessed using numerical standard errors. Methods for computing such standard errors are described in Chapter 8 of this volume by Geweke et al. Because the draws produced by MCMC algorithms are correlated, time-series methods are used to compute the standard errors; also larger samples are required to achieve a given level of accuracy relative to a situation involving independent draws.

Although the above remarks on convergence and numerical standard errors were made in the context of the Gibbs sampler for β and Σ, they also apply to other MCMC algorithms including the Gibbs sampler for β and the Metropolis–Hastings algorithm described below.

3.6 Gibbs Sampling with β

If the number of equations is large, making Σ of high dimension, then if may be preferable to use a Gibbs sampler based on the conditional posterior p.d.f.'s for the β_i from each equation. Note, however, that this alternative is not feasible if cross-equation restrictions on the β_i, as discussed in Sections 5 and 7, are present.

To proceed with this Gibbs sampler, we begin with starting values for all coefficients except the first, say $(\beta_2^{(0)}, \beta_3^{(0)}, \ldots, \beta_M^{(0)})$ and then sample iteratively using the following steps for the ℓth draw:

1. Draw $\beta_1^{(\ell)}$ from $f(\beta_1 | \beta_2^{(\ell-1)}, \ldots, \beta_M^{(\ell-1)})$
2. Draw $\beta_2^{(\ell)}$ from $f(\beta_2 | \beta_1^{(\ell)}, \beta_3^{(\ell-1)}, \ldots, \beta_M^{(\ell-1)})$

⋮

i. Draw $\beta_i^{(\ell)}$ from $f(\beta_i | \beta_1^{(\ell)}, \ldots \beta_{i-1}^{(\ell)}, \beta_{i+1}^{(\ell-1)}, \ldots, \beta_M^{(\ell-1)})$

⋮

M. Draw $\beta_M^{(\ell)}$ from $f(\beta_M | \beta_1^{(\ell)}, \ldots, \beta_M^{(\ell-1)})$

The conditional posterior p.d.f.'s are multivariate t-distributions from which we can readily draw values (see Appendix). Ordinary least squares estimates are adequate for starting values.

3.7 Metropolis-Hastings Algorithm

An alternative to Gibbs sampling is a Metropolis–Hastings algorithm that draws observations from the marginal posterior p.d.f. $f(\beta | y)$. As we shall see, this algorithm is particularly useful for an inequality-restricted prior, or if the equations are nonlinear. The algorithm we describe is a random-walk algorithm; it is just one of many possibilities. For others see, e.g., Chen et al. (2000).

The Metropolis–Hastings algorithm generates a candidate value β^* that is accepted or rejected as a draw from the posterior p.d.f. $f(\beta | y)$. When it is rejected, the previously accepted draw is repeated as a draw. Thus, rules are needed for generating the candidate value β^* and for accepting it. Let V be the covariance matrix for the distribution used to generate a candidate value. The maximum likelihood covariance matrix is usually suitable. For the linear SUR model this matrix is $[X'(\hat{\Sigma}^{-1} \otimes I_T)X]^{-1}$. Choose a feasible starting value $\beta^{(0)}$. The following steps can be used to draw the $(\ell + 1)$th observation in a random-walk Metropolis–Hastings chain:

1. Draw a candidate value β^* from an $N(\beta^{(\ell)}, cV)$ distribution where c is a scalar set such that β^* is accepted approximately 40–50% of the time.
2. Compute the ratio of the posterior p.d.f. evaluated at the candidate draw to the posterior p.d.f. evaluated at the previously accepted draw:

   $$r = \frac{f(\beta^* | y)}{f(\beta^{(\ell)} | y)}$$

   Note that this ratio can be computed without knowledge of the normalizing constant for $f(\beta \mid y)$. Also, if any of the elements of β^* fall outside a feasible parameter region defined by an inequality-restricted prior (see Section 5), then $f(\beta^* \mid y) = 0$. When $r > 1$, β^* is a more likely value then $\beta^{(\ell)}$ in the sense that it is closer to the mode of the distribution. When $r < 1$, β^* is further into the tails of the distribution. If $r > 1$, β^* is accepted; if $r < 1$, β^* is accepted with probability r. Thus, more draws occur in regions of high probability and fewer draws occur in regions of low probability. Details of the acceptance–rejection procedure follow in step 3.
3. Draw a value u for a uniform random variable on the interval $(0, 1)$:

   If $u \leq r$, set $\beta^{(\ell+1)} = \beta^*$
   If $u > r$, set $\beta^{(\ell+1)} = \beta^{(\ell)}$
   Return to step 1 with ℓ set to $\ell + 1$.

Let $q(\beta^* | \beta^{(\ell)})$ be the distribution used to generate the candidate value β^* in step 1. In our case it is a normal distribution. In more general Metropolis–Hastings algorithms, where our choice of distribution is not necessarily utilized, r is defined as

$$r = \frac{f(\beta^* \mid y)}{f(\beta^{(\ell)} \mid y)} \times \frac{q(\beta^{(\ell)} \mid \beta^*)}{q(\beta^* \mid \beta^{(\ell)})}$$

In our case $q(\beta^{(\ell)} \mid \beta^*) = q(\beta^* \mid \beta^{(\ell)})$. Various alternatives for $q(\cdot)$ have been suggested in the literature.

4 NONLINEAR SUR

Many economic models are intrinsically nonlinear, or a nonlinear model may result from substituting nonlinear restrictions on β into a linear model. The Gibbs sampling algorithms that we described are no longer applicable for a nonlinear SUR model. However, we can still proceed with the Metropolis–Hastings algorithm.

Suppose that the nonlinear SUR model is given by

$$y_i = h_i(X, \beta) + e_i \qquad i = 1, 2, \ldots, M \tag{31}$$

where the h_i are the nonlinear functions. The dimensions of y_i, h_i, and e_i are $(T \times 1)$. In this context X represents a set of explanatory variables and β is the vector of all unknown coefficients. The omission of an i-subscript on X, β is deliberate; the same coefficients and the same explanatory variables can occur in different equations. The earlier assumptions about the e_i are retained.

With a nonlinear model, $f(\beta) \propto$ *constant* may no longer be suitable as a noninformative prior; consideration needs to be given to the type of nonlinear function and to whether particular values for some parameters need to be excluded. Thus, we give results for a general prior on β, denoted by $f(\beta)$. We retain the noninformative prior $f(\Sigma) \propto |\Sigma|^{-(M+1)/2}$, and assume *a priori* independence of β and Σ. Thus, the prior p.d.f. is given by

$$f(\beta, \Sigma) \propto |\Sigma|^{-(M+1)/2} f(\beta) \tag{32}$$

The likelihood function can be written as

$$\begin{aligned} f(y|\beta, \Sigma) &= (2\pi)^{-MT/2}|\Sigma|^{-T/2} \\ &\quad \times \exp\{-\tfrac{1}{2}(y - h(X, \beta))'(\Sigma^{-1} \otimes I_T)(y - h(X, \beta))\} \\ &= (2\pi)^{-MT/2}|\Sigma|^{-T/2} \exp\{-\tfrac{1}{2}\mathrm{tr}(A\Sigma^{-1})\} \end{aligned} \tag{33}$$

where $y = (y_1', y_2', \ldots, y_M')'$, $h = (h_1', h_2', \ldots, h_M')'$, and now, A is an $(M \times M)$ matrix with the (i,j)th element given by

$$[A]_{ij} = [y_i - h_i(X, \beta)]'[y_j - h_j(X, \beta)] \tag{34}$$

The joint posterior p.d.f. for (β, Σ) is

$$f(\beta, \Sigma \mid y) \propto f(\beta)|\Sigma|^{-(T+M+1)/2} \exp\{-\tfrac{1}{2}\mathrm{tr}(A\Sigma^{-1})\} \tag{35}$$

and, integrating out Σ, the marginal posterior p.d.f. for β is

$$f(\beta \mid y) \propto f(\beta)|A|^{-T/2} \tag{36}$$

Thus, the posterior for β in the nonlinear SUR model involves the same determinant of sums of squares and cross-products of residuals as it does in the linear model. A more general prior has been added. (Of course, it could also have been included in the linear model.)

The Metropolis–Hastings algorithm described in Section 3.7 can be readily applied to the posterior p.d.f. in Eq. (36). Because the earlier results on conditional posterior p.d.f.'s for β and the β_i no longer hold, the draws need to be used directly to estimate posterior p.d.f's and their moments.

5 IMPOSING LINEAR EQUALITY RESTRICTIONS

Economic applications of SUR models frequently involve linear restrictions on the coefficients. For example, the same coefficient may appear in more than one equation, and the Slutsky symmetry conditions in demand models lead to cross-equation restrictions, or one might want to hypothesize that all equations have the same coefficients vector. Under the existence of cross-equation linear restrictions, the Gibbs sampler using β and Σ, and the Metropolis–Hastings algorithm, can still be used. However, the Gibbs sampler involving only β is no longer applicable. If the restrictions are all within equation restrictions, all three algorithms are possible.

Suppose a set of J linear restrictions is written as

$$R\beta = (R_1 \quad R_2)\begin{pmatrix} \eta \\ \gamma \end{pmatrix} = r \tag{37}$$

where R_1 is $(J \times J)$ and nonsingular, R_2 is $[J \times (K-J)]$, and η and γ are J and $(K-J)$ dimensional subvectors of β, respectively. To make this partition, it may be necessary to reorder the elements in β. Correspondingly, we can reorder the columns of X and partition it so that the linear SUR model can be written as

$$y = X\beta + e = (X_1 \quad X_2)\begin{pmatrix} \eta \\ \gamma \end{pmatrix} + e \tag{38}$$

This reordering may destroy the block-diagonal properties of X. From Eq. (37), we can solve for η as

$$\eta = R_1^{-1}(r - R_2\gamma) \tag{39}$$

Substituting Eq. (39) into Eq. (38) and rearranging yields:

$$y - X_1R_1^{-1}r = (X_2 - X_1R_1^{-1}R_2)\gamma + e$$

or

$$z = Z\gamma + e \tag{40}$$

where $z = y - X_1 R_1^{-1} r$ and $Z = X_2 - X_1 R_1^{-1} R_2$ represent new sets of "observations."

In general, Z and γ can no longer be partitioned unambiguously into M separate equations. However, the stochastic properties of e remain the same. Thus, all the results in Sections 2 and 3 that did not rely on partitioning of X and β can still be applied to the model in Eq. (40). In particular, a Gibbs sampler can be used to draw γ and Σ, and the Metropolis–Hastings algorithm can be used to draw γ, from their respective posterior p.d.f.'s.

6 IMPOSING INEQUALITY RESTRICTIONS

Possible inequality restrictions on the coefficients range from simple ones such as a sign restriction on a single coefficient to more complex ones such as enforcing the eigenvalues of a matrix of coefficients to be nonpositive. Letting the feasible region defined by the inequality constraints be denoted by S, and defining the indicator function:

$$I_S(\beta) = \begin{cases} 1 \text{ for } \beta \in S \\ 0 \text{ for } \beta \notin S \end{cases} \tag{41}$$

the inequality restrictions can be accommodated by setting up the otherwise noninformative prior p.d.f.:

$$f(\beta, \Sigma) \propto |\Sigma|^{-(M+1)/2} I_S(\beta) \tag{42}$$

Using Bayes' theorem to combine this prior with the likelihood function in Eq. (5), we obtain the joint posterior p.d.f.:

$$\begin{aligned} f(\beta, \Sigma \mid y) &\propto f(y \mid \beta, \Sigma) f(\beta, \Sigma) \\ &\propto |\Sigma|^{-(T+M+1)/2} \exp\{-\tfrac{1}{2}(y - X\beta)'(\Sigma^{-1} \otimes I_T)(y - X\beta)\} I_S(\beta) \\ &= |\Sigma|^{-(T+M+1)/2} \exp\{-\tfrac{1}{2}\mathrm{tr}(A\Sigma^{-1})\} I_S(\beta) \end{aligned} \tag{43}$$

From this result we can derive the following conditional and marginal posterior p.d.f.'s.

The conditional posterior p.d.f. for $(\beta \mid \Sigma)$ is the truncated multivariate normal distribution:

$$f(\beta \mid \Sigma, y) \propto \exp\{-\tfrac{1}{2}(\beta - \hat{\beta})' X' (\Sigma^{-1} \otimes I_T) X (\beta - \hat{\beta})\} I_S(\beta) \tag{44}$$

The conditional posterior p.d.f. for $(\Sigma \mid y)$ is the same inverted Wishart distribution as was given in Eq. (20). The marginal posterior p.d.f. for β is

$$f(\beta \mid y) \propto |A|^{-T/2} I_S(\beta) \tag{45}$$

The posterior p.d.f. for β_1 conditional on the remaining β_i is the truncated multivariate t-distribution:

$$f(\beta_1 \mid y, \beta_2, \beta_3, \ldots, \beta_M) \propto \left[\nu_1 + \frac{(\beta_1 - \tilde{\beta}_1)' X_1' Q_{(1)} X_1 (\beta_1 - \tilde{\beta}_1)}{\tilde{s}_1^2} \right]^{-(K_1+\nu_1)/2} I_S(\beta) \tag{46}$$

Of interest is how best to use these p.d.f's to draw observations on β, and possibly Σ, from their respective posterior p.d.f.'s. The conditional posterior p.d.f.'s for $(\beta \mid \Sigma)$ and $(\Sigma \mid \beta)$ can be used within a Gibbs sampler providing that the inequality restrictions are sufficiently mild for a simple acceptance–rejection algorithm to be practical when sampling from the truncated multivariate normal distribution. By a "simple acceptance–rejection algorithm," we mean that a draw is made from a nontruncated multivariate normal distribution and, if it lies outside the feasible region, it is discarded and replaced by another draw. This procedure will not be practical if the probability of obtaining a draw within the feasible region is small, which will almost always be the case if the number of inequality restrictions is moderate to large. Thus, we are using the term "mild" inequality restrictions to describe a situation where the maximum number of draws necessary before a feasible draw is obtained is not excessive.

If the inequality restrictions are not mild, then a Metropolis–Hastings algorithm can be employed. In the steps we described in Section 3, if a candidate value β^* is infeasible, then $r = 0$, and the retained draw is automatically the last accepted feasible draw. That is, $\beta^{(\ell+1)} = \beta^{(\ell)}$.

If the inequality restrictions are not mild, but are linear, then using a Gibbs sampler on subcomponents of β might prove successful. For example, using the truncated multivariate t-distributions for each of the β_i, as specified in Eq. (46), could be useful. Also, within different contexts, sampling from truncated multivariate t and multivariate normal distributions has been broken down into sampling from univariate conditional distributions by Geweke (1991) and Hajivassiliou and McFadden (1990). See also the Appendix.

7 THREE APPLICATIONS

7.1 Wheat Yield

In Griffiths et al. (2001) the following model was used for predicting wheat yield in five Western Australian shires:

$$\begin{aligned} Y_t = \beta_1 + \beta_2 t + \beta_3 t^2 + \beta_4 t^3 + \beta_5 G_t + \beta_6 G_t^2 \\ + \beta_7 D_t + \beta_8 D_t^2 + \beta_9 F_t + \beta_{10} F_t^2 + e_t \end{aligned} \tag{47}$$

Yield (Y_t) depends on a cubic time trend to capture technological change and on quadratic functions of rainfall during the germination period (G_t), the development period (D_t), and the flowering period (F_t). The rainfalls are measured relative to their sample means. Inequality restrictions are imposed to ensure that the response of yield to rainfall, at average rainfall, is positive. That is, for germination rainfall, for example, $\partial Y/\partial G = \beta_5 + 2\beta_6 > 0$. Thus, the feasible region for this example is

$$S(\beta) = \{\beta \mid \beta_5 + 2\beta_6 > 0, \beta_7 + 2\beta_8 > 0, \beta_9 + 2\beta_{10} > 0\} \tag{48}$$

Although Griffiths et al. (2001) used separate single equation estimation for the five shires and focused on several forecasting issues, investigation within a five-equation SUR model has started. Given that the inequality restrictions within each equation are relatively mild, but in total they are not, a Gibbs sampler using the truncated t densities in Eq. (46) seems a profitable direction to follow. Also, some preliminary work involving the Metropolis–Hastings algorithm on the complete β vector has proved effective.

7.2 Cost and Share Equations

In a second application, a translog cost function (constant returns to scale) and cost-share equations for merino woolgrowers (310 observations over 23 years) was estimated by Griffiths et al. (2000) using, as inputs, land, capital, livestock, and other factors. In the equations that follow c is cost, q is output, the w_i are input prices, and the S_i are input shares:

$$\log\left(\frac{c}{q}\right) = \beta_0 + \beta_T T + \sum_{i=1}^{4} \beta_i \log(w_i) + 0.5 \sum_{i=1}^{4}\sum_{j=1}^{4} \beta_{ij} \log(w_i)\log(w_j) + e_1$$

$$S_i = \beta_i + \sum_{j=1}^{4} \beta_{ij} \log(w_j) + e_i \quad i = 2, 3, 4$$

This SUR model has the following characteristics:

1. The equations are linear.
2. There are a number of linear equality restrictions that need to be imposed. Specifically, the β_{ij}s in the cost function are equal to the β_{ij}s in the share equations, and, furthermore, to satisfy homogeneity and symmetry, we require that

$$\sum_{i=1}^{4} \beta_i = 1 \qquad \sum_{j=1}^{4} \beta_{ij} = 0 \qquad \beta_{ij} = \beta_{ji}$$

3. Inequality restrictions are required for the functions to satisfy concavity and monotonicity. These restrictions are:
   - Monotonicity $0 < S_i < 1$
   - Concavity $B - S + ss'$ is negative semidefinite where

$$B = \begin{bmatrix} \beta_{11} & \beta_{12} & \beta_{13} & \beta_{14} \\ \beta_{21} & \beta_{22} & \beta_{23} & \beta_{24} \\ \beta_{31} & \beta_{32} & \beta_{33} & \beta_{34} \\ \beta_{41} & \beta_{42} & \beta_{43} & \beta_{44} \end{bmatrix}$$

$$S = \begin{bmatrix} S_1 & & & \\ & S_2 & & \\ & & S_3 & \\ & & & S_4 \end{bmatrix} \qquad s = \begin{bmatrix} S_1 \\ S_2 \\ S_3 \\ S_4 \end{bmatrix}$$

   Note that $B - S + ss'$ is negative semidefinite if and only if its largest eigenvalue is nonpositive.

   Since S_i depend on the input prices, a decision concerning the input prices at which S_i are evaluated, and the inequality restrictions imposed, needs to be made. The inequality restrictions were imposed at average input prices for each of the 23 years.
4. Given the severe inequality restrictions that were imposed, the Metropolis–Hastings algorithm was used.
5. The quantities of interest are nonlinear functions of the parameters. They are the elasticities of substitution:

$$\sigma_{ij} = \frac{\beta_{ij}}{S_i S_j} + 1 \quad i \neq j$$

and input demand elasticities:

$$\eta_{ii} = \frac{\beta_{ii}}{S_i} + S_i - 1$$

$$\eta_{ij} = \frac{\beta_{ij}}{S_i} + S_j$$

7.3 Expenditure Functions

Our third example involves two expenditure functions estimated from a sample of 1834 Bangkok households, and deflated by an "equivalence scale" measure of household size (Griffiths and Chotikapanich, 1997). For the tth observation, the functions are

$$w_{1t} = \frac{\alpha_1 m_{1t}}{x_t} + \frac{\beta_1 m_{1t}(x_t - \alpha_1 m_{1t} - \alpha_2 m_{2t} - \alpha_3 m_{3t})}{x_t(\beta_1 m_{1t} + \beta_2 m_{2t} + \beta_3 m_{3t})} + e_{1t}$$

$$w_{2t} = \frac{\alpha_2 m_{2t}}{x_t} + \frac{\beta_2 m_{2t}(x_t - \alpha_1 m_{1t} - \alpha_2 m_{2t} - \alpha_3 m_{3t})}{x_t(\beta_1 m_{1t} + \beta_2 m_{2t} + \beta_3 m_{3t})} + e_{2t}$$

$$m_{1t} = 1 + \delta_{21} n_{2t} + \delta_{31} n_{3t}$$

$$m_{2t} = 1 + \delta_{22} n_{2t} + \delta_{32} n_{3t}$$

$$m_{3t} = 1 + \delta_{23} n_{2t} + \delta_{33} n_{3t}$$

where

w_{jt} = expenditure proportion for commodity j,
x_t = total expenditure,
m_{jt} = equivalence scale for commodity j,
n_{2t} = number of extra adults (each household has at least one adult),
n_{3t} = number of children.

The unknown parameters are

$$(\alpha_1, \alpha_2, \alpha_3, \beta_1, \beta_2, \beta_3, \delta_{31}, \delta_{22}, \delta_{32}, \delta_{23}, \delta_{33})$$

This SUR model has the following characteristics:

1. The equations are nonlinear in the parameters.
2. A number of inequality restrictions were imposed, namely,
   $0 < \beta_1,\ \beta_2 < 1$ additional expenditure from a one-unit increase in supernumerary income must lie between zero and one.
   $0 \le \delta_{3j} \le \delta_{2j} \le 1$ expenditure requirements for extra adults are less than those for the first adult but greater than those for children.

$$\alpha_i < \min_t \left\{ \frac{e_{it}}{1 + \delta_{2i} n_{2t} + \delta_{3i} n_{3t}} \right\} \qquad i = 1, 2$$

The smallest level of consumption in the sample must be greater than subsistence expenditure, a constraint from the utility function.

3. Given the nonlinear equations and the inequality constraints, the Metropolis–Hastings algorithm was used.
4. Two nonlinear functions of the parameters are of interest. They are the general scale or "household size":

$$m_0 = \frac{x \sum_k \beta_k m_k}{x - \sum_k \alpha_k m_k + \sum_k \alpha_k \sum_k \beta_k m_k}$$

and the elasticities. Expressions for the latter can be found in Griffiths and Chotikapanich (1997).

8 FORECASTING

Suppose that we are interested in forecasting dependent variable values in the next period. The shire-level wheat yield application in the Section 7.1 is an example of where such a forecast would be of interest. In that case the objective is to forecast yield for each of the five shires. Since the yields are correlated via the stochastic assumptions of the SUR model, a joint forecast is appropriate. We can write the next period's observation as

$$y_* = X_* \beta + e_* \tag{49}$$

where y_* is an M-dimensional vector, X_* is an $(M \times K)$ block diagonal matrix with the ith block being a $(1 \times K_i)$ row vector containing next period's explanatory variables for the ith equation:

$$X_* = \begin{bmatrix} x_{1*} & & & \\ & x_{2*} & & \\ & & \ddots & \\ & & & x_{M*} \end{bmatrix} \tag{50}$$

and $e_* \sim N(0,\Sigma)$ is next period's $(M \times 1)$ random error vector. The conventional Bayesian forecasting tool is the predictive p.d.f. $f(y_* \mid y)$. Graphing marginal predictive p.d.f.'s from this density function, and computing its means, standard deviations, and probabilities of interest are the standard ways of reporting results.

The procedure for deriving the predictive p.d.f. is to begin with the joint p.d.f. $f(y_*,\beta,\Sigma \mid y)$ and then to integrate out Σ and β, either analytically or via a numerical sampling algorithm. Now,

$$\begin{aligned} f(y_* \mid \beta, \Sigma) &= (2\pi)^{-M/2}|\Sigma|^{-1/2}\exp\{-\tfrac{1}{2}(y_* - X_*\beta)'\Sigma^{-1}(y_* - X_*\beta)\} \\ &\propto |\Sigma|^{-1/2}\exp\{-\tfrac{1}{2}\mathrm{tr}(A_*\Sigma^{-1})\} \end{aligned} \tag{51}$$

where $A_* = [y - X_*\beta][y - X_*\beta]'$. Thus, using the posterior p.d.f. in Eq. (12) (no inequality restrictions), we have

$$\begin{aligned} f(y_*\beta, \Sigma \mid y) &= f(y_* \mid \beta, \Sigma)f(\beta, \Sigma \mid y) \\ &\propto |\Sigma|^{-(T+M+2)/2}\exp\{-\tfrac{1}{2}\mathrm{tr}[(A + A_*)\Sigma^{-1}]\} \end{aligned}$$

Using properties of the inverted Wishart distribution to integrate out Σ yields:

$$f(y_*, \beta \mid y) = \int f(y_*, \beta, \Sigma \mid y)d\Sigma \propto |A + A_*|^{-(T+1)/2} \tag{52}$$

Because analytical integration of β out of Eq. (52) is not possible, we consider the conditional predictive p.d.f. $f(y_* \mid \beta, y)$. It turns out that this p.d.f. is a multivariate student t. Thus, $f(y_* \mid y)$ and its moments can be estimated by averaging quantities from $f(y_* \mid \beta, y)$ over draws of β obtained using one of the MCMC algorithms described earlier.

To establish that $f(y_* \mid \beta, y)$ is a multivariate t-distribution, we first note that (see, e.g., Dhrymes 1978, p. 458)

$$|A + A_*| = |A|(1 + (y_* - X_*\beta)'A^{-1}(y_* - X_*\beta)) \tag{53}$$

Thus,

$$\begin{aligned} f(y_*|\beta, y) &\propto [1 + (y_* - X_*\beta)'A^{-1}(y_* - X_*\beta]^{-(T+1)/2} \\ &\propto \left[\nu_* + (y_* - X_*\beta)'\left(\frac{A}{\nu_*}\right)^{-1}(y_* - X_*\beta)\right]^{-(M+\nu_*)/2} \end{aligned} \tag{54}$$

where $\nu_* = T - M + 1$. Equation (54) is a multivariate t-distribution with mean:

$$E(y_*|\beta, y) = X_*\beta \tag{55}$$

covariance matrix:

$$V(y_*|\beta, y) = \frac{A}{\nu_* - 2} \tag{56}$$

and degrees of freedom ν_*. Given draws $\beta^{(\ell)}, \ell = 1, 2, \ldots, N$ from an MCMC algorithm, one can average the quantities in Eqs. (54)–(56) over these draws to estimate the required marginal predictive p.d.f.'s and their moments. Marginal univariate t distributions from Eq. (54) are averaged and the formulas are analogous to those in Eqs. (22), (26), and (27) except, of course, that our random variable of interest is now an element of y_*, say y_{*i}, not β_{ik}.

Percy (1992) describes an alternative Gibbs sampling approach where y_*, β, and Σ are recursively generated from their respective conditional p.d.f.'s. With our approach, it is not necessary to generate draws on y_*. Also, because we have derived the predictive p.d.f. conditional on β, the introduction of inequality restrictions on β does not change the analysis. The range of values of β over which averaging takes place is restricted, but that is accommodated by the way in which β is drawn, and the result in Eq. (54) still holds.

An interesting extension, and one that is of concern to Griffiths et al. (2001), is capturing the extra uncertainty created by not knowing the value of one or more regressors in X_*. We have this problem if a wheat-yield forecast is made prior to all rainfalls having been observed. The effect can be captured by modeling rainfall and averaging the predictive p.d.f. for yield conditional on rainfall over rainfalls draws made from its predictive p.d.f.

9 SOME EXTENSIONS

Consider estimating β in the SUR model when there are missing observations on one or more of the dependent variables. This problem was considered in the context of expenditure functions by Supat (1996). For the moment, assume that the observations are truly missing and that they are missing at random; they are not zeros created by negative values of an unobserved latent variable, as in the case with the Tobit model. Writing y^O to denote observed components and y^U to denote unobserved components, estimation can proceed within a Gibbs sampling framework using the conditional posterior p.d.f.'s $f(\beta|\Sigma, y^O, y^U)$, $f(\Sigma|\beta, y^O, y^U)$, and $f(y^U|\beta, \Sigma, y^O)$. The conditional posterior p.d.f.'s for β and Σ are the normal and inverted Wishart p.d.f.'s given in Eqs. (14) and (20). To investigate how to draw observations from $f(y^U|\beta, \Sigma, y^O)$, we write the $(M \times 1)$ tth observation $y_{(t)}$ as

$$y_{(t)} = X_{(t)}\beta + e_{(t)} \tag{57}$$

The subscript t has been placed in parentheses to distinguish the $(M \times 1)$ tth observation on all equations $y_{(t)}$ from the $(T \times 1)$ observations on the ith equation y_i. The structure of $X_{(t)}$ is similar to that of X_* defined in Eq. (50). We wish to consider Eq. (57) for all values of t where $y_{(t)}$ has one or more unobserved components. Reordering the elements if necessary, we can partition Eq. (57) as

$$\begin{pmatrix} y_{(t)}^{U} \\ y_{(t)}^{O} \end{pmatrix} = \begin{pmatrix} X_{(t)}^{U} \\ X_{(t)}^{O} \end{pmatrix} \beta + \begin{pmatrix} e_{(t)}^{U} \\ e_{(t)}^{O} \end{pmatrix} \tag{58}$$

where we write

$$E[e_{(t)} e'_{(t)}] = \begin{bmatrix} \Sigma^{UU} & \Sigma^{UO} \\ \Sigma^{OU} & \Sigma^{OO} \end{bmatrix} \tag{59}$$

The conditional posterior p.d.f. $f(y_{(t)}^{U} | \beta, \Sigma, y_{(t)}^{O})$ is a mulativariate normal distribution with mean:

$$E(y_{(t)}^{U} \mid \beta, \Sigma, y_{(t)}^{O}) = X_{(t)}^{O}\beta + \Sigma^{UO} \Sigma^{OO^{-1}} (y_{(t)}^{O} - X_{(t)}^{O}\beta) \tag{60}$$

and covariance matrix:

$$V(y_{(t)}^{U} | \beta, \Sigma, y_{(t)}^{O}) = \Sigma^{UU} - \Sigma^{UO} \Sigma^{OO^{-1}} \Sigma^{OU} \tag{61}$$

Furthermore, $(y_{(t)}^{U} | \beta, \Sigma, y_{(t)}^{O})$, $t = 1, 2, \ldots, T$ are independent. Thus, for generating $y_{(t)}^{U}$ within the Gibbs sampler, we use the conditional normal distributions defined by Eqs. (60) and (61) for all observations where an unobserved component is present.

Suppose, now, that the unobserved components represent negative values of a Tobit-type latent variable. In this case we have the additional posterior information that the elements of $y_{(t)}^{U}$ are negative. The conditional posterior p.d.f. for $(y_{(t)}^{U} | \beta, \Sigma, y_{(t)}^{O})$ becomes a truncated (multivariate) normal distribution with a truncation that forces $y_{(t)}^{U}$ to be negative. Its location vector and scale matrix (no longer the mean and covariance matrix) are given in Eqs. (60) and (61). A convenient algorithm for drawing from this truncated normal distribution is described in the Appendix.

For extensions into Probit models, see Geweke et al. (1997) and references cited therein. The literature on simultaneous equation models with Tobit and Probit variables can be accessed through Li (1998). Sets of SUR expenditure functions with a common parameter and with unobserved expenditures that result from infrequency of purchase are considered by Griffiths and Valenzuela (1998). Smith and Kohn (2000) study Bayesian estimation of nonparametric SURs.

10 CONCLUDING REMARKS

With the recent explosion of literature on MCMC techniques, Bayesian inference in the SUR model has become a practical reality. However, it is the author's view that, prior to the writing of this chapter, the relevant results have not been collected and summarized in a form convenient for applied researchers to implement. It is my hope that this chapter will facilitate and motivate many more applications of Bayesian inference in the SUR model.

ACKNOWLEDGMENTS

I am grateful to Michael Chua for research assistance, and to Denzil Fiebig and Chris O'Donnell for comments on an earlier draft.

REFERENCES

1. Albert, J.H. and Chib, S., 1996. Computation in Bayesian econometrics: An introduction to Markov chain Monte Carlo, In: Hill, R.C. ed. *Advances in Econometrics*. vol. 11A: *Computational Methods and Applications*. JAI Press, Greenwich, Connecticut, 3–24.
2. Chen, M.-H., Shao, Q.-M. and Ibrahim, J. G., 2000. *Monte Carlo Methods in Bayesian Computation*. Springer, New York.
3. Chib, S. and Greenberg, E., 1995. Hierarchical analysis of SUR models with extensions to correlated serial errors and time-varying parameter models. *Journal of Econometrics*, **68**, 339–360.
4. Chib, S. and Greenberg, E., 1996. Markov chain Monte Carlo simulation methods in econometrics. *Econometric Theory*, **12**, 409–431.
5. Cowles, M.K. and Carlin, B.P., 1996. Markov chain Monte Carlo convergence diagnostics: A comparative review. *Journal of the American Statistical Association*, **91**, 883–904.
6. Dhrymes, P.J., 1978. *Introductory Econometrics*, Springer, New York.
7. Drèze, J.H. and Morales, J.A., 1976. Bayesian full information analysis of simultaneous equations. *Journal of the American Statistical Association*, **71**, 919–923.
8. Fiebig, D., 2001. Seemingly unrelated regression. In: Baltage, B. ed. *Companion in Econometrics*. Blackwell, London, Ch. 5.
9. Fiebig, D.G. and Kim, J.H., 2000. Estimation and inference in SUR models when the number of equations is large. *Econometric Reviews*, **19**, 105–130.
10. Geweke, J., 1988. The secular and cyclical behavior of real GDP in nineteen OECD countries. *Journal of Business and Economic Statistics*, **6**, 479–486.
11. Geweke, J., 1991. Efficient simulation from the multivariate normal and student t-distributions subject to linear constraints. In: Keramidas E.M., ed. *Computing Science and Statistics: Proceedings of the Twenty-third Symposium on the Interface*. Interface Foundation of North America, Fairfax, pp. 571–578.
12. Geweke, J., Keane, M.P., and Runkle, D.E., 1997. Statistical inference in the multinomial multiperiod probit model. *Journal of Econometrics*, **80**, 127–166.
13. Gilks, W.R., Richardson, S. and Spielgelhalter, D.J., eds., 1996. *Markov Chain Monte Carlo in Practice*. Chapman and Hall, London.

14. Griffiths, W.E., 1999. Estimating consumer surplus: comment on using simulation methods for Bayesian econometric models: inference development and communication. *Econometric Reviews*, **18**, 75–88.
15. Griffiths, W.E. and Chotikapanich, D., 1997. Bayesian methodology for imposing inequality constraints on a linear expenditure function with demographic factors. *Australian Economic Papers*, **36**, 321–341.
16. Griffiths, W.E. and Valenzuela, M.R., 1998. Missing data from infrequency of purchase: Bayesian estimation of linear expenditure system. In: T.B. Fomby, and R.C. Hill, eds., *Advances in Econometrics*. vol. 13: *Messy Data, Missing Observations, Outliers and Mixed-Frequency Data*. JAI Press, Greenwich, Connecticut, pp. 75–102.
17. Griffiths, W.E., O'Donnell, C.J. and Tan-Cruz, A., 2000. Imposing regularity conditions on a system of cost and factor share equations. *Australian Journal of Agricultural and Resource Economics*, **44**, 107–127.
18. Griffiths, W.E., Newton, L.S. and O'Donnell, C.J., 2001. Predictive densities for shire level wheat yield in Western Australia. Paper contributed to the *Australian Agricultural and Resource Economics Society Conference*, Adelaide.
19. Griffiths, W.E., Skeels, C.J. and Chotikapanich, D., 2002. Sample size requirements for estimation in SUR models. In A. Ullah, A. Wan, and A. Chaturvedi, eds., *Handbook of Applied Econometrics and Statistical Inference*. Marcel Dekker, New York, 575–590.
20. Hajivassiliou, V.A. and McFadden, D.L., 1990. The method of simulated scores for the estimation of LDV models with an application to external debt crises. Yale Cowles Foundation Discussion Paper No. 967.
21. Judge, G.G., Griffiths, W.E., Hill, R.C., Lütkepohl, H. and Lee, T.-C., 1985. *The Theory and Practice of Econometrics*, 2nd ed. Wiley, New York.
22. Kloek, T. and van Dijk, H.K., 1978. Bayesian estimates of equation system parameters: An application of integration by Monte Carlo. *Econometrica*, **46**, 1–19.
23. Li, K., 1998. Bayesian inference in a simultaneous equation model with limited dependent variables. *Journal of Econometrics*, **85**, 387–400.
24. Lütkepohl, H., 1996. *Handbook of Matrices*. Wiley, Chichester, U.K.
25. Percy, D.F., 1992. Prediction for seemingly unrelated regressions. *Journal of the Royal Statistical Society B*, **54**, 243–252.
26. Percy, D.F., 1996. Zellner's influence on multivariate linear models. In: D.A. Berry, K.M. Chaloner, and J.K. Geweke, eds., *Bayesian Analysis in Statistics and Econometrics: Essays in Honor of Arnold Zellner*. Wiley, New York, Ch. 17.
27. Richard, J.-F. and Steel, M.F.J., 1988. Bayesian analysis of systems of seemingly unrelated regression equations under a recursive extended natural conjugate prior density. *Journal of Econometrics*, **38**, 7–37.
28. Richard, J.-F. and Tompa, H., 1980. On the evaluation of poly-*t* density functions. *Journal of Econometrics*, **12**, 335–351.
29. Smith, M. and Kohn, R., 2000. Nonparametric seemingly unrelated regression. *Journal of Econometrics*, **98**, 257–282.

30. Srivastava, V.K. and Dwivedi, T.D., 1979. Estimation of seemingly unrelated regression equations: A brief survey. *Journal of Econometrics*, **10**, 15–32.
31. Srivastava, V.K. and Giles, D.E.A., 1987. *Seemingly Unrelated Regression Equations Models: Estimation and Inference*. Marcel Dekker, New York.
32. Steel, M.F.J., 1992. Posterior analysis of restricted seemingly unrelated reression equation models: A recursive analytical approach. *Econometric Reviews*, **11**, 129–142.
33. Supat, K., 1996. Seemingly unrelated regression with missing observations on dependent variables. B.Ec. Honours dissertation, University of New England, Armidale, NSW, Australia.
34. Tanner, M.A., 1996. *Tools for Statistical Inference: Methods for the Exploration of Posterior Distributions and Likelihood Functions*, 3rd ed. Springer, New York.
35. Tierney, L., 1994. Markov chains for exploring posterior distributions. (With discussion and rejoinder.) *Annals of Statistics*, **22**, 1701–1762.
36. Zellner, A., 1962. An efficient method of estimating seemingly unrelated regressions and tests of aggregation bias. *Journal of American Statistical Association*, **57**, 500–509.
37. Zellner, A., 1971. *An Introduction to Bayesian Inference in Econometrics*. Wiley, New York.

APPENDIX: DRAWING RANDOM VARIABLES AND VECTORS

A.1 Multivariate Normal Distribution

To draw a vector y from a $N(\mu, \Sigma)$ distribution:

1. Compute the Cholesky decomposition H such that $HH' = \Sigma$
2. Generate z from $N(0, I)$
3. Calculate $y = \mu + Hz$

A.2 Multivariate t-Distribution

Consider the multivariate k-dimensional t-distribution with p.d.f.:

$$f(x \mid \mu, V) \propto [\nu + (x - \mu)' V^{-1} (x - \mu)]^{-(k-\nu)/2}$$

It has ν degrees of freedom, mean μ, and covariance matrix $(\nu/(\nu - 2))V$. (Assume $\nu > 2$.) To draw a vector x from this p.d.f.:

1. Compute the Cholesky decomposition H such that $HH' = V$
2. Generate the $(k \times 1)$ vector z_1 from $N(0, I_k)$
3. Generate the $(\nu \times 1)$ vector z_2 from $N(0, I_\nu)$
4. Calculate $x = \mu + Hz_1/\sqrt{z_2' z_2/\nu}$

A.3 Inverted Wishart Distribution

Let Σ have an m-dimensional inverted Wishart distribution with parameter matrix S and degrees of freedom ν. It has p.d.f.:

$$f(\Sigma|S) \propto |\Sigma|^{-(\nu+m+1)/2} \exp\{-\tfrac{1}{2}\operatorname{tr}(S\Sigma^{-1})\}$$

To draw observations on Σ:

1. Compute the Cholesky decomposition H such that $HH' = S^{-1}$.
2. Draw independent $(m \times 1)$ normal random vectors $z_1, z_2, \ldots, z_\nu$ from $N(0, I_m)$.
3. Calculate $\Sigma = \left(H\sum_{i=1}^{\nu} z_i z_i' H'\right)^{-1}$.

A.4 Univariate Truncated Normal Distribution

Suppose that x is a truncated normal random variable with location μ, scale σ, and truncation $a < x < b$. To draw x:

1. Draw a uniform (0, 1) random variable U.
2. Calculate:

$$x = \mu + \sigma\Phi^{-1}\left[\Phi\left(\frac{a-\mu}{\sigma}\right) + U\left(\Phi\left(\frac{b-u}{\sigma}\right) - \Phi\left(\frac{a-\mu}{\sigma}\right)\right)\right] \quad \text{(A.1)}$$

where Φ is the standard normal cumulative distribution function.

A.5 Multivariate Truncated Normal Distribution

Suppose that x is an m-dimensional multivariate truncated normal distribution such that $a_1 < x_1 < b_1$, $a_2 < x_2 < b_2, \ldots, a_m < x_m < b_m$.

1. Use Eq (A.1) to draw x_1.
2. Find the location and scale parameters for the truncated conditional normal distribution $(x_2 \mid x_1)$ conditional on x_1 drawn in step 1.
3. Apply Eq. (A.1) to the distribution $(x_2|x_1)$.
4. Find location and scale parameters for the distribution of $(x_3 \mid x_2, x_1)$ conditional on the draws made in steps 1 and 3.
5. Apply Eq. (A.1) to the distribution $(x_3 \mid x_2, x_1)$.
6. And so on.

10

Computationally Intensive Methods for Deriving Optimal Trimming Parameters

Marco van Akkeren
University of California at Berkeley, Berkeley, California, U.S.A.

1 INTRODUCTION

The information recovery process envisions sample data that are often limited, partial, or ill-conditioned, and the corresponding statistical models that form the basis for estimation and inference may in some cases be underdetermined or ill-posed. The ill-posed aspect may arise because nonstationary or other model specification reasons may cause the number of parameters to exceed the number of data points or the moment conditions to be less than or greater than the number of unknown parameters. In addition, the design matrix implied by nature or society may be ill-conditioned. Consequently, if traditional estimation procedures are used, the solution may be undefined or the estimates that are obtained may be highly unstable giving rise to high variance and low precision of the recovered parameters.

To have a formal basis for the analysis of these problems, consider the following probability model: suppose we make n observations on k random variables $y_i, x_{2i}, x_{3i}, \ldots, x_{ki}$. We can specify a relationship between the response variable y_i and the conditioning variables $x_{2i}, x_{3i}, \ldots, x_{ki}$

using the statistical model:

$$\mathbf{y} = \mathbf{X}\boldsymbol{\beta} + \mathrm{e} \tag{1}$$

where $x_{1i} = \mathbf{i}_n$ to indicate inclusion of the intercept term, $\mathbf{y} = (y_1, \ldots, y_n)'$ is the vector of observed sample values, $\mathbf{X}$ is a known $(n \times k)$ nonstochastic design matrix, and $\beta \in \mathcal{B} \subset \mathbb{R}^k$ represents the vector of unknown parameters. The n-dimensional noise vector $\mathbf{e}$ is assumed to represent a drawing from an unknown multivariate distribution with zero mean and covariance matrix $\Sigma_{\mathbf{e}} = \sigma^2\mathbf{I}_n$. This is the usual linear statistical model. For notation purposes, we adhere to the convention where bold capital letters refer to matrices and bold lower case letters refer to column vectors.

Given the linear statistical model, one plausible criterion is to find an estimator $\boldsymbol{\delta}(y) \in \mathcal{D}$ of the unknown parameter vector $\boldsymbol{\beta} \in \mathcal{B}$ that yields small expected squared error loss (SEL) relative to conventional estimators:

$$R(\boldsymbol{\delta}(\mathbf{y})|\boldsymbol{\beta}) = \mathbb{E}_y\|\boldsymbol{\delta}(\mathbf{y}) - \boldsymbol{\beta}\|^2 \tag{2}$$

If $\boldsymbol{\beta} \in \mathcal{B} \subset \mathbb{R}^k$ and under the SEL measure, the least squares (LS) estimator $\mathbf{b}(\mathbf{y})$ is best linear unbiased with constant risk equal to $\sigma^2\mathrm{tr}(\mathbf{X}'\mathbf{X})^{-1}$. If $\mathbf{e}$ is a multivariate normal distributed random variable then the maximum likelihood estimator $\boldsymbol{\delta}^0(\mathbf{y})$ of $\boldsymbol{\beta}$ equals the LS estimator $\mathbf{b}(\mathbf{y})$:

$$\boldsymbol{\delta}^0(\mathbf{y}) = \mathbf{b}(\mathbf{y}) = (\mathbf{X}'\mathbf{X})^{-1}\mathbf{X}\mathbf{y} \sim N_k(\boldsymbol{\beta}, \sigma^2(\mathbf{X}'\mathbf{X})^{-1}) \tag{3}$$

where $\boldsymbol{\delta}^0(\mathbf{y})$ is best unbiased and minimax under SEL and which is denoted by the notation $\boldsymbol{\delta}^*(\mathbf{y})$:

$$\sup_{\beta \in \mathcal{B}} R(\boldsymbol{\delta}^*(\mathbf{y})|\beta) = \min_{\delta(\mathbf{y}) \in \mathcal{D}} \sup_{\beta \in \mathcal{B}} R(\boldsymbol{\delta}(\mathbf{y})|\boldsymbol{\beta}) \tag{4}$$

The linear statistical model formulation Eq. (1), and the corresponding estimation rules, $\mathbf{b}(\mathbf{y})$ and $\boldsymbol{\delta}^0(\mathbf{y})$, form the basis for a range of statistical models involving sample observations from both discrete and continuous distributions.

In order to evaluate the performance of Eq. (3) more effectively, we should note that Stein (1955) and James and Stein (1961) showed the inadmissibility of $\boldsymbol{\delta}^0(\mathbf{b})$ for $k > 2$ using the traditional Stein estimator:

$$\boldsymbol{\delta}^S(\mathbf{b}) = \left[1 - \frac{(k-2)\sigma^2}{\|\mathbf{b}\|^2}\right]\mathbf{b} \tag{5}$$

which implies that $R(\boldsymbol{\delta}^S(\mathbf{b})|\boldsymbol{\beta}) \leq R(\boldsymbol{\delta}^0(\mathbf{b})|\boldsymbol{\beta})$ for all $\boldsymbol{\beta} \in \mathcal{B}$ and with strict inequality for at least one $\beta \in \mathcal{B}$. Baranchik (1970) proved that the traditional Stein estimator itself is dominated by Stein's positive-rule estimator:

$$\delta^{+}(\mathbf{b}) = \left[1 - \frac{(k-2)\sigma^2}{\|\mathbf{b}\|^2}\right]_{+} \mathbf{b} \tag{6}$$

where $[u]_+$ denotes max[0, u], (Judge and Bock, 1978), which has been shown to be inadmissible by Efron and Morris (1973). Investigation of the above formulations shows that the Stein estimator shrinks each element of **b** by the same amount towards zero. This does not need to be the case. Suppose a priori information is available that reduces the parameter space and allows for specification of a constant target vector $\mathbf{c} \in \mathbb{R}^k$. The Stein's modified estimator results:

$$\delta^{m}(\mathbf{b},\mathbf{c}) = \left[1 - \frac{(k-2)\sigma^2}{\|\mathbf{b}-\mathbf{c}\|^2}\right](\mathbf{b}-\mathbf{c}) + \mathbf{c} \tag{7}$$

which shrinks **b** towards the target **c** and is itself also dominated by a positive-rule modified Stein estimator. It can be shown that if **c** is appropriately chosen, i.e., $k^{-1}\mathbf{i}\mathbf{i}'\mathbf{b}$ where **i** is a vector of ones, then $\delta^{m}(\mathbf{b},\mathbf{c})$ is minimax for $k > 3$. For the modified Stein estimator to be effective, sufficient a priori information must be available to reduce the parameter space and accurately conjecture the target vector. If $\boldsymbol{\beta}$ is close to **c** then significant improvement is made in $R(\delta^{m}(\mathbf{b},\mathbf{c})|\boldsymbol{\beta})$, which dissipates quickly the larger is $\|\mathbf{c}-\boldsymbol{\beta}^2\|$ or the farther is **c** from a neighborhood around $\boldsymbol{\beta}$. Consequently, the quality of nonsample information is of importance to the sampling performance of Stein-like estimators.

More recently, Zellner (1994) introduced the balanced loss function as a means to incorporate both goodness of fit and precision of estimation in the estimator performance criterion. The general form of this loss function is

$$L_w(\boldsymbol{\delta}(\mathbf{y})|\boldsymbol{\beta}) = w\|\mathbf{y} - \mathbf{X}\boldsymbol{\delta}(\mathbf{y})\|^2 + (1-w)\|\boldsymbol{\delta}(\mathbf{y}) - \boldsymbol{\beta}\|^2 \tag{8}$$

where $0 \leq w \leq 1$ is the relative weight associated with prediction. Dey et al. (1999) have shown that if the estimator of the location parameters, expressed as $\boldsymbol{\delta}(\mathbf{y}) = \mathbf{b}(\mathbf{y}) + \mathbf{g}(\mathbf{y})$ dominates the LS estimator **b** where $\mathbf{b}(\mathbf{y})$ is defined by Eq. (3) and $g(\mathbf{y})$: $\mathbb{R}^n \to \mathbb{R}^k$, then $\boldsymbol{\delta}_w(\mathbf{y}) = \mathbf{b}(\mathbf{y}) + (1-w)\mathbf{g}(\mathbf{y})$ dominates **b** under $L_w(\boldsymbol{\delta}(\mathbf{y}|\boldsymbol{\beta})$ for all w. The Stein estimator consequently may be represented in this form and hence can be shown also to dominate the LS estimator under Zellner's balanced loss function. It is under this criterion that van Akkeren (1999) and van Akkeren, Judge, and Mittelhammer (2002) developed the data-based information theoretic (DBIT) estimator, which has attractive sampling properties in the case of ill-conditioned linear statistical model. With these concepts in mind, Section 2 develops the formulations for the DBIT estimator along with the ridge and principal components estimators where the optimal tuning parameter is chosen to minimize an unbiased estimator of the risk.

2 INFORMATION PROCESSING RULES

Given the framework of ill-conditioned inverse problems, we first review the DBIT estimator and state the properties that characterize its sampling behavior. Sections 2.2 and 2.3 will describe the ridge and principal components estimators, which are contained in the more general class of method of regularization estimators. Our review will be fairly general and is intended to give an overview of these methods. For more specific details, the reader is encouraged to examine the original papers.

2.1 The DBIT Estimator

In developing an information theoretic alternative estimation procedure, we generalize the traditional estimating function assumption $\mathbf{g}(\mathbf{y}, \boldsymbol{\beta}) = \mathbf{X}'(\mathbf{y} - \mathbf{X}\boldsymbol{\beta}) = \mathbf{0}$ and make use of information about the unknown distribution and parameters in the form of the following estimating function:

$$\mathbf{h}(\mathbf{y}, \boldsymbol{\beta}, \mathbf{e}) = \mathbf{X}'(\mathbf{y} - \mathbf{X}\boldsymbol{\beta} - \mathbf{e}) = \mathbf{0} \tag{9}$$

or equivalently,

$$n^{-1}[\mathbf{X}'\mathbf{y} - \mathbf{X}'\mathbf{X}\boldsymbol{\beta} - \mathbf{X}'\mathbf{e}] = \mathbf{0} \tag{10}$$

The k-dimensional structural data constraint contains $(k+n)$ unobservables, $(\boldsymbol{\beta}, \mathbf{e})$, and thus is ill-posed. Consequently, traditional matrix procedures can not yield a unique solution that satisfies the equation.

The structural data constraints, Eqs. (9) and (10), express one representation of the observable information that underlies the unknown and unobservable $\boldsymbol{\beta}$ and $\mathbf{e}$. Traditional information-processing rules are usually expressed as some function of the sample information either in a data or moment form and belong to the general class of estimators $\boldsymbol{\Psi} = \{\boldsymbol{\psi}: \delta(\mathbf{y}) = \sum_{i=1}^{n} (\boldsymbol{\psi}_i \odot y_i)\}$. For example, in the case of the LS estimator, we have $\boldsymbol{\psi}_i = (\mathbf{X}'\mathbf{X})^{-1}\mathbf{x}'_i$ which reduces to $\boldsymbol{\psi}_i = \mathbf{x}_i$ for the orthonormal linear model. Also, note that in the case of IV($h = k$) we have $\boldsymbol{\psi}_i = (\mathbf{Z}'\mathbf{X})^{-1}\mathbf{z}_i$ or $\boldsymbol{\psi}_i = [(\mathbf{X}'\mathbf{Z})(\mathbf{Z}'\mathbf{Z})^{-1}(\mathbf{X}'\mathbf{Z})(\mathbf{Z}'\mathbf{Z})^{-1}\mathbf{z}_i$ for IV($h > k$). Consequently, all are some weighted sum of y_i values, and all belong to the same estimator class given appropriate choices of $\boldsymbol{\psi}_i$. To solve the problem, we seek estimates of $\boldsymbol{\delta}(\mathbf{y})$ and $\boldsymbol{\gamma}(\mathbf{y})$ that satisfy Eqs. (9) and (10). In particular, expressing the observables $\mathbf{y}$ and $\mathbf{X}$ in terms of deviations about the mean, we seek $\mathbf{p}$ and $\mathbf{w}$

that satisfy

$$n^{-1}[\mathbf{X}'\mathbf{y} - \mathbf{X}'\mathbf{X}(\mathbf{X}' \odot \mathbf{P})\mathbf{y} - \mathbf{X}'(\mathbf{I}_n \otimes \mathbf{y}')\mathbf{w}] = 0 \tag{11}$$

such that $\boldsymbol{\delta}(\mathbf{y}) = (\mathbf{X}' \odot \mathbf{P})\mathbf{y}$ with the $(k \times n)$ matrix $\mathbf{P} = [\mathbf{p}_1, \mathbf{p}_2, \ldots, \mathbf{p}_k]'$ and $\gamma(y) = (I_n \otimes y')w$, where $\mathbf{w}' = [\mathbf{w}'_1, \mathbf{w}'_2, \ldots, \mathbf{w}'_n]$ and $\mathbf{p}' = [\mathbf{p}'_1, \mathbf{p}'_2, \ldots, \mathbf{p}'_k]$. We make use of the Hadamard product $\odot$ and let the weights $\mathbf{p}$ and $\mathbf{w}$ represent $(k+n)$ univariate probability distributions, one for each element of $\boldsymbol{\delta}$ and $\boldsymbol{\gamma}$. Thus, in our DBIT formulation, we apply the sample weights $\boldsymbol{\pi}$ marginally (coordinate wise) and specify a univariate probability distribution $\mathbf{p}_k$ on the sample observations for each of the estimates $\delta_k(\mathbf{y})$. Correspondingly, for the estimates of the disturbance terms $\gamma(\mathbf{y})$, a univariate probability distribution $\mathbf{w}_i$ on the sample observations is specified for each $\gamma_i(\mathbf{y})$, implying each e_i is contained in the convex hull of the centered sample points. Consistent with moment information in the form of Eqs. 9–11, and the Kullbaek-Leibler (KL) pseudo-distance measure, we have the extremum problem proposed by van Akkeren and Judge (2001):

$$\min_{\mathbf{p},\mathbf{w}} I(\mathbf{p}, \mathbf{q}, \mathbf{w}, \mathbf{u}) = \mathbf{p}' \log(\mathbf{p}/\mathbf{q}) + \mathbf{w}' \log(\mathbf{w}/\mathbf{u}) \tag{12}$$

subject to

$$\frac{\mathbf{X}'\mathbf{y}}{n} = \frac{(\mathbf{X}'\mathbf{X})}{n}(\mathbf{X}' \odot \mathbf{P})\mathbf{y} + \frac{(n^{-1})}{n}\mathbf{X}'(\mathbf{I}_n \otimes \mathbf{y}')\mathbf{w} \tag{13}$$

$$\mathbf{i}_k = (\mathbf{I}_k \otimes \mathbf{i}'_n)\mathbf{p} \tag{14}$$

$$\mathbf{i}_n = (\mathbf{I}_n \otimes \mathbf{i}'_n)\mathbf{w} \tag{15}$$

where $\mathbf{q}$ and $\mathbf{u}$ are the reference distributions composed of uniform probabilities in the case of noninformative priors. In this constrained optimization problem, the objective is to recover the $\mathbf{p}$ and $\mathbf{w}$ and thereby derive the optimal estimates of $\boldsymbol{\beta}$ and $\mathbf{e}$.

Given the objective of minimizing KL information that is inherent to the definition of the estimator, there is a tendency to draw the noise component vector to the centroid of the polyhedron formed from the centered $\mathbf{y}$ values. That is, KL information, which is based on an extension of Laplace's principle of insufficient reason or indifference, takes its global maximal value when all convexity weights are n^{-1}, which would draw the error estimates to their central tendency of the zero vector. Thus, imposing convexity weights on the feasible space of errors would seem a natural restriction. In particular, without further information, errors are estimated to be equal to their expected values. Regarding the normalization of the weights to have unit sum, note that traditional estimators also impose restrictions on the weights applied to the y_i values when determining the feasible values of the errors. For example, in the least squares case, the weights are defined by the positive semidefinite nature of the idempotent matrix, $(\mathbf{I}-\mathbf{X}(\mathbf{X}'\mathbf{X})^{-1}\mathbf{X}')$, that premultiplies $\mathbf{y}$ in the definition of the estimate of the error vector. In this case, each of the row sums of the idempotent matrix, which represents the sum of the weights applied to the $\mathbf{y}$ vector in defining each element of $\gamma(\mathbf{y})$ within the LS estimator context, equals 1 when the data is centered as it is here.

In terms of the $\mathcal{B}$ space, the feasible values are defined by $(\mathbf{X}'\otimes\mathbf{P})\mathbf{y}$. Under the KL criterion function Eq. (12), and based on arguments analogous to those used in motivating the behavior of the error component definition, the objective relative to the $\boldsymbol{\beta}$ component has a tendency to draw all the convexity weights to n^{-1} and seek the weights that are most maximally uninformative and could have arisen the greatest number of ways. Thus, there is a tendency to draw the estimate of $\boldsymbol{\beta}$ towards $n^{-1}\mathbf{X}'\mathbf{y}$. Note also that the row sums of the matrix, $(\mathbf{X}'\mathbf{X})^{-1}\mathbf{X}'$, weighting the $\mathbf{y}$ elements each equal zero. In comparison, the row sums of the weights applied to the $\mathbf{y}$ elements in defining the LS estimator of the $\boldsymbol{\beta}$ vector also equal zero when the data are centered.

Solving the constrained optimization problem defined by Eqs. (12)–(15) using the method of Lagrange multipliers, we obtain the optimal weights:

$$\hat{p}_{km}=\frac{q_{km}\exp\left[\frac{1}{n}\sum_{i=1}^{n}\sum_{l=1}^{K}\hat{\lambda}_l x_{il}x_{ik}x_{km}y_m\right]}{\sum_{m=1}^{n}q_{km}\exp\left\{\frac{1}{n}\sum_{i=1}^{n}\sum_{l=1}^{K}\hat{\lambda}_l x_{il}x_{ik}x_{km}y_m\right\}}$$

$$\equiv\frac{q_{km}\exp(.)}{\Omega_k(\hat{\lambda})}\quad\forall k,m \tag{16}$$

$$\hat{w}_{ij}=\frac{u_{ij}\exp\left\{\frac{1}{n}\sum_{l=1}^{K}\hat{\lambda}_l x_{il}y_j\right\}}{\sum_{j=1}^{n}u_{ij}\exp\left\{\frac{1}{n}\sum_{l=1}^{K}\hat{\lambda}_l x_{il}y_j\right\}}\equiv\frac{u_{ij}\exp(.)}{\Psi_i(\hat{\lambda})}\quad\forall i,j \tag{17}$$

that are used to obtain the estimates $\delta(\mathbf{y})$ and $\gamma(\mathbf{y})$. The strict convexity of the objective function and linear constraints ensure that the optimal solution exists and is uniquely defined when weights are strictly positive. In addition, further examination of Eqs. (16) and (17) reveals that $\hat{\mathbf{p}}$ and $\hat{\mathbf{w}}$ are both continuous and monotonic functions of $\boldsymbol{\lambda}$. These observations, in addition to certain regularity conditions, are used to show that the DBIT estimator is consistent and asymptotically normal. Consequently, the usual asymptotic tests may be applied for inference, and confidence regions can be estimated. Finite sampling properties of the DBIT estimator are investigated in van Akkeren, Judge and Mittelhammer (2002) and van Akkeren (1999) to show that DBIT estimates dominate the LS and traditional method of regularization (MOR) estimators under the SEL measure for different levels of ill-conditioning.

2.2 Method of Regularization

Traditionally, a restricted LS or penalized likelihood approach has been proposed to deal with ill-conditioned inverse problems. As such, a MOR estimator is used where the data are fit using a distance measure of goodness of fit $\psi(\mathbf{y},\mathbf{x};\boldsymbol{\beta})$ with penalty function $\phi(\boldsymbol{\beta})$ (Tikhonov and Arsenin, 1977; Titterington, 1985; O'Sullivan, 1986). It is clear that the properties of the resulting estimator can be derived using the theory of extremum estimation and estimating equations. Specifically, the ridge regressor $\boldsymbol{\delta}^r(\mathbf{y};\kappa)$ is defined as

$$\boldsymbol{\delta}^r(\mathbf{y};\kappa) = \arg\min_{\beta\in\mathcal{B}}[\psi(\mathbf{y},X;\boldsymbol{\beta}) + \kappa\phi(\boldsymbol{\beta})] \tag{18}$$

when $\kappa\to\infty$, the resulting ridge regressor is driven by the penalty function, while a value of $\kappa=0$ implies an estimate based on goodness of fit. Hoerl and Kennard (1970a,b; 1976) propose the ridge estimator to counter problems of singularity and near singularity of $\mathbf{X}'\mathbf{X}$ by adding matrix $\kappa\mathbf{I}$, where $\kappa\in\mathbb{R}^+$ such that the inverse of the sum is well defined. It follows that the normal equations:

$$(\mathbf{X}'\mathbf{X} + \kappa\mathbf{I})\boldsymbol{\delta} = \mathbf{X}'\mathbf{y} \tag{19}$$

can be uniquely solved for $\boldsymbol{\delta}^r(\mathbf{y};\kappa)$. Taking expectations of $\boldsymbol{\delta}^r(\mathbf{y};\kappa)$ reveals the estimator is biased while $\mathrm{Cov}(\mathbf{b}(\mathbf{y})) - \mathrm{Cov}(\boldsymbol{\delta}^r(\mathbf{y};\kappa))$ is shown to be positive semidefinite. The trade-off between bias and efficiency is employed to show that $R(\boldsymbol{\delta}^r(\mathbf{y};\kappa^*)|\boldsymbol{\beta}) < R(\mathbf{b}(\mathbf{y})|\boldsymbol{\beta})$ for optimal tuning parameter k^*, which is dependent of $\boldsymbol{\beta}$ and as such represents an impracticality to estimation.

Several ad hoc methods exist for selecting the optimal tuning parameter. One simple alternative sets $\kappa^* = \hat{\sigma}^2(k-2)/\|\mathbf{b}(\mathbf{y})\|^2$, which can be improved upon by applying an iterative ridge procedure (Hoerl and Kennard, 1976) that updates κ^* until $\boldsymbol{\delta}^r(\mathbf{y};\kappa)$ stabilizes. More recent developments have led to κ^*, which is the argument that minimizes an unbiased estimator of the risk (Beran, 1998). If precision of estimation is considered, algebraic manipulation allows us to derive the following:

$$R_1(\boldsymbol{\delta}(\mathbf{y};\kappa)|\boldsymbol{\beta}) = \mathbb{E}\|\boldsymbol{\delta}(\mathbf{y};\kappa) - \boldsymbol{\beta}\|^2 \tag{20}$$

$$= \mathrm{tr}\left[\mathbf{G}_\kappa(\sigma^2\mathbf{X}'\mathbf{X} + \kappa^2\boldsymbol{\beta}\boldsymbol{\beta}')\mathbf{G}_\kappa\right] \tag{21}$$

$$= \sigma^2 \sum_{i=1}^{k} \frac{\lambda_i}{(\lambda_i + \kappa)^2} + \kappa^2 \sum_{i=1}^{k} \frac{\tilde{\beta}_i^2}{(\lambda_i + \kappa)^2} \tag{22}$$

$$= \sum_{i=1}^{k} \frac{\sigma^2 \lambda_i + \kappa^2 \tilde{\beta}_i^2}{(\lambda_i + \kappa)^2} \tag{23}$$

where tr[.] indicates the standard notation for the trace of [.], $\mathbf{G}_\kappa = (\mathbf{X}'\mathbf{X} + \kappa\mathbf{I})^{-1}$, $\tilde{\boldsymbol{\beta}} = \mathbf{P}'\boldsymbol{\beta}$, and the matrix $\mathbf{P}$ is obtained from spectral decomposition of $\mathbf{X}'\mathbf{X} = \mathbf{P}\Lambda\mathbf{P}'$ with λ_i being the ith characteristic root of $\mathbf{X}'\mathbf{X}$. The above formulation clearly contains two unknowns; the variance, σ^2, and the transformed vector of parameters $\tilde{\boldsymbol{\beta}} = \mathbf{P}'\boldsymbol{\beta}$. The variance parameter can be easily estimated using an unbiased estimator such as $s^2 = \|\mathbf{y} - \mathbf{Xb}(\mathbf{y})\|^2 \|$ $(n-k)$ whereas $\tilde{\beta}_i^2$ may be estimated by $\tilde{b}_i^2 - s^2/\lambda_i$. Hence, by substituting these estimates for the unknowns in Eq. (23) an unbiased estimator of the risk [Eq. (20)] is obtained:

$$\hat{R}_1(\hat{\beta}, \hat{\sigma}^2) = \sum_{i=1}^{k} \frac{s^2 \lambda_i + \kappa^2 (\tilde{b}_i^2 - s^2/\lambda_i)}{(\lambda_i + \kappa)^2} \tag{24}$$

As stated, this empirical risk function addresses precision of estimation in terms of the SEL measure. If goodness of fit is considered, an alternative empirical prediction risk is obtained from the following algebraic manipulations:

$$R_2(\boldsymbol{\delta}(\mathbf{y};\kappa)|\boldsymbol{\beta}) = \mathbb{E}\|\mathbf{X}\boldsymbol{\delta}(\mathbf{y};\kappa) - \mathbf{X}\boldsymbol{\beta}\|^2 \tag{25}$$

$$= \mathbb{E}[(\boldsymbol{\delta}(\mathbf{y};\kappa) - \boldsymbol{\beta})'\mathbf{X}'\mathbf{X}(\boldsymbol{\delta}(\mathbf{y};\kappa) - \boldsymbol{\beta})] \tag{26}$$

$$= \mathrm{tr}[\mathbf{X}'\mathbf{X}\mathbb{E}(\boldsymbol{\delta}(\mathbf{y};\kappa) - \boldsymbol{\beta})(\boldsymbol{\delta}(\mathbf{y};\kappa) - \boldsymbol{\beta})'] \tag{27}$$

$$= \mathrm{tr}\left[\mathbf{X}'\mathbf{X}\left[\mathbf{G}_\kappa(\sigma^2\mathbf{X}'\mathbf{X} + \kappa^2\boldsymbol{\beta}\boldsymbol{\beta}')\mathbf{G}_\kappa\right]\right] \tag{28}$$

$$= \sum_{i=1}^{k} \frac{\sigma^2 \lambda_i^2 + \kappa^2 \lambda_i \tilde{\beta}_i^2}{(\lambda_i + \kappa)^2} \tag{29}$$

The unknown quantities, σ^2 and $\tilde{\beta}_i^2$, in Eq. (29) can be replaced using the same unbiased estimators as before. To consider both prediction and precision of estimation in the estimation objective, a linear combination of $\hat{R}_1$ and $\hat{R}_2$ yields Zellner's balanced loss function of the form:

$$\hat{R}_w(\hat{\beta}, \hat{\sigma}^2) = w\hat{R}_1(\hat{\beta}, \hat{\sigma}^2) + (1 - w)\hat{R}_2(\hat{\beta}, \hat{\sigma}^2) \tag{30}$$

for a particular value of $w \in [0,1]$. The value of $\kappa^* \geq 0$ is selected to minimize the objective [Eq. (30)] and consequently is entirely dependent on the data and not on nonsample information or a priori support space restrictions. While the approach for minimizing an unbiased estimator of the risk leads to a data-based procedure for selecting the optimal tuning parameter, the resulting estimator, $\boldsymbol{\delta}(\mathbf{y}; k^*)$, is difficult to evaluate and the corresponding sampling properties are analytically intractable. Simulation and resampling methods may, however, represent a computationally convenient means to evaluate the point estimates.

2.3 Principal Components Estimator

In addition to ridge regression, one frequently used method for dealing with ill-conditioned data is to reduce the informational demands by considering a subset of the parameter space $\mathcal{B}$. In principal components regression this amounts to removing the conditioning variables that contribute little to the explanatory power of the model under a specific criterion. Several procedures have been proposed for variable elimination (Hill et al., 1997; Fomby and Hill, 1978; Mundlak, 1981; Hill and Judge, 1987); however, these methods tend to lack rigorous theoretical justification and produce estimates with poor sampling properties.

Beran (1998) proposes an alternative formulation that addresses the identification problem of which specific column vectors of $\mathbf{X}$ to remove. This method uses an approach that miminizes an unbiased estimator of the risk in a fashion similar to that of Section 2.2. The criterion under which to evaluate estimator performance simply becomes the squared error prediction loss (SEPL), or the balanced loss function $L_w(\boldsymbol{\delta}(y)|\boldsymbol{\beta})$, where $w = 1$. As a starting point for this model, denote by $\mathbf{P} = (\mathbf{p}_1, \ldots, \mathbf{p}_k)$ that $(k \times k)$ orthogonal matrix whose columns $\mathbf{p}_i$ are the eigenvectors of $\mathbf{X'X}$ numbered according to the magnitude of characteristic roots $\lambda_1 \geq \lambda_2 \geq \cdots \geq \lambda_k$. The linear statistical model is then reparameterized to the following:

$$\mathbf{y} = \mathbf{XPP'}\boldsymbol{\beta} + \mathbf{e} = \tilde{\mathbf{X}}\tilde{\boldsymbol{\beta}} + \mathbf{e} \tag{31}$$

which is the principal components regression model. Here, the ith principal component is $\tilde{\mathbf{x}}_i = \mathbf{Xp}_i$ and it follows that $\tilde{\mathbf{x}}_i'\tilde{\mathbf{x}}_i = \lambda_i$. The principal components estimator is consequently formulated as $\tilde{\mathbf{b}}_\kappa(\mathbf{y}) = (\tilde{\mathbf{X}}'_\kappa\tilde{\mathbf{X}}_\kappa)^{-1}\tilde{\mathbf{X}}'_k\mathbf{y}$ where the subscript indicates the first κ column vectors of $\tilde{\mathbf{X}}$. The concept behind this method becomes clear in that one would like to remove the $\tilde{\mathbf{x}}_i$ associated with a small characteristic root λ_i, yet the major question of how to determine the correct cut-off point remains.

In addressing this problem, we consider the canonical modulation estimator proposed by Beran (1998) and exploit the singular value decomposition of $\mathbf{X}=\mathbf{ULV}'$ with the properties such that $\mathbf{U}'\mathbf{U}=\mathbf{I}_k$, $\mathbf{V}'\mathbf{V}=\mathbf{I}_k$ and $\mathbf{L}=\text{diag}(l_1, l_2, \ldots, l_k)$ are the singular values. It can be shown using a few algebraic steps that the candidate estimator for $\mathbf{X}\boldsymbol{\beta}$ can be expressed as $\mathbf{Xb}_f(\mathbf{y})=\mathbf{UFU}'\mathbf{y}$ where the $(k \times k)$ matrix $\mathbf{F}$ consists of a string of κ ones followed by $(k-k)$ zeros and all off-diagonal elements are zero as well. The prediction risk of this estimator where $\mathbf{Z}=\mathbf{U}'\mathbf{y}$ and $\boldsymbol{\xi}=\mathbf{LU}\boldsymbol{\beta}$ is derived as follows:

$$R(\mathbf{b}_f(\mathbf{y})|\boldsymbol{\beta}) = \mathbb{E}\|\mathbf{UFU}'\mathbf{y} - \mathbf{ULV}'\boldsymbol{\beta}\|^2 \tag{32}$$

$$= \mathbb{E}\|\mathbf{FZ} - \boldsymbol{\xi}\|^2 \tag{33}$$

$$= \mathbb{E}\text{tr}\{[\mathbf{F}(\mathbf{Z}-\boldsymbol{\xi}) - (\mathbf{I}-\mathbf{F})\boldsymbol{\xi}][\mathbf{F}-\boldsymbol{\xi}) - (\mathbf{I}-\mathbf{F})\boldsymbol{\xi}]'\} \tag{34}$$

$$= \text{tr}\{\mathbf{F}^2\mathbb{E}(\mathbf{Z}-\boldsymbol{\xi})(\mathbf{Z}-\boldsymbol{\xi})' + (\mathbf{I}-\mathbf{F})^2\boldsymbol{\xi}\boldsymbol{\xi}^2\} \tag{35}$$

$$= \text{tr}\{\sigma^2\mathbf{F}^2 + \boldsymbol{\xi}\boldsymbol{\xi}'(\mathbf{I}-\mathbf{F})^2 \tag{36}$$

$$= \sum_{i=1}^{k} [\sigma^2 f_i^2 + \xi_i^2(1-f_i)^2] \tag{37}$$

where the unknowns, (σ^2, ξ_i^2), can be estimated by the unbiased estimators s^2 and $z_i^2 - s^2$ since it is observed that $\mathbb{E}[z_i^2] = \sigma^2 + \xi_i^2$. Minimization of the empirical representation of Eq. (37) leads to the data-based estimate of $\mathbf{X}\boldsymbol{\beta}$, but derives an estimator functional form that is difficult to evaluate analytically for finite sample properties. Large sample properties of the resulting estimator represent one means to evaluate the estimates whereas resampling and Monte Carlo methods could provide a computationally simple alternative.

3 EXAMPLES

To give the discussed estimators operational content, this section investigates two sets of data that are frequently studied in the statistics literature because of particular problems that characterize the data. Estimates derived using the DBIT method are compared to usual LS results as well as those computed using principal components, ridge regression, and nonparametric smoothing techniques. One should keep in mind that while the discussion has focussed on small sample estimator performance in terms of precision of estimation and prediction, this section concentrates on goodness of fit as the criterion for comparison.

3.1 Motorcycle Data

The first example is a data set composed of 133 observations of acceleration against time for a simulated motorcycle accident that may be found in Silverman (1985) and is analyzed by Härdle (1991) and Venables and Ripley (1994). For each observation the measurement time is between 0 and 60 sec. Accordingly, we postulate that acceleration y_i against time t_i satisfies the trigonometric regression model:

$$y_i = \beta_0 + \sum_{j=1}^{12} \left[\alpha_j \cos(2\pi j t_i/60) + \beta_j \sin(2\pi j t_i/60)\right] + e_i \quad \text{for } 1 \le i \le 133 \tag{38}$$

Here, as usual, we assume the vector of unknown disturbances $\mathbf{e}$ to represent drawings from an unknown multivariate distribution with mean $\mathbf{0}$ and covariance matrix $\Sigma_{\mathbf{e}} = \sigma^2\mathbf{I}$ in addition to standard assumptions of a nonstochastic design matrix $\mathbf{X}$ and unknown vector of coefficients $\boldsymbol{\beta} \in \mathcal{B}$ as stated in Section 1. The degree to which one column vector of $\mathbf{X}$ lies in the subspace spanned by the other is reflected in the condition number $\kappa(\mathbf{X}'\mathbf{X})$, which is defined as the ratio of the largest to the smallest characteristic root of $\mathbf{X}'\mathbf{X}$, and equals about 189.08 for this particular data set, implying that collinearity represents a substantial problem.

3.2 DBIT Regression Estimates

Given the large sample size of the data, $n = 133$, the limiting properties proposed by van Akkeren, Judge, and Mittelhammer (2001) suggest DBIT estimates should be fairly close to the true parameter β_0 and moreover, they should be normally distributed with close approximation to LS estimates. Solving the optimization problem, Eqs. (12)–(15), leads to a computational difficulty due to the many constraints and unknown variables of the model. As a practical solution, we propose reducing the number of targets for the kth parameter from $\{\gamma x_{ki} y_i\}_{i=1}^{p}$ to $\{\gamma x_{ki} y_i\}_{i=1}^{j}$, where j refers to the number of ordered quantiles and $\gamma \in \mathbb{R}^+$ is a scale parameter that increases or tightens the distance between the empirical target points. For example, we may choose to select $j = 5$, leaving us with the minimum, maximum, the interquartile range, and the median. Likewise, we reduce the number of targets of the ith error term from $\{y_i - (1/n)\mathbf{i}'\mathbf{y}\}_{i=1}^{n}$ to $\{y_i - (1/n)\mathbf{i}'\mathbf{y}\}_{i=1}^{j}$, which sets the bounds of the support space as well as giving less weight to outlying target points.

TABLE 1 Regression Results for the Silverman Motorcycle Data

Parameter	$\gamma=0.1$	$\gamma=1$	$\gamma=10$	LS
β_0	−22.24872	−13.89796	−13.23652	−13.25907
β_1	1.27064	8.71495	9.95335	9.06142
β_2	7.06728	14.61594	15.40384	15.41838
β_3	−8.04954	−23.49255	−23.17484	−23.11846
β_4	4.55280	14.32972	15.20501	15.20574
β_5	−0.43147	−4.51787	−4.06511	−4.04139
β_6	−0.90008	−2.65975	−2.16492	−2.15621
β_7	−1.24315	−1.55327	−1.03614	−1.03692
β_8	0.58241	0.78432	1.19926	1.19661
β_9	0.21283	−0.62523	−0.32104	−0.31714
β_{10}	−1.58363	−3.98569	−3.82590	−3.81345
β_{11}	−0.41283	−1.64735	−1.49423	−1.49148
β_{12}	−0.27299	−0.66754	−0.42911	−0.43592
β_{13}	−6.33800	−28.36732	−29.22203	−29.09952
β_{14}	9.16865	30.81417	30.92620	30.86495
β_{15}	−3.80181	−12.58201	−12.90998	−12.89735
β_{16}	−2.68503	−6.30882	−6.40717	−6.41988
β_{17}	0.32144	3.01598	2.98387	2.95566
β_{18}	2.07148	0.74785	0.44230	0.45266
β_{19}	−0.00556	−0.71955	−0.87236	−0.87333
β_{20}	−2.25243	−2.48885	−2.58998	−2.60282
β_{21}	2.02036	−2.14160	−2.59033	−2.57853
β_{22}	0.64224	2.80111	2.81314	2.81422
β_{23}	−1.09331	−2.00492	−2.02609	−2.01187
β_{24}	−0.74670	0.19485	0.24022	0.24519

Table 1 displays the DBIT regression estimates for $j=11$ and 3 selected values of the scale parameter γ with Fig. 1 showing the graphical representation of these estimates. It is observed that the data display an unusual nonlinear shape where the dispression of points around the regression line varies from very wide to tight. Of the selected values of γ, it is clear that 0.1 sets supports that are too tight and hence not able to span the entire parameter space, resulting in the poor estimates found in column 2 of Table 1. A value of 1 or 10 for this parameter leads to more acceptable results with closer approximation to LS estimates. For parameter estimates under $\gamma=1$ and 10, the variance estimates are accordingly 546.5530 and 545.9179 compared to LS estimate of 545.9045. Next, we compare the resulting regression fits with principal components and ridge regression techniques.

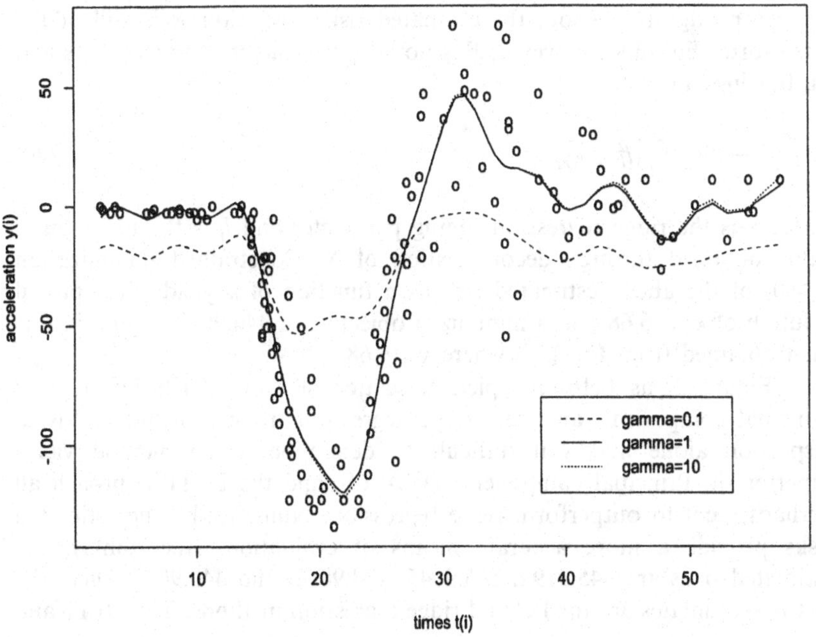

FIGURE 1 DBIT regression fits for $\gamma = 0.1$, 1, and 10.

3.3 Alternative Estimates

Applying Beran's canonical modulation estimator for principal components regression amounts to estimating $\mathbf{X}\boldsymbol{\beta}$ by $\mathbf{X}\mathbf{b}_f(\mathbf{y}) = \mathbf{U}\mathbf{F}\mathbf{U}'\mathbf{y}$, where $\mathbf{F}$ is a diagonal matrix whose diagonal consists of a string of 1s followed by a (possibly empty) string of 0s, and $\mathbf{U}$ is obtained through singular value decomposition of $\mathbf{X}$. The optimal number of 1s or principal components is found by minimization of an objective function, which is the estimated quadratic risk of estimating $\mathbf{X}\boldsymbol{\beta}$ by $\mathbf{X}\delta(\mathbf{F})$. This criterion is of the form:

$$\hat{R} = \frac{1}{k}\sum_{i=1}^{k}[s^2 f_i^2 + (z_i^2 - s^2)(1 - f_i)^2] \tag{39}$$

where $\mathbf{z} = \mathbf{U}'\mathbf{y}$ is a k-dimensional vector of which z_i is its ith element, f_i is either a $\mathbf{1}$ or 0, and σ^2 is the LS unbiased estimator of the variance. Computation of the optimal number of principal components reveals that the empirical risk, Eq. (37), is minimized for the square matrix $\hat{\mathbf{F}}$ with a string of 16 diagonal 1s followed by 9 diagonal 0s, and all off-diagonal elements are 0. The value of the empirical risk corresponding 16 principal components is 178.9882.

For ridge regression, the estimated risk expression $\hat{R}$ is still of the same form, Eq. (3.9); however, $\mathbf{F}$ is no longer a diagonal matrix of 1s and 0s, but instead,

$$\mathbf{F} = \mathrm{diag}[l_i^2/(l_i^2 + \kappa)] \tag{40}$$

where κ is the ridge regression tuning parameter and l_i is the ith singular value obtained through decomposition of $\mathbf{X}$. Constrained minimization ($\kappa \geq 0$) of the above estimated risk as a function of κ yields the optimal solution of $\kappa^* = 6.68$ and a minimized objective of 440.90. The matrix $\hat{\mathbf{F}}$ is then obtained from Eq. (40) where $\kappa = 6.68$.

Figure 2, as before, depicts a scatter plot on which DBIT, best principal component, and best ridge regression fits are included. Visual inspection alone makes it difficult to determine which method yields superior fit. Principal components (PC), LS, and the DBIT approach all perhaps seem to outperform ridge regression. Numerically, the estimated risks provide a more accurate means of evaluation. Accordingly, the estimated risks are 545.9197, 545.9045, 178.9882, and 440.9017 where the last two estimates are for PC and ridge regression methods. The DBIT and

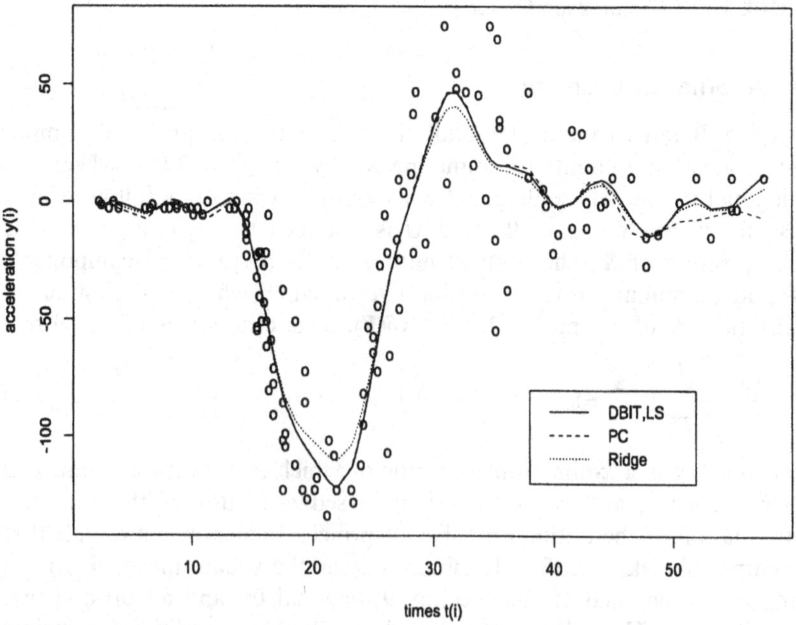

FIGURE 2 Comparative fits for the simulated motorcycle data.

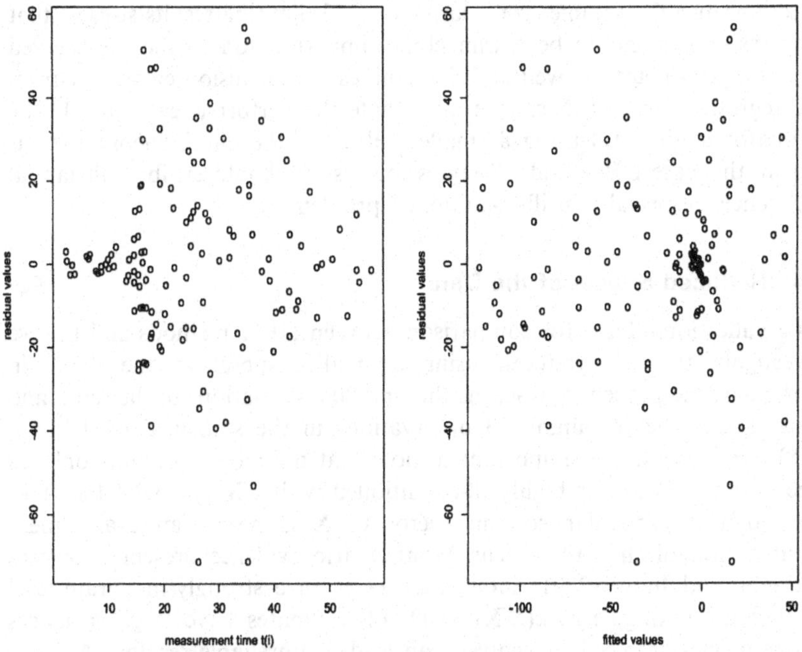

FIGURE 3 Regression diagnostics for DBIT estimates.

LS risks are close indeed, reflecting the nearly identical parameter estimates.

To study further the fit produced by the DBIT method, we refer to Fig. 3, which contains two plots; the first shows the unadjusted residuals against the times t_i while the second shows the unadjustes residual values against the fitted means $\hat{y}_i$. The first plot indicates that the residual values are very close to zero for measurement times up to 18, confirming a very close fit of the regression line. From times of 18 to 50, the dispersion around zero increases as the fit of the regression worsens. The outliers are indicated by a poor fit and large residual value, e.g., greater than 50 in absolute value. The second plot shows the residual values against the fitted values. No apparent systematic pattern can by discerned from mere visual inspection. The plot appears as a blur centered around the zero axis. However, outliers can also be identified by residual values greater than 50 in absolute value, which corresponds to Figs. 1 and 2.

From the preceding analysis it becomes evident that the DBIT method fits the motorcycle data reasonably well under the discussed criteria. Since the sample size is large, $n=133$, limiting results suggest that these estimates should be within close approximation to the unobserved location parameters as well as LS estimates, a conclusion evidenced from the regression results. Next, we investigate the performance of the DBIT estimator against LS using a smaller subset of the data. Theory tells us that in this case DBIT and LS regression results should exhibit substantial differences, especially in ill-conditioned problems.

3.4 Reduced Subset of the Data

As an additional basis for comparison between DBIT methods and LS, we investigate the fit produced using a smaller subset of the data. In particular, the procedure discards the first 80 observations of the data, and instead uses the remaining 53 observations in the statistical model, Eq. (3.8) (see Fig. 4). It is important to note that the model contains only 28 free parameters and is highly ill-conditioned with $\kappa(\mathbf{X}'\mathbf{X})=3.260466e+17$ such that a particular column vector of $\mathbf{X}$ is nearly an exact linear combination of the others. The Monte Carlo evidence presented by van Akkeren and Judge (2001) shows that, even for a strongly moderate level of ill-conditioning, e.g., $\kappa(\mathbf{X}'\mathbf{X})=100$, LS estimates have large variances where a one-sample data analysis can lead to unreliable results.

Table 2 compares LS and DBIT estimates computed with $\gamma=1$ and $j=5$, and shows that the estimates are strongly divergent. In contrast, the fitted means $\hat{y}_i$ are roughly similar, but with LS significantly "overfitting" the scatter plot. The estimated variances are 958.8862 and 1062.606 with a lower value for LS as expected. If we consider the information that has been omitted when using only a subset of our sample, it becomes evident that the DBIT method produces estimates significantly closer to the results of Table 1 than does LS. In this aspect, the DBIT method performs better than LS in that the results produced seem reasonable in comparison to the larger data set. Since the true location parameters are unobservable, we rely on asymptotic theory and consistency of the above estimates to conclude that the estimates in Table 1 are more accurate than those produced in Table 2.

3.5 Canadian Earnings Data

The following set of data is from Ullah (1985) and is analyzed by Chu and Marron (1991). Figure 5 shows the scatter plot of 205 data points of log (income) against age where an obvious upward sloping trend followed by a decreasing bend is observable. To form a relationship between earnings

TABLE 2 Regression Results for the Reduced Silverman Motorcycle Data

Parameter	DBIT($\gamma=1$)	LS
β_0	3.30360	−2.30566e+9
β_1	−5.36008	−1.78241e+9
β_2	5.64011	2.45064e+9
β_3	−1.53831	2.64337e+9
β_4	2.34618	2.20861e+8
β_5	−3.41809	−9.25645e+8
β_6	1.18621	−4.58583e+8
β_7	−3.98931	3.79388e+7
β_8	6.98036	8.46492e+7
β_9	−5.03064	1.73084e+7
β_{10}	−1.49742	−2.18760e+6
β_{11}	2.12458	−8.84852e+5
β_{12}	−9.39358	−3.72874e+4
β_{13}	−3.21095	−3.97916e+9
β_{14}	7.38785	−2.74960e+9
β_{15}	−7.52927	8.32280e+8
β_{16}	4.86897	1.83482e+9
β_{17}	−7.64678	5.63959+8
β_{18}	5.77725	−3.08928e+8
β_{19}	−3.03776	−2.38434e+8
β_{20}	−0.16285	−2.45786e+7
β_{21}	1.07573	1.93529e+7
β_{22}	−1.60745	5.39118e+6
β_{23}	6.89803	7.20995e+4
β_{24}	5.02754	−5.90341e+4

and age, the following polynomial regression model is proposed:

$$y_i = \beta_0 + \sum_{j=1}^{k} \beta_j x_i^j + e_i \quad \text{for} \quad 1 \le i \le 205 \tag{41}$$

The usual assumptions regarding the linear statistical model are valid (Section 1): the unknown errors e_i are independent with mean 0 and variance σ^2, the design matrix $\mathbf{X}$ is considered nonstochastic, and the unknown location parameters $\beta \in \mathcal{B}$ are finite valued.

3.6 Estimation Results

In applying the DBIT estimation procedure to the Canadian earnings data set, the same computational difficulty arises as in the previous example in that the number of unknown variables associated with 205

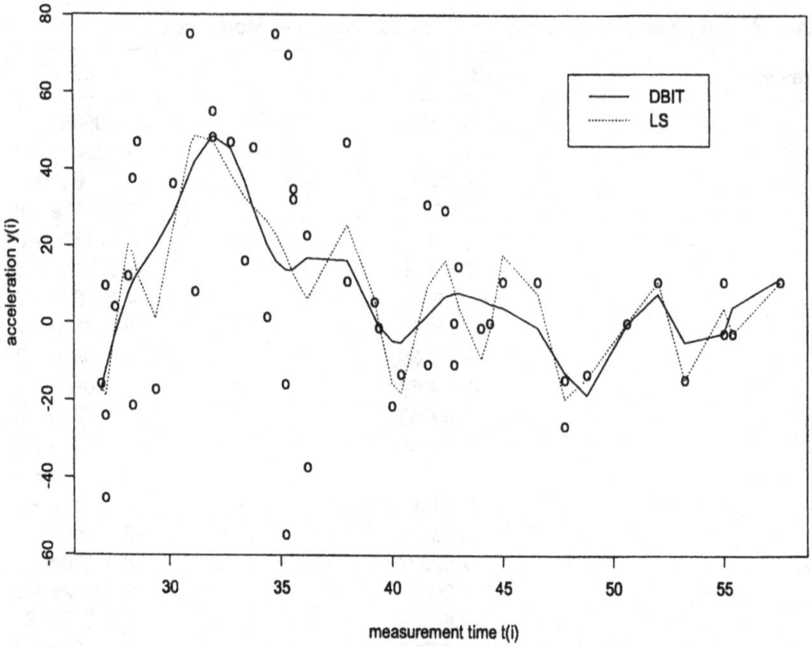

FIGURE 4 Comparative fits for the reduced motorcycle data, *n*: 81-133.

data points creates a formidable optimization problem. The computational burden is similarly eased through application of ordered quantiles for which $j=5$ is selected. To address the model selection problem of finding the optimal number of polynomial regressors, the Mallows' C_p (Mallows, 1973) criterion is applied, which penalizes the residual sum of squares by twice the number of parameters times the residual mean square for the initial model (Venables and Ripley, 1994, Chap. 10). The final model is chosen through a sequence of forward or backward steps, which minimizes this criterion. Specifically, we have

$$C_p = RSS_k + 2ps^2 \tag{42}$$

for p parameters of the submodel, and s^2 is the unbiased estimate of the variance in the full model.

Applying backward model selection, LS under the above C_p criterion derives an optimal model with five polynomial regressors, whereas the DBIT approach produces a final model with four explanatory variables. This result may very well be a consequence of rounding error of using the GAMS optimization package in conjunction with the large values

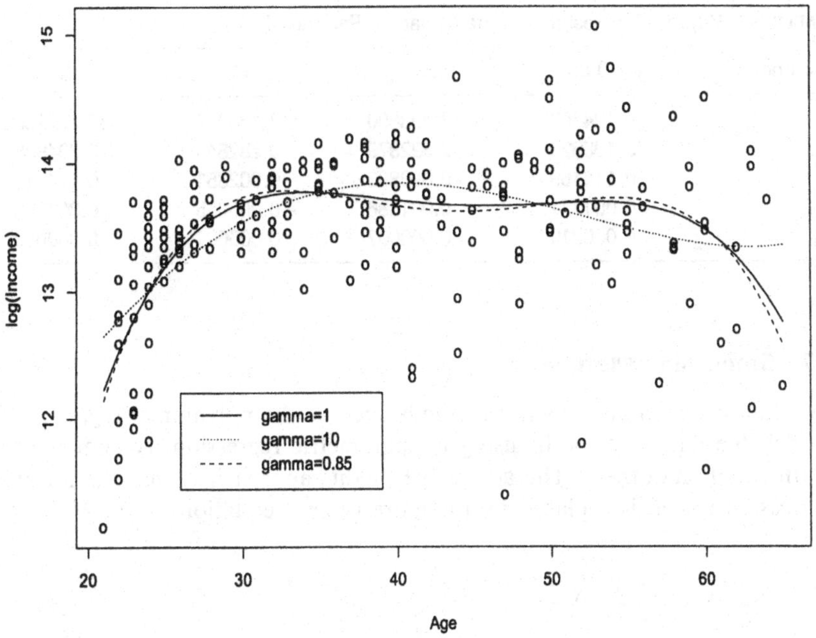

FIGURE 5 DBIT fits for the Canadian earnings data.

produced in fourth to fifth degree polynomials. Table 3 summarizes the main results. It is observed that DBIT estimates under $\gamma = 0.85$ are very close indeed to LS results. While coefficients derived under the standard scale parameter of $\gamma = 1$ are significantly different, a reasonable fit is produced as can be seen from Fig. 5. The large sample size of the data set leads us to suggest that the DBIT and LS estimates should be fairly similar, a result supported by Table 1.

Figure 5 shows several fits attained under different values of the scale parameter γ. While each of the three fits seems reasonable, the coefficients corresponding to $\gamma = 0.85$ and $\gamma = 10$ are significantly different. From visual inspection, it appears that the DBIT fitted values under $\gamma = 0.85$ describe the variation of the data best, accounting for the dips and turning points in the data. The dip in fitted values between age of 33 to 50 represents a decrease in income found by Ullah (1985) to be significant and cannot be readily observed from visual inspection. From a labor economics perspective, this observation represents an interesting issue in whether this phenomenon is truly due to some underlying economic structure or merely a coincidental sampling artifact.

TABLE 3 Regression Results for the Canadian Earnings Data

Parameter	$\gamma=0.85$	$\gamma=1$	$\gamma=10$	LS
β_0	−19.124700	−13.498690	8.126709	−19.177150
β_1	3.198929	2.592379	0.282846	3.204908
β_2	−0.114169	−0.090809	−0.002351	−0.114413
β_3	0.001771	0.001386	−0.000055	0.001775
β_4	−0.000010	−0.000007	0.000001	−0.000010

3.7 Smoothing Methods

In addition to semiparametric methods such as the polynomial regression model, the data may be fit using nonparametric regression techniques or scatter plot smoothers. The advantage of this approach is that the linear regression model is replaced by the more general equation:

$$\mathbf{y} = \beta_0 + \sum_{j=1}^{k} f_j(\mathbf{x}_j) + \mathbf{e} \tag{43}$$

where f is a twice-differentiable function that includes the linear regression model. Generally, these methods produce estimates that result from a compromise between goodness of fit and degree of smoothness. A variety of nonparametric regressors exist, but for comparison purposes to the DBIT approach, we will present only the smoothing spline and the kernel regression method.

The cubic smoothing spline, S-Plus command *smooth.spline*, selects the function f in Eq. (43), which minimizes a weighted functional between goodness of fit and smoothness of the form:

$$S_\lambda(f) = \sum_{i=1}^{n} [y_i - f(x_i)]^2 + \lambda \int [f''(x)]^2 \, \mathrm{d}x \tag{44}$$

where λ is a degree of freedom parameter that regulates the amount of smoothness and $\int [f''(x)]^2 \mathrm{d}x$ is the roughness penalty used to quantify local variation. The minimization of $S_\lambda(f)$ over the class of twice-differential functions defined on the interval $[a, b] = [x_1, x_n]$ has the unique solution, $\hat{m}_\lambda(x)$, which is a cubic spline (polynomial) due to the particular choice of the roughness penalty. Several methods exist for selecting an optimal parameter, but as before in our case visual inspection is used to determine λ. Another nonparametric method, the kernel regression estimator, Splus command *ksmooth*, has the specific functional form:

$$\hat{y}_i = \sum_{j=1}^{n} y_i K\left(\frac{x_i - x_j}{b}\right) \Big/ \sum_{j=1}^{n} K\left(\frac{x_i - x_j}{b}\right) \tag{45}$$

Here, as in the case of density estimation, K is the kernel function for which we shall use the standard normal density $N(0, 1)$, and b is the bandwidth parameter that determines the amount of smoothness. The kernel can be any continuous, symmetric, and bounded function that must integrate to one. Choice of optimal bandwidth amounts of finding the optimal balance between bias versus variance of Eq. (45), often called the Nadaraya–Watson estimator (Nadaraya, 1964; Watson, 1964). Again, the optimal value of b is selected using mere visual inspection.

Figure 6 displays the two smoothing methods compared to the DBIT results attained using optimal polynomial regression under the Mallows' C_p criterion. The smoothing spline parameter λ is set equal to 15 with the kernel regression parameter b equal to 3.5. The plot at first glance indicates that the curves are fairly similar with the characteristics

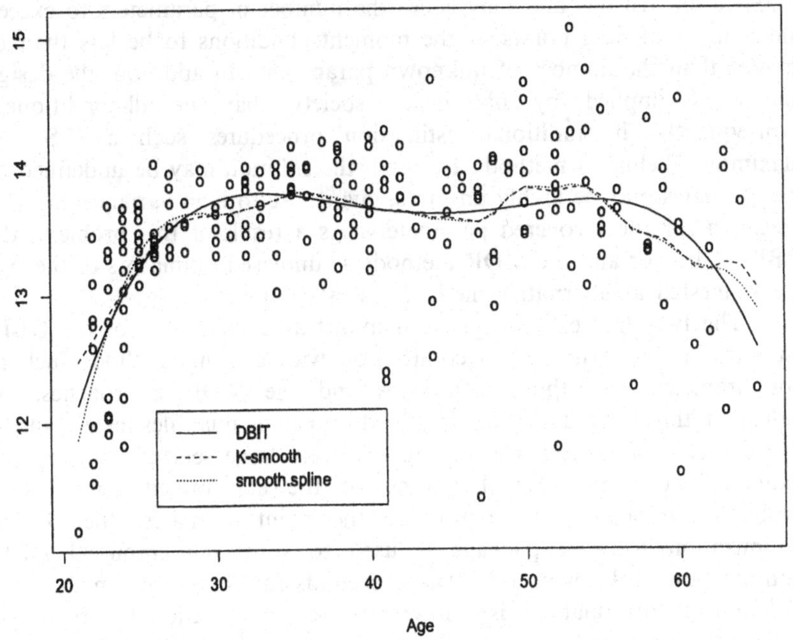

FIGURE 6 Scatter plot and smooths for Canadian earnings data.

of the data well reflected by each fit. The DBIT method seems to pick up the initial increase in earnings from age 20 to 30 better than *ksmooth* but not as well as the cubic smoothing spline. The dip in earnings is well observed by each method; however, the DBIT curve seems to weight the final data point in excess, producing a steeper slope then with alternative methods.

4 CONCLUDING REMARKS

Economic data are to a large extent nonexperimentally generated. Since the data, the statistical model, and the corresponding estimation and inference procedure are interdependent, we are restricted to the process of nonexperimental model building. This means there is uncertainty relative to the probability model that best describes the sampling process for the observed data and little basis for specifying the underlying sampling distribution. In addition, since the data are nonexperimental in nature, there may not be enough information contained in the design matrix $\mathbf{X}$ and the noisy data $\mathbf{y}$ to permit the recovery of the k-dimensional $\boldsymbol{\beta}$ vector by traditional LS or maximum likelihood estimation methods. The ill-posed aspect may arise because of nonstationary or other model specification reasons and may cause the number of parameters to exceed the number of data points or the moment conditions to be less than or greater than the number of unknown parameters. In addition, the design matrix $\mathbf{X}$ implied by nature or society may be ill-conditioned. Consequently, if traditional estimation procedures such as LS and maximum likelihood methods are used, the solution may be undefined or the estimates may be highly unstable, giving rise to high variance and low precision for the recovered parameters. As a result of this problem, the DBIT estimator and the MOR methods as unbiased estimators of the risk are suggested as alternative methods of estimation.

The two data examples have demonstrated the ability of the DBIT estimator to perform well in comparison with existing methods such as nonparametric smoothing techniques and the MOR approaches. As indicated, this method may be simplified to include quantiles instead of the entire vector of targets when using relatively large sets of data, hereby reducing the computational burden of the constrained optimization problem significantly. In computing the point estimates, the GAMS optimization software package is used to solve the empirical DBIT optimization problem in a matter of seconds for the given sample sizes. The DBIT formulation is shown to be computationally tractable, especially in the quantile formulation. In large samples, DBIT estimates are shown to approach LS results. However, LS tends to "overfit" and produce unreliable point estimates in contrast to our method, especially when ill-conditioning of the design matrix is a considerable problem as evidenced by the analysis of the reduced Silverman motorcycle data set. One possible implication of the MOR approach is that one may consider the case of applying a biased Stein's estimator of the risk, a formulation that could lead to desirable sampling properties when the data exhibit collinearity problems.

ACKNOWLEDGMENTS

George Judge and Rudy Beran provided valuable insights and feedback during the course of this research.

REFERENCES

Baranchik, A.J., 1970. A family of minimax estimators of the mean of a multivariate normal distribution. *Annals of Mathematical Statistics* **41**, 642–645.

Beran, R.J., 1998. React fits to linear models and scatterplots. Unpublished working paper. Dept. of Statistics, University of California of Berkeley (1–34).

Chu, C. and Marron, J., 1991. Choosing a kernel regression estimator. *Statistical Science* **6**, 404–436.

Dey, D.K., Ghosh, M., and Strawderman, W.E., 1999. On estimation with balanced loss functions. Statistics & Probability Letters, **45**, 97–101.

Efron, B. and Morris, C., 1976. Families of minimax estimators of the mean of a multivariate normal distribution. *Annals of Statistics* **34**, 11–21.

Fomby, T.B. and Hill, R.C., 1978. Multicollinearity and the minimax conditions for the Bock Stein-like estimator. *Econometrica* **47**, 211–213.

Härdle, W., 1991. *Smoothing Techniques with Implementation in S.* Springer Verlag, New York.

Hill, R.C. and Judge, G.G., 1987. Improved prediction in the presence of multicollinearity. *Journal of Econometrics* **35**, 83–100.

Hill, R.C., Fomby, T.B., and Johnson, S.R., 1997. Component selection norms for principal components regression. *Communications in Statistics* A **5**, 309–333.

Hoerl, A. and Kennard, R., 1970a. Ridge regression: Biased estimation for nonorthogonal problems. *Technometrics* **12**, 55–67.

Hoerl, A. and Kennard, R., 1970b. Ridge regression: Iterative estimation of the biasing parameter. *Technometrics* **12**, 69–82.

Hoerl, A. and Kennard, R., 1976. Ridge regression: Iterative estimation of the biasing parameter. *Communications in Statistics* A **4**, 77–88.

James, W. and Stein, C., 1961. Estimation with quadratic loss. *Proceedings of the Fourth Berkeley Symposium on Mathematical Statistics and Probability* vol. 1, 361–379.

Judge, G.G. and Bock, M., 1978. *The Statistical Implications of Pre-Test and Stein-Rule Estimators in Econometrics*. North-Holland, New York.

Mallows, C.L., 1973. Some comments on Cp. *Technometrics* **15**, 661–676.

Mundlak, Y., 1981. On the concept of nosignificant functions and its implications for regression analysis. *Journal of Econometrics* **16**, 139–150.

Nadaraya, E.A., 1964. On estimating regression. *Theory of Probability and Its Applications*, **10**, 186–190.

O'Sullivan, F., 1986. A statistical perspective on ill-posed inverse problems. *Statistical Science* **1**, 502–527.

Silverman, B., 1985. Some aspects of the spline smoothing approach to non-parametric regression curvefitting. *Journal of the Royal Statistical Society* B **47**, 1–52.

Stein, C., 1955. Inadmissibility of the usual estimator for the mean of a multivariate normal distribution. *Proceedings of the Third Berkeley Symposium on Mathematical Statistics and Probability*, pp. 197–206.

Tikhonov, A.N. and Arsenin, V.Y., 1977. *Solutions to Ill-Posed Problems*. Winston, Washington, DC.

Titterington, D., 1985. Common structures of smoothing techniques in statistics. *International Statistical Review* **53**, 141–170.

Ullah, A., 1985. Specification analysis of econometric models. *Journal of Quantitative Economics* **2**, 187–209.

van Akkeren, M., 1999. Data Based Information Theoretic Estimation, Ph.D. Dissertation, University of California at Berkeley.

van Akkeren, M., Judge, G.G., and Mittelhammer, R., 2002. Generalized moment based estimation. *Journal of Econometrics*, **107**, 127–148.

Venables, W.N. and Ripley, B.D., 1994. *Modern Applied Statistics with S-Plus*. Springer Verlag, New York.

Watson, G.S., 1964. Smooth regression analysis. *Sankkhyz, Ser. A.*, **26**, 359–372.

Zellner, A., 1994. Bayesian and non-bayesian estimation using balanced loss functions. In: S. Gupta and J. Berger, eds. *Statistical Decision Theory and Related Topics*. Springer Verlag, New York.

11

Estimating and Testing Fundamental Stock Prices: Evidence from Simulated Economies

R. Glen Donaldson
University of British Columbia, Vancouver, British Columbia, Canada
Mark J. Kamstra
Federal Reserve Bank of Atlanta, Atlanta, Georgia, U.S.A.

1 INTRODUCTION

A core tenet of financial economics states that, in a market populated by rational investors, the fundamental price of an asset equals the expected discounted present value of its future cashflows. This implies that in a rational and efficient market, and in the absence of price bubbles,* stock price movements are driven by forecasted changes in dividends and discount rates and not by the "irrational exuberance" of traders. Interest in the extent to which stock prices are fundamentally driven has been particularly high of late, with everyone from academics to the Chairman of the U.S. Federal

*The term "bubble" is formally defined below, but for now can be thought of as a situation in which some force other than a "fundamental factor," such as dividends and discount rates, drives stock prices.

Reserve Board offering opinions. Research into fundamental valuation is, therefore, particularly timely and, at least for the type of computer-intensive research we undertake in this chapter, is made possible by recent increases in computing power available to financial econometricians.

Several different procedures have been developed to estimate fundamental stock prices and to test whether market prices deviate in significant ways from the estimated fundamental price. Many studies have reported what appear to be important deviations from fundamentals, especially around market crashes such as those which occurred in 1929 and 1987. The resulting belief that financial markets may be "excessively volatile," and may potentially contain "price bubbles" that push market prices away from fundamental valuations, has contributed to the institution of policies such as trading halts in the face of large price moves. Whether such policies help or hurt financial markets depends in important ways on the extent to which market prices are driven by fundamental factors as opposed to "irrational exuberance," and the answer to this question crucially depends on the accuracy of the fundamental price estimates and tests used in the analysis.

To the best of our knowledge, no one has yet undertaken a thorough and systematic investigation of the accuracy of various fundamental price estimating and testing procedures.* Thus, while a fundamentals-estimation exercise might value a share of stock at $\$X$, we cannot be sure how confident to be in the estimate. For example, the estimate could be biased up or down and/or be very imprecise (i.e., have a large variance). With a small bias and low variance the estimate might be very accurate and therefore of considerable use. Conversely, a significantly biased estimates with very low precision might be wildly inaccurate and thus of little value. Establishing the properties of various fundamentals estimation and testing procedures would, therefore, be of significant benefit to academics, investors, and policy makers who often employ fundamental price estimates in their work. The accuracy of fundamental price estimates can be obtained analytically under only very restrictive and special circumstances

---

*Bollerslev and Hodrick (1995) provide a survey of much of this literature, and themselves use simulation techniques to investigate tests of market efficiency. Bollerslev and Hodrick (1995) focus much of their simulation study on the constant discount rate case and consider only one ad hoc model for time-varying discount rates. This model, like the constant discount rate case, allows an analytic solution for the market price. We consider formal discount rate processes that admit serial dependence and hence rule out a general analytic solution. Allowing serial dependence in discount rates considerably complicates the analysis, but also considerably enriches the range of possible dynamics for prices and returns.

(such as assuming that dividend growth rates and interest rates will never change in the future, which is of course highly unrealistic). To circumvent this analytical intractability, we develop and test in this chapter a simulation-based method for calculating the properties of various fundamental price estimating and testing procedures. We are particularly interested in determining the statistical properties of fundamental price and return estimates commonly used in both industry and academia and in investigating the effects that estimation inaccuracies may have on the variety of volatility tests commonly employed in the literature.

Our strategy in this chapter is as follows: (1) use financial market data to estimate time-series models for dividend growth and discount rates, (2) use these models to simulate dividend growth and discount rate paths for a variety of possible economies that do not contain bubbles, (3) calculate fundamental prices for these bubble-free economies based on the simulated dividend growth and discount rates (which prices we call "market prices" since they are fair-value prices for the simulated market economies), (4) use various fundamental valuation models to estimate fundamental prices for each of the simulated economies, and (5) compare market price versus fundamental price and investigate statistical properties of common tests for excess volatility and bubbles in stock prices using the data we simulated under the (true) null hypothesis that there are no price bubbles. We apply this procedure to S&P 500 stock price data.

Results produced using our computer-aided techniques suggest that, while stock prices are indeed volatile, they are not more volatile than one would reasonably find in an economy driven by fundamentals, thereby suggesting that market prices are not "excessively volatile." We also find that traditional tests for excess volatility and bubbles overreject a true null of no bubbles in samples of the size traditionally employed in the literature. Indeed, most tests we investigate find overwhelming evidence of bubbles in data series we construct under the conditions that there are no bubbles. In other words, we demonstrate that traditional tests for price bubbles frequently find bubbles in data that do not in fact contain bubbles.

In Section 2, we describe a variety of common fundamental pricing models and tests for excess volatility and bubbles. In Section 3 we present our Monte Carlo procedure for estimating fundamental prices/returns and investigating test performance. We discuss the results of our efforts in Section 4. Section 5 concludes.

2 FUNDAMENTAL PRICING METHODS AND TESTS

Define P_t^M as a stock's beginning-of-period-t market price, r_t as the rate used to discount payments received during period t, and $\mathcal{E}_t\{\cdot\}$ as the conditional expectation operator, with the conditioning information being the set of information available to investors at the beginning of period t. Investor rationality requires that the current market price of a stock, which will pay a dividend D_{t+1} at the beginning of period $t+1$ and then immediately sell for the ex-dividend market price P_{t+1}^M, satisfies Eq. (1):

$$P_t^M = \mathcal{E}_t\left\{\frac{P_{t+1}^M + D_{t+1}}{1+r_t}\right\} \tag{1}$$

We can solve Eq. (1) forward to period T (where $T > t$) and substitute realized dividends and discount rates in for their expected values to produce Eq. (2):

$$P_t^X = \sum_{i=1}^{T-t}\left(\prod_{k=1}^{i}\left[\frac{1}{1+r_{t+k-1}}\right]\right)D_{t+i} + \left(\prod_{k=1}^{T-t}\left[\frac{1}{1+r_{t+k-1}}\right]\right)P_T^M \tag{2}$$

The superscript X on the left-hand-side price indicates that this is the ex-post rational price; i.e., the price that an investor would have rationally paid for the stock had she known that the market price of the stock in period T was going to be P_T^M and that the stock was going to pay the sequence of dividends that it actually paid between periods t and T. The ex-post rational price is not a fundamental price, nor is it a price that would ever be observed in the marketplace. However, for reasons explained below, it is nonetheless useful in analyzing price behavior.

2.1 Fundamental Pricing Models

To calculate the fundamental price of stock, we need to solve Eq. (1) forward into infinite time under the transversality condition that the expected present values of the market price P_{t+i}^M falls to zero as i goes to infinity: i.e., there are no price "bubbles." This produces the familiar result that today's fundamental stock price, P_t^F, equals the expected present value of future dividends, i.e.,

$$P_t^F = \mathcal{E}_t\left\{\sum_{i=1}^{\infty}\left(\prod_{k=1}^{i}\left[\frac{1}{1+r_{t+k-1}}\right]\right)D_{t+i}\right\} \tag{3}$$

Note from Eq. (3) that the fundamental price is a function only of dividends, discount rates, and time, and not of the market price.

Defining the growth rate of dividends during period t as $g_t \equiv (D_{t+1} - D_t)/D_t$ allows the preceding equation to be rewritten as

$$P_t^F = D_t\mathcal{E}_t\left\{\sum_{i=1}^{\infty}\left(\prod_{k=1}^{i}\left[\frac{1+g_{t+k-1}}{1+r_{t+k-1}}\right]\right)\right\} \tag{4}$$

or, defining the discounted dividend growth rate as $y_t \equiv (1+g_t)/(1+r_t)$.

$$P_t^F = D_t \mathcal{E}_t \left\{ \sum_{i=1}^{\infty} \prod_{k=1}^{i} y_{t+k-1} \right\} \tag{5}$$

This can be rewritten as shown in Eq. (6) (note that y_t is *not* in the time t information set since it depends on D_{t+1} through g_t):

$$P_t^F = D_t \sum_{i=1}^{\infty} \mathcal{E}_t \left\{ \prod_{k=1}^{i} y_{t+k-1} \right\} \tag{6}$$

In practice, fundamental valuation calls for forecasting future discounted dividend growth rates (i.e., dividends and discount rates) to solve for the sum product in Eq. (4) [or, equivalently, Eq. (6)]. Since Eq. (4) cannot be solved analytically in general, this usually requires that some restrictive assumptions be made concerning the time-series processes driving dividends and discount rates. The simplest approach, introduced by Gordon (1962), is to assume that discount rates and dividend growth rates will be constant, at r and g, respectively, for all future time so that Eq. (4) reduces to Eq. (7) in which the superscript G denotes the Gordon-model fundamental price estimate:

$$P_t^G = D_t \left[\frac{1+g}{r-g} \right] \tag{7}$$

The Gordon model is certainly convenient, but its extremely restrictive assumptions of constant r and g do not tend to produce the most accurate valuations possible. Several attempts have therefore been made to relax Gordon's original restrictions and yet still retain an analytical solution to Eq. (4). The literature in this area is rather large and includes papers by Malkiel (1963), Fuller and Hsia (1984), Brooks and Helms (1990), Hurley and Johnson (1994, 1998) and Yao (1997), among others. Early literature in this area broke future time into several "chunks," with dividend growth and discount rates constant within each time chunk, but different between chunks. For example, dividends might be forecasted to grow at a "high rate" for the first 3 years, and then grow at some "normal rate" for the rest of time, so that the sum in Eq. (4) would be broken into two parts, each part solved separately, and then added together (see, e.g., Brooks and Helms, 1990).

Recent work on extending the Gordon model has largely focused on allowing for more complicated time-series behavior of dividends and discount rates, while retaining the ability to solve Eq. (4) analytically. Two particularly good examples, found in Yao (1997), are the additive Markov model (Eq. 1 of Yao, 1997) and the geometric Markov model

(Eq. 2 of Yao, 1997). These appear below as Eq. (8) and Eq. (9), respectively:

$$P_t^{ADD} = D_{t-1}/r + [1/r + (1/r)^2](q^u - q^d)\Delta \tag{8}$$

$$P_t^{GEO} = D_{t-1}\left[\frac{1 + (q^u - q^d)\Delta^{\%}}{r - (q^u - q^d)\Delta^{\%}}\right] \tag{9}$$

in which q^u is the proportion of the time the dividend increases, q^d is the proportion of the time the dividend decreases, $\Delta = \sum_{t=2}^{T} |D_t - D_{t-1}|/(T-1)$ is the average absolute value of the level change in the dividend payment, and $\Delta^{\%} = \sum_{t=2}^{T} |(D_t - D_{t-1})/D_{t-1}|/(T-1)$ is the average absolute value of the percentage rate of change in the dividend payment.

Donaldson and Kamstra (1996, 2000) develop an alternative approach (hereafter referred to as the DK procedure) which does not require that Eq. (4) be solved analytically. The DK procedure instead uses Monte Carlo methods to solve this equation numerically. The DK procedure eases the conditional constancy restrictions on dividend growth rates and discount rates of the Gordon procedure by modeling the discounted dividend growth rate y in Eq. (6) explicitly as conditionally time varying. The basic idea of the DK procedure is to estimate an econometric model for the time-series behavior of y_{t+i} (the discounted dividend growth rate) in Eq. (6), which is equivalent to Eq. (4), use this model and randomly drawn innovations to simulate time paths for the possible future evolution of y_{t+i}, then take the present value of the forecasted time paths to find a fundamental price. This is done thousands (even millions) of times, with a different sequence of randomly drawn innovations each time, so as to integrate out the expectation in Eq. (6) and thus produce a numerical (as opposed to analytical) estimate of the fundamental price.

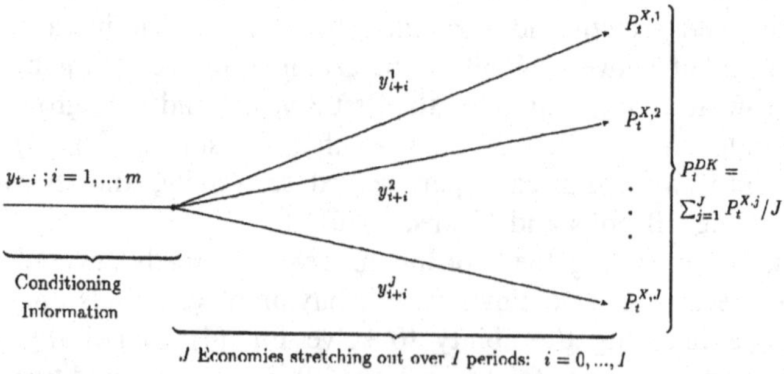

EXHIBIT 1 Diagram of DK Monte Carlo integration.

The sequence of steps employed to produce a DK fundamental price estimate are described below and depicted in Exhibit 1.

Step A: Use in-sample data to specify and estimate an econometric model for conditionally time-varying y_t.

Step B: Use the estimated model from Step A and data up to period t to simulate, out-of-sample, possible realizations of y_{t+i}: $i=1,\ldots,I$, where I is chosen to be very large (infinity, were it practical),* for a cross-section of J different possible economies, so that we follow the evolution of y across a panel of J simulated economies over I periods of time. We accomplish this by using the estimated model from Step A to make a conditional mean forecast of y, denoted $\hat{y}_t$, then simulate a population of J possible shocks around the mean and add these shocks to $\hat{y}_t$ to produce $y_t^j : j = 1,\ldots,J$ for the cross-section of j economies at time t. We then repeat this procedure through time for each of the J economies, conditioning on y_t^j, to form $y_{t+i}^j : i = 1,\ldots,I$. We therefore produce an out-of-sample panel of values of y for J economies stretching out I periods, all of which have been simulated based on the in-sample data.

Step C: Calculate the fundamental DK price based on this panel of simulated y using Eq. (6) as follows:

$$P_t^{DK} = D_t \sum_{j=1}^{J} \left(\sum_{i=1}^{I} \prod\nolimits_{k=1}^{i} y_{t+i}^{j} \right) \Big/ J \tag{10}$$

Note that the preceding equation can be rearranged by bringing the dividend level D_t inside the parentheses so that

$$P_t^{DK} = \sum_{j=1}^{J} \left(D_t \sum_{i=1}^{I} \prod\nolimits_{k=1}^{i} y_{t+k}^{j} \right) \Big/ J$$

It is interesting to note here that the parenthesized term in the preceding equation—i.e., $D_t \sum_{i=1}^{I} \prod_{k=1}^{i} y_{t+k}^{j}$ —is for economy j the ex-post rational price of the stock from Eq. (2), with the terminal period pushed out to infinity as I goes to infinity. In other words, if an investor stood at the end of time and looked back over history to see that the stock had paid a stream of dividends that had grown at rate g_{t+i}^{j} and which were discounted

*Note that as I increases, the product $\prod_{i=1}^{I} y_{t+i}$ converges to zero since y is less than unity in steady state. We found that, in practice, $I=400$ was easily sufficient to have our simulations converge to the point where increasing I had no impact on the fundamental price. In other words, a dividend received 400 years in the future has essentially no impact on today's price and thus 400 years in the future is equivalent to infinity for practical purposes in our simulations.

at rate r^j_{t+i} such that the discounted dividend growth rate had been y^j_{t+i} in this jth economy, the investor would rationally feel that she should have been willing to pay $P^{X,j}_t = D_t \sum_{i=1}^{\infty} \prod_{k=1}^{i} y^j_{t+k}$ to have purchased the stock back in time t where $P^{X,j}_t$ is commonly referred to by Shiller (1981) and others as the "ex-post rational price" [although when applied to data it is truncated at the end of the data period, as shown in Eq. (2)].

The representation of a single fan of the DK simulation as an ex-post rational price is interesting because it highlights the relationship between the ex-post rational price, the Gordon price, and the DK technique, and therefore provides some insight into the likely properties of the price estimates produced by various methods. For example, it is clearly seen that all Gordon-based methods are restricted versions of the DK technique. The basic Gordon model, for example, imposes conditionally constant dividend growth rates and discount rates such that each of the j economies in the simulation are identical and based on a constant y. If these restrictions are invalid and there is good reason to believe that they are indeed invalid—then the Gordon estimator will be biased and inefficient relative to DK.

The ex-post rational price calculation performed by Shiller (1981) and others is also problematic because they estimate a fundamental price based only on one realization of dividends and discount rates: the realization actually observed in the true market. In other words, Shiller's method stands at the end of time and asks what an investor with perfect foresight would have paid back at date t for a share of stock, had she known the dividends and discount rates that were to occur in the future.* Conversely, a real-life investor looks into an uncertain future when making purchasing decisions and therefore considers a universe of possible economies that might unfold when valuing a share of stock—this is the DK method. The ex-post rational price, based on only one time path, will, therefore, provide an imprecise picture of the thousands of possible economies, and their associated dividend time paths, that were considered when forming the market price at which real-life traders buy and sell stock. The extent to which this imprecision might affect our view on important asset-pricing questions, such as whether market prices are excessively volatile or not, is studied below.

*Of course, if the investor had perfect foresight, and therefore faced no uncertainty, she would not require an equity risk premium so the discount rate she would use would be lower than the rate r used in the ex-post price calculation. This subtlety is typically ignored in the literature that uses ex-post price calculations. For consistency we will also follow the literature here.

2.2 Some Tests Using Fundamental Prices

One of the key features of analytically based fundamental pricing procedures is that the fundamental prices they produce behave differently than observed market prices in important ways. For example, the time path of such fundamental prices is typically much smoother than that of observed market prices, which has led many researchers to conclude that market prices are excessively volatile. Indeed, there is an entire literature devoted to the study of excess volatility in financial markets [reviewed by Camerer (1989) and Cochrane (1992)]. Conversely, Donaldson and Kamstra (1996, 2000) find that the DK procedure produces fundamental stock prices that behave substantially the same as market prices in terms of return volatility. Of course, the evidence presented in all of these papers is based on the comparison of one estimate of the fundamental price (the estimate produced by the model in question) with one realization of the market price (the price series we see in the true market data). Given that the results come from only one price series, there is no way to be truly sure how accurate (biased, precise, etc.) the fundamentals estimate might be in general. The use of computer-aided financial econometric techniques in an investigation of the accuracy of various fundamentals-estimation procedures is, therefore, one important objective of this chapter.

Another objective of this chapter is to investigate the properties of various tests applied to fundamental and actual stock price data, in particular tests for excess volatility and price bubbles. Tests we investigate in this chapter include the following standard tests for price bubbles.

First, Camerer (1989) observes that market prices would boom upward and crash downward with bubbles expanding and collapsing, which would lead market prices to be excessively volatile relative to fundamental prices if the market price contained bubbles. This suggests a test with a no-bubbles null hypothesis that the variance of the percentage rate of change in market prices and the percentage rate of change in fundamental prices are equal, and an alternative hypothesis that the market price contains bubbles and thus that the variance in the percentage rate of change in market prices is greater than the percentage rate of change in fundamental prices.

Second, a direct test of bubbles asks whether the market and fundamental prices are cointegrated (e.g., Campbell and Shiller, 1987). Bubbles in the market price will lead to a nonstationary difference between the market price and the fundamental price estimate; a standard unit root test on this difference therefore tests the null of cointegration. See Dickey and Fuller (1981) and Said and Dickey (1984) for a description of unit root tests.

Third, Mankiw *et al.* (1985) (MRS) develop two tests for bubbles. They exploit a decomposition of the difference between the ex-post rational price, P_t^X [calculated by inserting realized dividends and discount rates into the present value equation as in Shiller (1981), as seen in Eq. (2)], and the fundamental price estimate P_t^F, relative to the observed market price P_t^M:

$$(P_t^X - P_t^F) = (P_t^X - P_t^M) + (P_t^M - P_t^F)$$

Since the total volatility of the two sides of the preceding equation are equal by definition, the volatility of each individual component on the right-hand side should be less than or equal to the volatility of the left-hand side. The test that compares the volatility of the first term on the right-hand side with the volatility of the left-hand side will be labeled the MRS1 test, while the test which compares the volatility of the second term on the right-hand side with the volatility of the left-hand side will be labeled the MRS2 test. In both cases the null hypothesis of no bubbles (i.e., no excess volatility) is that the term on the left-hand side is more volatile than the right-hand side term to which it is being compared. In considering these tests it is worth asking whether any of them have proper size, or whether they over-reject as some researchers argue. Note that these tests involve a joint null hypothesis that the market price does not contain a bubble and that the fundamental price *estimate*, produced by some fundamental pricing model, shares similar properties with the market's true unobserved fundamental price. A finding of systematic over-rejection across simulated market economies (where the null of no bubbles is imposed) would therefore indicate that the model used to produce fundamental price estimates is mis-specified.

In this chapter we employ Monte Carlo procedures to investigate the performance of various fundamentals-estimation methodologies and of simulated market prices, and also of various procedures commonly used to test for excess volatility and bubbles in stock prices. Although our particular application of Monte Carlo methods is new, there is already a significant literature that uses Monte Carlo methods in asset-pricing applications. Indeed, there is body of work that (directly or indirectly) simulates stock prices and dividends, under various assumptions, to investigate price and dividend behavior (e.g., Scott, 1985; Kleidon, 1986; West, 1988a,b; Campbell, 1991; Mankiw *et al.*, 1991; Hodrick, 1992; Timmermann, 1993, 1995; Campbell and Shiller, 1998). However, these studies typically impose restrictions on the dividend and discount rate processes so as to obtain fundamental prices from some variant of the Gordon (1962) model discussed above and/or some log-linear approximating framework.

Rather than impose approximations to solve Eq. (4) analytically, we will instead simulate the dividend growth and discount rate processes directly, and evaluate the expectation through Monte Carlo integration techniques. This approach is computationally burdensome since it requires us to perform a Monte Carlo simulation of a Monte Carlo simulation, but it is the only way to evaluate Eq. (4) without approximation error.* We also take care to calibrate our models to the time-series properties of the data. Dividend growth, for instance, is strongly autocorrelated in the S&P500 stock market data, in contrast to the assumption of a log random walk for dividends often imposed in this literature.

3 SIMULATING FUNDAMENTAL STOCK PRICES

Consider again the pricing relationship of Eq. (4), rewritten below for convenience, in which P is the price (for notational convenience we drop the superscript "F"), D is the dividend, g is the dividend growth rate, and r is the discount rate:

$$P_t = D_t \mathcal{E}_t \left\{ \sum_{i=1}^{\infty} \left(\prod_{k=1}^{i} \left[\frac{1 + g_{t+k-1}}{1 + r_{t+k-1}} \right] \right) \right\}$$

We now execute our strategy of (1) calibrating models to market dividends and interest rates, (2) simulating bubble-free economies using these calibrated models, (3) calculating fundamental prices for these bubble-free economies, (4) forming fundamental price *estimates*, including the Gordon models and the DK model, and (5) evaluating these fundamental price estimates and tests for bubbles.

3.1 Dividends and Discount Rates

The first step is to estimate time-series models for dividend growth and interest rates so that the Monte Carlo simulations generate dividends and discount rates that match real-world dividends and discount rates. This will allow us to generate nonbubble prices and returns and compare them to real-world prices and returns [e.g., a finding that the real-world prices and returns look significantly different from the simulated (nonbubble)

*There is still Monte Carlo simulation error, but that is random, unlike most types of approximation error, and it can also be measured explicitly. This simulation error is analogous to simple cases such as the simulation error associated with Monte Carlo experiments on the size of a test statistic.

prices and returns would provide evidence of non-fundamental movements in real-world prices and returns; possibly bubbles]. This will also allow us to evaluate conventional fundamental price estimation methods, including variants of the Gordon growth model, and conventional tests for bubbles, such as the MRS tests, to explore the properties of the estimators and test statistics when there are no bubbles in the price.

Our dividend process is calibrated to the S&P 500 stock index annual dividend data, 1952–1998, collected as described in Shiller (1989). The discount rate is defined to be the risk-free interest rate plus a constant equity premium, where the risk-free rate is the interest rate on a 1-year U.S. T-bill as constructed by the U.S. Federal reserve, 1952–1998.* A constant equity premium of 5.77% is added to the risk-free interest rate to produce a discount rate consistent with the stock price data.†

S&P 500 dividends grew at an average rate of 5.5% per year, over 1952–1998, with a standard deviation of 3.7%. As dividend growth rates have a minimum value of −100% and no theoretical maximum, a natural choice for their distribution is the log normal in the S&P 500 data. The logarithm of 1 plus the dividend growth rate has mean 0.0531 and standard deviation 0.035 over 1952–1998. We estimated simple autoregressive moving average (ARMA) time-series models for the logarithm of 1 plus the dividend growth rate and found the best model by the Bayesian information criterion to be a moving average model of order 1 (MA(1)) with the MA(1) coefficient equal to 0.60 Standard tests for normality of this error term do not reject the null of normality,‡ and standard tests for autocorrelation and autoregressive conditional heteroskedasticity (ARCH) fail to reject the null of homoskedasticity and no serial correlation.§,¶

---

*We choose this measure because, as Cochrane (1992) notes, "there is a long tradition in the volatility test and investment or capital-budgeting literature that measures time-varying discount rates from interest rates plus risk premiums that are constant over time."

†Finance theory requires that $\mathcal{E}\{(1+R)/(1+i+\pi)\}=1$, where R is the log of one plus the stock return, i is the risk-free rate, and π is the equity premium (i.e., $1/[1+i+\pi]$ is the "pricing kernel"). See, e.g., Campbell et al. (1997). Selection of the equity premium that produces the requisite pricing kernel was accomplished with a grid search, and this leads to a risk premium of 5.77% in the annual S&P 500 data over 1952–1998.

‡These tests include the Shapiro–Wilk. Kolmogorov–Smirnov, Cramer–von Mises, and Anderson–Darling tests. See SAS Procedures Guide (1999).

§See Engle (1982) for the seminal treatment of ARCH effects.

¶These are based on autocorrelations of the residuals and residuals squared.

As for interest rates, since economic theory admits a wide range of possible interest-rate processes there are a variety of models possible, from constant to autoregressive and highly nonlinear heteroskedastic forms. The autoregressive model of order 1 (AR(1)) of the logarithm of interest rates, as described in Hull (1993, p. 408), will be used here as it fits our data well and restricts nominal rates to be positive. Standard specification tests for normality, autocorrelation, and ARCH on the error term from an AR(1) model of the logarithm of interest rates do not reject the null of no misspecification. The 1-year T-bill rates have mean 0.059 and standard deviation 0.03 over 1952–1998. The AR(1) coefficient estimate in the regression of log interest rates on lagged log interest rates equals 0.83. Finally, the error terms from the MA(1) model of log dividend growth rates and log interest rates are correlated, with a correlation coefficient of 0.21.

Properties of fundamental prices and returns produced by Eq. (4) hinge delicately on the modeling of the dynamics of the dividend growth and interest-rate processes. For instance, fundamental prices will equal a constant times the dividend level and fundamental returns will be very smooth over time if dividend growth and interest rates are equal to constants plus independent innovations. However, modeling these data series to capture the serial dependence of dividend growth rates and interest rates observed in the data, as we have done, will typically lead to time-varying price-dividend ratios and variable returns of the sort we see in the S&P 500 stock market data.

3.2 Monte Carlo Experiment

We now detail the Monte Carlo experiment by which the price, P, is arrived at given conditioning information on the dividend level, D, dividend growth rate, g, and interest rate, r. That is, we detail for the eth economy (where $e = 1, \ldots, E$) the formation of the price P_t^e given D_t^e, g_t^e, and r_t^e. P_t^e is the market price that Eq. (4) states would obtain in economy e if the stock market in economy e is rational, efficient, and bubble-free. In terms of timing and information, recall that P_t^e is the stock's beginning-of-period-t market price based on fundamental factors, r_t^e is the rate used to discount payments received during period t and is known at the beginning of period t, D_t^e is paid at the beginning of period t, $g_t^e \equiv (D_{t+1}^e - D_t^e)/D_t^e$ and is not known at the beginning of period t since it depends on D_{t+1}^e, and $\mathcal{E}_t\{\cdot\}$ is the conditional expectation operator, with the conditioning information being the set of information available to investors at the beginning of period t. Finally, recall from Eq. (4) that investor rationality requires

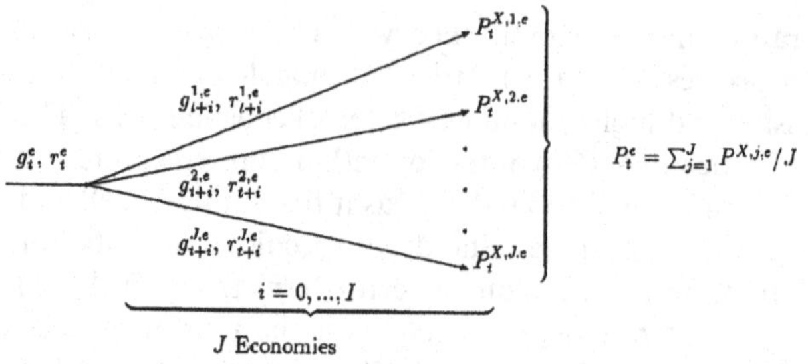

EXHIBIT 2 Diagram of a sample market price calculation for the *t*th observation of the *e*th economy (Steps 1 and 2).

$$P_t^e = D_t^e \mathcal{E}_t \left\{ \sum_{i=1}^{\infty} \left(\prod_{k=1}^{i} \left[\frac{1 + g_{t+k-1}^e}{1 + r_{t+k-1}^e} \right] \right) \right\}$$

Based on the preceding equation, we generate prices by generating a multitude of possible streams of dividends and discount rates, present-value discounting of the dividends with the interest rates, and averaging the results, i.e., by conducting a Monte Carlo integration. Hence, we produce P_t^e, $e = 1, \ldots, E$ utilizing only dividend growth rates and discount rates. The exact procedure is described below and summarized in Exhibit 2.

3.2.1 Simulating Stock Prices for a Rational, Efficient, Bubble-Free Market

Step 1: When forming P_t^e, the most recent fundamental information available to a market trader would be g_{t-1}^e, D_t^e, and r_t^e. The quantities g_{t-1}^e, D_t^e and r_t^e must, therefore, be generated directly in our Monte Carlo experiment, whereas P_t^e must be calculated based on these *g*, *r*, and *D* since this is how prices would be determined by a rational market participant in a bubble-free economy. In the steps below the risk free T-bill rate is indicated as *r* and the discount rate (i.e., the risk-free interest rate plus the risk premium of 5.77%) as r_*. The objective of Steps 1(a)–(c) is to produce dividend growth rates and interest rates that replicate the real-world dividend growth and discount rate data. That is, the simulated dividend growth rates and interest rates must have the same mean, variance, correlation structure, and autocorrelation structure as the real-world dividend growth rates and interest rates.

Step 1(a): Note that since the logarithm of one plus the dividend growth rate is modeled as a MA(1) process, $\log(1 + g_t^e)$ is a function of only innovations, labeled ε_g^e. Note also that since the logarithm of the interest rate is modeled as an AR(1) process, $\log(r_t^e)$ is a function of

$\log(r_{t-1}^e)$ and an innovation labeled ε_r^e. Set the initial dividend, D_1^e, equal to the S&P 500's dividend value for 1951 (observed at the end of 1951), and the lagged innovation of the logarithm of the dividend growth rates $\varepsilon_{g,0}^e$ to 0. To match the real-world interest rate data, set $\log(r_0^e) = -3.05$ (the mean value of log interest rates required to produce interest rates matching the mean and variance of observed T-bill rates). Set the standard deviation of the innovation to the log interest rate process to 0.242, and the standard deviation of the innovation of the log dividend growth rate process to 0.0305. Then, generate two independent standard normal random numbers, $\varepsilon_{1,1}^e$ and $\varepsilon_{2,1}^e$, and form two correlated random variables, $\varepsilon_{r,1}^e = 0.242(0.21\varepsilon_{1,1}^e + (1 - 0.21^2)^{0.5}\varepsilon_{2,1}^e)$ and $\varepsilon_{g,1}^e = 0.0305\varepsilon_{1,1}^e$. These are the simulated innovations to the interest rate and dividend growth rate processes, formed to have standard deviations of 0.242 and 0.0305, respectively, to match the data, and to be correlated with correlation coefficient 0.21 as we find in the S&P 500 and T-bill data. Next, form $\log(1 + g_1^e) = 0.0531 + 0.60\varepsilon_{g,0}^e + \varepsilon_{g,1}^e$ and $\log(r_1^e) = -0.18 + 0.94\log(r_0^e) + \varepsilon_{r,1}^e$.* Also, form $D_2^e = D_1^e(1 + g_1^e)$.

Step 1(b): Produce two correlated normal random variables, $\varepsilon_{r,2}^e$ and $\varepsilon_{g,2}^e$ as in Step 1(a), and conditioning on $\varepsilon_{g,1}^e$ and $\log(r_1^e)$ from Step 1(a) produce $\log(1 + g_2^e) = 0.0531 + 0.60\varepsilon_{g,1}^e + \varepsilon_{g,2}^e$, $\log(r_2^e) = -0.18 + 0.94\log(r_1^e) + \varepsilon_{r,2}^e$ and $D_3^e = D_2^e(1 + g_2^e)$.

Step 1(c): Repeat Step 1(b) to form $\log(1 + g_t^e)$, $\log(r_t^e)$ and D_t^e for $t = 3, 4, 5, \ldots, T$ and for each economy $e = 1, 2, 3, \ldots, E$, then calculate the dividend growth rate g_t^e and the discount rate $r_{*,t}^e = r_t^e + 0.0577$.

Step 2: For each time period $t = 1, 2, 3, \ldots, T$ and economy $e = 1, 2, 3, \ldots, E$ we must calculate present value prices, P_t^e. In order to do this we must solve for the expectation of the infinite sum of discounted future dividends conditional on time $t-1$ information for economy e. That is, we must produce a cross-section of dividends and interest rates that might be observed in periods $t, t+1, t+2, \ldots$ given what is known at period $t-1$ and use these to solve the expectation of Eq. (4). The counter j below indexes the cross-section of future economies that could possibly evolve from the current state of the economy.

*Notice that the AR(1) parameter for the log interest rate process is estimated to be 0.83, but we have set it to 0.94 in the simulations. It is well known that the coefficient estimate in an AR(1) OLS regression is biased downwards; see, for instance, Kennedy (1992, p. 147). Monte Carlo experiments were employed to determine the appropriate correction for our data, as in Orcutt and Winokur (1969), and this led to the setting of 0.94. The intercept term had to be adjusted as well to reflect this new setting.

Step 2(a): Set $\varepsilon_{g,t-1}^{j,e} = \varepsilon_{g,t-1}^{e}$ and $\log(r_{t-1}^{j,e}) = \log(r_{t-1}^{e})$ for $j = 1, 2, 3, \ldots, J$. Generate two independent standard normal random numbers, $\varepsilon_{1,t}^{j,e}$ and $\varepsilon_{2,t}^{j,e}$ and form two correlated random variables $\varepsilon_{r,t}^{j,e} = 0.242(0.21\varepsilon_{1,t}^{j,e} + (1 - 0.21^2)^{0.5}\varepsilon_{2,t}^{j,e})$ and $\varepsilon_{g,t}^{j,e} = 0.0305e_{1,t}^{j,e}$ for $j = 1, 2, 3, \ldots, J$.* These are the simulated innovations to the interest rate and dividend growth rate processes, respectively. Form $\log(1 + g_t^{j,e}) = 0.0531 + 0.60\varepsilon_{g,t}^{j,e} + \varepsilon_{g,t}^{j,\varepsilon}$ and $\log(r_t^{j,e}) = -0.18 + 0.94\log(r_{t-1}^{j,e}) + \varepsilon_{r,t}^{j,e}$.

Step 2(b): Produce two correlated normal random variables $\varepsilon_{r,t+1}^{j,e}$ and $\varepsilon_{g,t+1}^{j,e}$ in Step 2(a), and conditioning on $\varepsilon_{g,t}^{j,e}$ and $\log(r_t^{j,e})$ from Step 2(a) produce $\log(1 + g_{t+1}^{j,e}) = 0.0531 + 0.60\varepsilon_{g,t}^{j,e} + \varepsilon_{g,t+1}^{j,e}$, and $\log(r_{t+1}^{j,e}) = -0.18 + 0.94\log(r_t^{j,e}) + \varepsilon_{r,t+1}^{j,e}$, for $j = 1, 2, 3, \ldots, J$.

Step 2(c): Repeat Step 2(b) to form $\log(1 + g_{t+i}^{j,e})$ and $\log(r_{t+i}^{j,e})$ for $i = 2, 3, 4, \ldots, I$, $j = 1, 2, 3, \ldots, J$, and economies $e = 1, 2, 3, \ldots, E$. Solve for the dividend growth rate $g_{t+i}^{j,e}$, the dividends $D_{t+i}^{j,e}$, and the discount rate $r_{*,t+i}^{j,e} = r_{t+i}^{j,e} + 0.0577$ for $i = 0, 1, 2, \ldots, I$.

Step 2(d): The present discounted value of each of the individual, J streams of dividends is now taken in accordance with Eq. (4). Note that for each of the j streams the present value price so calculated for that stream is the ex-post rational price for the jth economy; the price a rational investor would pay for the stock if she knew for certain that the jth economy would obtain. We can, therefore, call the jth present value $P_t^{X,j,e}$, where the superscript X denotes ex-post rational.

In considering these prices, note that according to Eq. (4) the stream of discount and dividend growth rates should be infinitely long, while in our simulations we extend the stream only a finite number of periods, I. Since the ratio of gross dividend growth rates to gross discount rates, i.e., the y's in Eq. (6) are less than one in the steady state, the individual product elements in the infinite sum in Eq. (6) [equivalently, Eq. (4)] eventually converge to zero as I increases. Indeed, this convergence to zero is exactly what is required for the absence of price bubbles! We therefore set I large enough in our simulations so that the truncation does not materially effect our results. As mentioned previously, we found that setting $I = 400$ (years) accomplishes this conservatively. That is, the discounted value today of a dividend payment received 400 years in the future is essentially zero. Also note that the steps above are required to produce P_t^e, D_t^e, g_{t-1}^e, and $r_{*,t}^e$ for $e = 1, \ldots, E$; the intermediate terms

*For our random number generation we made use of a variance reduction technique, stratified sampling. This technique has us drawing pseudorandom numbers, ensuring that $q\%$ of these draws come from the qth percentile, so that our sampling does not weight any grouping of random draws too heavily.

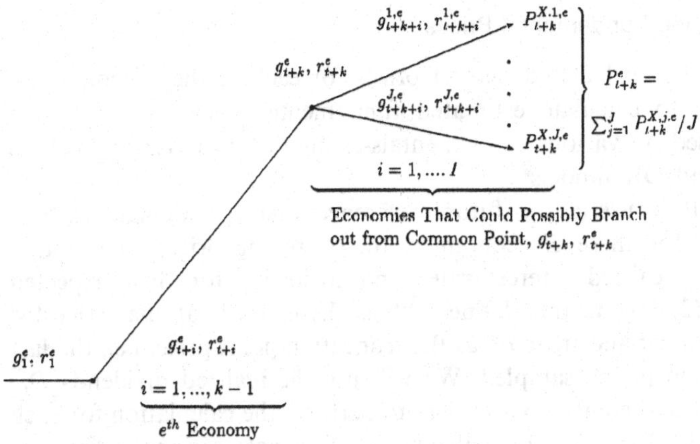

EXHIBIT 3 Diagram of the *e*th economy of *E* economies and calculation of the *e*th economy market price at time $t+k$ [Step 2(e)].

superscripted with a j are required only to perform the numerical integration that yields P^e_t.

Step 2(e): Perform Steps 1(a)–(c) and 2(a)–(d) for $t = 1, \ldots, T$, rolling out E independent economies for T periods. The length of the time series T is chosen to be 47 to imitate the 47 years of annual data we have available from the S&P 500 from 1952–1998.* This produces D^e_t, r^e_t, and P^e_t, for $t = 1, 2, 3, \ldots, T$ and $e = 1, 2, 3, \ldots, E$. We can also construct the market returns for these economies:

$$R^e_t = (P^e_{t+1} + D^e_{t+1} - P^e_t)/P^e_t$$

and the equity premium, π^e, that agents in the *e*th economy would observe. The equity premium satisfies the pricing-kernel condition:†

$$E\left\{\frac{1 + R^e_t}{1 + r^e_t + \pi^e}\right\} = 1$$

Exhibit 3 summarizes this procedure by which we calculate market prices for the E economies in Step 2(e).

*To avoid initial conditions contaminating the simulations, the dividend growth rates and interest rates are simulated for over 100 periods before dividends and prices are calculated.

†As with determining the equity premium for the S&P 500 data, this requires a nonlinear estimation problem to be solved and was accomplished with a grid search. The equity premium is restricted to be positive and thus is set to 0 if $\pi^e < 0$.

3.2.2 Calculating Fundamental Prices

Now that we have calculated market prices for each of the E economies, we next move to calculate estimated fundamental prices for the same economies based on various fundamentals-estimation procedures, such as the Gordon and DK models.

Step 3: It is useful for future reference first to calculate ex-post rational prices for these E economies; these are the prices that obtain by substituting realized interest rates and dividends for their expected values in Eq. (2). To do this define $P_t^{X,e}$ as the ex-post rational price for economy e at time t and treat P_T^e as the truncation point price, i.e., the last price observation in the sample.* We will use the realized dividends D_t^e, interest rates r_t^e and equity premium π^e to perform the calculation for each economy $e = 1, 2, 3, \ldots, E$. This calculation of ex-post rational prices for each of the E economies and $t = 1, \ldots, T$ time periods produces E *time series* of ex-post rational prices produced by our simulations.

Step 4: We must also form Gordon-model prices for each of these E economies. That is, for each of the $e = 1, 2, 3, \ldots, E$ simulated economies for which market prices were calculated in the previous section, we now calculate a fundamental price using the Gordon model.

Step 4(a): Set $\bar{g}^e = \sum_{t=1}^{T} g_t^e/T$ and $\bar{r}_*^e = \sum_{t=1}^{T} r_t^e/T + \pi^e$, and form the Gordon price:

$$P_t^{G,e} = D_t^e\left[\frac{1+\bar{g}^e}{\bar{r}_*^e - \bar{g}^e}\right]$$

Step 4(b): In addition to the classic Gordon model, there have been a variety of extensions to make the Gordon model more realistic by allowing dividend growth rates to vary over time. We implement the trinomial dividend models of Yao (1997), the additive Markov model, and the geometric Markov model specified in Eqs. (8) and (9).

The additive model has us estimating for each $e = 1, 2, 3, \ldots, E$ economy the average absolute value of the change in the dividend payment. $\Delta^e = \sum_{t=2}^{T} |D_t^e - D_{t-1}^e|/(T-1)$, the proportion of the time the dividend increases, $q^{e,u}$, and the proportion of the time the dividend decreases, $q^{e,d}$, in order to form the price estimate:

---

*In the literature (e.g., Shiller, 1981), when calculating ex-post rational prices, a truncation point is chosen to be either fixed, at say the last observed market price, or to be a moving point 20 or more years ahead of the date for which we are calculating the price. With a fixed date, the ex-post converges to the market price at the terminal date.

$$P_t^{ADD,e} = D_{t-1}^e/\bar{r}_*^e + [1/\bar{r}_*^e + (1/\bar{r}_*^e)^2](q^{e,u} - q^{e,d})\Delta^e$$

The geometric model has us estimating for each economy $e = 1, 2, 3, \ldots, E$ the average absolute value of the percentage change in the dividend payment, $\Delta^{e,\%} = \sum_{t=2}^{T} |(D_t^e - D_{t-1}^e)/D_{t-1}^e|/(T-1)$, to form the price estimate:

$$P_t^{GEO,e} = D_{t-1}^e \left[\frac{1 + (q^{e,u} - q^{e,d})\Delta^{e,\%}}{\bar{r}_*^e - (q^{e,u} - q^{e,d})\Delta^{e,\%}} \right]$$

Step 5: Finally, we also consider the Donaldson and Kamstra (1996) procedure, which focuses on the ratio of dividend growth to discount rates, $\log(y_t) \equiv \log((1+g_t)/(1+r_t+\pi))$, rather than on dividend growth and interest rates separately. The DK procedure calls for evolving many possible streams of y's into the future with a Monte Carlo simulation and then taking the present value as in Eq. (10). Donaldson and Kamstra (1996) argue that y_t is better behaved than g_t or discount rates alone and that it makes more sense to forecast y_t since this is the object of investor interest in Eq. (6). In other words, investors care about the ratio of gross dividend growth to the gross discount rate, not about each variable individually, so it makes more sense to forecast the ratio y. The DK procedure is described below and summarized in Exhibit 4.

Step 5(a): To implement the DK procedure, we start by estimating a model for $\log(y)$ that captures its dynamic evolution over time. For each of the $e = 1, 2, 3, \ldots, E$ simulated economies for which market prices were calculated in the previous section, we now use the Bayesian information criterion to select the "best" model from the set of ARMA (p, q) models, $(p, q) = (1,0)$, $(1,1)$, and $(2,0)$ for $\log(y_t)$. We find that $(p, q) = (1,0)$ is

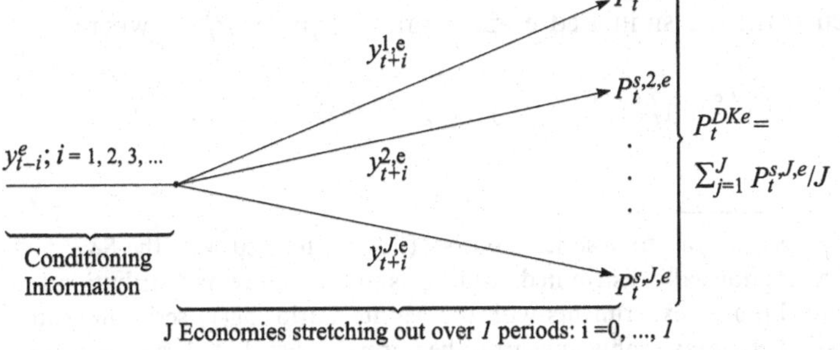

EXHIBIT 4 Diagram of DK Monte Carlo integration of the eth economy.

usually, but not always, chosen. For each of the E economies we then use the estimated model to make conditional mean forecasts $\widehat{\log(y_t^e)}$, $t=1,\ldots,T$, conditional on only data observed before period t for that economy. We also estimate the sample standard deviation ($\hat{\sigma}^e$) of the residual of the model within each economy. With $E=1000$ economies we would do this step 1000 times, once for each of the $e=1,\ldots,E$ economies.

Step 5(b): Now simulate discounted dividend growth rates for each of the $e=1,2,3,\ldots,E$ simulated economies. That is, produce $\log(y^e)$ that might be observed in period t in economy e given what is known at period $t-1$ in economy e. To do this for a given period t and economy e, simulate a population of J independent possible shocks (draws from a normal distribution,* mean 0, and standard deviation equal to $\hat{\sigma}^e$), which we label $\varepsilon_t^{j,e}$, $j=1,\ldots,J$, and add these shocks separately to the conditional mean forecast $\widehat{\log(y_t^e)}$ from Step 5(a) so as to produce $\log(y_t^{j,e})=\widehat{\log(y_t^e)}+\varepsilon_t^{j,e}$, $j=1,\ldots,J$. This is a simulated cross-section of J possible realizations of $\log(y_t^e)$ considered at time $t-1$ for economy e, i.e., different paths that economy e may take next period. Note here that we are performing a Monte Carlo simulation on each of the e economies that were themselves generated by a Monte Carlo simulation, so that if we generate $E=1000$ economies in Steps 1–2, and $J=1000$ economies in Step 5, we are in total generating one million economies for the DK simulation (which is precisely what we have done for the investigations in this chapter).

Step 5(c): Next, for each economy e, use the estimated model from Step 5(a) to make the conditional mean forecast $\widehat{\log(y_{t+1}^{j,e})}$, conditional on only the jth realization for period t, $\log(y_t^{j,e})$, and $\varepsilon_t^{j,e}$, and the data known at period $t-1$, and simulate a population of J independent shocks $\varepsilon_{t+1}^{j,e}$, $j=1,\ldots,J$ as in Step 5(b) to form $\log(y_{t+1}^{j,e})$.

Step 5(d): Repeat Step 5(c) to form $\log(y_{t+2}^{j,e})$, $\log(y_{t+3}^{j,e})$, $\ldots$, $\log(y_{t+I}^{j,e})$ for each of the J economies, where I is the number of periods into the future the simulation is run ($I=400$) in our simulations, as explained above). Then form the simulated ex-post rational price, $P_t^{s,j,e}$, where

$$P_t^{s,j,e}=D_t^e(y_t^{j,e}+y_t^{j,e}y_{t+1}^{j,e}+y_t^{j,e}y_{t+1}^{j,e}y_{t+2}^{j,e}+\cdots):\quad j=1,\ldots,J$$

*The regression error from time-series models of $\log(y)$, formed with the S&P 500 and T-bill data, are normally distributed, leading us to this choice of distribution for the shocks here. Limited experiments with the Monte Carlo $\log(y)$ series indicates that the regression error invariably appears to be normally distributed. As y is a ratio of log normal random variables, this is not surprising.

corresponding to the J possible economies based on the previously simulated economy e.

Step 5(e): Calculate the DK fundamental price for each time period $t=1,\ldots,T$ and each economy $e=1,2,3,\ldots,E$:

$$P_t^{DK,e}=\sum_{j=1}^{J}P_t^{s,j,e}/J$$

The DK procedure outlined above is represented diagrammatically in Exhibit 4. For the experiments in this chapter we set $E=1000$, $J=1000$, and $I=400$.*

3.2.3 Sensitivity of the Monte Carlo Results

Careful analysis of any Monte Carlo simulation must include a discussion of the simulation error itself. In a world of unlimited resources, the simulation error can be driven down to negligible scales by increasing the number of replications, in our case increasing the number of simulated economies, E, from 1000 to several million, and increasing the fans, J, used in the calculation of each economy's market price from 1000 to several million. This chapter's Monte Carlo experiment involves a simulation of a numerical integration (to calculate market prices) as well as the method of Donaldson and Kamstra (1996) which is also a numerical integration (to calculate the DK prices). The scale of the simulation quickly reaches frightening proportions as we increase the number of replications (economies), E, or fans, J, in the numerical integrations. The choice of 1000 economies, each stretching out for 47 years, and then the choice of 1000 fans used to calculate the market prices for each year of each economy, led to roughly one month of CPU time on an SUN UltraSparc II 400 MHz machine, the practical limit for this experiment given competing demands for this resource.

To determine the simulation error, we must conduct a simulation of the simulations. Unlike some Monte Carlo experiments (such as those estimating the size of a test statistic under the null) the standard error of the simulation error for most of our estimates (returns, prices, etc.) are themselves analytically intractable, and must be simulated. In order to

*Less than 2% of the simulations (economies) yielded DK models that were not stable, or that produced rollouts that were not stable, and were excluded from the analysis. These economies were not otherwise remarkable, and the majority of the experimental results we present are qualitatively unchanged whether these simulations are included or not.

estimate the standard error of the simulation error in estimating market prices, we estimated a single market price 1000 times, each time independent of the other, and from this set of prices computed the mean and variance of the price estimate. If the experiment had no simulation error, each of the 1000 price estimates would be identical. With the number of fans, J, equal to 1000 we find that the standard deviation of the simulation error is only 0.28% of the price, which is sufficiently small as to not be a source of concern for our study.*

4 RESULTS OF THE MONTE CARLO EXPERIMENTS

We now explore results based on our simulated market economies and the fundamental prices calculated from them. The first set of results focus on summary statistics for returns, dividend–price (D/P) ratios, dividend growth rates, interest rates, and risk premia produced by our simulated economies as compared to data from real-life financial markets. The second set of results focuses on tests of bubbles in the simulated economies, as well as on some related statistics.

Recall that no experiment presented here is calibrated to actual S&P 500 *prices*. All experiments are calibrated only to dividend growth rates and discount rates, and then prices are calculated based on the underlying fundamental factors. Great care was taken to ensure that our experiments faithfully replicated the mean, variance, cross-correlation, and autoregressive structure of log dividend growth rates and log discount rates. If S&P 500 prices contain bubbles, we would expect that S&P 500 returns and D/P ratios would be excessively volatile relative to our simulated economies, which do not themselves contain bubbles. Furthermore, since we have no bubbles in our simulated economies, the properties of the fundamental price estimates and the ex-post rational price under the null of no bubbles can be obtained from our simulations. This allows us to investigate the size properties of tests for bubbles based on these bubble-free price estimates.

Recall from the discussion above that we have 47 years of data from the S&P 500 and that each of our 1000 simulated economies therefore also produce 47 years of data. For each of the 1001 economies under

*We further investigated the sensitivity of selected results by increasing the number of simulated economies, E, or the number of fans, J, to as large as 10,000 (restricting to 2 years the number of periods each economy was followed out), but found no qualitative changes to our experimental results. Of course, this did reduce the simulation error by a factor of square root 10, but was not of much practical gain since the simulation error was already extremely small.

investigation (one actual economy and 1000 simulated economies) we calculate the mean, standard deviation, skewness, and kurtosis of a variety of interesting variables, including dividend growth rates, interest rates, stock returns, and D/P ratios. For example, we calculate the mean return of the 47 years of the actual S&P 500, which appears in the top-left cell of the results shown in Table 1. We also calculate the mean return for each of our 1000 simulated economies, which produces 1000 mean return estimates, the distribution of which is summarized in the remaining cells of the first row of results shown in Table 1 (the *X*th percentile reports the return for the economy with the *X*th largest return).

We begin our analysis of the results in Table 1 by considering results for dividend growth rates and discount rates. Given that we calibrated the discount rates and dividend growth rates to the true data, we would expect that discount and dividend growth rates from our simulated economies would appear quite similar to the true market data. And indeed, we see

TABLE 1 Statistics on the S&P 500 Index and Simulated Market Index

		Data					
			Mean of simulated data	Percentiles of simulated data			
Statistic		S&P 500		1%	5%	95%	99%
Returns	Mean	0.11571	0.10065	0.07300	0.07984	0.12517	0.14214
	Std. dev.	0.13539	0.09137	0.05361	0.06057	0.13068	0.14969
	Skewness	−0.39741	0.13638	−0.66906	−0.48990	0.71935	1.29074
	Kurtosis	2.75816	3.02994	1.87952	2.09439	4.49553	6.92236
D/P ratios	Mean	0.03629	0.04872	0.03321	0.03654	0.06724	0.08030
	Std. dev.	0.00946	0.00955	0.00266	0.00364	0.02062	0.02653
	Skewness	0.44283	0.71078	−0.31723	−0.00336	1.64431	2.11081
	Kurtosis	2.68087	3.12383	1.51714	1.81999	5.77567	8.06559
Dividend growth rates	Mean	0.05513	0.05517	0.03731	0.04262	0.06746	0.07141
	Std. dev.	0.03649	0.03680	0.02651	0.02918	0.04435	0.04828
	Skewness	0.25694	0.09863	−0.71792	−0.45213	0.66111	1.03167
	Kurtosis	3.29200	2.73179	1.82880	2.00100	3.82972	4.51803
Interest rates	Mean	0.05887	0.06008	0.01588	0.02391	0.11970	0.16291
	Std. dev.	0.02981	0.02984	0.00626	0.00949	0.06851	0.08978
	Skewness	0.95245	0.86858	−0.16511	0.12500	1.82812	2.30151
	Kurtosis	3.69054	3.42080	1.55605	1.85578	6.74406	8.93784
Risk premia	Mean	0.05774	0.05102	0.00000	0.01141	0.07599	0.08382

means, standard deviations, skewness, and kurtosis for our simulated interest rates and dividend growth rates very nearly identical to the true interest rate and dividend growth rate data.* Furthermore, each and every statistic for dividend growth rates and interest rates lies near the center of the distribution of the simulated series. Given that we generated dividend growth rates and interest rates to match the true data this is not surprising and is only presented as a verification check. The true challenge, which may provide some insight into the existence of bubbles in the S&P 500 data, is to see if S&P 500 returns and prices also look like returns and prices from economies that share dividend and interest rate characteristics with the true economy, but have been simulated to have no bubbles.

From the first row of Table 1 we see that, on average, our simulated economies produce mean returns very close to returns from the actual S&P 500. Indeed, the S&P 500's return lies near the middle of the distribution of returns from our simulations. The skew and kurtosis of S&P 500 returns also fall well within the 90% confidence interval formed by the simulated economies. Only the standard deviation of market returns is (barely) outside the 90% confidence interval formed by our simulated economies. Price levels themselves can of course not be compared directly since price is a random walk and thus the unconditional mean, standard deviation, skewness, and kurtosis do not even exist. Financial studies therefore typically focus on the D/P ratio, since prices and dividends are cointegrated. From Table 1 we see that the D/P ratio of the S&P 500 is similar to our simulated data, with all statistics measured for the S&P 500 data within a 90% confidence interval formed by the simulated economies' statistics (with the sole exception of the mean D/P ratio, which is outside the 90% confidence interval but within the 98% confidence interval, the 5 percentile range). Again, similar to returns, the S&P 500 D/P ratios do not stand out as wildly unusual, given the behavior of returns we see in no-bubble economics.

In sum, the results for dividend growth rates, interest rates, returns, and D/P ratios shown in Table 1 suggest that the behavior of the true S&P 500 stock index over the past 50 years is not out of line with how we might expect a financial market to behave if that market priced stocks in a rational, efficient, and bubble-free manner, given interest rate and dividend fundamentals. What other studies have argued to be anomalies in price behavior, such as volatile return series, seem in our study to be entirely

*For instance, the S&P 500 dividend growth rates have a slight positive skew of 0.257, while the 5 and 95 percentiles of the simulated series of dividend growth rates are −0.45 and 0.66, respectively.

consistent with prices based on dividend and interest rate fundamentals. The key difference between our investigation and previous work is our calibration to the dividend and interest rate characteristics of the economy with no simplifying assumptions when simulating market prices, returns, and various statistics and their distributions derived from these prices and returns. Conversely, previous studies, even those that have used simulation techniques, can only make (educated) guesses at what the empirical distribution of various tests and statistics might be without the simplifying assumptions imposed. To the extent that these guesses and simplifying assumptions (e.g., independent dividend innovations) are inaccurate, or to the extent that small sample properties of the financial time series are not accounted for, misleading results can be obtained and "anomalies" can be found where none truly exist. Thus, the approach we adopt leads to a conclusion opposite to many studies in the literature, starting with Shiller (1981), which conclude that stock prices are excessively volatile relative to fundamentals.

To investigate further the summary statistics reported in Table 1, we provide a series of figures that show the bivariate distributions of the economy-by-economy statistics that went to comprise Table 1. The S&P 500 economy could conceivably be relatively usual compared to the simulated economies, statistic by statistic, but have a very unusual combination (pairing) of statistics, which would become obvious from a bivariate plot of these statistics for all the economies (e.g., the standard deviation of returns and the mean of returns for the S&P 500 might each appear similar to simulated data, but the mean-standard deviation combination seen in the S&P 500 data might be very different from the mean-standard deviation combinations produced by the simulations). The plots in Figs. 1–5 display data points for each of our simulated economies and for the S&P 500, with the actual S&P 500 data represented by a large black circle and the simulated data by smaller circled points (which can, on occasion, appear in the plots to be small black dots because of several circles partially overlying each other).*

Figure 1 presents standard deviations versus means in a group of four panels for returns (Panel A), D/P ratios (Panel B), dividend growth rates (Panel C), and interest rates (Panel D). We would expect the dividend growth rate and interest rate plots to have the actual S&P 500 data in the center of the cloud of points associated with the simulated

*For presentation, data points that completely (or very close to completely) overlay each other were condensed down to a single point.

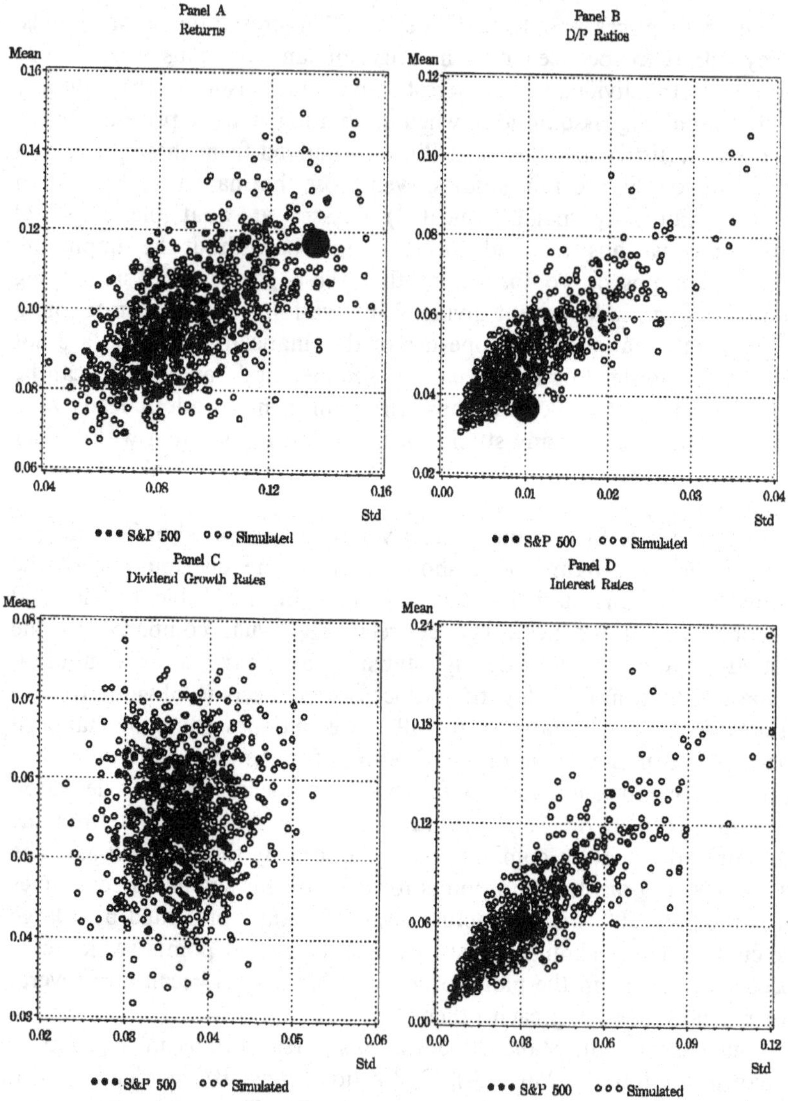

FIGURE 1 Mean versus standard deviation.

economies since we calibrated the experiments to these two quantities, and this is indeed what we observe in Fig. 1, Panels C and D. As for returns in Panel A, we see that economies generated by calibrating to only discount rates and dividend growth rates produce mean/variance pairs for returns that do not look terribly dissimilar to the S&P 500 return data. The plot of

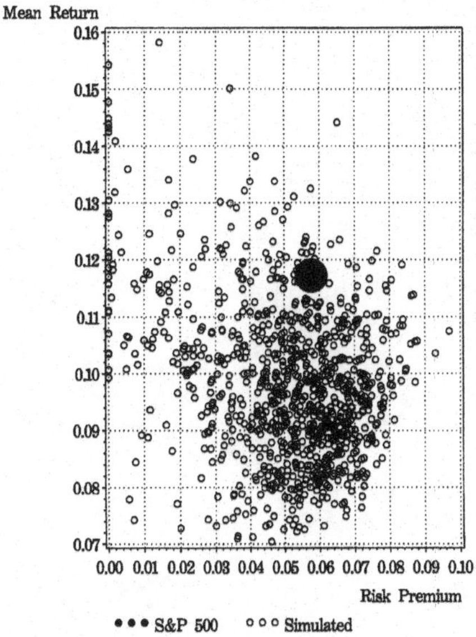

FIGURE 2 Return mean versus risk premium.

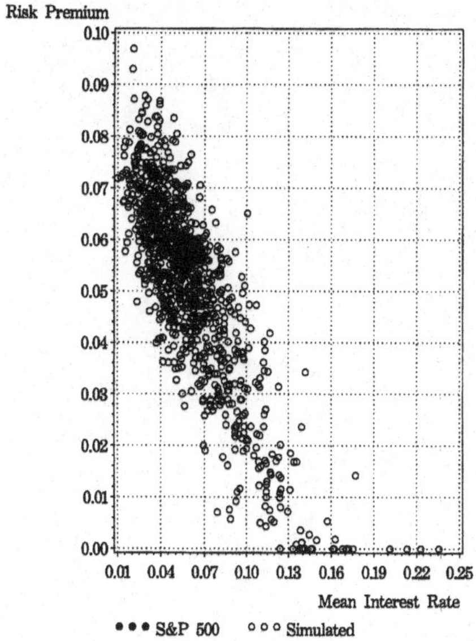

FIGURE 3 Interest rate mean versus risk premium.

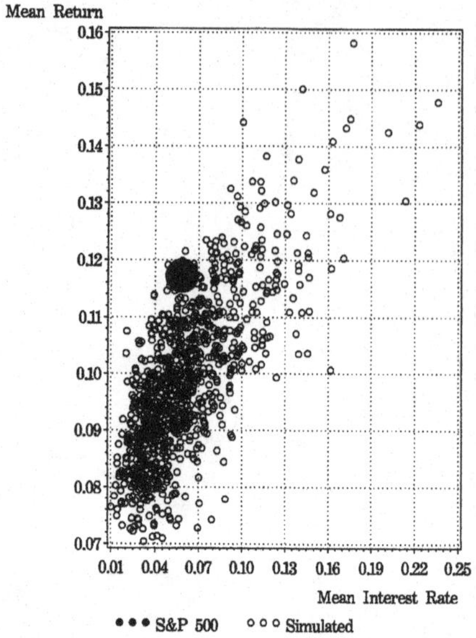

FIGURE 4 Return mean versus interest rate mean.

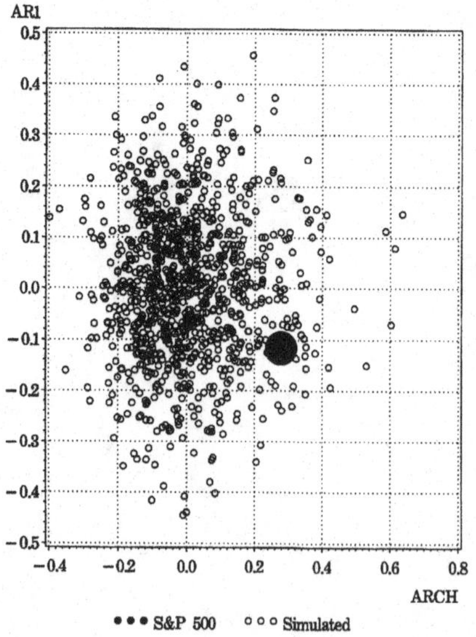

FIGURE 5 Return AR(1) coefficient versus ARCH(1) coefficient.

the D/P ratios means versus standard deviations in Panel B also shows that the first two moments of the S&P 500 data are not entirely out of the bounds of possibility, although the S&P 500 data do stray to the very edge of the joint distribution of means and standard deviations for the D/P ratios from the simulated economies. In other words, prices in the S&P 500 data seem higher relative to dividends than we typically see in our simulated economies. This suggests that, while the S&P 500's D/P ratio is not low enough to claim with any degree of confidence that current market prices contain a bubble of irrational exuberance, our evidence does suggest that actual market prices do seem rather high relative to what we observe in our simulated economies. In other words, S&P 500 stock prices may indeed be suspiciously high as some people suggest, but they are not so high as to allow us to conclude that they are not being driven by fundamentals.

Turning our attention back to Table 1, in the last row of Table 1 we present equity premia from the true data and equity premia from our simulated economies. Note that since we have only one equity premium value per economy we cannot present a standard deviation, skew, or kurtosis measure for the S&P 500 data; there is only the one equity premium estimate of 5.77%. Similarly, in each of the simulated economies we have a single measure of the equity premium and thus cannot present a standard deviation, skew, or kurtosis measure for the simulated equity premia either. However, we do see from Table 1 that the risk premia possible in a simulated economy, with dividend and interest rate processes set to mimic actual dividend and interest rate processes, can range from 0 to over 8%, even if the true equity premium is below 6%. This suggests that the equity premium we see in the actual S&P 500 data could be an imprecise estimate of the market's true underlying risk premium and leads us to wonder whether findings of so-called anomalies, such as the Mehra and Prescott (1985) "equity premium puzzle," are really puzzling, or whether such findings are simply small sample flukes.*

To examine further the equity premium issue, we plot in Fig. 2 the relationship between stock returns and the equity premium. Recall that the actual S&P 500 data are represented by a large black circle and the simulated data by smaller circled points. Although the high equity premium of the last half-century has been much discussed, and argued to be excessive, we see that the distribution of possible equity premia and

*For an initial identification of the equity premium puzzle see Mehra and Prescott (1985). This seminal work spawned a significant body of research on the equity premium puzzle, much of which is reviewed in Campbell *et al.* (1997).

mean market returns over a 47-year period is quite spread out and that the S&P 500's experience is not terribly unusual, at least if one allows discount rates and dividend growth rates to be autocorrelated random variables as we do. In fact, in a bubble-free economy with fundamentals (i.e., dividend growth rates and discount rates) calibrated to match the true economy, it is possible to observe measured equity premia very nearly equal to mean returns, even when we would expect that over hundreds of years the mean return would converge to 11% (the mean return of the S&P 500 over the past 47 years) and the mean equity premium to 6% (the equity premium of the S&P 500 over the past 47 years). The variance of the S&P 500 discount rates and market returns is so large that there is very little we can draw from the so-called equity premium puzzle in the short time periods available to us in the true economy.

Some authors have argued that the equity premium is "too large," not because stock returns are too high, as originally argued, but rather because risk-free interest rates are too low (Kocherlakota, 1996). Figure 3 therefore presents mean interest rates versus the equity premia. Again we find the experience of the true data over the last 47 years to be unremarkable relative to our simulated economies. It is not uncommon to find in our simulations very low mean interest rates and high risk premia. In fact, low average interest rates are virtually a prerequisite to finding high equity premia, as the sum of the two is restricted to add up (approximately) to the mean return, and the mean return rarely strays above 14% over a 47-year period, according to our simulations.

For completeness, in Fig. 4 we present mean returns versus interest rates. This combination shows the actual S&P 500 data to have a relatively high return relative to the interest rate, but even in this case the S&P 500 data is within the envelope of the simulated economies.

For all the data and statistics we explore, including results on returns, D/P ratios, equity premia, etc., the S&P 500 data do not seem to be wildly unusual among economies we have generated. These economies have been generated to be bubble-free and calibrated to the S&P 500 dividend and discount rate experience. Taken together, this evidence therefore suggests that the S&P 500 index does not contain price bubbles. Given this evidence that the stock market does not contain bubbles, why is the financial econometric literature replete with tests for bubbles that in general support the argument that stock prices are excessively volatile? It is to this question that we now turn.

Table 2 presents the results of four common statistical tests for stock price bubbles, presented in rows 1–4 of the table, with the columns indicating the pricing method (the ex-post rational price and four different fundamental price estimates) the test was based on. Each test for bubbles

TABLE 2 Statistics and *p* Values from Time-Series Tests for Bubbles, Fads, and Excess Volatility

	Fundamental forecasting model				
Statistic or p-value ↓	Ex post	Gordon growth	Gordon additive	Gordon geometric	DK
Camerer test[a]	1.00000	1.00000	1.00000	1.00000	0.59021
MRS1 test[a]	—	0.28746	0.05505	0.24873	0.21305
MRS2 test[a]	—	0.31091	0.32314	0.35474	0.27013
Campbell-Shiller test	0.31600	0.10296	0.02446	0.13354	0.21101
Mean % Error[b]	0.07682	0.08475	0.17549	0.00187	−0.06629
% Error (σ)	0.17728	0.17842	0.34676	0.17842	0.15557

[a] *p* Values are from a paired-sample *t*-test, applied as follows. Let $x_{1,t}$ and $x_{2,t}$ be the two series we are interested in testing for difference in variance, let x_1 and x_2 be the sample means of $x_{1,t}$ and $x_{2,t}$, respectively, and define $z_t \equiv (x_{1,t}-x_1)^2-(x_{2,t}-x_2)^2$, i.e., the difference in two paired samples. Then, under standard regularity conditions, a test of the null-hypothesis that the first moment of z_t is greater than 0 can be interpreted as a test of $x_{1,t}$ having greater variance than $x_{2,t}$ and implemented as a standard paired-sample *t*-test for the mean of z_t being greater than 0. The test statistic has an asymptotic normal distribution under fairly weak conditions, permitting non-normality as well as some dependence and heterogeneity [see Lehmann (1975) and Gastwirth and Rubin (1971)].

[b] Percentage error calculated as price forecast minus market price all divided by price forecast.

was applied at the 10% significance level for each pricing method (where applicable). In Table 2 we present the actual rejection rates across our simulated economies for each of these tests and for each of the fundamental price estimation techniques. If each test is well-behaved, i.e., has proper size, and if these fundamentals estimation techniques are reliable guides to the true fundamental value, then we would expect approximately 10% rejection rates in each entry for the first four rows of the table, as all these tests are being applied under the null of no bubbles.*

From the first four rows of Table 2 we see that the most commonly applied tests for bubbles, the Camerer test and the two MRS tests, are

*The rejection rate we calculate with these simulations is a random variable, centered on 10% with a standard deviation derived from the binomial distribution, equal to the square root of 0.10 times 0.90 divided by the number of simulated economies, or 0.0095 approximately. Thus, we would expect the rejection rates to differ from 10% by no more than approximately 0.02 at the 5% level of significance.

extremely unreliable. If the tests have correct size, the *p*-value entries in Table 2 would all be 0.10. However, in all but one case the empirical *p* values far exceed their theoretical values, which implies that these tests typically grossly over-reject. In other words, common tests for stock price bubbles find bubbles where none actually exist. Indeed, the Camerer test finds bubbles 100% of the time using ex-post and Gordon fundamental price series and 59% of the time using the DK fundamental price series, even though the market price series does not contain bubbles. Only the Campbell–Shiller test procedure (the Dickey–Fuller unit root test) comes close to having the correct size.

The last two rows of Table 2 report results on the fundamental pricing error, defined as the fundamental price minus the simulated market price, all divided by the simulated market price. Of all the fundamentals estimation techniques, the DK procedure is the most reliable forecaster of market prices, with the smallest forecast error variance and among the smallest mean errors. The superiority of the DK procedure over other simpler methods is not very great, however. All the fundamental price estimation methods we employ suffer from the same basic problem, which is the difficulty of estimating key parameters (the mean dividend growth rate and discount rate in the case of the basic Gordon growth method, the parameters of the ARMA process for discounted dividend growth rates in the case of the DK method, etc.). Clearly, based on the results of Table 2, there is scope for improvement in fundamentals estimation techniques.

Finally, we explore derived statistics from our experiments, including statistics such as the sample autocorrelation at one lag in market stock returns and the ARCH effect in stock returns. The fashion in which our data are constructed should lead to returns that are independent and identically distributed, but the distribution of the sample autocorrelation at one lag and the ARCH(1) coefficient estimate may not be well approximated with the usual asymptotic distributions. Figure 5 presents the universe of pairs of AR(1)–ARCH(1) coefficient estimates from our simulated economies' time series of returns and also plots the coefficient estimates from the S&P 500 returns data. From Fig. 5 we see that, once again, the S&P 500 data do not appear wildly at odds with the cloud of data points from our simulated economies. Furthermore, we see from our simulated economies that by random chance we can find economies that exhibit extremely strong annual ARCH effects, with the ARCH(1) coefficients achieving values ranging from −0.4 to 0.6, and equally strong autoregressive effects, with positive and negative magnitudes of roughly 0.50 on the AR(1) coefficient, even though neither of these effects are actually present in the data.

5 CONCLUSIONS

The purpose of this chapter has been to use advanced simulation and Monte Carlo techniques to learn more about the behavior of fundamental stock prices and the properties of statistical tests commonly used to test for price bubbles in stock market data. We produce simulated stock price series based on dividend growth and discount rate fundamentals so as to simulate rational, efficient, and bubble-free economies that share dividend and interest rate characteristics with the true economy. While much conventional wisdom suggests that stock prices are "too volatile" and may be influenced by the "irrational exuberance" of traders, our evidence does not support this conclusion. On the contrary, we find that actual stock prices do not behave significantly differently from our simulated prices from rational, efficient, and bubble-free economies. We also find that traditional tests for excess volatility and bubbles over-reject the null of no bubbles in samples of the size traditionally employed in the literature. Indeed, most tests we investigate find overwhelming evidence of price bubbles even through we have constructed that data to have no bubbles. In other words, we demonstrate that traditional tests for bubbles frequently find bubbles in data that do not in fact contain bubbles.

The key difference between our investigation and previous work is our calibration to the dividend and interest rate characteristics of the economy with no simplifying assumptions when simulating market prices, returns, and various statistics and their distributions derived from these prices and returns. Conversely, previous studies, even those that have used simulation techniques, can only make (educated) guesses at what the empirical distribution of various tests and statistics might be without the simplifying assumptions imposed. To the extent that these guesses and simplifying assumptions (e.g., independent dividend innovations) are inaccurate, or to extent that small sample properties of the financial time series are not accounted for, misleading results can be obtained and "anomalies" can be found where none truly exist. While more work remains to be done in this area, we believe that the results of this chapter at least raise some concerns with some previously employed approaches and suggest an avenue for further investigation.

ACKNOWLEDGMENTS

We thank Lisa Kramer for helpful comments and the Social Sciences and Humanities Research Council of Canada for financial support. The views expressed are those of the authors and not necessarily those of the Federal Reserve Bank of Atlanta or the Federal Reserve System.

REFERENCES

Bollerslev, T. and Hodrick, R.J., 1995. Financial market efficiency tests. In: M.H. Pesaran and M.R. Wickens, eds. *Handbook of Applied Econometrics*, vol. 1. Blackwell, Oxford, pp. 415–458.

Brooks, R. and Helm, B., 1990. An *n*-stage fractional period, quarterly dividend discount model. *Financial Review* **25**, 651–657.

Camerer, C., 1989. Bubbles and fads in asset prices. *Journal of Economic Surveys* **3**, 3–38.

Campbell, J.Y., 1991. A variance decomposition for stock returns. *Economic Journal* **101**, 157–179.

Campbell, J.Y. and Shiller, R.J., 1987. Cointegration tests of present value models. *Journal of Political Economy* **95**, 1062–1088.

Campbell, J.Y. and Shiller, R.J., 1998. Valuation ratios and the long-run stock market outlook. *Journal of Portfolio Management* **24**, 11–26.

Campbell, J.Y., Lo, A.W. and MacKinlay, A.C., 1997. *The Econometrics of Financial Markets*. Princeton University Press, Princeton, NJ.

Cochrane, J. 1992. Explaining the variance of price–dividend ratios. *Review of Financial Studies* **5**, 243–280

Dickey, D.A. and Fuller, W.A. 1981. Likelihood ratio statistics for autoregressive time series with a unit root. *Econometrica* **49**, 1057–1072.

Donaldson, R.G. and Kamstra, M.J., 1996. A new dividend forecasting procedure that rejects bubbles in asset prices. *Review of Financial Studies* **9**, 333–383.

Donaldson, R.G. and Kamstra, M.J., 2000. Forecasting fundamental stock price distributions. Unpublished manuscript, Simon Fraser University, Burnaby, BC, Canada.

Engle, R.F., 1982. Autoregressive conditional heteroskedasticity with estimates of the variance of United Kingdom inflation *Econometrica* **50**, 987–1008.

Fuller, R.J. and Hsia, C., 1984. A simplified common stock valuation model. *Financial Analysts Journal* **40**, 51–54.

Gastwirth, J.L. and Rubin, H., 1971. The behavior of robust estimators on dependent data. Purdue University Statistics Mimeo, Number 197.

Gordon, M., 1962. *The Investment, Financing and Valuation of the Corporation*. Irwin, Homewood, IL.

Hodrick, R.J. 1992. Dividend yields and expected stock returns: Alternative procedures for inference and measurement. *Review of Financial Studies* **5**, 357–386.

Hull, J., 1993. *Options, Futures and Other Derivative Securities*. Prentice-Hall, Englewood Cliffs, NJ.

Hurley, W.J. and Johnson, L.D., 1994. A realistic dividend valuation model. *Financial Analysts Journal* **50**, 50–54.

Hurley, W.J. and Johnson, L.D., 1998. Generalized Markov dividend discount models. *Journal of Portfolio Management* **24**, 27–31.

Kennedy, P., 1992. *A Guide to Econometrics*. MIT Press, Cambridge, MA.

Kleidon, A.W., 1986. Variance bound tests and stock price valuation models. *Journal of Political Economy* **94**, 953–1001.

Kocherlakota, N.R., 1996. The equity premium: still a puzzle, *Journal of Economic Literature* **34**, 42–71.

Lehmann, E.L., 1975. *Non-Parameterics*. Holden-Day, San Francisco.

Malkiel, B.G., 1963. Equity yields, growth, and the structure of share prices. *American Economic Review* **53**, 1004–1031.

Mankiw, N.G., Romer, D. and Shapiro, M.D., 1985. An unbiased reexamination of stock market volatility. *Journal of Finance*. **40**, 667–687.

Mehra, R. and Prescott, E., 1985. The equity premium: A puzzle. *Journal of Monetary Economics* **15**, 145–161.

Orcutt, G.H. and Winokur, H.S., 1969. First order autoregression: Inference, estimation, and prediction. *Econometrica* **37**, 1–14.

Said, E.S. and Dickey, D.A., 1984. Testing for unit roots in autoregressive moving average models of unknown order. *Biometrika* **71**, 599–607.

SAS Institute Inc., 1999. SAS Procedures Guide, Version 8, Cary, NC: SAS Institute Inc.

Scott, L.O., 1985. The present value model of stock prices: regression tests and Monte Carlo results, *Review of Economics and Statistics* **67**, 599–605.

Shiller, R., 1981. Do stock prices move too much to be justified by subsequent changes in dividends? *American Economic Review* **71**, 421–436.

Shiller, R., 1989. *Market Volatility*. MIT Press, Cambridge, MA.

Timmermann, A.G., 1993. How learning in financial markets generates excess volatility and predictability in stock prices. *Quarterly Journal of Economics* **108**, 1135–1145.

Timmermann, A.G., 1995. Tests of present value models with a time-varying discount factor. *Journal of Applied Econometrics* **10**, 17–31.

West, K.D., 1988a. Bubbles, fads and stock price volatility tests: A partial evaluation. *Journal of Finance* **43**, 639–656.

West, K.D., 1988b. Dividend innovations and stock price volatility. *Econometrica* **56**, 37–61.

Yao, Y., 1997. A Trinomial dividend valuation model. *Journal of Portfolio Management* **23**, 99–103.

12

Neural Networks: An Econometric Tool

Johan F. Kaashoek
Econometric Institute, Erasmus University Rotterdam, Rotterdam, The Netherlands

Herman K. van Dijk
Econometric Institute, Erasmus University Rotterdam, Rotterdam, and Tinbergen Institute, Rotterdam, The Netherlands

1 INTRODUCTION

In recent decades one witnesses a substantial increase in the interest of econometricians for nonlinear models and methods. This is due to: (1) advances in the processing power of personal computers, (2) increased and successful research on algorithms for fast numerical optimization methods, and (3) the availability of large data sets. One of the nonlinear models that received much attention from applied researchers is a neural network, also known as a neural net. The basic idea behind a neural net is the tremendous data-processing capability of the human brain. Human brains consist of an enormous number of cells, labeled neurons. These neurons are connected and signals are transmitted from one cell to another through the connections. These connections are, however, not all equally strong. When a signal is transmitted through a strong connection it arrives more strongly in the receiving neuron. One may argue that there is a particular weight associated with each connection, which varies with the strength of the connection. Neurons may also receive signals from outside the brains. There are then transformed within the brains and returned to the outside world. The whole structure of signal processing between many (unobserved) cells can be described by a particular mathematical model that is therefore known as an *artificial* neural network model. For convenience we delete, henceforth, the qualification *artificial*. A more detailed description of the analogy between the mathematical neural network models and the working of the human brain is given by, e.g., Simpson (1990).

Neural networks are used in many sciences like biology, informatics, and econom(etr)ics. Within the latter field neural nets are, in particular, applied for the description and prediction of complex data patterns in economic time series. The field is very extensive and empirical illustrations are many. This chapter is not intended to give a complete survey. Instead, we start with a brief introduction on neural nets and their flexibility. Our focus is on the following two applications of neural network analysis with the aim of showing that neural nets are a convenient econometric tool:

1. Recovery of the unobserved dynamics; in particular, stability of a nonlinear system from a low-dimensional data set.
2. Specification of a neural network where a time-varying component is included.

In the first topic a neural net is used to recover the dynamic properties of a nonlinear system, in particular, its stability, by making use of the Lyapunov exponent. We use one simulated series from a structurally unstable chaotic model and some data from real exchange rates to illustrate the methods. Second, a two-stage network is introduced where the usual nonlinear model is combined with time transitions that may be handled by neural nets. The connection with time-varying smooth transitions models is indicated. The procedures are illustrated on a time-varying Philips curve using U.S. data from 1960–1997. We discuss connections with the existing literature, but refer for a general introduction to neural network to Hertz *et al.* (1991) and Bishop (1995) and the references cited therein.

2 A SIMPLE INTRODUCTION TO NEURAL NETWORKS

There exist many classes of neural networks, see, e.g., Hertz. *et al.* (1991) and Bishop (1995). In this chapter we restrict attention to a simple class that is known as the three-layer feed forward neural network, also labelled the Rumelhart–Hinton–Williams multilayer network after Rumelhart *et al.*

(1986). For expository purpose we describe and interpret this network as a generalization of the well known linear model from basic econometrics, see, e.g., Theil (1971, Chap. 3). Suppose that the cells of the network are partitioned into particular groups or layers and suppose further that there exist three such layers: the "input" layer, the "hidden" layer, and the "output layer." The cells of the input layer correspond to the "regressors" or "explanatory variables" in the standard linear regression model. The cell in the output layer correspond to the dependent variables in the linear model. The hidden layer contains cells that transmit the signals from the input layer to the output layer. These cells may be interpreted as unobserved components built into the linear model. A graph of a neural network with three cells in the input layer, two cells in the hidden layer and two cells in the output layer is shown in Fig. 1.

The network transmits signals as follows. A weighted sum of the signals of the input cells are sent to the hidden layer cells. Within the cells of this layer the values of the signals received are transformed by a so-called "activation function." A weighted sum of the transformed signals is then sent to the cells of the output layer. We note that the weights in the neural network correspond to unknown parameters in the linear model.

Henceforth, we make use of the following (standard) notation. A neural network with I cells in the input layer, H cells in the hidden layer, and O cells in the output layer is denoted as $nn(I,H,O)$. In Fig. 1 the network is given as $nn(3,2,2)$.

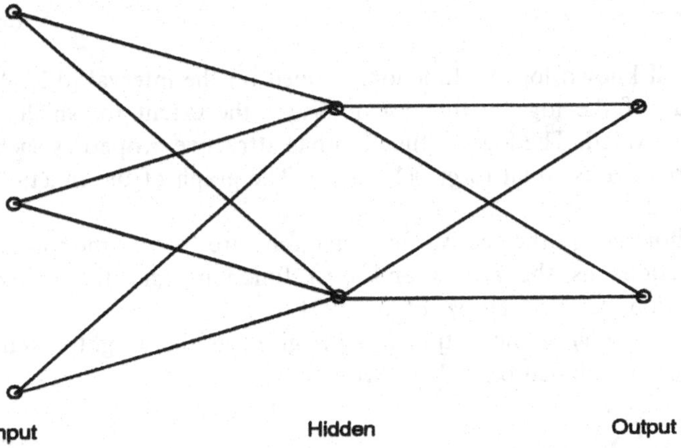

FIGURE 1 Graph of a neural network.

Next, we discuss the mathematical structure of a neural net. We make use of the following notation for cells, signals, and weights:

i	index of input cells, $i = 1, \ldots, I$
h	index of hidden layer cells, $h = 1, \ldots, H$
j	index of output cells, $j = 1, \ldots, O$
$g(\cdot)$	activation function
x_i	value of input cell i
y_j	value of output cell j
a_ih	weight of the signal from input cell i to hidden cell h
b_h	constant input weight for hidden cell h
c_jh	weight of the signal from hidden cell h to output cell j
d_j	constant weight for output cell j

The value of the signal that arrives in hidden cell h is given as

$$\text{Input for hidden cell } h = \sum_{i=1}^{I} (a_{ih} x_i) + b_h \tag{1}$$

The hidden cell h transforms the value of this signal with the activation function $g(\cdot)$ as follows:

$$\text{Output from hidden cell } h = g\left(\sum_{i=1}^{I} (a_{ih} x_i) + b_h \right) \tag{2}$$

where the activation function is a monotone increasing and bounded function given as

$$g(x) = \frac{1}{1 + e^{-x}} \tag{3}$$

which is the well known logistic function, defined on the interval [0,1]. A particular value of the logistic function indicates the extent to which a hidden cell is activated. The logistic function has attractive properties such as that the derivative is equal to $g(x)(1 - g(x))$. The graph of the function is given in Fig. 2.

Other choices of the activation function are other monotone squashing functions as the *arctan* and *tanh* functions and the *cosine* squashing function, see e.g., Hertz. *et al.* (1991).

Next, the value y_j of the output cell j is given as the weighted sum of the output of the hidden cells. It is equal to

$$y_j = \sum_{h=1}^{H} c_{jh} g\left(\sum_{i=1}^{I} a_{ih} x_i + b_h \right) + d_j \tag{4}$$

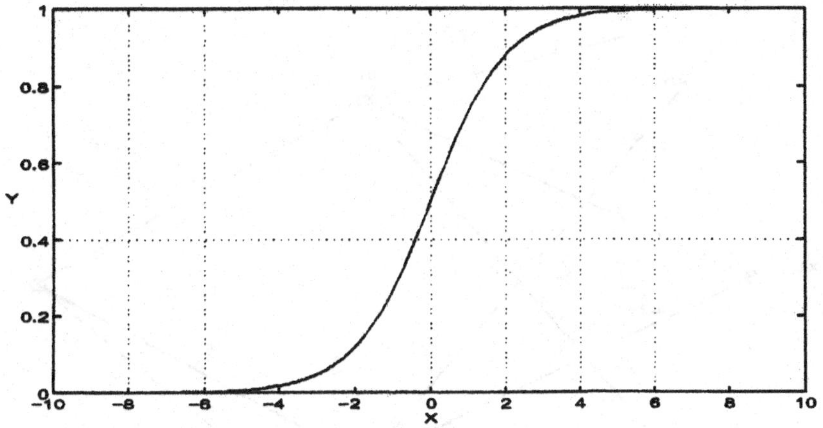

FIGURE 2 Graph of logistic activation function.

In matrix notation one can write:

$$y = C\mathcal{H} + d \tag{5}$$

$$\mathcal{H} = G(xA + b) \tag{6}$$

where

$x \in \mathbb{R}^I$ $y \in \mathbb{R}^O$
$A = (a_{ih}), I \times H$ matrix $b \in \mathbb{R}^H$
$\mathcal{H} = (h_1, \ldots, h_H)$, the vector of hidden cell outputs
$G: \mathbb{R}^H \to \mathbb{R}^H$ is the vector function, given by
$G(v) = [g(v_1, \ldots, g(v_H)]'$
$C = (c_{jh}), O \times H$ matrix $d \in \mathbb{R}^O$

The neural network $nn(\cdot)$ describes the situation at one moment in time and it indicates a deterministic relation. By adding a subscript t and an error term ε one obtains the system:

$$\begin{aligned} y_t &= C\mathcal{H}_t + d + \varepsilon_t \\ \mathcal{H}_t &= G(x_t A + b) \end{aligned} \tag{7}$$

Figure 3 is a representation of an $nn(3, 2, 1)$. The mathematical specification of this model is equal to

$$y_t = d + \frac{c_1}{1 + \varepsilon^{-a_{11}y_{t-1} - a_{12}y_{t-2} - b_1}} + \frac{c_2}{1 + e^{-a_{21}y_{t-1} - a_{22}y_{t-2} - b_{12}}} \tag{8}$$

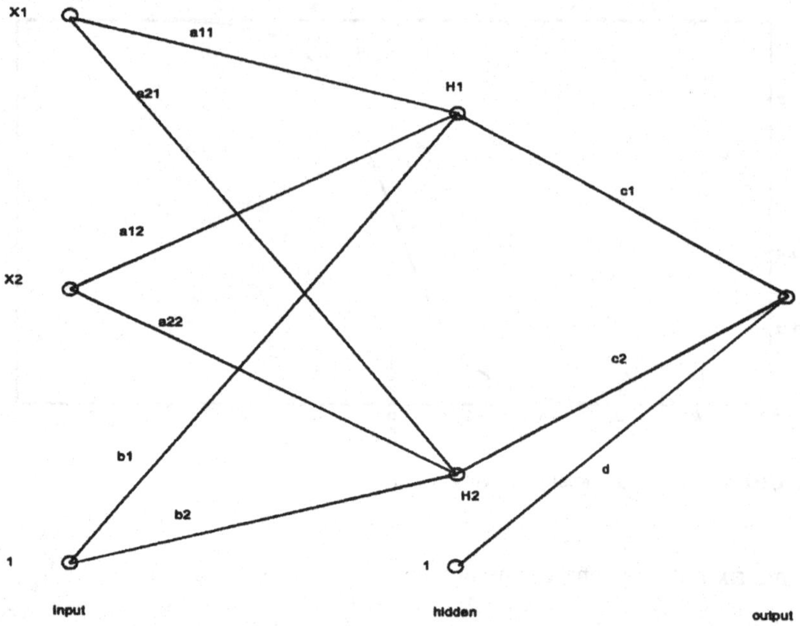

FIGURE 3 Graph of a neural network.

We note that this model is closely related to a threshold autoregessive model: for details see Granger and Teräsvirta (1993) and Van Dijk (1999).

2.1 Flexibility of Neural Networks

The flexibility of three-layer feed forward neural nets is well documented. It is summarized by its so-called "universal approximation" property. Most of this approximation theory starts with Kolmogorov's representation theorem (1957). This provides the background for the Hecht-Nielsen article (1987). From the point of view of a neural network user, the Kolmogorov theorem provides, however, a justification for the existence of approximations in the reverse way. That is to say, the number of layers and cells are given but not the functional form of the one-dimensional activation functions. In a neural network one encounters the opposite case: the activation functions g are given (to some extent) but, at least, the number of hidden layer cells is unknown.

The articles of Gallant and White (1989) (with the revealing title: "There exists a neural network that does not make avoidable mistakes"), Cybenko (1989), Funahashi (1989), Hecht-Nielsen (1989, 1990) and

Hornik *et al.* (1989). Lesho *et al.* (1993) and White (1989), provide the theoretical background for the statement: "A (three layer) feed forward network is an universal approximator." This general statement should be interpreted in the sense that any square integrable function can be approximated arbitrary close in L_2 norm.

Further, the articles of Hornik *et al.* (1990) and Gallant and White (1992) extended the approximation capabilities of the network to the derivative of a function.

2.2 Estimation of Parameters of Neural Networks

A generally accepted optimization principle is to minimize the norm of

$$\min_{\theta} \sum_{t} \|y_t - \hat{y}_t(\theta)\|^2 \tag{9}$$

where $\|.\|$ is the Euclidean norm. For the case of a neural net it follows that one minimizes the criterion function:

$$\min_{A,b,C,d} \sum_{t} \|y_t - CG(x_t A + b) - d\|^2 \tag{10}$$

Well known methods for numerical optimization are the simplex method and the BFGS method, which is a gradient method (see Press *et al.*, 1988).

Initial values for the parameters A, b, C, and d are chosen randomly from the uniform distribution on $[-0.5, 0.5]$. We emphasize that the logistic function is insensitive for large and small values. Therefore one should scale down the range of data before processing through the network. Since the output range of a hidden cell lies between 0 and 1, an appropriate range for the data should be $[0.1, 0.9]$.

2.3 Determining the Size of a Neural Network

Neural nets are flexible, but the price of increased flexibility is the danger of "overfitting." This statement may be explained as follows.

In empirical econometric models one assumes that an observed economic time series consists of a part that can be explained and a part that is labelled unexplained or "residual noise." With "overfitting" this noise is also "fitted." One then obtains a wrong picture of the real data-generating process and the quality of the forecasts may be badly affected. "Overfitting" with neural nets may occur by increasing the number of hidden cells, which increases the number of parameters, without increasing the number of explanatory variables or inputs. Because of this possibility neural nets are more sensitive to "overfitting" than other classes of models like autoregressive models. Therefore, it is important to develop methods that determine the optimal size of a neural net. Pruning methods apply to the reduction of large neural networks to smaller ones. Two methods can be distinguished: weight (interconnection) reduction or node reduction. Examples can be found in Hertz *et al.* (1991) and Bishop (1995) (see also Mozer and Smolensky, 1980).

Below we summarize a descriptive method to reduce the size of a neural net. Descriptive methods are useful for exploratory data analysis. This type of analysis is more and more needed since in recent decades large data sets for economic variables have become available. In these data, complex patterns of the economic variables may occur. For more details we refer to Kaashoek and van Dijk (1998).

2.3.1 Pruning a Network: Incremental Contribution Method

The method we follow is labeled incremental contribution method." It looks for each cell separately how much the specific cell contributes to the overall performance of the network. When this contribution is considered to be low then such a cell is a candidate for excluding from the existing network (and all its connections). Re-estimating the reduced network may confirm this exclusion. To measure the contribution of a cell we look at two quantities.

First, the square of the correlation coefficient (R^2) between y and $\hat{y}$, the neural network output where

$$R^2 = \frac{(\hat{y}'y)^2}{(y'y)(\hat{y}'\hat{y})} \tag{11}$$

and y as well as $\hat{y}$ are taken in deviation of the means. The procedure applies to hidden layer cells as well as to input layer cells, but here we restrict ourselves to hidden layer cells.

The contribution of cell h can now be measured by leaving out this cell and its connection from the network, and again calculating the square of the correlation coefficient; we denote network estimates with cell h left out by $\hat{y}_{-h}$, and the corresponding R^2 is defined as

$$R^2_{-h} = \frac{(\hat{y}_{-h}y)^2}{(y'y)(\hat{y}'_{-h}\hat{y}_{-h})} \tag{12}$$

The incremental contribution R^2_{incr} (i) is now given as

$$R^2_{\text{incr}}(h) = R^2 - R^2_{-h} \tag{13}$$

In the group of hidden layer cells, cells with a low R^2_{incr} are candidates for exclusion.

The second quantity involves the idea of principal components (Malinvaud, 1970; Theil, 1971). Let again $\hat{y}_h$ be the network output with exclusion hidden layer cell h. Construct the vector e_{-h} of residuals:

$$e_{-h} = y - \hat{y}_{-h} \tag{14}$$

and the matrix E_{-H}:

$$E_{-H} = (e_{-1}, \ldots, eH_{-H}) \tag{15}$$

The matrix $E_{-H'}E_{-H}/T$ is an estimate of the covariance matrix of the e_{-h}'s. The principal component of $E_{-H'}E_{-H}$ is the eigenvector at maximal eigenvalue in the absolute sense. Hence, the principal components provide a linear combination of $(e_{-1}, \ldots, e_{-H})$, which explains the largest part of the variance.

Which fraction is explained by each of the eigenvectors is given by the relative weight w_i with

$$w_h = \frac{\lambda_h}{\sum_{h=1}^{H} \lambda_h} \quad w_i = \frac{\lambda_i}{\sum_{h=1}^{H} \lambda_h} \tag{16}$$

where λ_h are the eigenvalues of $E_{-H'}E_{-H}$. In the case of the principal component, λ_H is the largest eigenvalue.

Now one can look at the components of the principal component itself: cell with low incremental contribution will have a low (in the absolute sense) coefficient in the eigenvector composing the principal component. The exclusion of those cells will cause a relatively small increase in the residuals; these cells are again candidates for exclusion.

Finally, one may apply a graphical analysis. Assuming the graph of $\hat{y}$ fits well with the graph of y, the graph of $\hat{y}_{-h}$ may differ less from the graph of y for those cells with a "low contribution."

The procedure involves the contributions of one cell only. It is a "feature" of neural networks that sometimes pairs of cells do have a similar contribution with the output of the cells having reverse sign. Such a "behavior" can be detected by graphical analysis and by observing that in the principal component analysis, explained above, such types of cells do have (almost) equal coefficients. In that case one has to look at the incremental contribution of both cells together.

We end this section with a remark on the descriptive nature of our procedure. As stated before, large data sets have become available in the economic sciences. Important examples are data set on household behavior and labour employment at the individual level. Further, data sets on financial variables in the stock and exchange rate markets. This appears as only a beginning. The new scanner data in marketing constitute great and fascinating challenges in data analysis. Because of its flexibility, we believe that a "pruned" neural network may be a useful instrument for exploratory data analysis in order to find possibly complex patterns of economic variables.

Our method is descriptive and as such has the same limitations as all techniques of data summarizations like the construction of histograms and/or the plotting of time series together with a list of summary statistics. However, using neural networks one can perform a dynamic analysis and compute the long-run properties (see Section 3). For a statistical approach to neural network analysis in econometrics, we refer to White (2000).

3 STABILITY ANALYSIS OF COMPLEX NONLINEAR SYSTEMS

A linear autoregressive system of equations of the n-vector of variables $\boldsymbol{x}(\boldsymbol{t})$ can be written as

$$x_{t+1} = Ax_t, x_t \in \mathbb{R}^N \tag{17}$$

The stability of fixed or equilibrium points depends on the eigenvalues λ (real or complex) of the matrix A. Taking the absolute value or modulus of the eigenvalues, $\|\lambda\|$, the fixed point will be unstable if for some eigenvalue λ, $\|\lambda\| > 1$, which is equivalent to $\ln\|\lambda\| > 0$. The same holds for the stability of orbits or time series $x_0, \ldots, x_t, \ldots$. Note that in a point $x_t \in \mathbb{R}^N$, the logarithm of the local expansion rate in the direction of a vector $v \in \mathbb{R}^N$ is given as

$$\ln\left\| A\frac{v}{\|v\|}\right\| \tag{18}$$

Lyapunov exponents are a generalization of the above concept for nonlinear systems. They are defined as the (spatial or time) means of the logarithm of local expansion rates. In the case of time means, the expansion rates are calculated in the time series $x_0, \ldots, x_t, \ldots$ where in a point $x_t \in \mathbb{R}^N$, the logarithm of the local expansion rate in the direction of a vector $v \in \mathbb{R}^N$ is now given as

$$\ln\left\| D_x F(x_t)\frac{v}{\|v\|}\right\| \tag{19}$$

where $F\colon \mathbb{R}^N \to \mathbb{R}^N$ is the data-generating function:

$$x_{t+1} = F(x_t) \tag{20}$$

and D_xF is the Jacobian. Then, the Lyapunov exponent λ_{v_0}, with start direction vector v_0 and start value x_0, is defined as

$$\lambda_{v_0} = \lim_{T\to\infty} \frac{1}{T} \sum_{t=1}^{T} \ln \left\| D_xF(x_{t-1}) \frac{v_{t-1}}{\|v_{t-1}\|} \right\| \tag{21}$$

$$v_t = D_xF(x_{t-T})v_{t-1}, \|v_0\| = 1 \tag{22}$$

The dependence of λ on v_0 and on x_0 seems to indicate an infinite number of Lyapunov exponents. However, this is not the case. In general, there are as many Lyapunov exponents as the dimension N of the system; see (Guckenheimer and Holmes, 1983). Above all, for an ergodic system, with the space mean being equal to time mean, it follows that for almost all start directions v_0, and for almost all initial values x_0, the value of λ_{v_0} will be the largest Lyapunov exponent (Arnold and Avez, 1988; Guckenheimer and Holmes, 1983).

Since the mean of logarithms is the logarithm of the geometric mean, one can also write

$$\lambda_{v_0} = \lim_{T\to\infty} \frac{1}{T} \ln \prod_{t=1}^{T} \left\| D_xF(x_{t-1}) \frac{v_{t-1}}{\|v_{t-1}\|} \right\| \tag{23}$$

$$= \lim_{T\to\infty} \frac{1}{T} \ln \|D_xF^T(x_0)v_0\| \tag{24}$$

If some (or all) of the Lyapunov exponents are positive then one has so-called "sensitivity on initial values," a characteristic of chaotic series. Hence, in particular, the largest Lyapunov exponent is of interest: if positive, then the series is unstable, if negative then the series is stable. This result can be applied directly if F, the data-generating function, is known. However, in practice, one observes only a one-dimensional, finite time series $\{x_t, t = 0, \ldots, T-1\}$. So the question is how to extract from the series $\{x_t\}$ the dynamic properties, especially the value of the largest Lyapunov exponent of the original (unknown) model, Eq. (20).

Although only the series x_t is given, one can use the embedding theorem of Takens (1981) to reconstruct from the one-dimensional series x_t the original deterministic and smooth model. This theorem says that if the data x_t have a *deterministic explanation*, which means that the data-generating process is smooth and deterministic, there exists a finite embedding m, such that the dynamic system $\Psi: \mathbb{R}^m \to \mathbb{R}^m$ given by

$$\Psi : (x_t, x_{t-1}, \ldots, x_{t-m+1}) \to (x_{t+1}, x_t, \ldots, x_{t-m+2}) \tag{25}$$

has the same dynamical properties as the original one. Moreover, if the original system is N-dimensional then an embedding dimension $m = 2N + 1$ will be sufficient to have "Ψ reconstruct the original system."

Note that the only unknown component of Ψ is the first component function $\Psi_1 : \mathbb{R}^m \rightarrow \mathbb{R}$ with

$$\Psi_1 : (x_t, x_{t-1}, \ldots, x_{t-m+1}) \rightarrow x_{t+1} \tag{26}$$

For the other components, $i > 1$, yields $\Psi_i(x_t, x_{t-1}, \ldots, x_{t-m+1}) = x_{t-i+2}$. This function Ψ_1 should be approximated, and a neural network seems to be a proper candidate to do the job because of its "universal approximation" property (Kaashoek and van Dijk, 1994).

Once Ψ is found, the Lyapunov exponents can be calculated using Eq. (23). This means the calculation of $D_x\Psi$ in each point of the time series $(x_t, x_{t-1}, \ldots, x_{t-m+1})$. Note that $D_x\Psi$ has the form of a companion matrix. However, to avoid overflow in calculating the product of $D_x\Psi$'s the more stable method of Eckmann–Ruelle is used. In this case, a QR decomposition of $D_x\Psi$ is calculated at each point and the Lyapunov exponents are now simply the products of diagonal elements of the matrices R (Eckmann and Ruelle, 1985).

We apply this procedure to two data sets, one simulated and one economic time series.

3.1 Simulated Data Experiment

In this experiment we use simulated data. The data are generated by the model:

$$\begin{aligned} x_{t+1} &= \alpha x_t + \varepsilon_t - 0.5 \\ \varepsilon_{t+1} &= \gamma \varepsilon_t (1 - \varepsilon_t) \end{aligned} \tag{27}$$

where only the series x_t is observed. The data, called $CH95$, are generated with $\alpha = 0.95$ and $\gamma = 4$. Although completely deterministic, this model has some nice features:

- The series ε_t is chaotic: for initial values in [0, 1], the series ε_t is bounded between 0 and 1 but has the "sensitivity of initial values" property characteristic for chaotic series.
- The model is structurally unstable ($\gamma = 4$) e.g., a small change in the coefficient value 4 will cause a dynamically different data series. For instance, a γ value greater than 4 will for almost all initial values (between 0 and 1) generate diverging (exploding) data ε_t; for values less than 4, periodic data are possible: the system is *structurally* unstable.

Suppose that only the one-dimensional data set x_t is observed. Our goal should be to extract from this series, the dynamical properties of the original data-generating process: only Lyapunov exponents are considered.

Note that model equation (27) is equivalent to a second-order nonlinear difference equation in x_t:

$$x_{t+2} = \alpha x_{t+1} + \gamma(x_{t+1} - \alpha x_t + 0.5)(1 - x_{t+1} + \alpha x_t - 0.5) \qquad (28)$$

where the right-hand side is a (nonlinear) *continuous* function in (x_{t+1}, x_t). So this case is well suited for a neural network approximation. Moreover, since the neural network is intrinsically nonlinear, the chaotic aspect of the data could be grasped by such an approximation while a simple linear autoregressive model would certainly fail as an "accurate" approximation.

The original data generating model, Eq. (27), has two Lyapunov exponents, which can be calculated analytically. In a point (x_i, ε_i), the Jacobian of the system function F is given by

$$\begin{pmatrix} 0.95 & 1 \\ 0 & 4(1-\varepsilon_i) \end{pmatrix} \qquad (29)$$

Since the Jacobian $D_xF^T(x_0, \varepsilon_0) = \prod_{t=1}^{T} D_xF(x_{t-1}, \varepsilon_{t-1})$, one has

$$D_xF^T(x_0, \varepsilon_0) = \begin{pmatrix} 0.95^T & a_{12} \\ 0 & a_{22}(T) \end{pmatrix} \qquad (30)$$

where $a_{22}(T) = \prod_{t=1}^{T} 4(1 - \varepsilon_{t-1})$. It is obvious that D_xF^T has two eigenvalues: namely 0.95^T and $a_{22}(T)$. With $v_0 = (1, 0)$, $D_xF^T(x_0, \varepsilon_0)v_0 = [(0.95)^T, 0]$, which results in a Lyapunov exponent value of ln(0.95). It is well known that the system $\varepsilon_l = 4\varepsilon_{l-1}(1 - \varepsilon_{l-1})$ has Lyapunov exponent ln(2). Since $\ln(2) > \ln(0.95)$, for any start vector $v_0 \neq (1,0)$, the resulting Lyapunov exponent will be ln(2). Writing both Lyapunov exponents in base 2 logarithm, the values will be $\ln(0.95)/\ln(2) \sim 0.074$ and 1.

Before starting with the neural network computations of the function Ψ_1 see Eq. (26), first the scatter diagram $\{(x_{t-1}, x_t\}$ of the series *CH*95 is given. The sample size T is 200. Since e_t is bounded between 0 and 1, the graph of (x_{t-1}, x_t) lies between the lines $y = 0.95x + 0.5$ and $y = 0.95x - 0.5$. (See Fig. 4.)

As said above, in order to extract from the one-dimensional observed data $\{x_t\}$, the dynamic properties, especially the value of the largest Lyapunov exponent of the original model, Eq. (27), only the function:

$$\Psi_1(x_t, x_{t-1}, \ldots, x_{t-m+1}) = x_{t+1}$$

should be approximated by a neural network. This implies a choice for the value of m, or in neural terms, the size of the input layer. Since the original system has dimension 2, based on Takens embedding theorem, it will be

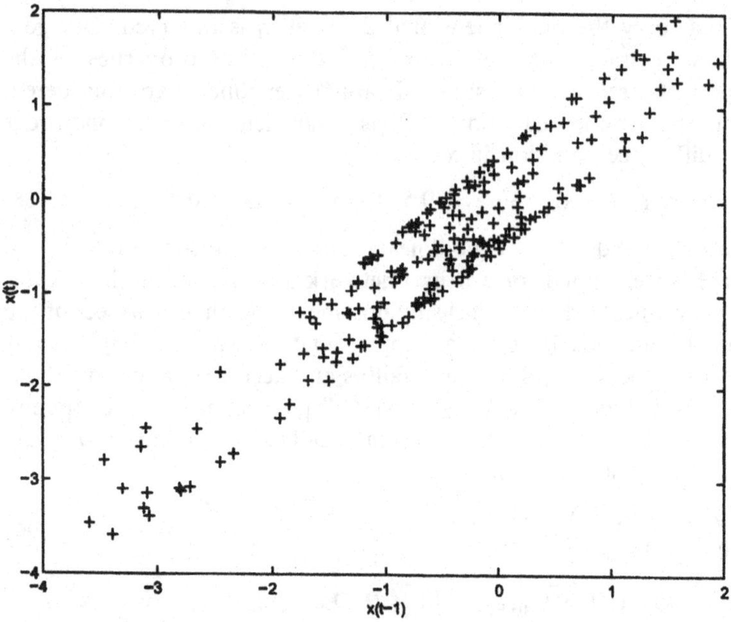

FIGURE 4 Scatter diagram of series *CH*95.

(more than) sufficient to take as neural network input variables $(x_t, x_{t-1}, \ldots, x_{t-5})$: adding an additional constant, the dimension of the input layer will be 6.

The initial number of hidden layers will be 5, while the output layer has only one cell, the target value being x_{t+1}. Before processing the data through the neural network, the data are scaled down to a range of [0.1, 0.9]. Linear scaling does not affect the value of the Lyapunov exponents. Initial parameter values are randomly chosen from [−0.5, 0.5].

To summarize the performance of this network, the quantities:

$$R^2 = \frac{(\hat{y}'y)^2}{(y'y)(\hat{y}'\hat{y})} \tag{31}$$

$$MSSR = \frac{1}{T}(y - \hat{y})'(y - \hat{y}) \tag{32}$$

$$SIC = \ln(MSSR) + \frac{n_p}{2T}\ln(T) \tag{33}$$

are calculated. In Table 1 the results for the *nn* (6,5,1) are reported.*

*All values reported are rounded-off at the last reported digit.

In Table 2 the incremental contributions of the hidden layer nodes are given: both R^2 and principal component vector (row with label PrincComp. in Table 2 indicate that two hidden layer nodes can be removed.

The incremental contributions of inputs are shown in Table 3: the inputs x_{t-4}, x_{t-3}, x_{t-2} should be removed. Applying node removal, the network is reduced to two inputs plus a constant and three hidden layer nodes (plus constant). This $nn(3, 3, 1)$ performs as well as the larger network (see Table 4). Both input nodes do have the same contributions: no further reduction at this level is applied.

The incremental contributions of hidden layer nodes is shown in Table 5. Based on the principal component vector (with weight 99%) hidden node 1 could be removed (see again Table 4 for the results on this network). The performance is just slightly worse compared to the larger

TABLE 1 Results of $nn(6, 5, 1)$

R^2	*MSSR*	*SIC*
1.00	1.4×10^{-9}	−9.70

TABLE 2 Contribution of Hidden Layer Nodes

Hidden cell	1	2	3	4	5
R^2_{incr}	0.000	0.000	0.517	0.880	0.940
PrincComp.	−0.000	−0.000	−0.081	−0.733	0.675

TABLE 3 Contribution of Input Layer Nodes

Input cell	x_{t-4}	x_{t-3}	x_{t-2}	x_{t-1}	x_t
R^2_{incr}	0.000	0.000	0.000	0.124	0.241
PrincComp.	0.000	0.000	−0.000	0.613	0.790

TABLE 4 Results of $nn(3, 3, 1)$ and $nn(3, 2, 1)$

Network	R^2	*MSSR*	*SIC*
$nn(3, 3, 1)$	1.00	1.5×10^{-9}	−10.04
$nn(3, 2, 1)$	1.00	6.3×10^{-9}	−9.31

TABLE 5 Contribution of Hidden Layer Nodes in *nn*(3, 3, 1) Network

Hidden cell	1	2	3
R^2_{incr}	0.696	0.890	0.943
PrincComp.	0.114	−0.745	0.657

TABLE 6 Lyapunov Exponents $\gamma = 4$

Series	Actual x_t	Orbit $\hat{x}_t$
Lyapunov exponent	0.9951	1.0155
Lyapunov exponent	−0.0735	−0.0729

network. No further reduction is applied. So we end up with a network of two inputs x_{l-1}, x_l (plus constant) and two hidden layer nodes.

Based on this $nn(3, 2, 1)$ network, the Lyapunov exponents of Ψ are calculated. This can be done in two ways: either along the actual data (x_t, x_{t-1}), or along a series $(\hat{x}_{t-1}, \hat{x}_t)$, where $\hat{x}_t$ is a series generated by the neural network function $nn(3, 2, 1)$:

$$\hat{x}_{t+1} = nn(3, 2, 1)(\hat{x}_{t-1}, \hat{x}_t) \tag{34}$$

The series $\hat{x}_t$ is the dynamic forecast given some initial values $\hat{x}_0 = x_0$, $\hat{x}_1 = x_1$*; here and in the following, such a series will be denoted as *orbit* in contrast to the actual data series $\{x_t\}$.

If the function Ψ is indeed a proper approximation of the original model then one should expect that the dynamic properties along the actual data and along the orbit data should be similar.

The results for the Lyapunov exponents are given in Table 6; they are in both cases near the theoretical values. Note that for the orbit data the largest Lyapunov exponent is greater than 1; this would correspond in the original model, Eq. (27), with a coefficient of $\varepsilon_t(1 - \varepsilon_t)$ larger than 4: for almost all initial values, the series ε_t would diverge. However, the orbit data $\hat{x}_t$ converge to a large amplitude periodic pattern: in Fig. 5 the graph of $(t, \hat{x}_t)$ for the $t = 1, \ldots, 150$ (left panel in Fig. 5) and for $t = 1, \ldots, 300$ (right panel in Fig. 5) is compared to the graph of the actual data (t, x_t). Since the original system is chaotic, one should not expect that both series, actual and orbit data, fall together but one should expect that the graph of both series shows a similar pattern; even in this case (and also in the larger

*The initial values are taken from the actual data.

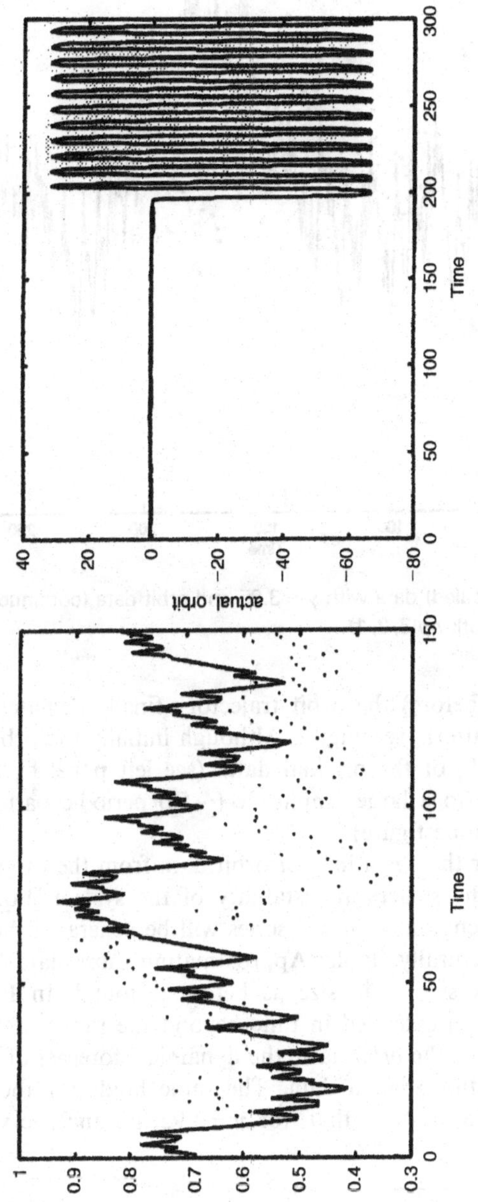

FIGURE 5 Actual (scaled) data *CH95*, $\gamma = 4$ and orbit data (continuous line) generated by neural network *nn*(3, 2, 1). Left panel: time index from 1 to 150; right panel: time index from 1 to 300.

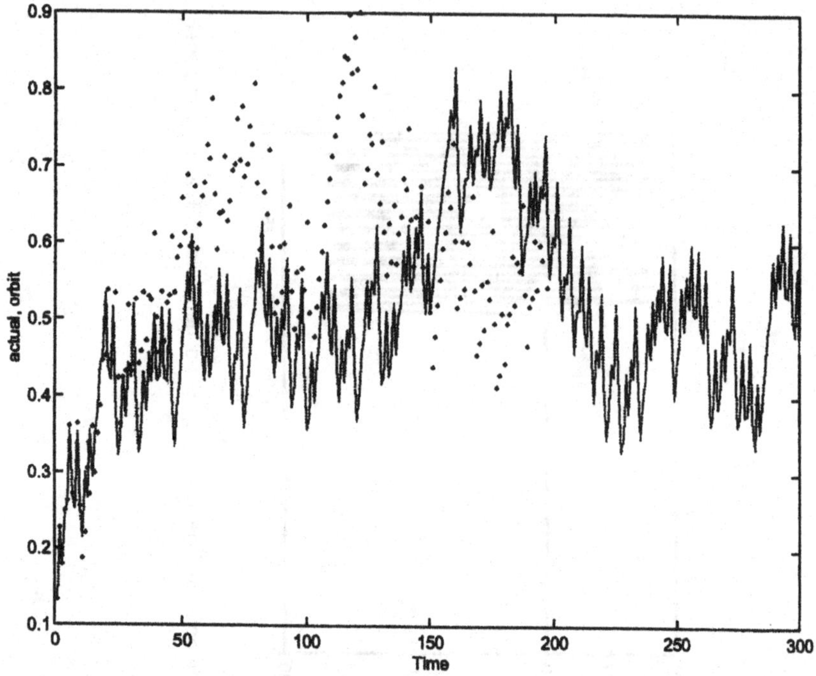

FIGURE 6 Actual (scaled) data with $\gamma = 3.95$ and orbit data (continuous line) generated by neural network *nn*(3, 2, 1).

networks obtained before) the orbit trajectory finally deviates essentially from the original pattern; see Fig. 5. Although initially the orbit data stay in the range [0.1, 0.9] of the original data, (see left panel of Fig. 5), the orbit data converge to a large amplitude (~ 50) periodic pattern (see the right panel of the same figure).

The reason for the deviations of orbit data from the original pattern is to be found in the structural instability of the system [Eq. (27)] with $\gamma = 4$. If $\gamma = 3.95$ then still a chaotic series will be generated; however, the system itself is structurally stable. Approximating those data by a neural network, a network similar in size as before is found. In Fig. 6 again actual and orbit (even extended in time beyond the range of the original data) are shown. Now the *orbit* data, the dynamical forecast of given value x_0, "behaves" like the original data. The same holds for the Lyapunov exponents (see Table 7). Note that, for $\gamma = 3.95$, no analytic value of the

*The initial values are taken from the actual data.

TABLE 7 Lyapunov Exponents $\gamma = 3.95$

Series	Actual x_t	Orbit $\hat{x}_t$
Lyapunov exponent	0.8618	0.8614
Lyapunov exponent	−0.0782	−0.0727

largest Lyapunov exponent is available. Analytical values of Lyapunov exponents can be calculated if the data-generating process is ergodic and if the space distribution of the data is known; see, for more details, Arnold and Avez (1988). For the process $\varepsilon_{t+1} = 3.95\varepsilon_t(1 - \varepsilon_t)$ no analytical expression is known for the distribution of the data ε_t. The second Lyapunov exponent will still be ~ -0.074.

Finally, we look how the network $nn(3, 2, 1)$ has approximated the original data. For both cases, $\gamma = 4$, $\gamma = 3.95$, none of the parameters A, b, C, and d are very large. For instance in the case of $\gamma = 4$, the input vector of hidden nodes $xA + b$, is given as

$$\begin{aligned}(xA + b)_1 &= -3.1557x_{t-1} + 3.3197x_t - 1.3688 \\ &= 3.3197(-0.9506x_{t-1} + x_t - 0.4123) \\ (xA + b)_2 &= -0.7453x_{t-1} + 0.8095x_t + 0.0516 \\ &= 0.8095(-0.9207x_{t-1} + x_t + 0.0637)\end{aligned} \tag{35}$$

Note that reported parameter values are estimation results using scaled data.

3.2 Nonlinear Trends in Real Exchange Rates

The economic time series are monthly observations of the natural logarithm of real exchange rates. We report those between the yen and the dollar (January 1957 to March 1998). This series is denoted by *JPUS* (see Fig. 7). Since the original data process is unknown, a proper embedding dimension m, e.g., delay vector $(x_t, \ldots, x_{t-m+1})$, is also unknown. One way out is to extract from the correlation dimension the embedding dimension (Grassberger and Procaccia, 1983; Kaashoek and van Dijk, 1991).

Another approach would be to start with a rather large network and prune this network until further reduction would essentially corrupt the performance. In this case, the initial network was taken to be $nn(6, 10, 1)$ with input variables $(1, x_t, x_{t-1}, \ldots, x_{t-4})$. The performance of this network, and the successively reduced networks, are summarized in Table 8. At the same time, in the column "λ" the largest Lyapunov exponent along the actual data set is reported while in column "$\hat{\lambda}$" the

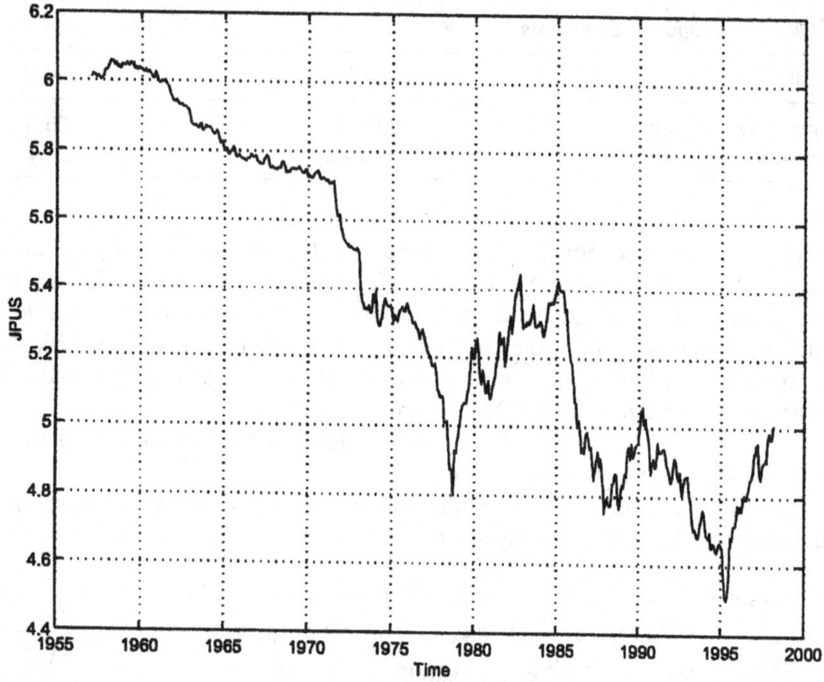

FIGURE 7 Time series *JPUS*: Yen-Dollar exchange rates.

TABLE 8 Results of Neural Network Approximation of *JPUS* Data

Network	R^2	*MSSR*	*SIC*	λ	$\hat{\lambda}$	Pruning
$nn(6,10,1)$	0.997	0.16×10^{-3}	-3.880	0.058	-0.013	3 redundant hidden cells
$nn(6,7,1)$	0.997	0.16×10^{-3}	-4.041	0.033	-0.038	2 redundant hidden cells
$nn(6,5,1)$	0.997	0.16×10^{-3}	-4.130	0.048	-0.143	2 redundant input cells
$nn(4,5,1)$	0.997	0.16×10^{-3}	-4.194	0.048	-0.032	2 redundant hidden cells
$nn(4,3,1)$	0.996	0.18×10^{-3}	-4.188	0.076	-0.034	3 redundant input cells
$nn(2,3,1)$	0.996	0.19×10^{-3}	-4.215	-0.011	-0.019	

largest Lyapunov exponent along the *orbit* is reported. The applied reduction can be found in the column "Pruning."

The input variables of the resulting network $nn(2,3,1)$ are $\{(1, x_t)\}$. In Fig. 8 the *orbit* based on the $nn(2,3,1)$ network is shown (the data are scaled down to the range $[0.1, 0.9]$): the orbit follows nicely the pattern of the actual data converging to a scaled value of 0.26 compatible with the unscaled value of -0.74.

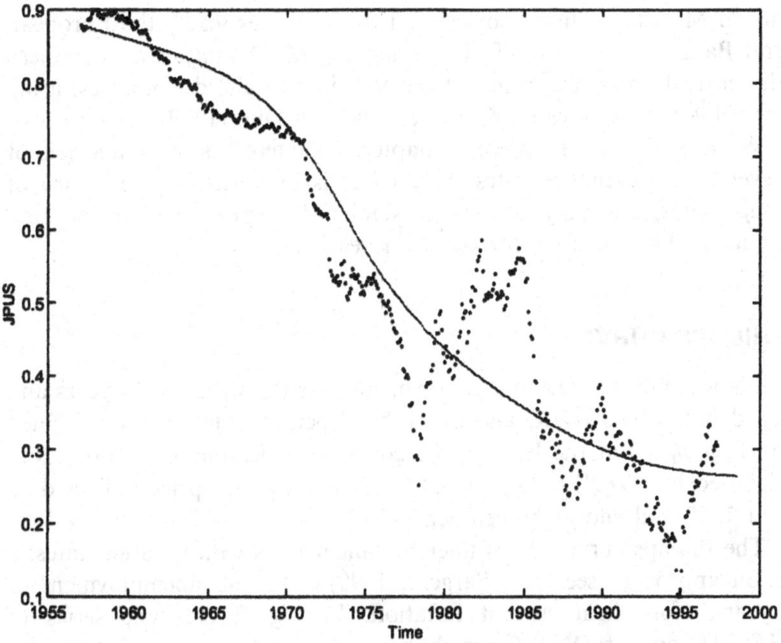

FIGURE 8 *nn*(2, 3, 1) orbit compared to actual (scaled) *JPUS* data.

In all cases the orbit is stable (negative largest Lyapunov exponent) while along the actual data, the Lyapunov exponent is positive indicating an unstable series except for the final *nn*(2, 3, 1) network. The differences between the networks with respect to statistics are marginal. Note also that the Lyapunov exponent based on actual data still encompasses stochastic elements if present in the data itself. The smallest network is preferable.

Our empirical analysis indicates that nonlinear trends may become an important part of a modeling strategy. So far, the empirical literature on real exchange rate analysis makes mostly use of autoregressive models with possible unit root behavior; see, e.g., Schotman and van Dijk (1991) and the references cited therein, in particular, Dornbusch (1976), Frankel (1979), and Meese and Rogoff (1988).

An issue in that literature is whether or not the hypothesis of purchasing power parity (*PPP*) holds. Nonrejection of a unit root has led researchers to conclude that *PPP* does not hold. The results presented in this chapter indicate that the *PPP* hypothesis should be investigated on a nonlinear model.

Our empirical analysis may also be used to verify whether or not the long-run equilibrium values of several exchange rates of countries of the

European Monetary Union correspond to the values set by the European Central Bank in January 1999. For a more detailed analysis on nonlinear trends in real exchange rates of several industrialized countries using neural networks we refer to Kaashoek and van Dijk (1999).

We note that in the present chapter our interest is in an analysis of the level of the exchange rates. When one is interested in the return of exchange rates, one may perform a similar descriptive analysis on first differences. This is a topic for further research.

4 PHILLIPS CURVE

The data are monthly U.S. unemployment rates (all workers, 16 years and older), denoted by *LHUR*, and monthly 12-period inflation rates defined by $100\ln(p_t/p_{t-12})$ at the level of the consumer price index p_t: those data are denoted by *INFR*12. The unemployment rates and price indices data start at 1960 and end at November 1997.

The Phillips curve relates unemployment rates with inflation rates: a common approach (see, e.g., Sargent, 1999) is to link unemployment of one year before with current inflation. In Fig. 9 the time series of *LHUR*(−1)* and of *INFR*12 are shown.

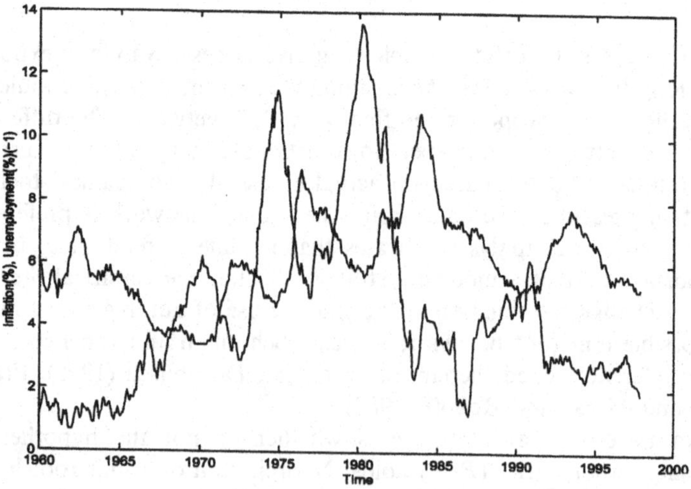

FIGURE 9 Time series of inflation rates *INFR*12 and unemployment rates *LHUR*(−1).

*Here, and in the following, 1-year delayed data are denoted by an −1 argument.

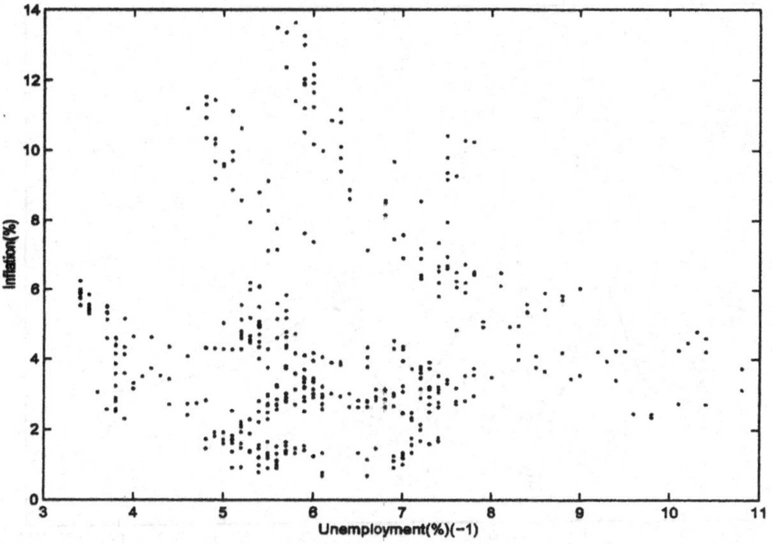

FIGURE 10 Phillips curve data (*LHUR*(− 1), *INFR*12).

The Phillips curve data {(*LHUR*(−1)(*t*), *INFR*12(*t*))} are shown in Fig. 10. Two attempts are made to model the relation between *LHUR*(−1) and *INFR*12.

First: Let $nn(2,6,1)$ be the neural network with inputs a constant term and the variable *LHUR*(−1), consider six hidden layer cells and let the target output value be *INFR*12. The performance of this network, which is to fit the Phillips curve to the data of Fig. 10, is poor: $MUSSR=0.13$, $R^2=0.11$. This is rather obvious because with input variable *LHUR*(−1), the points of Fig. 10 can hardly be considered as generated by a single valued relation (function). The graph of the neural network function $x \rightarrow nn(2,6,1)(x)$ confirms this: it seems to consists of four descending functions (all four being standard Phillips curves) with smooth transitions in between (see Fig. 11). Networks with more hidden layers show similar patterns. Although a neural network is capable of generating step-functions, it cannot of course model "multi-level relations." The graph in Fig 10 is rather to be considered as generated by some explicit time-dependent relation (t, *LHUR*(−1))→*INFR*12.

Second attempt: Referring again to Fig. 11, a time-varying approach is tempting, as in time-varying smooth transition models (see, e.g., Van Dijk, 1999). In this section, we will add an additional argument to the notation of a neural network, either (x) or (t) depending on whether the input variables include explicitly data x or time indices t.

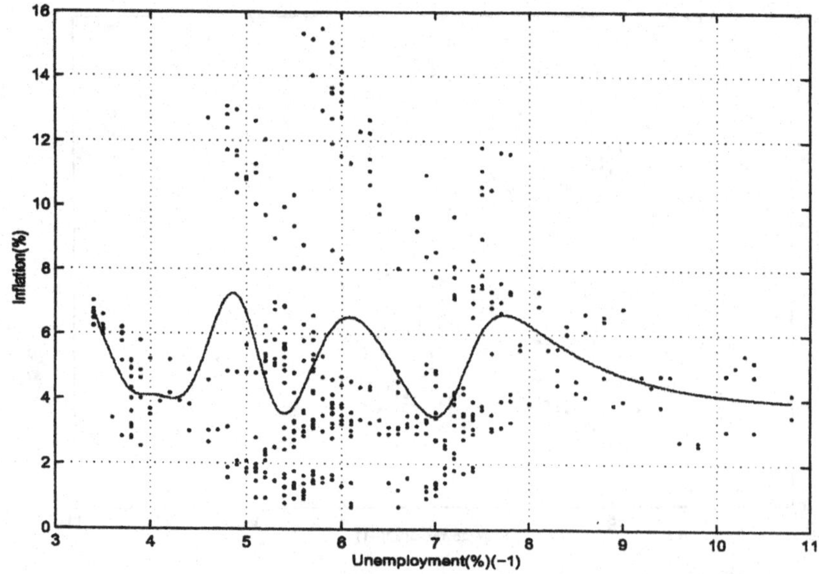

FIGURE 11 Graph of neural network function $nn(2, 6, 1)(x)$ and actual data.

Suppose four time periods, and for each period a neural network approximation (of unknown size) written as $nn_i(x)$, $i = 1, \ldots, 4$ with

$$nn_i(x) = c_i g(a_i x + b_i) + d_i, \forall \quad i = 1, \ldots, 4 \tag{36}$$

then a formulation of time-varying smooth transition model could be

$$nn_1(x)g_1(t) + nn_2(x)g_2(t) + nn_3(x)g_3(t) + nn_4 g_4(t) \tag{37}$$

subject to some normalization, say

$$g_1(t) + g_2(t) + g_3(t) + g_4(t) = \gamma, \forall t \tag{38}$$

Hence, Eq. (37) can be written as

$$\begin{aligned} &nn_1(x)(\gamma - g_2(t) - g_3(t) - g_4(t)) \\ &\qquad + nn_2(x)g_2(t) + nn_3(x)g_3(t) + nn_4 g_4(t) \end{aligned} \tag{39}$$

Collecting the transition functions, one obtains

$$\begin{aligned} &\gamma nn_1(x) + g_2(t)(nn_2(x) - nn_1(x)) \\ &\qquad + g_2(t)(nn_3(x) - nn_1(x)) + g_4(t)(nn_4(x) - nn_1(x)) \end{aligned} \tag{40}$$

Now assume

$$nn_i(x) = (1+\gamma)nn_1(x), \quad i = 2, \ldots, 4 \tag{41}$$

then the model equation can be written as

$$\gamma\{1 + g_2(t) + g_3(t) + g_4(t)\}nn_1(x) \tag{42}$$

Note that the condition (41) assumes that for all four regimes, the neural network approximation differs only by a multiplicative constant. Assuming

$$g_i(t) = c_i/(1 + \exp(-a_i t - b_i)) \tag{43}$$

then in full neural network notation, the formulation of Eq. (42) is

$$nn(2, H_t, 1)(t) \times nn(I, H, 1)(x) \tag{44}$$

Assuming four times transitions ($H_t = 3$), our neural network model is given by

$$(t, x) \rightarrow nn(2, 3, 1)(t) \times nn(I, H, 1)(x) \tag{45}$$

with

$$nn(2, 3, 1)(t) : t \rightarrow 1 + \sum_{i=1}^{3} \frac{c_i}{1 + e^{-a_i t - b_i}} \tag{46}$$

$$nn(I, H, 1)(x) : x \rightarrow \delta + \sum_{i=1}^{H} \frac{\gamma_i}{1 + e^{-a_i x - \beta_i}} \tag{47}$$

The *nn*(2, 3, 1) (*t*) will be denoted as the *t*-network; as the size of this network is fixed, we will leave out from the notation the size-arguments, and write just *nn*(*t*). The function *nn*(*I*, *H*, 1) (*x*) will be called the *x*-network; like the *t*-network, the *x*-network is denoted as *nn*(*x*). However, in this case we shall explicitly mention the size of this network, especially the value of *H*.

Hence, instead of Eq. (44), we write in short:

$$nn(t) \times nn(x) \tag{48}$$

The input variable x will be (again) the data series $LHUR(-1)$; hence, the x-network has $I = 2$, and the input data t will be a time-index series {1,2,...}. As before, x data are scaled down to [0.1, 0.9]. However, for convenience, in graphs the data are scaled back to their original range.

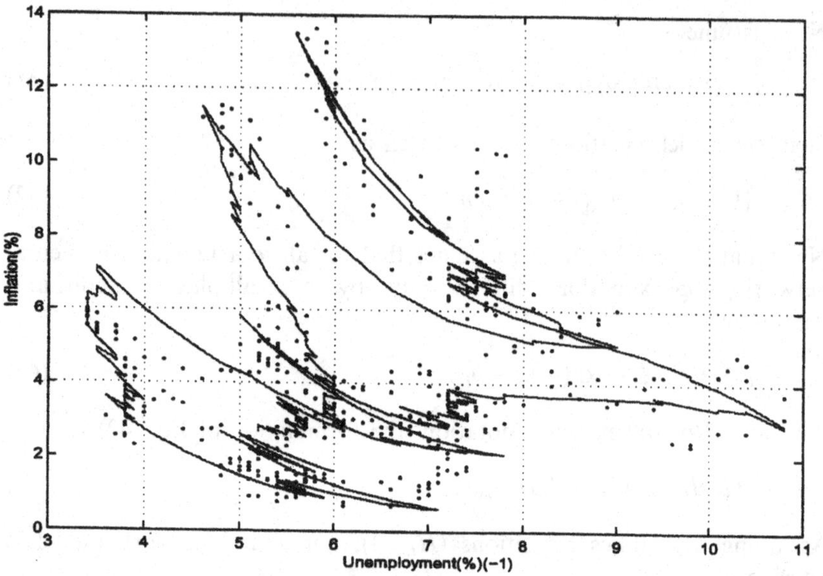

FIGURE 12 Estimated data of model $nn(t) \times nn(x)$, with $H=3$ and actual data (dots).

The number of hidden layer cells H in the x-network $nn(x)$ has to be defined. We report on the results for $H=3$ and $H=1$. Networks with more hidden layer cells do not give other results in the sense that the neural networks generate outputs similar to the one depicted in Fig. 12; additional hidden layer cells have a low incremental contribution.

Results on $H=3$.

Statistical values: $MUSSR=0.002$, $R^2=0.96$.

In Fig. 12 the estimated- and actual data are shown.

The performance is much better (as expected) than the foregoing network approximation.

In order to find out how the time network performs and whether an elementary "Phillips curve" is found by this model, the time-series outputs of the t-network $nn(t)$ and x-network $nn(x)$ are graphed separately (see Fig. 13).*

It seems that the x-network has indeed found some basic Phillips-curve structure. However, the t-network shows only three levels while

*All data in the figures are on the same scale: all data are scaled back to the original range.

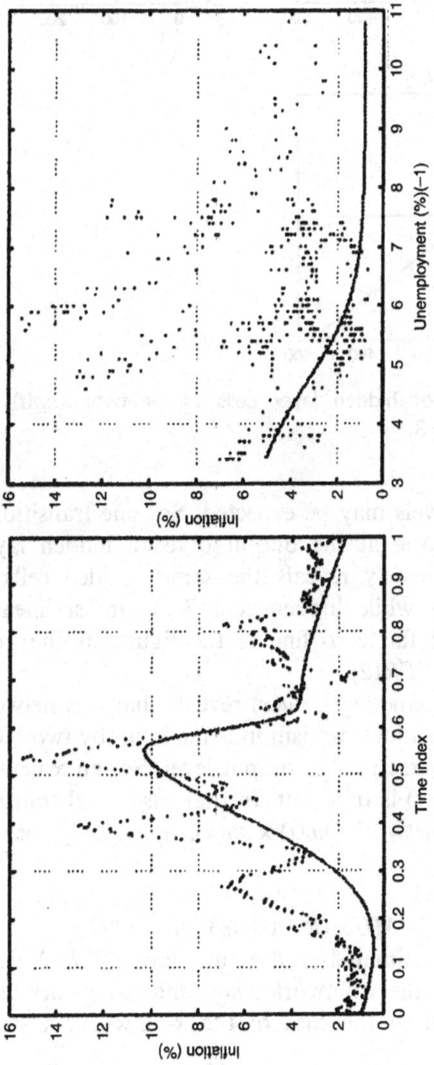

FIGURE 13 Actual data *INFR*12 (dots) compared to output of *t*-network $nn(t)$ (left) and *x*-network $nn(x)$ (case $H = 3$) (right).

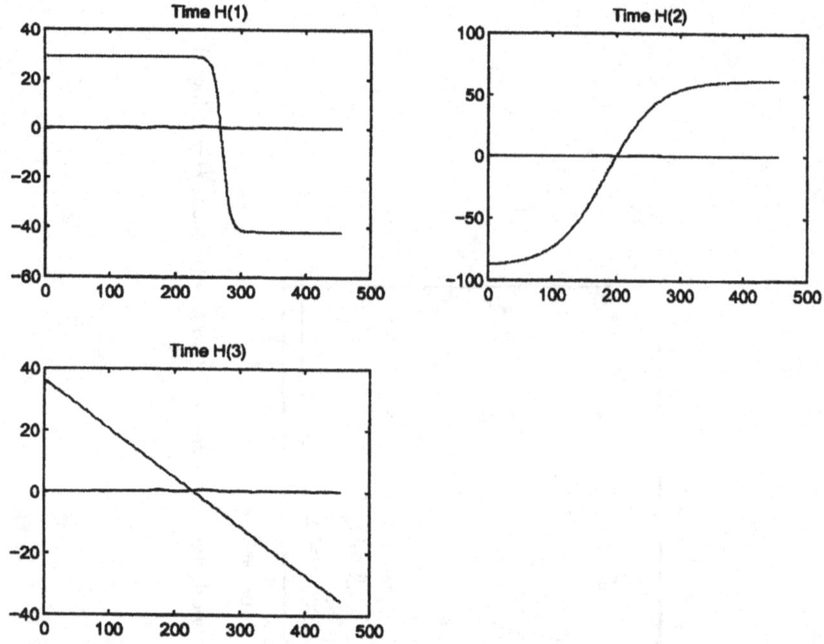

FIGURE 14 Output of hidden layer cells of *t*-network *nn*(*t*) in the case of the *x*-network with $H=3$.

from Fig. 11 four levels may be expected. So, one transition is missing in the *t*-network. Looking at the output of each hidden layer cell of the *t*-network *nn*(*t*) separately reveals the same: hidden cells 1 and 2 will generate three levels while hidden cells 3 is almost linear in time (see Fig. 14) (the almost flat zero line in the figures is the graph of actual inflation rate data *INFR*12).

Examining the *x*-network *nn*(*x*) reveals that this network has a level transition itself; such a level transition is made up by two hidden cells with almost equal but reverse in sign output level. So, an *x*-network with only one hidden cell ($H=1$) is tried out. In that case, level transitions, if found by the combined network $nn(t) \times nn(x)$, may only be found in the *t*-network.

Results on $H=1$.

Statistical values: $MUSSR=0.0023$, $R^2=0.93$.

In Fig. 15 the estimated- and actual data *INFR*12 are shown.

The output of the *t*-network *nn*(*t*) and *x*-network *nn*(*x*) show the three transitions and in the case of the *x*-network, a smooth "Phillips curve" (see Fig. 16).

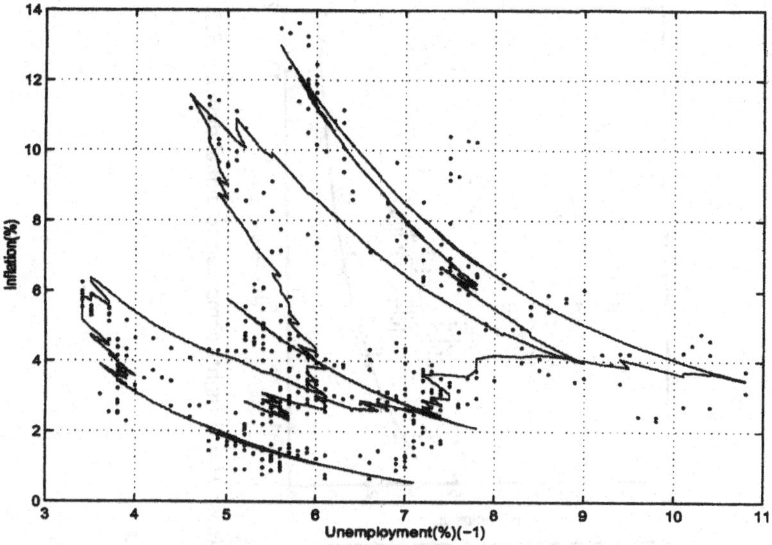

FIGURE 15 Estimated data of model $nn(t) \times nn(x)$, with $H=1$ and actual data *INFR*12.

It seems that the x-network has indeed found (again) some basic Phillips-curve structure and now the time network $nn(t)$ indeed generates four levels. Approximately, those time transitions occur where the input to a hidden cell will be zero, or otherwise stated, in the point of symmetry of the logistic function. In the case of the t-network, this will be at $t_i = -b_i/a_i$ [see Eq. (46)]. In this case the transitions take place at

$$\begin{aligned} t_1 &= 171.93 \sim \text{April 1974} \\ t_2 &= 300.82 \sim \text{February 1982} \\ t_3 &= 425.13 \sim \text{June 1995} \end{aligned} \tag{49}$$

The data t_1 corresponds more or less to the first oil crises, data t_2 corresponds to a business cycle slowdown in the U.S. economy while the third data t_3 coincides with the beginning of a period of low inflation and low unemployment. Looking at the output of each hidden layer cell of the t-network $nn(t)$ separately reveals the same (see Fig. 17). We repeat that our method is descriptive. For a statistical approach which enables one to compute confidence intervals around these dates, we refer to White (2000).

We end this section with a similar remark as made at the end of Section 3 on nonlinear trends in real exchange rates. That is, our empirical analysis on the Phillips curve indicates that the unemployment–inflation

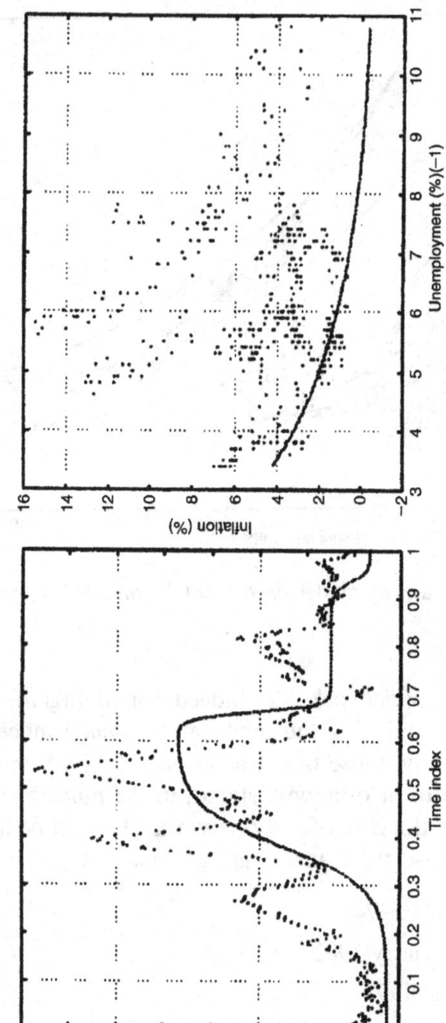

FIGURE 16 Output of $nn(t)$ (left) and $nn(x)$ (case $H=1$) (right) compared to actual data *INFR*12.

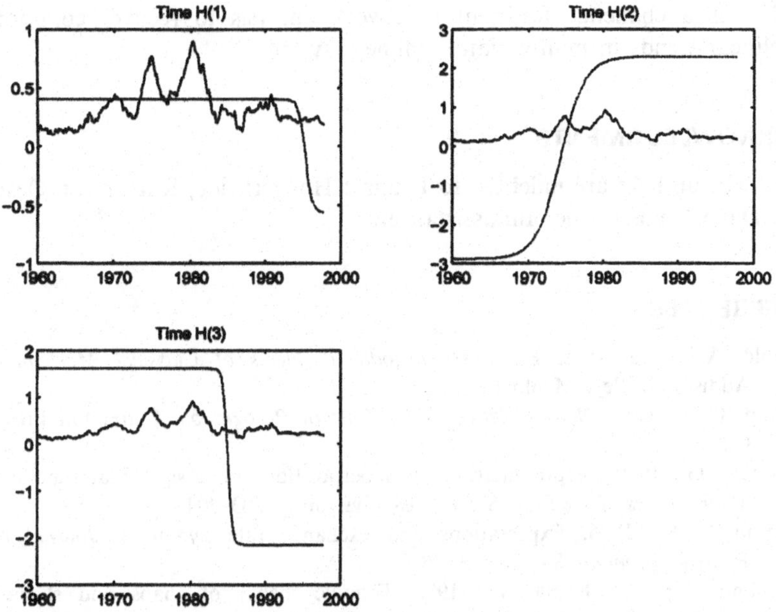

FIGURE 17 Output of hidden layer cells of $nn(t)$ (x-network case with $H=1$).

trade-off is shifting over time. We have also presented an approximation of the dates of these transitions. It is of considerable interest to find an economic explanation of these time transitions. This is a topic which is, however, beyond our exploratory data analysis.

5 CONCLUSION

In this chapter we gave a brief exposition of neural networks and their flexibility in handling complex patterns in economic data. A descriptive method to prune the size of neural nets in order to avoid overfitting is summarized. It is shown how a neural network is used to recover the dynamic properties of a nonlinear system, in particular, its stability by making use of the Lyapunov exponent. A two-stage network has been introduced where the usual nonlinear model is combined with time transitions, which may be handled by neural networks. The empiricial examples on nonlinear trends in real exchange rates and a time-varying Philips curve using U.S. data indicate the applicability of the proposed procedures. Further research is needed to allow for more than one output and it is a challenge for neural network analysis to recover common nonlinear trends in multivariate nonlinear systems.

ACKNOWLEDGMENTS

The authors are indebted to Lennart Hoogerheide, Rutger van Oest, and Daan Ooms for helpful assistance.

REFERENCES

Arnold, V.I. and Avez, Z., 1988. *Ergodic Problems of Classical Mechanics.* Addison-Wesley, Menlo Park, CA.

Bishop, C.M., 1995. *Neural Networks for Pattern Recognition.* Clarendon Press, Oxford.

Cybenko, G., 1989. Approximation by superpositions of a sigmoidal function. *Mathematics of Control, Signals and Systems* **2**, 303–304.

Dornbusch, R., 1976. Expectations and exchange rate dynamics. *Journal of Political Economy* **84**, 1161–1176.

Eckmann, J.P. and Ruelle, D., 1985. Ergodic theory of chaos and strange attractors. *Reviews of Modern Physics* **57**, 617–656.

Frankel, J.A., 1979. On the mark: A theory of floating exchange rates based on real interest differentials. *American Economic Review* **69**, 610–622.

Funahashi, K., 1989. On the approximate realization of continuous mapping by neural networks. *Neural Networks* **2**, 183–192.

Gallant, A.R. and White, H., 1989. There exists a neural network that does not make avoidable mistakes. *Proceedings of the International Conference on Neural Networks*, San Diego, 1988, IEEE Press, New York.

Gallant, A.R. and White, H., 1992. On learning the derivatives of an unknown mapping with multilayer feedforward networks. *Neural Networks* **5**, 129–138.

Granger, C.W.J. and Teräsvirta, T., 1993. *Modelling Nonlinear Economic Relationships.* Oxford University Press, New York.

Grassberger, P. and Procaccia, I., 1983. Measuring the strangeness of strange attractors. *Physica* **9D**, 189–208.

Guckenheimer, J. and Holmes, P., 1983. *Nonlinear Oscillations, Dynamical Systems, and Bifurcations of Vector Fields.* Springer-Verlag, New York.

Hecht-Nielsen, R., 1987. Kolmogorov mapping neural network existence theorem. *Proceedings of the IEEE First International Conference on Neural Networks*, San Diego, pp 11–13.

Hecht-Nielsen, R., 1989. Theory of the backpropagation neural network. *Proceedings of the International Joint Conference on Neural Networks*, Washington, D.C., pp 593–605.

Hecht-Nielsen, R., 1990. *Neurocomputing.* Addison-Wesley, Menlo Park, CA.

Hertz, J. Krogh, A., and Palmer, R.G., 1991. *Introduction to the Theory of Neural Computation.* Addison-Wesley, Reading, MA.

Hornik, K., Stinchcombe, M., and White, H., 1989. Multilayer feedforward networks are universal approximators. *Neural Networks* **2**, 359–366.

Hornik, K., Stinchcombe, M., and White, H., 1990. Universal approximation of an unknown mapping and its derivatives using multilayer feedforward networks. *Neural Networks* **3**, 551–560.

Kaashoek, J.F. and van Dijk, H.K., 1991. A note on the detection of chaos in medium sized time series. *International Series of Numerical Mathematics*, vol. **97**. Birkhäuser Verlag, Basel.

Kaashoek, J.F. and van Dijk, H.K., 1994. Evaluation and application of numerical procedures to calculate Lyapunov exponents. *Econometric Reviews* **13**, 123–137.

Kaashoek, J.F. and van Dijk, H.K., 1998. A simple strategy to prune neural networks with an application to economic time series. Report 9854/A, Econometric Institute, Erasmus University Rotterdam.

Kaashoek, J.F. and van Dijk, H.K., 1999. Neural network analysis of varying trends in real exchange rates. Report EI9915/A, Econometric Institute, Erasmus University Rotterdam.

Kaashoek, J.F. and van Dijk, H.K., 2002. Neural network pruning applied to res. exchange rate analysis, *Journal of Forecasting* **21**, 559–577.

Kolmogorov, A.N., 1957. On the representation of continuous functions of many variables by superposition of continuous functions of one variable and addition. *American Mathematical Monthly Translation* **28**, 953–956. (Russian original in *Doklady Akademii Nauk SSSR* **144**, 55–59.)

Leshno, M., Lin, V.Y., Pinkus, A., and Schocken, S., 1993. Multilayer feedforward networks with a nonpolynomial activation function can approximate any function. *Neural Networks* **6**, 861–867.

Malinvaud, E., 1970. *Statistical Methods of Econometrics.* North-Holland, Amsterdam.

Meese, R.A. and Rogoff, K., 1988. Was it real? The exchange rate–interest differential relation over the modern floating rate period. *Journal of Finance* **43**, 933–948.

Mozer, M.C. and Smolensky, P., 1989. Skeletonization: A technique for trimming the fat from a network via relevance assessment. In: D.S. Touretzky ed. *Advances in Neural Information Processing Systems*, vol. 1, Morgan Kaufmann, San Mateo, CA.

Press, W.H., Flannery, B.P., Teukolsky, S.A. and Vetterling, W.T., 1988. *Numerical Recipes*. Cambridge University Press, Cambridge, UK.

Rumelhart, D.E., Hinton, G.E., and Williams, R.J., 1986. Learning representations by error propagation. In: D.E. Rumelhart, J.L. McClelland, and the PDP Research Group eds, *Parellel Distributed Processing*, vol. 1. MIT Press, Cambridge, MA.

Sargent, T.J., 1999. *The Conquest of American Inflation*. Princeton University Press, Princeton, NJ.

Schotman, P. and van Dijk, H.K., 1991. A Bayesian analysis of the unit root in real exchange rates. *Journal of Econometrics* **49**, 195–238.

Simpson, P.K., 1990. *Artificial Neural Systems: Foundations, Paradigms, Applications and Implementations*. Pergamon Press, Oxford.

Takens, F., 1981. Detecting strange attractors in turbulence. In: D.A. Rand and L.S. Young, eds. *Dynamical Systems and Turbulence*. Springer-Verlag, Berlin.

Theil, H., 1971. *Principle of Econometrics*. Wiley, New York.

Van Dijk, D., 1999. *Smooth Transition Models*: Extensions and Outlier Robust Inference. Tinbergen Institute Research Series, No. 200. Erasmus University Rotterdam.

White, H., 1989. Some asymptotic results for learning on single hidden layer feedforward network models. *Journal of the American Statistical Association* **84**, 1003–1013.

White, H., 2000. A reality check for data snooping. *Econometrica* **68**, 1097–1127.

13

Real-Time Forecasting with Vector Autoregressions: Spurious Drift, Structural Change, and Intercept Correction

Ronald Bewley
The University of New South Wales, Sydney, Australia

1 INTRODUCTION

In its simplest form, a vector autoregression (VAR) is an unrestricted reduced-form model that expresses each variable as a linear function of a constant and the lags of that and each other variable in the system. Since each equation in a VAR has the same regressors, they can be estimated separately by OLS. However, even in moderately sized systems with, say, six variables and four lags of each, and a constant term, there are 25 parameters to be estimated in each equation so that over-parameterization has often been cited as the main cause of poor forecasting performance.

Since their introduction in the late 1970s, one strand of research has focused on reducing the parameter space by imposing stochastic restrictions, which imply that each variable follows a random walk with drift (Litterman, 1980, 1986; Doan *et al.*, 1984). These restrictions are justified by appealing to the common observation that most economic time series are integrated to order one, $I(1)$. The combination of these priors with data-determined parameter estimates produces the variant commonly known as the BVAR, or Bayesian VAR, with the so-called Litterman priors after its originator. However, the estimation procedure is more properly known as mixed estimation (Theil and Goldberger, 1961) and it can trivially be implemented using OLS with a data-augmentation dummy variable procedure (Robertson and Tallman, 1999).

The original BVAR models have become commonplace in the literature with their relative success usually being attributed to the general reduction in the parameter space (e.g., Artis and Zhang, 1990; Funke, 1990; Holden and Broomhead, 1990; Shoesmith, 1995). In a dissenting note, Bewley (2000a) argues that this improvement is more likely due to the correction of the downward bias in the estimated unit roots (or near-unit roots) that has been shown to increase with the number of variables in the system (Abadir *et al.*, 1999). More recently, Sims and Zha (1998) have further developed the BVAR methodology with priors that have more general applicability and an estimation procedure that is more in the true Bayesian spirit.

A separate line of enquiry has focused on pretesting the time series for unit roots and, where appropriate, testing for cointegration in the spirit of Engle and Granger (1987) and Johansen (1988). While these methods typically result in only a minimal reduction in the parameter space, the imposition of unit roots have been shown to produce marked improvements in forecast accuracy over both an unrestricted VAR and a BVAR with Litterman priors (Bewley, 2000a, 2000b). The main justification for this result is that forecasts of $I(1)$ variables from vector error correction (VEC) models rapidly approach linear time trends. Therefore, it is the precision of the estimate of the constant term, rather than short-run dynamics, that is the key issue in forecasting with $I(1)$ time series (Clements and Hendry, 1995, 1998b; Bewley, 2000b).

Clements and Hendry's (1995, 1998b) taxonomy of the sources of forecast errors shows that, if it is assumed that the data-generating process (dgp) is stable over time, and a consistent estimator is used, there are only three important sources of forecast error in a VAR in levels with $I(1)$ data: the failure to impose unit roots; over-parameterization arising from an unrestricted reduced form specification; and, importantly, the precision in estimation of the drift parameters. In the standard VAR framework, the constant terms are nonlinear functions of all of the drift parameters and all of the (many) coefficients on the lagged variables, so that it is a non-trivial matter to address the question of controlling the estimates of drift in that framework. However, Bewley (2000b) proposed a new representation which allows a simple, possibly Bayesian, direct estimation of the drift parameters in a VEC. Thus, in case where some variables contain drift while others do not, the so-called mixed drift case, substantial improvements in forecast accuracy can be made by imposing exact, or stochastic, restrictions on the drift parameters.

Since the estimation of drift is of such importance in VEC models, Bewley and Yang (2000) developed a number of statistics to test for structural breaks specifically in that parameter vector. While system tests previously existed (Andrews, 1993; Andrews and Ploberger, 1994), these new tests allow for a simple variable-by-variable test for parameter constancy that are more appropriate when only some of the variables are subject to structural breaks in drift.

The methods introduced in this chapter are designed to take a properly specified model that has been successfully subjected to appropriate diagnostic tests and use that model in a real-time setting. In practice, structural breaks are commonplace and potentially an extremely important contributor to poor forecasting performance. As a result, Hendry and Clements (1994), Clements and Hendry (1995, 1996, 1998a,b), and Hendry (1996, 1997) have suggested correcting the intercept at each forecast origin to realign the forecasts after a break has occurred within the sample. This correction is achieved by adding the most recent residual, or the average over the last few periods, to the one-step forecasts or to the forecasts at all lead times.

While Clements and Hendry advocate intercept corrections with VEC models, they argue that they have less merit in a VAR model in the differences (DVAR). Indeed, they consider a (non intercept corrected) DVAR, which results from a mis-specifying of a VEC model, as a viable alternative to intercept correcting a VEC model. In that sense, Clements and Hendry consider breaks in the long-run means of the equilibrium-(error-) correction terms as more important, and more likely, sources of forecast error than breaks in drift parameters. This view is supported by Eitrheim *et al.*'s (1990) experimentation with a macroeconometric model of the Norwegian Central Bank.

While there is clear merit in the Clements and Hendry prescription, the results in Bewley (2000b) suggest that a fruitful approach might be to allow the drift parameters, rather than the constant terms in a VEC, to change in the recent past in both a classical and a mixed-estimation sense. While these methods can be applied to either VEC or DVAR models, the emphasis in this chapter is on the latter. It is found that substantial improvements in forecast accuracy can be made by intercept correcting DVARs.

The proposed intercept correction methods are introduced in Section 2 for the general VEC model. The design of the simulation experiment is presented in Section 3 and the results are reported in Section 4. Conclusions are drawn in Section 4.

2 INTERCEPT CORRECTION

Consider an $n \times 1$ vector of $I(1)$ times series, y_t, that can be represented by a VAR with p lags, VAR(p):

$$y_t = a + \sum_{i=1}^{p} A_i y_{t-i} + u_t \tag{1}$$

If $\sum A_i - I$ has rank $r > 0$, Eq. (1) can be written as VEC(p):

$$\Delta y_t = a + \alpha[\beta' y_{t-1}] + \sum_{i=1}^{p-1} B_i \Delta y_{t-i} + u_t \tag{2}$$

where $B_j = -\sum_{i=j+1}^{p} A_i$ and $\sum_{i=1}^{p} A_i - I = \alpha\beta'$, with α and β being $n \times r$. If the time series are not cointegrated ($r = 0$), then a VAR in Δy_t with $p - 1$ lags is appropriate, DVAR($p - 1$).

Clements and Hendry (1998b) write Eq. (2) as

$$[\Delta y_t - \delta] = \alpha[\beta' y_{t-1} - \gamma] + \sum_{i=1}^{p-1} B_i[\Delta y_{t-i} - \delta] + 1 \tag{3}$$

where γ is the mean of the equilibrium-correction terms, $\beta' y_{t-1}$, and δ is the mean of Δy_t, the drift parameters. The forecasts of Eq. (3) approach linear time trends with slope δ. If follows from a comparison of Eqs. (2) and (3) that the constant term, a, in Eq. (2) is a nonlinear function of the drift, δ, and all of the B_j:

$$a = \left[\left(I - \sum_{i=1}^{p-1} B_i\right)\delta - \alpha\gamma\right] \tag{4}$$

Bewley (2000b) proposed an alternative representation of Eqs. (2) or (3) that allows direct estimation of the drift parameter vector, δ. This is achieved by applying the Bewley (1979) transformation to Eq. (3):

$$\Delta y_t = \delta + \zeta[\beta' y_{t-1} - \gamma] + \sum_{i=1}^{p-2} C_i \Delta^2 y_{t-i} + v_t \tag{5}$$

where the C_i are nonlinear functions of all of the B_j and ζ is a nonlinear function of all the B_j and α. Since each Equation in (5) is exactly identified, two-stage least squares (2SLS), conditional on a superconsistent estimate of β, and using a constant term, the equilibrium-correction terms and the $p-1$ lags of Δy_t as instruments, is equivalent to indirect least squares, implying an exact relationship between the estimates of the three basic forms of a VEC: Eqs. (2), (3), and (5). Bewley (2000b) introduced Eq. (5) to enable some of the drift parameters to be set to zero with exact or stochastic restrictions in the mixed-drift situation. Equation (5) is referred to as the mixed-drift VAR (MDVAR).

When restrictions are placed on the MDVAR, the equations are no longer exactly identified and a system estimation approach, such as three-stage least squares (3SLS), or iterated 3SLS, is appropriate.

If it is assumed that the means of the equilibrium-correction terms are constant over time, but that all of the drift parameters have changed in a recent time period, an appropriately defined dummy variable can be added to each equation of (5) or, equivalently, Eqs. (2) or (3). This is one notion of intercept correction. If only a subset of the drift parameters has changed, it is not possible to impose linear restrictions on the dummy variable augmentation to Eq. (2) or (3). However, this is not the case in the MDVAR, and so possible gains might be had by exactly or stochastically restricting the dummy variable parameters in that framework.

On the other hand, if the long-run means of the equilibrium correction terms, ζ, vary, it is not possible to distinguish between changes in the two long-run means, δ and ζ with a single dummy variable for each equation. Importantly, changes in any one element of ζ implies that all of the dummy variable coefficients are nonzero in either the VEC or MDVAR. In that sense, it is not clear that there is an advantage to pursuing the latter when equilibrium-correction terms are included and are possibly subject to structural breaks. The notion that the equilibrium-correction terms do not break, but the individual series do, has been termed co-breaking by Hendry and discussed in Clements and Hendry (1998b) and elsewhere.

Since it may be the case that only a subset of the time series has been subjected to a structural break, the mixed-break case, but that this is not known to the model builder, consideration of a Bayesian intercept-correction method, which introduces a dummy to the equations in (5), but shrinks the coefficients to zero, may prove fruitful. Such a procedure has two potential advantages. First, shrinking the intercept correction toward zero implies that the stochastically restricted MDVAR is less likely to over-react to noise. Second, those series without breaks would have dummy variable coefficients with large standard errors, resulting in restricted estimates that would be attracted toward zero more so than for series that did break. When a similarly defined Bayesian intercept correction is applied to a standard DVAR or VEC, the mixed-break case implies that all of the dummy variables coefficients are typically nonzero so that this differential shrinking to zero cannot easily be exploited.

In order to implement this stochastic restriction on a DVAR or VEC, it is useful to consider the *i*th equation of a DVAR as a standard regression equation:

$$w = X\pi + \varepsilon \tag{6}$$

using obvious notation (conditional on the equilibrium-correction terms having been defined) and assuming that the first column of X is the dummy variable introduced for intercept correction. The stochastic restriction can be written as

$$r = R\pi + \nu \tag{7}$$

where $r=0$ and R is a row vector of zeroes except for the first element, which is unity.

It is necessary to prespecify the variance of ν, which controls the so-called tightness of the prior; $\mathrm{Var}(\nu)=\lambda^2$. Assuming $\mathrm{Var}(\epsilon)=\sigma^2 I$, stacking Eqs. (6) and (7) produces the augmented model:

$$\begin{vmatrix} w \\ r \end{vmatrix} = \begin{vmatrix} X \\ R \end{vmatrix} \Pi + \begin{vmatrix} \epsilon \\ \nu \end{vmatrix} \tag{8}$$

which might be estimated by GLS since σ is not necessarily equal to λ. Alternatively, Eq. (7) can be multiplied through by $\hat{\sigma}/\lambda$ and this equation stacked with Eq. (6):

$$\begin{vmatrix} w \\ r \end{vmatrix} = \begin{vmatrix} X \\ R^* \end{vmatrix} \Pi + \begin{vmatrix} \epsilon \\ \nu^* \end{vmatrix} \tag{8$'$}$$

where the first element of R^* is $\hat{\sigma}/\lambda$ and the remaining elements are zero. The variance of ϵ and ν^* are now approximately the same. Thus, Eq. (8) can be estimated by OLS; the so-called dummy variable augmentation procedure. In keeping with the BVAR literature, this estimate of σ is taken from a $p-1$th order autoregression of $\Delta y_{i,t}$.

When exact or stochastic restrictions are applied to the MDVAR, certain modifications are necessary. The correlation of X with ϵ requires X to be replaced by its least squares predictions $\hat{X}$ in typical 2SLS fashion. However, as $\lambda \to 0$, the resulting estimates do not approach those that would have been computed if the dummy variables has been excluded from the system, as would be the case if this were applied in the DVAR/VEC situation. This difference arises because the instruments used in creating $\hat{X}$ include the dummy variable which is then being "excluded" by the restrictions. The difference might be expected to be very small when the restriction is true but, when a significant break has occurred, sizeable differences can emerge in parameter estimates and forecast accuracy. To rectify this problem, the first-stage regressions can be similarly augmented with the dummy variable procedure to shrink the first-stage dummy variable coefficients to zero. That is, the $T+1$th observations are all zero except for that corresponding to the dummy variable that is set to $\hat{\sigma}/\lambda$. In this way, the restricted estimates approach those from not including an intercept correction as $\lambda \to 0$ and the unrestricted estimates, including an intercept-correction term, as $\lambda \to \infty$. While $\hat{\sigma}$ is taken from autoregressions of $\Delta y_{i,t}$ in the case of a DVAR/VEC, $\hat{\sigma}$ is taken from the Bewley transformation of that autoregression in the MDVAR.

3 MONTE CARLO DESIGN

In a series of papers, Hendry, and Clements and Hendry, have argued that intercept corrections to VEC models perform on a par with DVARs without corrections and that corrections to the latter tend to reduce forecast performance due to an over-reaction to noise. Since the analysis of Section 2 suggests that there might be greater potential to make improvements in MDVARs than in DVARs or VEC models, the ensuing experiment has been designed to investigate the merits, if any, of intercept correcting an MDVAR when there is no cointegration. Thus, consider a two-equation mixed-break dgp with T observations and no cointegration:

$$\Delta y_t = \delta_1 + \phi_1(\Delta y_{t-1} - \delta_1) + \phi_2(\Delta x_{t-1} - \delta_2 - \delta^* D_{t-1}) + \varepsilon_{1,t} \tag{9}$$

$$\Delta x_t = \delta_2 + \delta^* D_t + \phi_3(\Delta x_{t-1} - \delta_2 - \delta^* D_{t-1}) + \varepsilon_{2,t} \tag{10}$$

where ε_i is n.i.d (0,1) and $E(\varepsilon_1\ \varepsilon_2)=0$. D_t is the dummy variable that introduces the structural break in the drift of x_t but not y_t. Note that this dummy is lagged when it operates on Δx_{t-1} to align it with the timing of the break.

Equivalently, (9) and (10) can be expressed as

$$\Delta y_t = \gamma_1 + \mu_1 D_{t-1} + \phi_1 \Delta y_{t-1} + \phi_2 \Delta x_{t-1} + \varepsilon_{1,t} \tag{9$'$}$$

$$\Delta x_t = \gamma_2 + \delta^* D_t + \mu_2 D_{t-1} + \phi_3 \Delta x_{t-1} + \varepsilon_{2,t} \tag{10$'$}$$

where $\gamma_1 = (1-\phi_1)\delta_1 - \phi_2\delta_2$; $\gamma_2 = (1-\phi_3)\delta_2$; $\mu_1 = -\phi_2\delta^*$; $\mu_2 = -\phi_3\delta^*$. Thus, both constants shift in a DVAR as shown in Eqs. (9$'$) and (10$'$), providing that $\phi_2 \neq 0$, but there is no observable shift in y_t because the shift in x_t is offset by the change in the drift term in Eq. (9). However, in a real-time forecasting situation, there is a delay of one period before the constant term in Eq. (9$'$) changes as, in the first period of the change, there is a single unit element in D_t corresponding to the last observation and, hence, in that special case $D_{t-1}=0$.

It is assumed that $D_t=0$ if $t<T-q$ and $D_t=1$ if $t\geq q$. That is, the break in drift is permanent (and continues throughout the forecasting period). Moreover, q is also assumed to be sufficiently small as to make the shift difficult to detect. Any previous changes in drift are assumed to have been modeled. Following the construction of the Andrews and Ploberger (1994) and Bewley and Yang (2000) Wald tests, breaks are assumed not to have occurred at the ends of the sample, say within 15% of the end of the sample. This serves as a guide to the value of q. Two values of q are considered 0 and 4. Clearly, even if a significant shock were to be detected in the last observation, corresponding to $q=0$, it would be impossible to test whether the shock was temporary or permanent. In two cases corresponding to $q=0$ and 4, it is assumed in estimation that the location of the hypothesized break is known. In a third case, the assumed location, q^*, is set to 4 while the true value is $q=0$. This serves two purposes. In reality, the location of a break would not be known and second, it is consistent with Clements and Hendry's approach of averaging recent errors so that overreaction to shocks is less prevalent.

Two basic dgps are considered. The first, referred to as the "diagonal dgp," has $\phi_1=\phi_3=0.5$ and $\phi_2=0$. This implies that the constant term in the first equation of the DVAR does not depend on the size of the break in the second, δ^*; i.e., there is no "transmitted" structural break. In the second dgp, $\phi_1=\phi_3=0.5$ so that the speed of adjustment to equilibrium is unchanged, but $\phi_2=-0.8$, implying that both constant terms in the DVAR depend on δ^*; i.e., there is a "transmitted" structural break in the first equation but not the first variable. Three values of δ^* are considered: 1, 2, and 4 and these compare to $\text{Var}(\Delta y_t)=\text{Var}(\Delta x_t)=1/(1-0.5^2)=4/3$. Thus, $\delta^*=4$ is unlikely to be encountered in many applications but is included to assess the impact of the proposed methods in extrema.

Since none of the forecasting comparisons depend on the drift parameters, both were arbitrarily set equal to one. These values, together with the other parameters chosen for the dgp, produce time series that visually have the characteristics of many macroeconomic time series. That is, they exhibit strong drift, but deviations from a linear trend are pronounced. A stylized example for a single replication of the triangular dgp is given in Fig. 1 for $T=85$ and $\delta^*=0,1,2,4$ with a break occurring at observation 71. In the case of the diagonal dgp, the y series has the same dgp as x but without a break.

Each experiment is based on 5000 replications and $T=75$, 150 with the forecast comparisons being based on the 1- and 10-step ahead mean

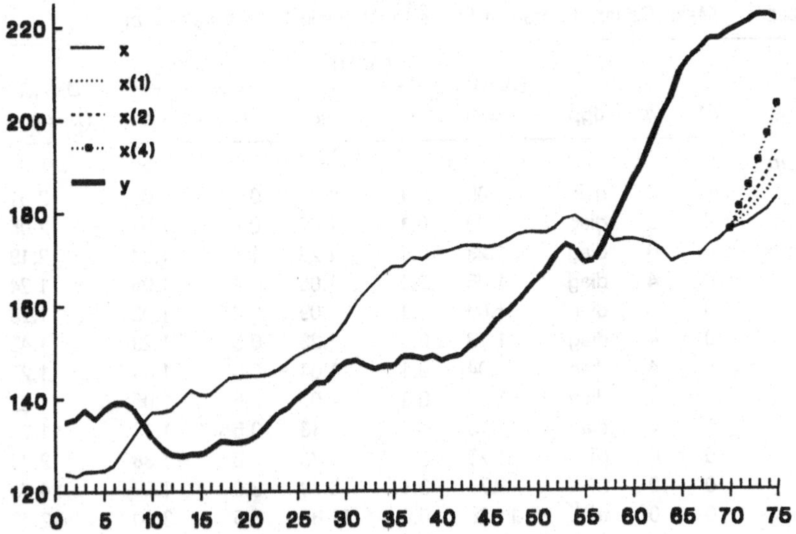

FIGURE 1 Stylized time series with breaks.

squared forecast error (msfe). In each case, the true lag length of one in first differences was chosen so as to abstract from any forecast inaccuracy due to selecting the wrong lag length. The DVAR was estimated with no correction and with an unrestricted intercept correction. The forecasts from these are necessarily the same as those from the comparable MDVARs. In addition, each form was estimated in its Bayesian form for 41 different values of $\lambda = (0.05, 0.1, 0.2, 0.3, 0.4, 0.5, 0.6, 0.7, 0.8, 0.9, 1.0, 1.1, 1.2, 1.3, 1.4, 1.5, 1.6, 1.7, 1.8, 1.9, 2.0, 2.5, 3.0, 3.5, 4.0, 4.5, 5.0, 6.0, 7.0, 8.0, 9.0, 10.0, 15.0, 20.0, 25.0, 30.0, 35.0, 40.0, 50.0, 75.0, 100.0)$.

4 RESULTS

The 1-step and 10-step-ahead msfes for the DVAR with no intercept correction ($\lambda = 0$) and with an unrestricted intercept correction ($\lambda \rightarrow \infty$), together with the Bayesian variant of the DVAR and MDVAR that produced the smallest msfe for that estimator and that equation, are presented in Tables 1 and 2 for the case $T = 75$.

When there is no transmitted drift, as there is in the diagonal dgp, the first variable has no structural break and, therefore, it would be expected that the no intercept correction model would produce the smallest msfe. On the other hand, the second variable does exhibit a break,

TABLE 1 Mean Squared Forecast Errors for Variable 1 (no break in drift)

δ^*	q^*	q	dgp	DVAR ($\lambda=0$)	DVAR λ	DVAR mse	MDVAR λ	MDVAR mse	DVAR ($\lambda=\infty$)
1-step									
1	0	0	diag	1.05	0.1	1.05	0.7	1.04	2.01
2	0	0	diag	1.09	0.1	1.09	0.8	1.07	2.05
4	0	0	diag	1.23	0.1	1.23	1.0	1.20	2.19
1	0	4	diag	1.05	0.1	1.05	0.4	1.05	1.24
2	0	4	diag	1.09	0.1	1.09	0.4	1.08	1.28
4	0	4	diag	1.23	0.1	1.23	0.5	1.23	1.43
1	4	4	diag	1.04	0.1	1.04	0.5	1.04	1.22
2	4	4	diag	1.07	0.1	1.07	0.5	1.06	1.22
4	4	4	diag	1.13	0.1	1.13	0.5	1.13	1.23
1	0	0	tri	1.73	0.1	1.73	1.8	1.38	2.73
2	0	0	tri	3.78	0.1	3.78	3.0	1.68	4.78
4	0	0	tri	11.88	0.0	11.88	3.5	3.10	12.90
1	0	4	tri	1.73	0.0	1.73	1.0	1.67	1.93
2	0	4	tri	3.78	0.0	3.78	1.7	3.25	4.00
4	0	4	tri	11.88	0.0	11.88	2.0	9.68	12.16
1	4	4	tri	1.58	1.5	1.24	2.5	1.16	1.25
2	4	4	tri	2.77	∞	1.33	3.5	1.22	1.33
4	4	4	tri	4.50	∞	1.57	∞	1.57	1.57
10-step									
1	0	0	diag	39.19	0.0	39.19	0.2	39.17	375.87
2	0	0	diag	39.73	0.0	39.73	0.3	39.70	405.19
4	0	0	diag	42.08	0.0	42.08	0.3	41.99	524.02
1	0	4	diag	39.19	0.0	39.19	0.2	39.16	100.67
2	0	4	diag	39.73	0.0	39.73	0.2	39.69	103.31
4	0	4	diag	42.08	0.0	42.08	0.2	42.02	114.32
1	4	4	diag	39.18	0.0	39.18	0.2	39.15	100.15
2	4	4	diag	39.75	0.0	39.75	0.2	39.73	101.01
4	4	4	diag	43.01	0.0	43.01	0.2	43.00	103.83
1	0	0	tri	122.88	0.0	122.88	1.2	119.81	1516.56
2	0	0	tri	160.11	0.0	160.11	1.5	143.18	3338.25
4	0	0	tri	311.47	0.0	311.47	1.5	251.77	10616.73
1	0	4	tri	122.88	0.0	122.88	0.7	121.79	316.26
2	0	4	tri	160.11	0.0	160.11	0.8	155.70	480.62
4	0	4	tri	311.47	0.0	311.47	0.8	296.54	1148.62
1	4	4	tri	121.26	0.0	121.26	1.1	112.95	281.68
2	4	4	tri	153.00	0.0	153.00	1.3	121.95	336.38
4	4	4	tri	268.94	0.0	268.94	1.6	192.50	497.74

TABLE 2 Mean Squared Forecast Errors for Variable 2 (break in drift)

δ^*	q^*	q	dgp	DVAR ($\lambda=0$)	DVAR λ	DVAR mse	MDVAR λ	MDVAR mse	DVAR ($\lambda=\infty$)
1-step									
1	0	0	diag	1.35	0.6	1.22	1.6	1.19	2.24
2	0	0	diag	2.16	0.9	1.32	2.5	1.24	2.92
4	0	0	diag	5.47	1.1	1.47	3.0	1.21	5.62
1	0	4	diag	1.35	0.4	1.32	0.9	1.32	1.42
2	0	4	diag	2.16	∞	1.82	6.0	1.82	1.82
4	0	4	diag	5.47	∞	3.45	∞	3.45	3.45
1	4	4	diag	1.29	0.5	1.17	1.3	1.16	1.29
2	4	4	diag	1.78	0.8	1.21	2.5	1.17	1.32
4	4	4	diag	2.66	1.2	1.24	4.5	1.15	1.39
1	0	0	tri	1.34	0.6	1.21	1.5	1.19	2.26
2	0	0	tri	2.13	0.9	1.32	2.5	1.25	2.96
4	0	0	tri	5.36	1.1	1.49	3.0	1.21	5.76
1	0	4	tri	1.34	0.4	1.31	1.0	1.30	1.40
2	0	4	tri	2.13	∞	1.78	5.0	1.78	1.78
4	0	4	tri	5.36	∞	3.34	∞	3.34	3.34
1	4	4	tri	1.28	0.5	1.16	1.4	1.15	1.27
2	4	4	tri	1.75	0.8	1.20	2.5	1.16	1.30
4	4	4	tri	2.51	1.2	1.22	4.5	1.14	1.38
10-step									
1	0	0	diag	114.37	0.6	78.97	1.5	76.39	427.98
2	0	0	diag	344.50	0.9	110.47	2.5	95.47	659.90
4	0	0	diag	1265.87	1.1	152.36	3.0	86.38	1587.59
1	0	4	diag	114.37	0.3	106.20	0.9	106.27	135.42
2	0	4	diag	344.50	∞	235.24	7.0	234.90	235.24
4	0	4	diag	1265.87	∞	629.90	∞	629.90	629.90
1	4	4	diag	103.84	0.5	66.69	1.3	64.01	102.80
2	4	4	diag	288.39	0.8	78.56	2.5	67.71	108.21
4	4	4	diag	849.37	1.1	86.82	5.0	60.73	123.44
1	0	0	tri	115.13	0.6	78.56	1.6	76.52	412.05
2	0	0	tri	347.46	0.9	110.48	2.5	94.38	635.81
4	0	0	tri	1277.64	1.1	159.79	3.5	89.84	1529.30
1	0	4	tri	115.13	0.4	106.45	0.9	106.78	134.29
2	0	4	tri	347.46	∞	238.43	7.0	238.13	238.43
4	0	4	tri	1277.64	∞	651.79	∞	651.79	651.79
1	4	4	tri	105.50	0.5	66.03	1.4	63.56	100.58
2	4	4	tri	298.05	0.8	77.46	2.5	66.79	106.22
4	4	4	tri	922.26	1.2	85.30	5.0	60.00	123.31

but it does not follow that an unrestricted intercept correction is to be preferred. This is the essence of the Clements and Hendry view that DVARs should not be intercept corrected. However, as the size of the break increases, unrestricted correction would become increasingly preferable.

From Table 1 it can be noted that both the 1-step and 10-step forecasts with the diagonal dgp do indeed have lower msfes when $\lambda=0$ than when $\lambda \to \infty$, and that the estimated λ is either zero or very small in the case of the DVAR. However, very small gains in msfe are apparent in the MDVAR with most of the estimated values of λ being in the approximate range 0.2–0.5 for 1-step ahead forecasts. Small values of λ are also estimated at the longer lead time.

Turning to Table 2, it can be noted that the ranking of the forecast performance for the second variable, and the diagonal dgp for the cases of $\lambda=0$ and $\lambda \to \infty$, depends on the size of the break, δ^*, the number of periods for which the break has been in operation, q, and the number of periods for which the dummy variable assumes a change, q^*. Not surprisingly, it is better to intercept correct (in the unrestricted sense) when δ^* is large, except when the break occurred in the last observation and residual averaging ($q^* > 0$) does not occur. Whether or not λ is estimated in the DVAR or the MDVAR form, the msfe is never lower in either extreme and, sometimes, the gains from the Bayesian procedure are very substantial and greater than that achieved by residual averaging.

In the case of the diagonal dgp and variable 2, it can be also noted from Table 2 that the estimated values of λ are typically larger than those in Table 1, reflecting the existence of the structural change and its accommodation with an intercept correction. Most values for the DVAR are in the range 0.5–1.1 while that range is approximately 1.5–5.0 for the MDVAR.

The results for the triangular dgp are quite different from those with a diagonal dgp. In the triangular case, both equations in the DVAR experience a structural break while only the second breaks in the MDVAR form. For variable 1, unrestricted intercept correction is preferable to no correction in the case $q^*=q=4$, but the reverse is true for the longer lead time. In most cases the Bayesian DVAR performed similarly to the no-correction DVAR and substantial gains can be found in the MDVAR variant. The estimated values of λ are mostly in the range 0.7–3.0 for the Bayesian MDVAR.

Not surprisingly, some of the greatest gains in msfe can be found in the triangular dgp and variable 2, and the Bayesian MDVAR is far superior to the Bayesian DVAR in many of the cases. Indeed, the Bayesian MDVAR produces the smallest (or equivalent) msfe in all but one of the 72 cases reported in Tables 1 and 2.

Two problems arise from the comparisons apparent in Tables 1 and 2. First, the value of λ is not restricted to be the same in the two equations. Unless the model builder has different information about the likelihood of a break in each variable, a common value of λ should be used. Second, it is not possible to ascertain the sensitivity of the msfe for λ in the neighborhood of the minimum.

The former problem can be addressed by combining the msfe for the two equations for a common value of λ. Naturally, the results depend on the weights attached to each series. In Table 3, the two msfes are simply added, reflecting the similar trends, scale, and variability of the two time series. In no case is the Bayesian DVAR, the unrestricted intercept correction, or the no-correction DVAR, preferred to the Bayesian MDVAR and, in some cases, the gains are very large indeed. The greatest gains are made when the break is very large, $\delta^*=4$, and the dgp is triangular. However, even when $\delta^*=1$, $q=q^*=4$, and the dgp is triangular, the gain in msfe for 10-step ahead forecasts of the Bayesian MDVAR is 20% over not correcting and 50% over intercept correction.

The sensitivity of the results to changes in λ is considered in Figs. 2–9 for a break of $\delta^*=2$ and $q=q^*=4$. In each case there are three pairs of lines corresponding to variable 1, with the suffix (1), variable 2, with the suffix (2), and the sum of two msfes, with the suffix (1+2). Within each pair, one line refers to the Bayesian DVAR and the other to the Bayesian MDVAR. A log scale is used for λ to allow for additional detail at smaller values of λ.

It can be seen by comparing Figs. 2–9 that there is a pronounced minimum sum of msfes for the Bayesian MDVAR. In some cases, such as in Fig. 7, the scale of the graph masks the full extent of the gain for that model. These minima also occur for variable 2, but it is the combination of the flatness of the curve for variable 1 over a wide range of λ, and the minimum of the curve for variable 2, that make the overall gain so pronounced.

The Bayesian MDVAR produces a lower sum of msfe, particularly at the longer lead time, and for a reasonably wide range of λ. In each case, an arbitrarily selected value of 1 would have produced a forecasting model preferable to either an unrestricted correction or no correction. While a more extensive set of experiments is warranted, the results of this design are sufficiently encouraging to explore the use of the Bayesian intercept correction MDVAR in practice.

The results for $T=150$ are not presented, but are available from http://economics.web.unsw.edu.au/people/rbewley, owing to their similarity with those given here. In essence, the additional observations improve the precision of the estimates of the basic DVAR but there are still

Table 3 Sum of Mean Squared Forecast Errors for Variables 1 and 2

δ^*	q^*	q	dgp	DVAR ($\lambda=0$)	DVAR λ	DVAR mse	MDVAR λ	MDVAR mse	DVAR ($\lambda=\infty$)
1-step									
1	0	0	diag	2.40	0.5	2.31	1.3	2.26	4.25
2	0	0	diag	3.25	0.8	2.55	2.0	2.44	4.97
4	0	0	diag	6.70	1.1	2.97	3.0	2.67	7.82
1	0	4	diag	2.40	0.2	2.38	0.6	2.37	2.66
2	0	4	diag	3.25	0.7	3.04	1.8	3.03	3.10
4	0	4	diag	6.70	∞	4.87	∞	4.87	4.87
1	4	4	diag	2.33	0.4	2.25	1.0	2.23	2.51
2	4	4	diag	2.85	0.7	2.35	2.0	2.31	2.54
4	4	4	diag	3.79	1.1	2.43	4.5	2.37	2.62
1	0	0	tri	3.07	0.5	2.97	1.7	2.57	4.99
2	0	0	tri	5.91	0.8	5.18	2.5	2.93	7.74
4	0	0	tri	17.23	1.1	13.53	3.5	4.36	18.66
1	0	4	tri	3.07	0.3	3.05	1.0	2.97	3.33
2	0	4	tri	5.91	0.7	5.68	1.9	5.07	5.78
4	0	4	tri	17.23	∞	15.50	2.5	13.46	15.50
1	4	4	tri	2.86	0.9	2.44	1.7	2.32	2.53
2	4	4	tri	4.51	3.5	2.63	3.0	2.39	2.63
4	4	4	tri	7.01	∞	2.95	7.0	2.86	2.95
10-step									
1	0	0	diag	153.57	0.4	132.40	1.1	129.50	803.85
2	0	0	diag	384.23	0.7	201.87	1.9	186.21	1065.09
4	0	0	diag	1307.95	1.0	316.36	3.0	246.07	2111.60
1	0	4	diag	153.57	0.2	149.47	0.6	149.33	236.09
2	0	4	diag	384.23	0.7	318.83	1.8	317.46	338.55
4	0	4	diag	1307.95	∞	744.23	∞	744.23	744.23
1	4	4	diag	143.02	0.4	118.49	1.0	116.24	202.95
2	4	4	diag	328.14	0.6	149.87	2.0	142.32	209.22
4	4	4	diag	892.38	1.0	174.36	4.5	156.44	227.27
1	0	0	tri	238.01	0.2	234.99	1.5	196.96	1928.61
2	0	0	tri	507.57	0.3	486.34	2.0	252.89	3974.06
4	0	0	tri	1589.10	0.4	1493.40	2.5	493.36	12146.03
1	0	4	tri	238.01	0.1	237.50	0.9	228.74	450.55
2	0	4	tri	507.57	0.2	499.65	1.5	427.80	719.05
4	0	4	tri	1589.10	0.4	1523.64	2.0	1203.70	1800.42
1	4	4	tri	226.76	0.3	215.36	1.3	177.24	382.26
2	4	4	tri	451.04	0.5	336.04	1.9	202.49	442.59
4	4	4	tri	1191.20	0.9	544.45	3.5	358.20	621.05

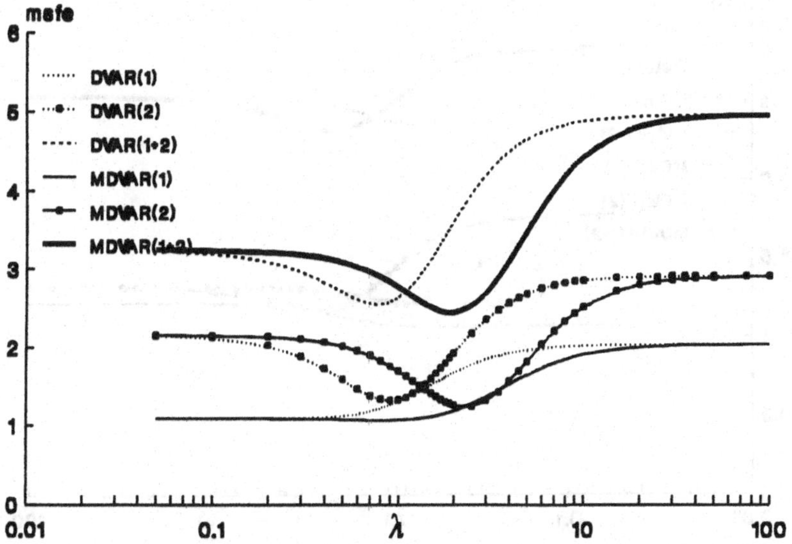

FIGURE 2 One-step-ahead msfe for diagonal dgp and $\delta^*=2$, $q^*=q=0$.

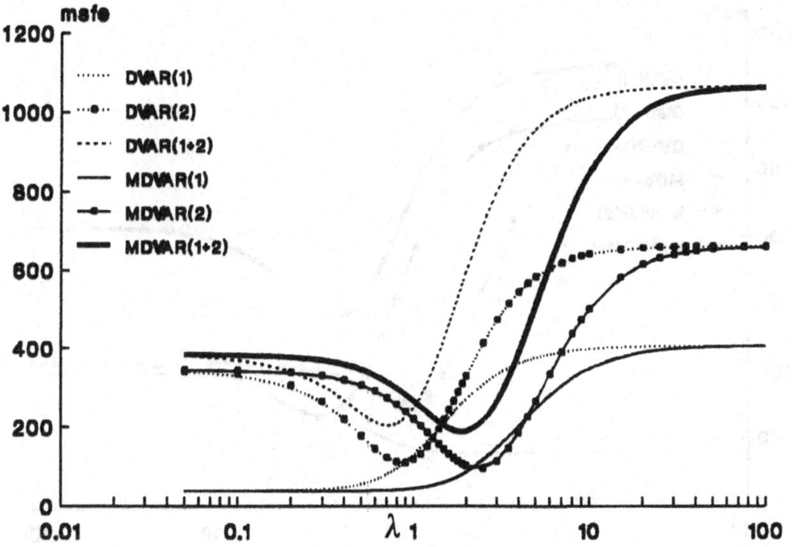

FIGURE 3 Ten-step-ahead msfe for diagonal dgp and $\delta^*=2$, $q^*=q=0$.

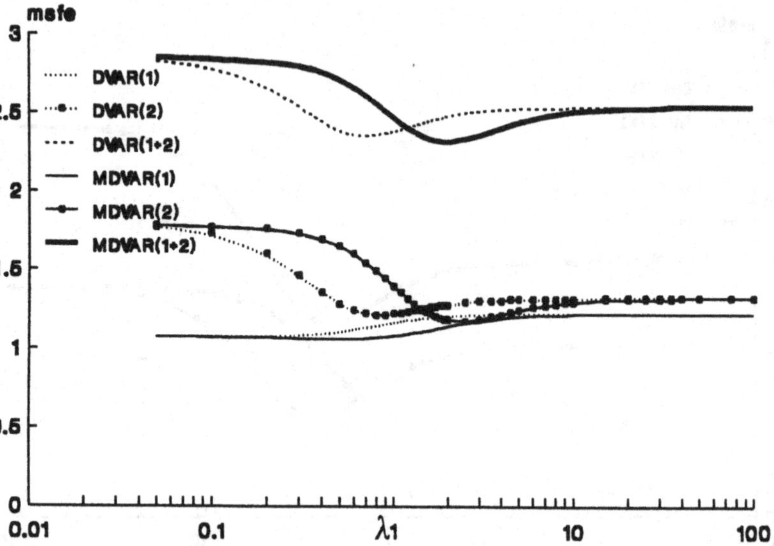

FIGURE 4 One-step-ahead msfe for diagonal dgp and $\delta^*=2$, $q^*=q=4$.

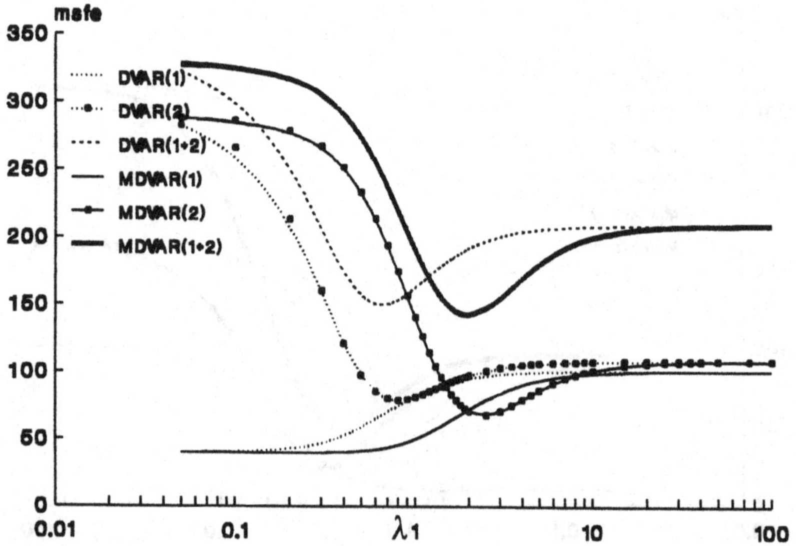

FIGURE 5 Ten-step-ahead msfe for diagonal dgp and $\delta^*=2$, $q^*=q=4$.

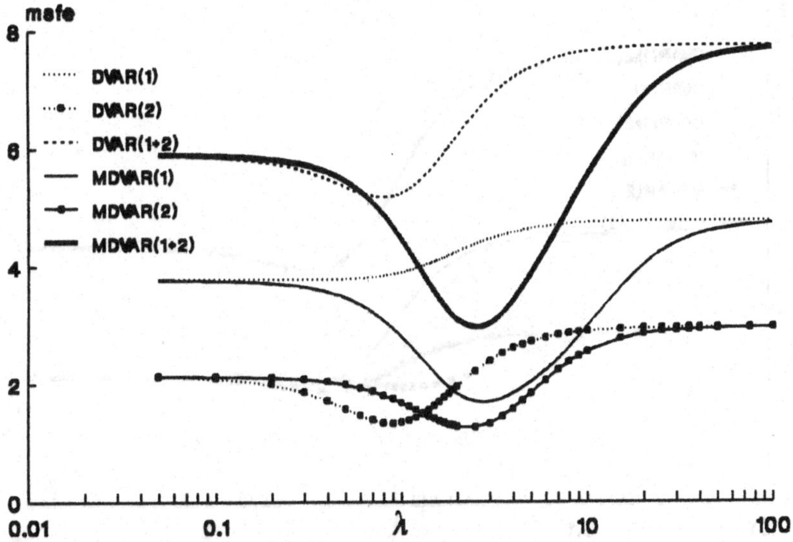

FIGURE 6 One-step-ahead msfe for triangular dgp and $\delta^* = 2$, $q^* = q = 0$.

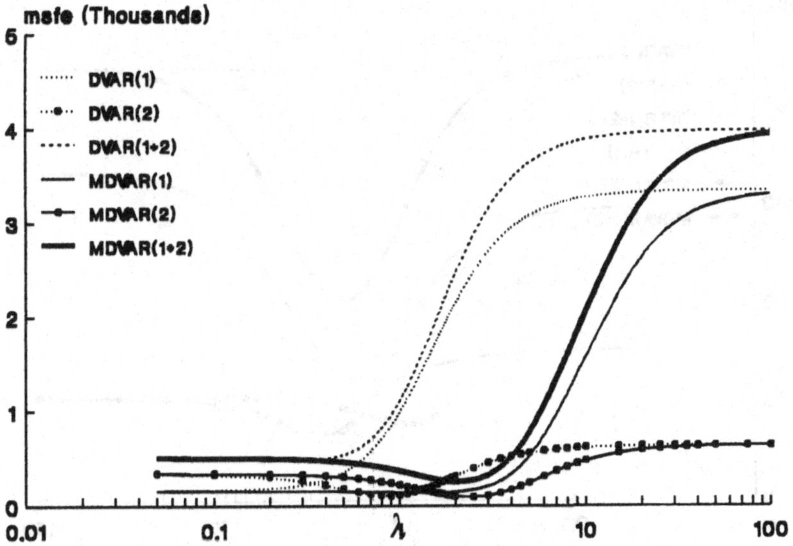

FIGURE 7 Ten-step-ahead msfe for triangular dgp and $\delta^* = 2$, $q^* = q = 0$.

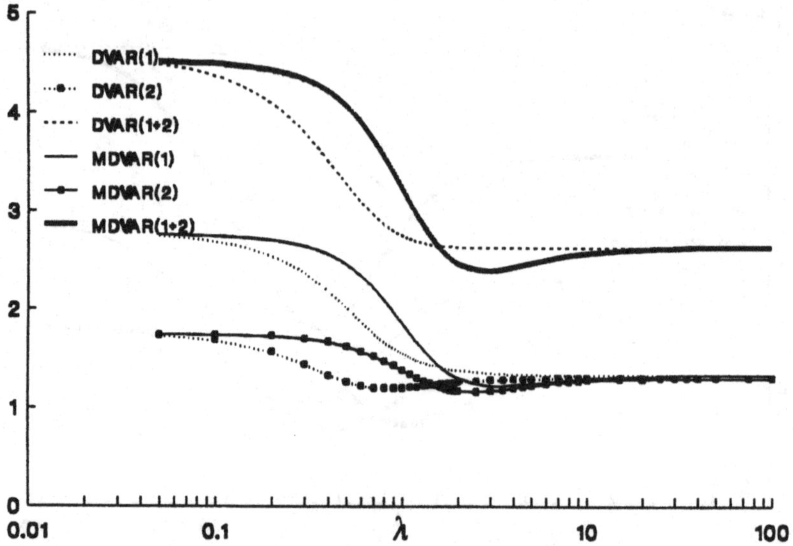

FIGURE 8 One-step-ahead msfe for triangular dgp and $\delta^* = 2$, $q^* = q = 4$.

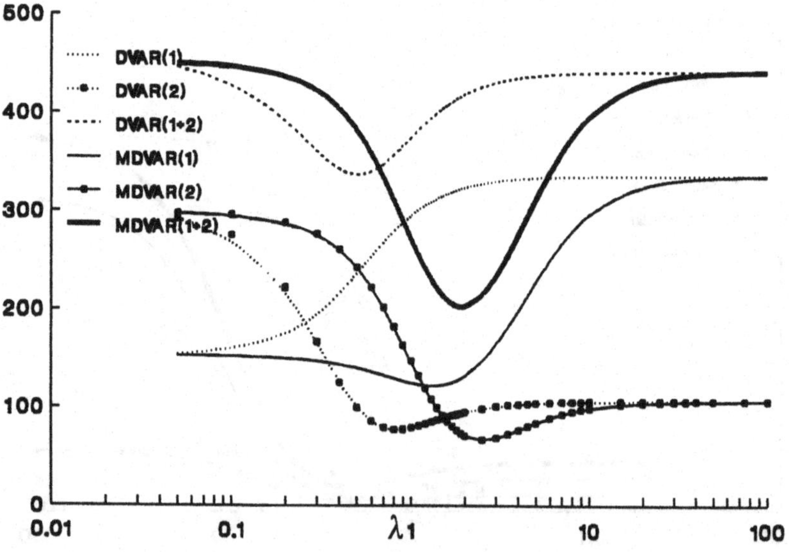

FIGURE 9 Ten-step-ahead msfe for triangular dgp and $\delta^* = 2$, $q^* = q = 4$.

only one ($q=0$) or five ($q=4$) observations in the new regime after the break. In that sense, the forecast relativities of the model do not depend on sample size.

5 CONCLUSIONS

A new strategy has been proposed for estimating and testing VAR forecasting models, including the possibility of structural change, in a series of papers: Bewley (2000a, 2000b) and Bewley and Yang (2000), and the present chapter. The essence of the approach is to focus attention on the drift parameters. Spurious drift, arising from poorly determined drift estimates of variables that would not be expected to trend, cause problems of forecast credibility and, while relevant restrictions could be placed on a standard VEC or DVAR model, the nonlinearity of the restrictions make such estimation cumbersome and prone to optimization problems. The proposed framework for estimating VARs—the mixed-drift VAR (MDVAR)—facilitates this estimation and has been shown to permit more accurate and more credible forecasts to be produced (Bewley, 2000b), using either exact or stochastic restrictions.

The computational time to estimate the proposed parameterization is trivial owing to there being an analytical solution. However, there is much scope for experimenting with different priors when stochastic restrictions are being contemplated, but even this is quite feasible. For example, 5000 replications, 41 sets of priors, 75–150 observations, and two equations took only 13 min with a Pentium II processor and a program written in FORTRAN.

In a natural extension to Bewley (2000b), Bewley and Yang (2000) proposed a set of statistics to test for structural change specifically in the drift parameters. The MDVAR framework simplifies the allocation of any structural change in the system to attribute it to possibly a subset of the variables. The present chapter considers possible structural change in the most recent observations, and an automatic Bayesian procedure for adapting to structural change in drift, without overacting to noise, is proposed. It has been found that, contrary to previous research (Clements and Hendry), significant gains in mean squared forecast error can be made by intercept correcting a VAR in differences. Indeed, when there is a significant degree of causality in the system, and structural change is confined to a subset of the variables, substantial gains in forecast accuracy can occur with the new estimator.

While the usual caveat of drawing conclusions from Monte Carlo experiments is at least as applicable here as elsewhere, the results are sufficiently encouraging to warrant further investigation. In particular, work is progressing on extending these results to discriminating between structural breaks in drift and in the long-run means of equilibrium-correction terms.

ACKNOWLEDGMENTS

This work was funded by an *Australian Research Council* grant (A10007132). This chapter has benefited greatly from many useful discussions with, and comments from, David Hendry, whom I thank. The procedures developed in this chapter were inspired by "Hans" Theil's repeated least squares and mixed estimation innovations. Sadly, Hans, my dear friend, mentor, and sometime coauthor passed away while I was writing this chapter. I would like to acknowledge Hans' support and encouragement over the years. I have also benefited from presentations at the UNSW time series group weekly workshops and the excellent research assistance provided by Glenn Anderson.

REFERENCES

Abadir, K.M., Hadri, K. and Tzavalis, E., 1999. The influence of VAR dimensions on estimator biases. *Econometrica* **67**, 63–181.

Andrews, D.W.K., 1993. Tests for parameter instability and structural change with unknown change point. *Econometrica* **61**, 825–856.

Andrews, D.W.K. and Ploberger, W., 1994. Optimal tests when a nuisance parameter is present only under the alternative. *Econometrica* 62, 1383–1414.

Artis, M.J. and Zhang, W., 1990. BVAR forecasts for the G-7. *International Journal of Forecasting* **6**, 349–362.

Bewley, R.A., 1979. The direct estimation of the equilibrium response in a linear dynamic model. *Economics Letters* **3**, 357–361.

Bewley, R., 2000a. Forecast accuracy, coefficient bias and Bayesian vector autoregressions. *Mathematics and Computers in Simulation*, forthcoming.

Bewley, R., 2000b. Controlling spurious drift in macroeconomic forecasting models. Unpublished manuscript. University of New South Wales, Sydney.

Bewley, R. and Yang, M., 2000. Testing for structural breaks in the long-run means of VARs. Unpublished manuscript, University of New South Wales, Sydney.

Clements, M.P. and Hendry, D.F., 1995. Forecasting in cointegrated systems. *Journal of Applied Econometrics* **10**, 127–146.

Clements, M.P. and Hendry, D.F., 1996. Intercept corrections and structural change. *Journal of Applied Econometrics* **11**, 475–494.

Clements, M.P. and Hendry, D.F., 1998a. Forecasting economic processes. *International Journal of Forecasting* **14**, 111–131.

Clements, M.P. and Hendry, D.F. 1998b. *Forecasting Economic Time Series.* Cambridge University Press, Cambridge, UK.

Doan, T., Litterman, R. and Sims, C., 1984. Forecasting and conditional projections using realistic prior distributions. *Econometric Reviews*, **3**, 1–100.

Eitrheim, Ø., Husbeø, T.A. and Nymoen, R., 1999. Equilibrium correction vs. differencing in macroeconomic forecasting. *Economic Modelling* **16**, 515–544.

Engle, R.F. and Granger, C.W.J., 1987. Cointegration and error correction: representations, estimation and testing. *Econometrica* **55**, 252–276.

Funke, M., 1990. Assessing forecast accuracy of monthly vector autoregressive models. *International Journal of Forecasting* **6**, 363–378.

Hendry, D.F., 1996. On the constancy of time series econometric equations. *Economic and Social Review* **27**, 401–422.

Hendry, D.F., 1997. The econometrics of macroeconomic forecasting. *Economic Journal* **107**, 1330–1357.

Hendry, D.F. and Clements, M.P., 1994. On a theory of intercept corrections in macroeconomic forecasting. In: Holly, S. ed., *Money, Inflation and Employment: Essays in Honour of James Ball.* Edward Elgar, Aldershot, UK, pp. 160–182.

Holden, K. and Broomhead, A., 1990. An examination of vector autoregressive forecasts for the U.K. economy. *International Journal of Forecasting* **6**, 11–23.

Johansen, S., 1988. Statistical analysis of cointegrating vectors. *Journal of Economic Dynamics and Control* **12**, 231–254.

Litterman, R.B., 1980. A Bayesian procedure for forecasting with vector autoregressions. Unpublished manuscript, Massachusetts Institute of Technology, Cambridge, MA.

Litterman, R.B., 1986. Forecasting with Bayesian vector autoregressions—five years of experience. *Journal of Business and Economic Statistics* **4**, 25–37.

Robertson, J.C. and Tallman, E.W., 1999. Vector autoregressions: forecasting and reality. *Federal Reserve Bank of Atlanta Economic Review*, First Quarter, 4–18.

Shoesmith, G.L., 1995. Multiple cointegrating vectors, error correction, and forecasting with Litterman's model. *International Journal of Forecasting* **11**, 557–567.

Sims, C.A. and Zha, T.A., 1998. Bayesian methods for dynamic multivariate models. *International Economic Review* 39, 949–968.

Theil, H. and Goldberger, A.S., 1961. On pure and mixed statistical estimation in economics. *International Economic Review* **2**, 65–78.

14

Econometric Modeling Based on Pattern Recognition via the Fuzzy C-Means Clustering Algorithm

David E. A. Giles
University of Victoria, Victoria, British Columbia, Canada

Robert Draeseke
Treaty Negotiations, Ministry of Attorney General, Government of British Columbia, Victoria, British Columbia, Canada

1 INTRODUCTION

The specification and estimation of any econometric relationship poses significant challenges. This is especially true when it comes to the choice of functional form, as the latter is not always suggested or prescribed by the underlying economic theory. Any mis-specification of the functional form of an econometric model can have serious consequences for statistical inference—e.g., the parameter estimates may be inconsistent. In response to the rigidities associated with explicit and rigid parametric functional relationships, and to avoid the consequences of their mis-specification, a more flexible nonparametric approach to their formulation and estimation is often very attractive (e.g., Pagan and Ullah, 1999).

Such nonparametric estimation does, however, have its drawbacks. Most notably, in order to perform well, multivariate kernel regression requires a reasonably large sample of data, and it founders when the number of explanatory variables grows. With regard to the latter point there are really two problems. First, there is evidence that the small-sample properties of the multivariate kernel estimator can be quite unsatisfactory (e.g., Silverman, 1986). Second, the rate of convergence of the kernel estimator decreases as the dimension of the model grows—this is the so-called "curse of dimensionality." Recently, Coppejans (2000) has dealt with this problem by using cubic B-spline estimation in conjunction with earlier results from Kolmogorov (1957) and Lorentz (1966) that enable a function of several variables to be represented as superpositions of functions of a single variable.

There is considerable appeal in the prospect of finding approaches to the formulation and estimation of econometric relationships that are highly flexible in terms of their functional form, make minimal parametric assumptions, perform well with either small or large data sets, and are computationally feasible even in the face of a large number of explanatory variables. In this chapter we use some of the tools of fuzzy set theory and fuzzy logic in the pursuit of just such an objective.

These tools have been applied widely in many disciplines since the seminal contributions of Zadeh (1965,1987) and his colleagues. These applications are numerous in such areas as computer science, systems analysis, and electrical and electronic engineering. The construction and application of "expert systems" is widespread in domestic appliances, motor vehicles, and commercial machinery. The use of fuzzy sets and fuzzy logic in the social sciences appears to have been limited mainly to psychology, with very little application to the analysis of economic problems and economic data. There are a few such examples in the field of social choice—e.g., Dasgupta and Deb (1996), Richardson (1998), and Sengupta (1999). In the area of econometrics, however, there have been surprisingly few applications of fuzzy set/logic techniques. Indeed, we are aware of only four such examples.

Josef et. al. (1992) used this broad approach in the context of modeling with panel data. Lindström (1998) used fuzzy logic in a rather different way to model fixed investment in Sweden on the basis of the level and variability in the real interest rate. More specifically, he generated an index for aggregate investment using fuzzy sets and operators to combine the imprecise information associated with the interest rate. He found that this approach provided a simple method of modeling inherent nonlinearities in a flexible (nonparametric) manner. It should be noted that Lindström's analysis involved no *statistical inference*, as such. This same technique was used by Draeseke and Giles (1999,2002) to generate an index of the New Zealand underground economy, and it has now been programmed into the SHAZAM (2000) econometrics package. The advantage of using this particular technique in that context was that the variable of interest is intrinsically unobservable.

Shepherd and Shi (1998) used fuzzy set/logic theory in a rather different manner in their analysis of U.S. wages and prices. Their modeling technique involved clustering the data into fuzzy sets, estimating the relationship of interest over each set, and then combining each submodel into a single overall model by using the membership functions associated with the clustered data together with the Takagi and Sugeno (1985) approach to fuzzy systems. Their analysis also highlighted the ability of such fuzzy analysis to detect and measure nonlinear relationships in a very flexible way, even though the underlying submodels may be linear. Indeed, this is a specific and attractive feature of the Takagi–Sugeno analysis. In this chapter we draw on the approach of Shepherd and Shi and adapt it to model a number of interesting relationships. We discuss some of the issues that arise when adopting this semiparametric approach in the specification and estimation of an econometric model, and by way of specific practical applications we illustrate some of its merits relative to standard parametric regression and conventional (kernel) nonparametric regression.

The use of fuzzy sets and fuzzy logic is just one of several possible approaches within the general area of expert systems and artificial intelligence. In the context that is of interest to us here, an obvious competitive approach is that based on (artificial) neural networks (e.g., White, 1992; White and Gallant, 1992; Bishop, 1995). Neural networks also facilitate the modeling of nonlinear relationships whose underlying form is not parametrically constrained at the outset. However, our adaptation of the Takagi–Sugeno/Shepherd–Shi (TSSS) methodology has an important and appealing advantage over the use of neural networks. Namely, the basic relationships that are estimated have a direct economic interpretation, and they can be related to the underlying economic theory in a useful way. The same is not generally true in the case of neural networks. Moreover, and as we shall see in more detail below, the TSSS methodology is computationally efficient, especially in as much as it requires only one "pass" through the data. In contrast, neural networks usually have to be "trained," and this involves more protracted computations.

In Section 2 we outline the background concepts in fuzzy set/logic theory that we will be using. A crucial component of our analysis is the implementation of the so-called "fuzzy c-means" algorithm to partition the data set in a flexible manner, and this algorithm is considered in some detail in Section 3. In Section 4 we describe our variant of the TSSS modeling procedure, and Section 5 presents the results associated with several illustrative empirical applications of this fuzzy technology. Some concluding remarks are given in Section 6.

2 FUZZY SETS AND FUZZY LOGIC

As was noted above, fuzzy logic relates to the notion of fuzzy sets, the theoretical basis for which is usually attributed to Zadeh (1965). Under regular set theory, elements either belong to some particular set or they do not. Another way of expressing this is to say that the "degree of membership" of a particular element with respect to a particular set is either unity or zero. The boundaries of the sets are hard, or "crisp." In contrast to this, in the case of fuzzy sets, the degree of membership may be any value on the continuum between zero and unity, and a particular element may be associated with more than one set. Generally, this association involves different degrees of membership with each of the fuzzy sets. Just as this makes the boundaries of the sets fuzzy, it makes the location of the centroid of the set fuzzy as well.

To consider an illustrative situation relating to a single economic variable, suppose we wish to distinguish between situations of excess supply and those of excess demand in relation to the price of some good. In traditional set theory we would have a situation such as

$$S_s = \{p : p > p^*\}$$
$$S_d = \{p : p < p^*\}$$
$$S_e = \{p^*\}$$

as the crisp sets representing those prices associated with excess supply, excess demand, and the equilibrium price (p^*), respectively. Any particular price, say \$5, would definitely be in one and only one of these sets. That is, if its degree of membership with S_i is denoted u_i, then $u_j = 1$ implies that $u_k = 0$, for all $k \neq j$; for $k, j = s, d, e$. In contrast to this, in a fuzzy set framework the sets S_s, S_d, and S_e would not have sharp boundaries and a particular price (such as \$5) would be associated to some degree or other with each of these sets. For instance, its degrees of membership might be $u_s = 0.2$, $u_d = 0.3$, $u_e = 0.5$. Note that although the u_i's sum to unity in this example, in general they need not, and they should not be equated with the "probabilities" that the price of \$5 lies in each set.

It should also be noted that in the example above, the concepts involved are defined in a crisp manner: "excess supply," "excess demand," and "equilibrium." More generally, fuzzy set theory is just as capable of handling vague linguistic concepts, such as "a rather high price," or "a very low demand." So, we could broaden the above example to allow for fuzzy sets involving prices associated with "a very high excess supply," "a moderate excess supply," "a small excess supply," and so on. Again, the boundaries of these sets would be fuzzy, and degrees of membership would map prices to the sets.

The application of the inductive premise to fuzzy concepts poses some difficulties—not all of the usual laws of set theory are satisfied. In particular, the "law of the excluded middle" is violated, so a different group of set operators must be adopted. So, for example, the "union" operator is replaced by the "max" operator, "intersection" is replaced by "min," and "complement" is replaced by subtraction from unity. Under these "fuzzy" operators the commutative, associative, distributive, idempotency, absorption, excluded middle, involution, and De Morgan's Laws are all satisfied in the context of fuzzy sets. For example, if the universal set is $U = \{a, b, c, d\}$, let the fuzzy sets S_1 and S_2 be defined as $S_1 = \{0.4/a, 0.5/b, 1/d\}$ and $S_2 = \{0.1/a, 0.2/b, 0.5/c, 0.9/d\}$. Here, the numbers are the "degrees of membership," which map the elements of U to S_1 and S_2. Then, $A \cup B = \{0.4/a, 0.5/b, 0.5/c, 1/d\}$ and $A \cap B = \{0.1/a, 0.2/b, 0/c, 0.9/d\}$, $S_1^c = \{0.6/a, 0.5/b, 1/c, 0/d\}$, and $S_2^c = \{0.9/a, 0.8/\mathrm{b}, 0.5/\ c, 0.1/d\}$.

Fuzzy sets can be used in conjunction with logical operators in a manner that will enable us to construct models in a flexible way. More specifically, we can use the degrees of membership that associate the values of one or more input variables with different fuzzy sets, together with logical "IF," "THEN," "AND" operations, to derive membership values that associate an output variable with one or more fuzzy sets. For example, in the case of the demand for a particular good, we might have a fuzzy rule of the form:

> "IF the own-price of the good is quite low, AND the price of the only close substitute good is rather high, THEN the demand for the good in question will be fairly high."

Or, to anticipate one of our empirical examples later in this chapter.

> "IF the interest rate is fairly high, AND income (output) is relatively low, THEN the demand for money will be quite low."

Of course, in general, there will be more than one fuzzy rule, and the output variable may have potential membership in more than one fuzzy set:

Rule 1: "IF the interest rate is fairly high, AND income (output) is relatively low, THEN the demand for money will be quite low."

Rule 2: "IF the interest rate is average, AND income (output) is relatively high, THEN the demand for money will be moderately high."

In such situations, a fuzzy outcome for the output variable could be inferred by taking account of the relevant membership values, and the

MAX/MIN operators associated with fuzzy sets. However, a fuzzy outcome is not usually adequate. For instance, in the last example, there is only limited interest in knowing that the fuzzy model predicts that in a certain period the demand for money will be "moderately high," or "quite low." It would be much more helpful to know, e.g., that the model predicted a demand of $10 million. In other words, we need a way of "defuzzifying" the predictions of the model.

The Takagi and Sugeno (1985) and Sugeno and Kang (1985) approach to dealing with this issue involves modifiying the above methodology to one that involves rules of the form:

Rule 1′: "IF the interest rate (r) is fairly high, AND income (Y) is relatively low, THEN the demand for money is $M = f_1(r, Y) = M_1$."

Rule 2′: "IF the interest rate (r) is average, AND income (Y) is relatively high, THEN the demand for money is $M = f_2(r, Y) = M_2$."

Here, f_1 and f_2 are crisp, and generally parametric, functions that yield numerical values for M. In their simplest form these functions would be linear relationships. The various values (M_1, M_2 etc.) emerging from these rules can then be combined by taking a weighted average, based on the degrees of membership associated with the input variables and the fuzzy input sets. This is described in more detail in Section 4. First, however, we need to give further consideration to the definition and construction of the fuzzy input sets, and this involves grouping or "clustering" the input data appropriately.

3 THE FUZZY C-MEANS ALGORITHM

3.1 Overview of the Algorithm

In our modeling analysis, which is described in detail in Section 4, we need to determine the partitioning of the sample data for each explanatory (input) variable into a number of clusters. These clusters have "fuzzy" boundaries, in the sense that each data value belongs to each cluster to some degree or other. Membership is not certain, or "crisp." Having decided on the number of such clusters to be used, some procedure is then needed to locate their midpoints (or more generally, their centroids) and to determine the associated membership functions and degrees of membership for the data points. To this end, Shepherd and Shi (1998) used a variant of the "fuzzy c-means" (FCM) algorithm. (The latter is sometimes termed the fuzzy k-means algorithm in the literature.) The FCM algorithm is really a generalization of the "hard" c-means algorithm. It appears to date from Ruspini (1970), although some of the underlying concepts were explored by MacQueen (1967). The FCM algorithm is closely associated with such early contributors as Bezdek (1973) and Dunn (1973, 1974, 1977), and is widely used in such fields as pattern recognition, for instance.

The algorithm provides a method of dividing up the n data points into c fuzzy clusters (where $c < n$), while simultaneously determining the locations of these clusters in the appropriate space. The data may be multidimensional, and the metric that forms the basis for the usual FCM is "squared error distance." The underlying mathematical basis for this procedure is as follows. Let x_k be the kth (possibly vector) data point ($k = 1, 2, \ldots, n$). Let v_i be the center of the ith (fuzzy) cluster ($i = 1, 2, \ldots, c$), let $d_{ik} = \|x_k - v_i\|$ be the distance between x_k and v_i, and let u_{ik} be the "degree of membership" of data point k in cluster i, where

$$\sum_{i=1}^{c} (u_{ik}) = 1$$

The objective is to partition the data points into the c clusters and simultaneously locate those clusters and determine the associated "degrees of membership," so as to minimize the functional:

$$J(U, v) = \sum_{i=1}^{c} \sum_{k=1}^{n} (u_{ik})^m (d_{ik})^2$$

There is no prescribed manner for choosing the exponent parameter, m, which must satisfy $1 < m < \infty$. In practice, $m = 2$ is a common choice. In the case of crisp (hard) memberships, $m = 1$.

In broad terms, the FCM algorithm involves the following steps:

1. Select the initial location for the cluster centers.
2. Generate a (new) partition of the data by assigning each data point to its closest cluster center.
3. Calculate new cluster centers as the centroids of the clusters.
4. If the cluster partition is stable then stop. Otherwise go to step 2 above.

In the case of fuzzy memberships, the Lagrange multiplier technique generates the following expression for the membership values to be used at step 2 above:

$$u_{ik} = 1 \Big/ \left\{ \sum_{j=1}^{n} [(d_{ik})^2 / (d_{jk})^2]^{1/(m-1)} \right\}$$

Notice that a singularity will arise if $d_{jk}=0$ in the above expression. This occurs if, at any point, a cluster center exactly coincides with a data point. This can be avoided at the start of the algorithm and generally will not arise subsequently in practice due to machine precision. If the memberships of data points to clusters are "crisp," then

$$u_{ik} = 0; \quad \forall i \neq j$$
$$u_{jk} = 1; \quad \text{js.t. } d_{jk} = \min\{d_{ik}, i = 1, 2, \ldots, c\}$$

The updating of the cluster centers at step 3 above is obtained via the expression:

$$v_i = \left[\sum_{k=1}^{n} (u_{ik})^m x_k\right] \bigg/ \left[\sum_{k=1}^{n} (u_{ik})^m\right]; \qquad i = 1, 2, \ldots, c$$

The fixed-point nature of this problem ensures the existence of a solution. See Bezdek (1981, Chap. 3) for complete and more formal mathematical details. As will be apparent, the FCM algorithm is simple to program. We have chosen to do this using standard commands in the SHAZAM (2000) econometrics package—this makes it convenient to compare our results with those from the other standard regression modeling techniques that are available in that package.

3.2 Some Computational Issues

Among the issues that arise in the application of the FCM are the following. First, a value for the number of clusters and the initial values for their centres have to be provided to start the algorithm. It appears that the results produced by the FCM algorithm can be sensitive to the choice of these start-up conditions. Second, being based on a "squared error" measure of distance, the algorithm can be sensitive to noise or outliers in the data.

In relation to the first of these issues, several proposals have been made. Among these are the following. Linde et al. (1980) proposed the binary splitting (BS) technique to initialise the clusters. The disadvantage of the BS technique is that it requires at least 10 or 20 data points in each cluster, and in practice this may not be attainable. Huang and Harris (1993) extended this approach to the "direct search binary splitting" technique. Tou and Gonzales (1974) proposed the "simple cluster-seeking" (SCS) method, which deals not only with the initial value issue, but also provides the clusters themselves. However, the SCS technique is *not* invariant to the order in which the data points are considered, and it also depends on certain "threshold" settings.

Yager and Filev (1992) introduced the so-called "mountain method" for deciding on the initial values. Their approach is apparently simple, but it becomes computationally burdensome in the context of multidimensional data. Consequently, Chiu (1994) proposed a modification of their approach, called "subtractive clustering," that does not suffer the same problem. This is the approach used by Shepherd and Shi (1998). Babu and Murty (1993) used genetic algorithms to initialize the clusters, and Katsavounidis et al. (1994) proposed a further method (KKZ). More recently, Al-Daoud and Roberts (1996) proposed two initialization methods that appear to be especially useful in the context of very large data sets, and which out-performed both the KKZ and SCS techniques with their test data.

In relation to the second of the above issues (noise and outliers in the data), there have again been a number of contributions. Among these are the following. Following on from earlier (computationally expensive) contributions by Weiss (1988) and Jolion and Rosenfeld (1989), Davé (1991) modified the FCM algorithm (still retaining a "squared error" distance concept) so that noisy data points are effectively allocated to a "noise cluster," rather than being allocated to other clusters, and so "contaminating" the latter. More recently, Keller (2000) suggested another related modification involving the use of a modified objective function with a weighting factor added for each data point—the objective being to assign a kind of influence factor to the single data points. Some authors have taken up the usual techniques used in the statistical literature to deal with the sensitivity of least squares to the problem of outliers. That is to say, methods of "robust estimation" have been incorporated into the FCM algorithm, and these have included "M-estimation," "trimmed least squares," and "least absolute deviations" alternatives to the least squares component of the FCM algorithm. A recent example of this approach is that of Frigui and Krishnapuram (1996), who integrate M-estimation into the FCM algorithm. They provide a computationally attractive generalization of the FCM algorithm that deals with *both* the identification of the number of clusters and the allocation of data points to these clusters simultaneously, and which is robust to noise and outliers.

4 FUZZY ECONOMETRIC MODELING

In this section we describe the TSSS methodology that we subsequently apply to a number of econometric model estimation problems. The discussion here is quite expository, and we begin by outlining the analysis in the very simple case where there is a single input variable (other than, perhaps, a constant intercept). So, the fuzzy relationship is of the form:

$$y = f(x) + \varepsilon$$

where the functional relationship will typically involve unknown parameters, and ε is a random disturbance term. While there is no need to make any distributional assumptions about the latter, if this is done then these can be taken into account in the subsequent analysis. If the disturbance has a zero mean, the fuzzy function represents the conditional mean of the output variable, y. To this extent, the framework is the same as that which is adopted in nonparametric kernel regression.

The identification and estimation of the fuzzy model then proceeds according to the following steps:

Step 1: Partition the sample observations for x into c fuzzy clusters, using the FCM algorithm. This generates the membership values for each x value with respect to each cluster, and implicitly it also defines a corresponding partition of the data for y.

Step 2: Using the data for each fuzzy cluster separately, fit the models:

$$y_{ij} = f_i(x_{ij}) + \varepsilon_{ij}; \quad j = 1, \ldots, n_i; \quad i = 1, \ldots, c$$

In particular, if the chosen estimation procedure is least squares, then

$$y_{ij} = \beta_{i0} + \beta_{i1} x_{ij} + \varepsilon_{ij}; \quad j = 1, \ldots, n_i; \quad i = 1, \ldots, c$$

Step 3: Model and predict the conditional mean of y using

$$\hat{y}_k = \left[\sum_{i=1}^{c} (b_{i0} + b_{i1} x_k) u_{ik}\right] \bigg/ \left[\sum_{i=1}^{c} u_{ik}\right]; \quad k = 1, \ldots, n$$

where u_{ik} is the degree of membership of the kth value of x in the ith fuzzy cluster, and b_{im} is the least squares estimator of $\beta_{im}(m = 0, 1)$ obtained using the ith fuzzy partition of the sample.

Step 4: Calculate the predicted "input–output relationship" i.e., the derivative) between x and the conditional mean of y:

$$(\partial \hat{y}_k / \partial x_k) = \left[\sum_{i=1}^{c} (b_{i1} u_{ik})\right] \bigg/ \left[\sum_{i=1}^{c} u_{ik}\right]; \quad k = 1, \ldots, n$$

So, the fuzzy predictor of the conditional mean of y is a weighted average of the linear predictors based on the fuzzy partitioning of the explanatory data, with the weights (membership values) varying continuously through the sample. This latter feature enables nonlinearities to be modeled effectively. In addition, it can be seen that the separate modeling over each fuzzy cluster involves the use of fuzzy logic of the form "IF the input data are likely to lie in this region, THEN this is likely to be the predictor of the output variable," etc. The derivative of the conditional mean with respect to the input variable also has this weighted average structure and the same potential for nonlinearity.

Note that this modeling strategy is essentially a semiparametric one. The parametric assumptions could be relaxed further by using kernel estimation to fit each of the cluster submodels at Step 2 above, in which case the estimated *derivatives* (rather than coefficients) would be weighted at Steps 3 and 4, instead of the parameter estimates. However, some limited experimentation with this variation of the modeling methodology, in the context of the empirical applications described in Section 5, yielded results that were inferior to those based on least squares.

Of course, in general we will be concerned with models that have more than one input (explanatory) variable:

$$y = f(x_1, x_2, \ldots, x_p) + \varepsilon$$

In such cases the steps in the above fuzzy modeling procedure are extended as follows, assuming a linear least squares basis for the analysis for expository purposes:

Step 1′: Separately partition the n sample observations for each x_r into c_r fuzzy clusters (where $r = 1, 2, \ldots, p$), using the FCM algorithm. This generates the membership values for each observation on each x variable with respect to each cluster.

Step 2′: Consider all c possible combinations of the fuzzy clusters associated with the p input variables, where

$$c = \prod_{r=1}^{p} c_r$$

and discard any for which the intersections involve negative degrees of freedom ($n_r < p$). Let the number of remaining cluster combinations be c'.

Step 3′: Using the data for each of these c' fuzzy clusters separately, fit the models:

$$y_{ij} = \beta_{i0} + \beta_{i1}x_{1ij} + \beta_{i2}x_{2ij} + \cdots + \beta_{ip}x_{pij} + \varepsilon_{ij};$$
$$j = 1, \ldots, n_i; \quad i = 1, \ldots, c'$$

Step 4′: Model and predict the conditional mean of y by using

$$\hat{y}_k = \left[\sum_{i=1}^{c'}(b_{i0} + b_{i1}x_{1k} + \cdots + b_{ip}x_{pk})w_{ik}\right] \Big/ \left[\sum_{i=1}^{c'} w_{ik}\right];$$
$$k = 1, \ldots, n$$

where

$$w_{ik} = \prod_{r=1}^{p} \delta_{ij}u_{rjk}; \quad i = 1, \cdots, c'$$

Here, δ_{ij} is a "selector" that chooses the membership value for the jth fuzzy cluster (for the rth input variable) if that cluster is associated with the ith cluster combination ($i = 1, 2, \ldots, c$), and b_{im} is the least squares estimator of β_{im} obtained using the ith fuzzy partition of the sample.

Step 5′: Calculate the predicted "input–output relationships" i.e., the derivatives) between the conditional mean of y and each input variable, x_r:

$$(\partial\hat{y}_k/\partial x_{rk}) = \left[\sum_{i=1}^{c'}(b_{ir}w_{rik})\right] \Big/ \left[\sum_{i=1}^{c'} w_{ik}\right];$$
$$r = 1, \ldots, p; \quad k = 1, \ldots, n$$

Comparing these steps with those in the case of a single input variable, it is clear that the computational burden associated with the fuzzy modeling increases at the same rate as in the case of multiple linear regression as additional explanatory variables are added to the model. Under very mild conditions on the input data and the random error term, fitting the submodels over each fuzzy cluster yields a weakly consistent predictor of the conditional mean of the output variable. The partitioning of the sample into fuzzy clusters, and the determination of the associated membership functions, involves using only the explanatory variable data in a nonstochastic manner. If the explanatory variables are exogenous then so will be the membership values that are used to construct the weighted averages of the least squares predictors in Step 4 above. Then, the fuzzy predictor of the conditional mean of y at Step 4 will be weakly consistent.

5 SOME APPLICATIONS

We have applied the fuzzy econometric modeling described above to a number of illustrative estimation and smoothing problems, and the associated results are described in this section. As was noted in the context of the FCM algorithm, the computation for these applications was undertaken by writing command code for the SHAZAM (2000) econometrics package. We have chosen a number of simple examples that involve both cross-section and time-series data, and which involve various degrees of model complexity in terms of the number of explanatory variables involved.

5.1 Modeling the Earnings-Age Profile

In our first application the relationship of interest is one that explains the logarithm of earnings as a function of age, the latter being a proxy for years of work experience. Typically, in the labor economics literature, this relationship has been modeled by using standard parametric regression, typically with both age and its square being included as regressors. This quadratic relationship allows for the fact that earnings would be expected to increase with age through much of the working life, but that the rate of increase declines with age (e.g., Heckman and Polachek, 1974; Mincer, 1974). Mincer relates the concave quadratic function to the behavior that is implied by the optimal distribution of human capital investment over an agent's life cycle, but more recently Murphy and Welsch (1990) have provided evidence that a quartic relationship performs more satisfactorily than a quadratic one.

Our own application involves a cross-section data set relating to the earnings and ages of a sample of 205 Canadian individuals, all of whom had the same number of years of schooling. These data come from the 1971 Canadian Census Public Use Tapes, and have been used in various parametric and nonparametric studies by Ullah (1985), Singh et al. (1987), Pagan and Ullah (1999, pp. 152–157), and van Akkeren (this volume, Chap. 10). As Pagan and Ullah discuss, simple polynomial models result in a smooth concave relationship, but when a nonparametric kernel estimator is used the fitted relationship exhibits a "dip" around the age of 40–45. Interestingly, the same effect is obtained by van Akkeren (this volume, Chap. 10, Fig. 5) via his data-based information theoretic (DBIT) estimation procedure. A possible explanation for this dip is offered by Pagan and Ullah (1999, p. 154):

> [it may be due to] the generation effect, because the cross section data represent the earnings of people at a point in time who essentially belong to different generations. Thus the plot of earnings represents the overlap of the earnings trajectories of different generations. Only if the sociopolitical environment of the economy has remained stable intergenerationally can we assume

these trajectories to be the same. But this is not the case; one obvious counterexample being the Second World War. Therefore, the dip in the nonparametric regression might be attributed to the generation between 1935 and 1945.

Figure 1 illustrates this effect with the Canadian earnings data. There, we show the results of fitting a quadratic relationship by the method of least squares, as well as a nonparametric kernel estimate. The latter uses a normal kernel with the approximately optimal bandwidth described by Silverman (1986, p. 45). Also shown in Fig. 1 is the result of applying our fuzzy regression analysis with three fuzzy clusters and $m=2$. (The results were not sensitive to the latter choice of value for the exponent parameter in the application of the FCM algorithm.) Some of the details associated with the clustering of the data and the estimation of submodels appear in Table 1, together with a comparison of the quality of the overall "fit" of the fuzzy model as compared with the nonparametric kernel and quadratic least squares models. In Table 1(a) we see that the submodel estimates (based on each of the three fuzzy clusters) are fundamentally different from each other. When these are combined on the basis of the associated membership functions (which are depicted in Fig. 2) we are able to fit a flexible nonlinear relationship. In Table 1(b) we see that the fuzzy model fits the data better than the nonparametric model and

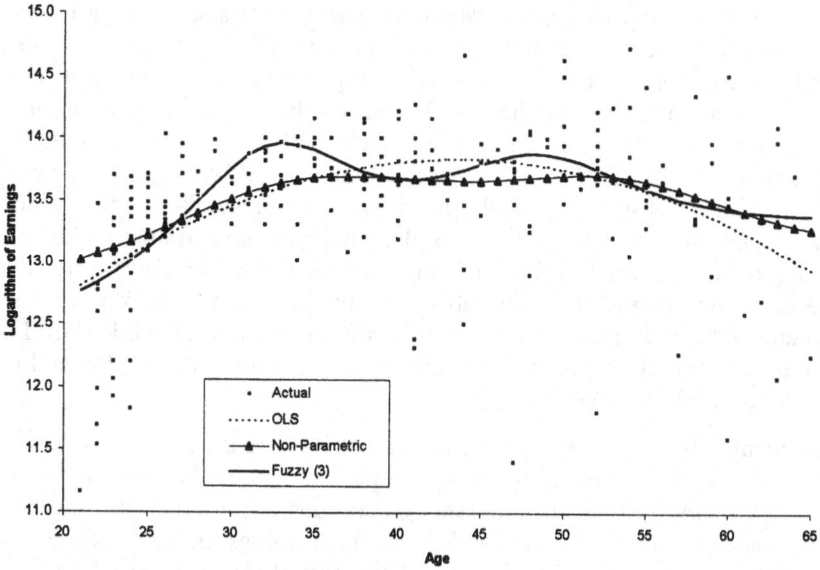

FIGURE 1 Alternative estimates of the earnings-age profile.

TABLE 1 Fuzzy Regression Results for Earning-Age Data ($c=3$; $m=2$)

(a) Submodel Results

Fuzzy cluster	Ranked observations	Cluster centre	Slope (t-value)	Intercept (t-value)
1	1–77 (21–32 years)	25.7421	0.1194 (6.74)	10.1199 (21.68)
2	78–148 (33–47 years)	39.7241	−0.0180 (−1.31)	14.3870 (26.22)
3	149–205 (48–65 years)	55.1273	−0.0475 (−2.53)	16.2073 (15.69)

(b) Comparative Model Performance

	Least squares	Nonparametric	Fuzzy
% RMSE[a]	4.403	4.459	4.406
% MAE[b]	3.017	3.014	2.992

[a]Percentage root mean squared error of fit.
[b]Percentage mean absolute error of fit.

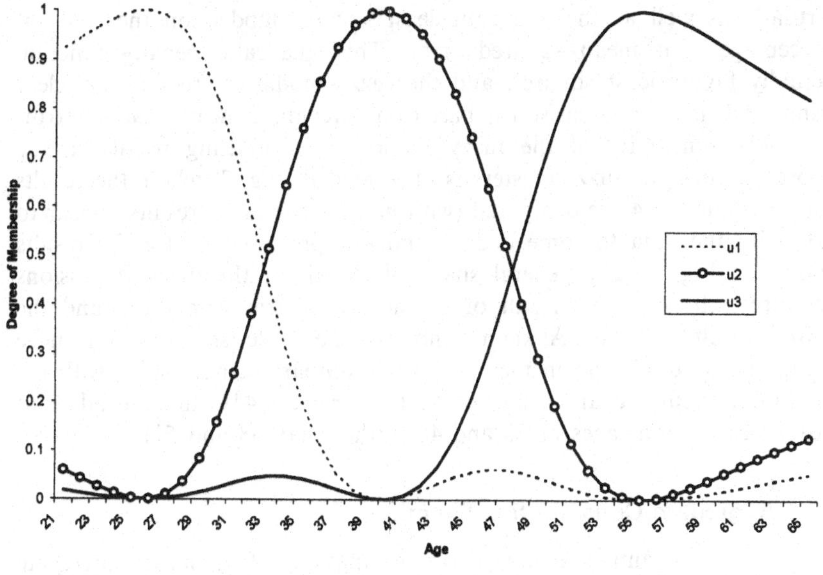

FIGURE 2 Membership functions for fuzzy regression ($c=3$; $m=2$).

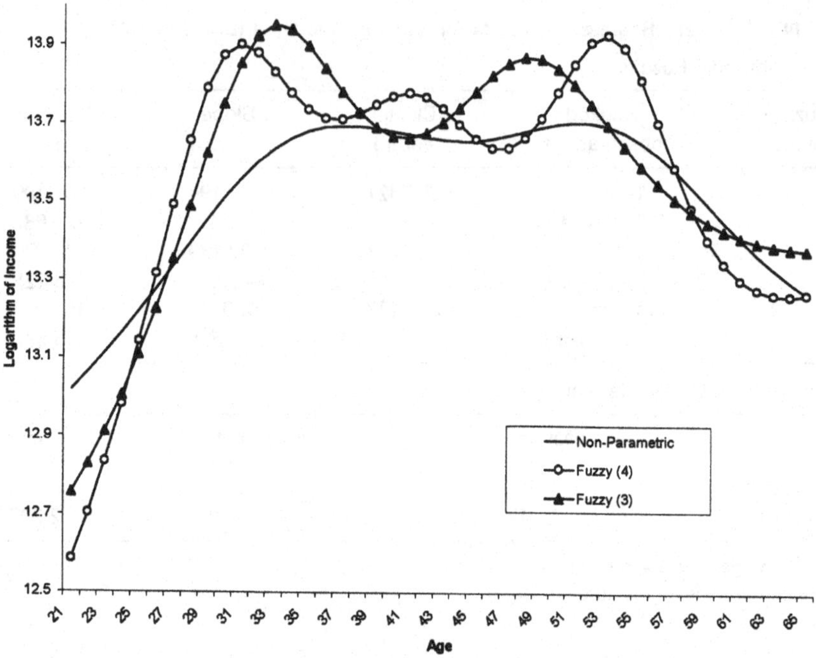

FIGURE 3 Sensitivity of fuzzy modeling to number of clusters.

virtually as well as the quadratic least squares model, on the basis of percentage root mean squared error. This quadratic penalty function actually favors least squares, and the fuzzy model emerges as the clear winner when the comparison is based on percentage mean *absolute* error.

The sensitivity of the fuzzy econometric modeling results to the chosen number of fuzzy clusters is illustrated in Fig. 3, where the results for $c = 3$ and $c = 4$ are compared (with $m = 2$). The latter results appear to "overfit" the data to some degree, and our preference is for the results shown in Fig. 1. The general shape of the fit of the fuzzy regressions reinforces the principal result of the nonparametric estimation (and the DBIT results of van Akkeren, this volume). We see an even more pronounced "dip" than in the case of the nonparametric results, with the minimum occurring at an age of 41 (rather than 44), and the adjacent peaks occurring at ages of 33 and 48 (rather than 38 and 51).

5.2 Aggregate Consumption Function

Our second example involves the estimation of a naive aggregate consumption function for the U.S.A., using monthly seasonally adjusted

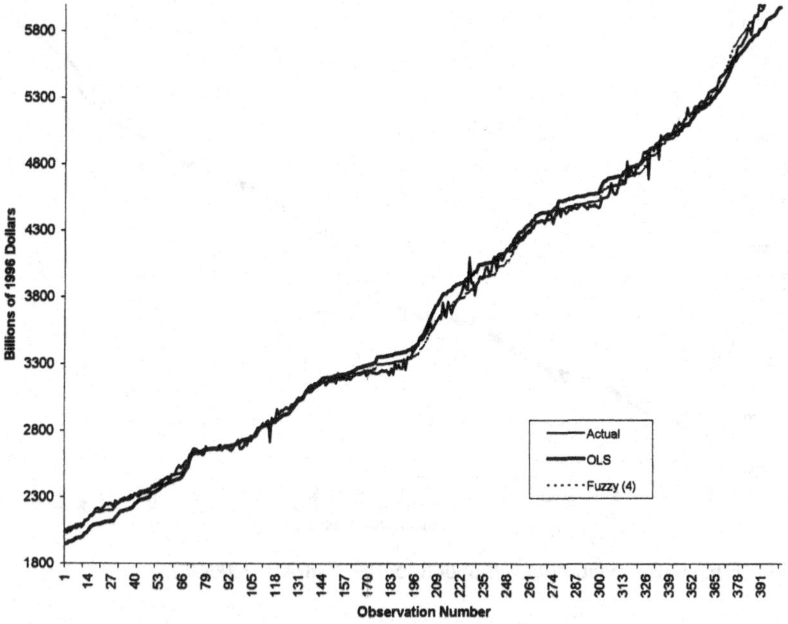

FIGURE 4 U.S. consumption model (monthly 1967*M*1-2000*M*6).

time-series data for personal consumption expenditures and real disposable personal income. The data are published by the Bureau of Economic Analysis at the U.S. Department of Commerce (2000), and are in real 1996 billions of dollars. The sample period is January 1967 to June 2000 inclusive. The model explains consumption expenditure simply as a function of disposable income. No dynamic effects or other explanatory variables are taken into account.

In this case we found that basing the analysis on four fuzzy clusters produced marginally better results than those based on three clusters. Figure 4 compares the "fit" of the fuzzy model with that obtained by simple least squares, and the corresponding comparison is made with a nonparametric kernel fit in Fig. 5. The latter was obtained with the same kernel and window choices as in the previous example, and $m=2$ was used again as the exponent parameter in the application of the FCM algorithm. As can be seen in these two figures, the fuzzy model "tracks" the data much more satisfactorily than do either of its competitors, and this is borne out by the percentage root mean square errors and percentage mean absolute errors that are reported in Table 2. The close similarity between the results based on three and four fuzzy clusters is clear in Fig. 6, and the % RMSE and % MAE values when $c=3$ are 1.327 and 1.033,

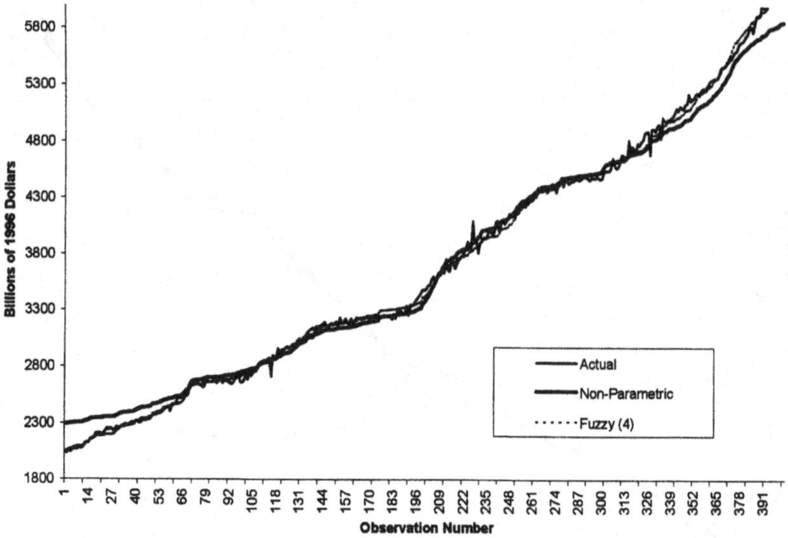

FIGURE 5 U.S. consumption model (monthly 1967*M*1-2000*M*6).

TABLE 2 Fuzzy Regression Results for Consumption-Income Data ($c=4$; $m=2$)

(a) Submodel Results

Fuzzy cluster	Ranked observations	Cluster centre	Slope (*t*-value)	Intercept (*t*-value)
1	1-108	2720.247	0.8140	184.45
			(92.18)	(7.55)
2	109-217	3681.513	0.8061	258.91
			(59.69)	(5.19)
3	218-332	4926.031	0.9740	−393.75
			(63.76)	(−5.26)
4	333-402	5973.417	1.1705	−1428.90
			(63.52)	(−13.05)

(b) Comparative Model Performance

	Least squares	Nonparametric	Fuzzy
% RMSE[a]	2.306	3.285	1.201
% MAE[b]	1.828	2.328	0.942

[a]Percentage root mean squared error of fit.
[b]Percentage mean absolute error of fit.

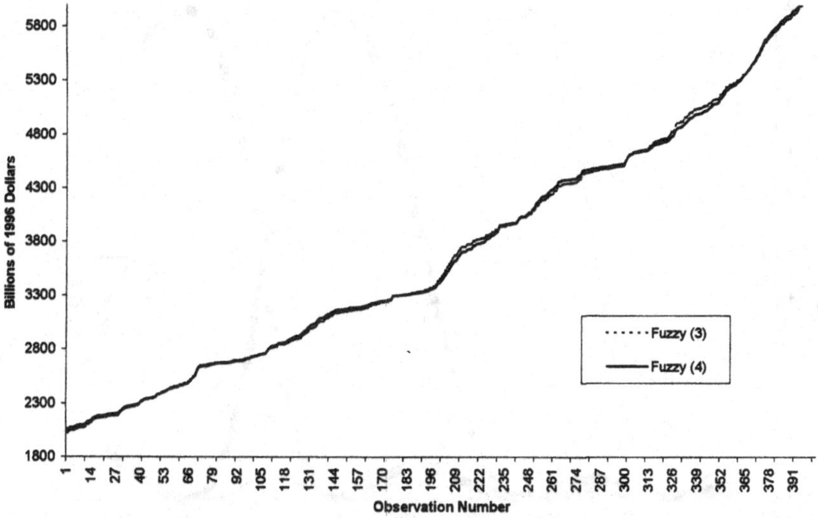

FIGURE 6 Sensitivity of fuzzy consumption model to number of fuzzy clusters.

respectively. The membership functions for the Fuzzy (4) model appear in Fig. 7.

The economic interpretation of the various fitted models is also interesting. In the linear least squares model the slope parameter is the (constant) marginal propensity to consume (m.p.c.), and is estimated to be 0.9598 ($t = 270.39$) from our sample. In the case of the nonparametric model the estimated derivatives (which represent the changing m.p.c.) range in value from 0.3134 to 1.2427 as the level of disposable income varies. This is shown in Fig. 8, where the income data have been ranked into ascending order. Values of the m.p.c. in excess of unity make no sense in economic terms, and the extreme and unusual pattern of the plot of the nonparametrically estimated m.p.c.'s in Fig. 8 suggests that these results should be treated with extreme skepticism. Also given in that figure are the corresponding results for the preferred Fuzzy (4) model. In this case the plot is much more reasonable than that for the nonparametric model, with the estimated m.p.c.'s ranging from 0.8061 to 1.1705 in value. The few estimates in excess of unity are still troublesome, of course, but apart from these, the derived m.p.c. values are more plausible economically than are those resulting form the nonparametric kernel estimation or from the (highly restrictive) least squares model.

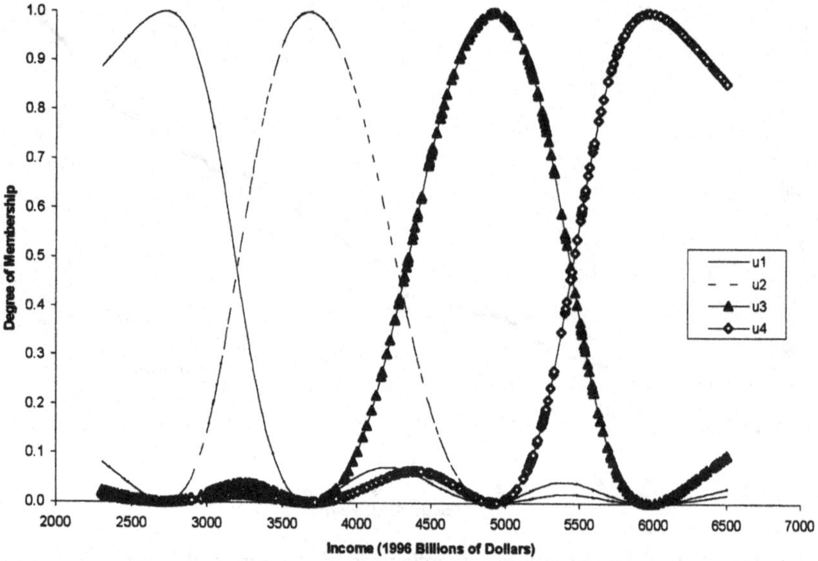

FIGURE 7 Membership functions for fuzzy regression ($c=4$; $m=2$).

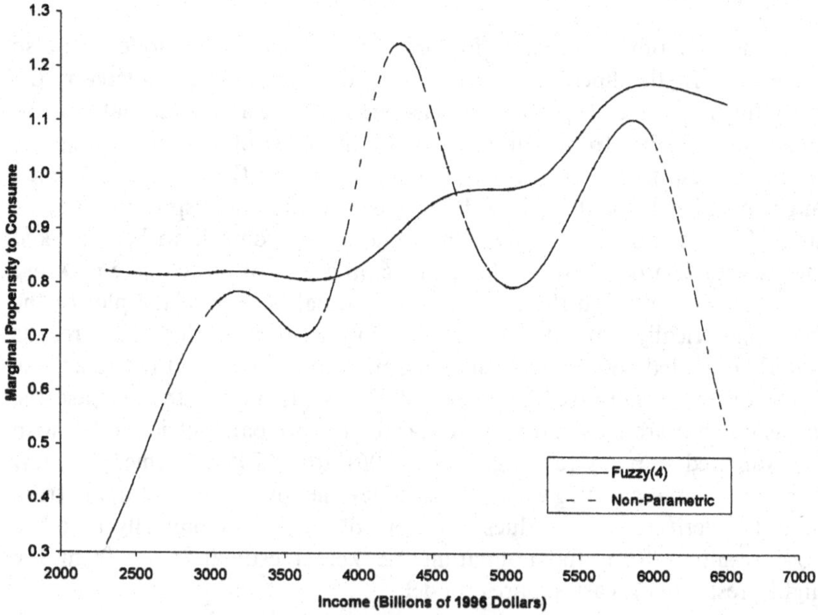

FIGURE 8 Predicted marginal propensities to consume.

5.3 Money Demand Model

Our third empirical application involves a simple demand for money model:

$$\log(M_t) = \beta_1 + \beta_2 \log(Y_t) + \beta_3 \log(r_t) + \varepsilon_t$$

where M is the money stock, Y is the output, and r is the rate of interest. We have used annual U.S. Department of Commerce (various years) data for the period 1960 to 1983 inclusive, as presented by Griffiths et al. (1993, p. 316). We used three fuzzy clusters to analyze each of the two input variables, and then considered all of the resulting nine combinations of the data partitioning. Of these, five resulted in empty or inadequately sized sets—either there were no sample points consistent with the intersection of the fuzzy sets for $\log(Y)$ and $\log(r)$, or there were insufficient to fit a regression. The sample points associated with the remaining four intersections are shown in Table 3, and the TSSS analysis was applied to this four-way partitioning of the data.

As Figs. 9 and 10 indicate, the fit of the fuzzy model clearly dominates that obtained from a nonparametric regression model, and it also dominates an OLS multiple regression analysis—especially at the end-points of the sample. This is confirmed by the % RMSE and % MAE figures that are shown in Table 3(b). Figures 11a and 11b provide details of the membership functions for the income and interest rate input variables, respectively, and Figs. 12a and 12b show the derivatives of the fuzzy model with respect to each of these input variables at each sample point. As the data are all in (natural) logarithms, these derivatives are elasticities, and in these figures they are compared with their counterparts from the nonparametric model. In each case, not only do the fuzzy elasticities evolve more plausibly with an increase in the input variable, but they also have the anticipated signs. In contrast, the elasticities derived from the nonparametric regression model have signs that conflict with the underlying economic theory for some values of income and the interest rate variable. In the case of the OLS regression model based on the full sample period, the (constant) income and interest rate elasticities are 0.7091 ($t = 44.41$) and -0.0533 ($t = -2.49$), respectively. The sample means of their fuzzy counterparts are quite similar in value, being 0.7368 and -0.0606, respectively, and the unsatisfactory nature of the non-parametric results is underscored when we note that the corresponding sample averages of those estimated elasticities are 0.3237 and 0.0666. (The sign of the latter value, of course, conflicts with the prior theory.) In short, the results of the fuzzy modeling dominate both OLS and nonparametric regression analysis in this example.

TABLE 3 Fuzzy Regression Results for Money Demand Model ($c_1 = c_2 = 3$; $m = 2$)

(a) SubModel Results

Fuzzy cluster	log(Income)		Fuzzy cluster	log(Interest rate)	
	Observations	Cluster centre		Observations	Cluster centre
Y_1	1-9	6.448	R_1	1-6	1.188
Y_2	10-18	7.122	R_2	7-19	1.740
Y_3	19-24	7.882	R_3	20-24	2.348

(b) Cluster Intersection Estimation Results

Fuzzy intersection	Observations	β_1 (t-value)	β_2 (t-value)	β_3 (t-value)
(Y1 ∩ R1)	1-6	1.4501	0.5638	−0.0053
		(3.81)	(8.43)	(−0.11)
(Y1 ∩ R2)	7-9	−1.3218	0.9976	−0.0819
		(n.a.)[a]	(n.a.)	(n.a.)
(Y2 ∩ R2)	10-18	0.8628	0.6632	−0.0320
		(5.63)	(32.94)	(−1.30)
(Y3 ∩ R3)	20-24	−0.4878	0.8718	−0.1465
		(−1.55)	(23.38)	(−5.26)

(c) Comparative Model Performances

	Least squares	Nonparametric	Fuzzy
% RMSE[b]	0.019	0.062	0.015
% MAE[c]	0.015	0.045	0.012

[a] Not available, as degrees of freedom are exactly zero.
[b] Percentage root mean squared error of fit.
[c] Percentage mean absolute error of fit.

5.4 Modeling Kuznets' "U-Curve"

In a seminal contribution, Kuznets (1955) postulated the existence of an inverted-U relationship between the degree of income inequality and economic growth. He argued that income inequality (perhaps as measured by the Gini coefficient) increases during the early stages of an economy's growth, reaches a maximum, and then declines as the economy matures. This hypothesis has been subjected to a substantial amount of empirical testing in the development economics literature. The results are somewhat mixed, depending on the type of data used (e.g., time-series, cross-section, or panel), the level of development of the

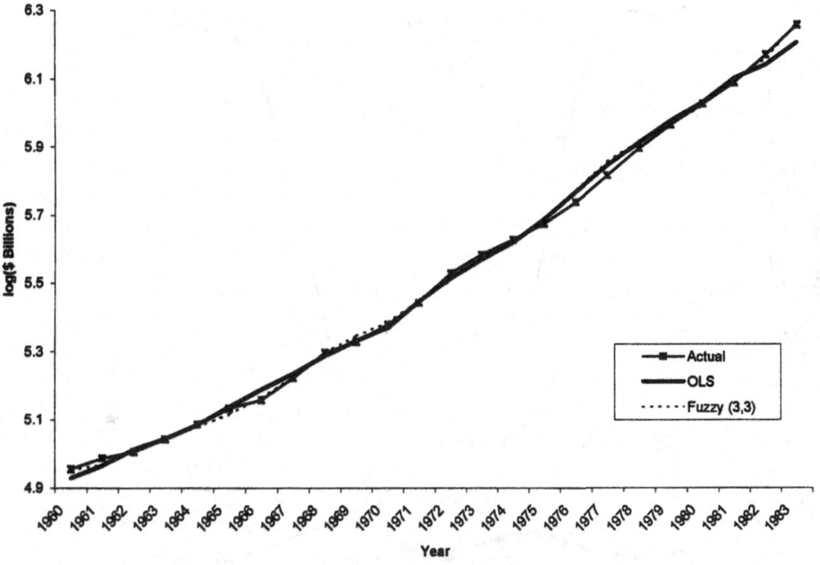

FIGURE 9 U.S. demand for money model.

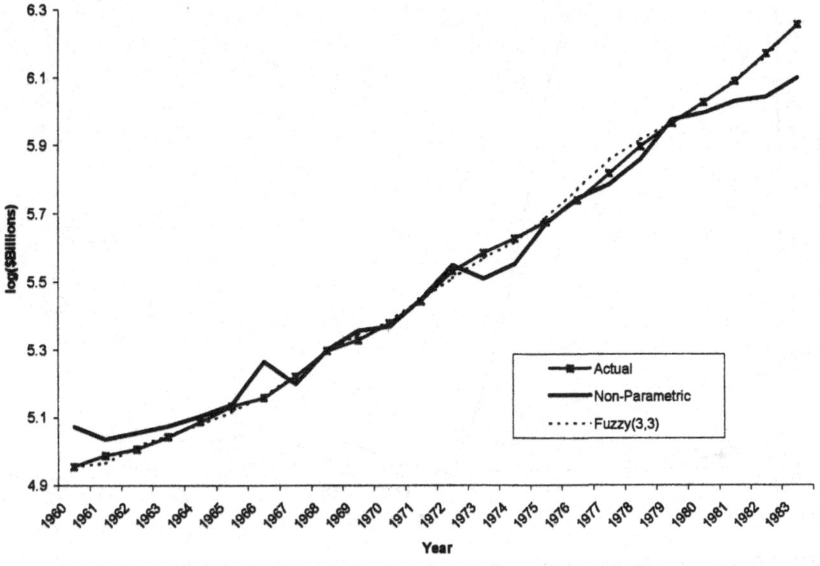

FIGURE 10 U.S. demand for money model.

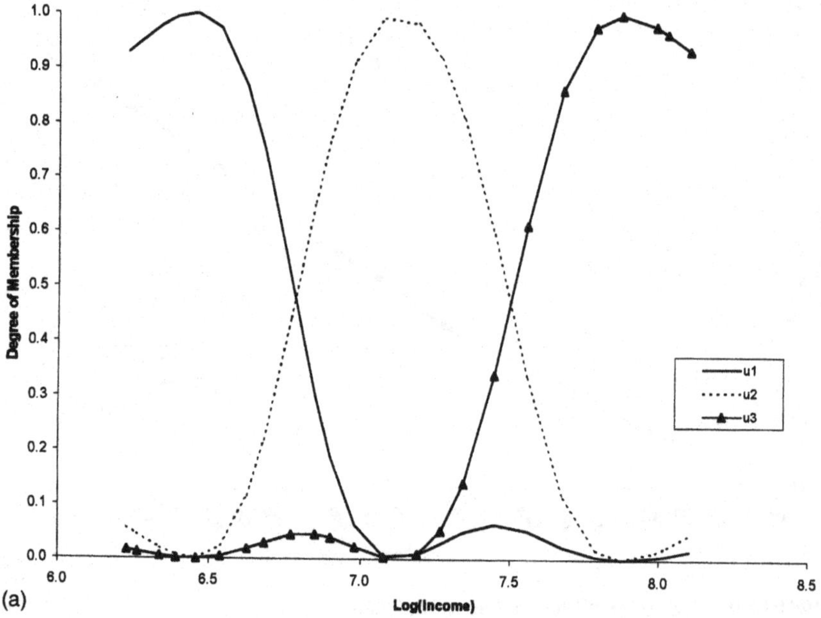

(a)

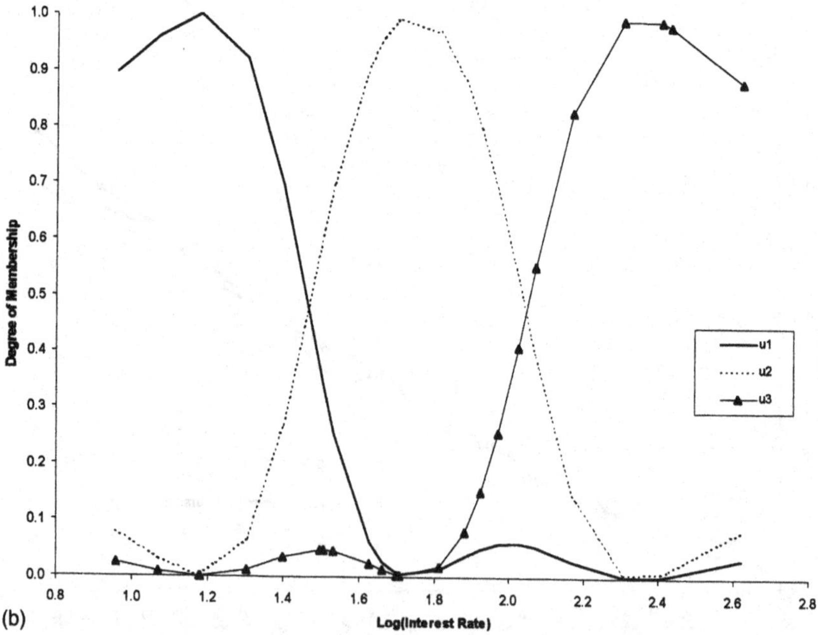

(b)

FIGURE 11 (a) Membership functions for income variable; (b) Membership functions for interest rate variable.

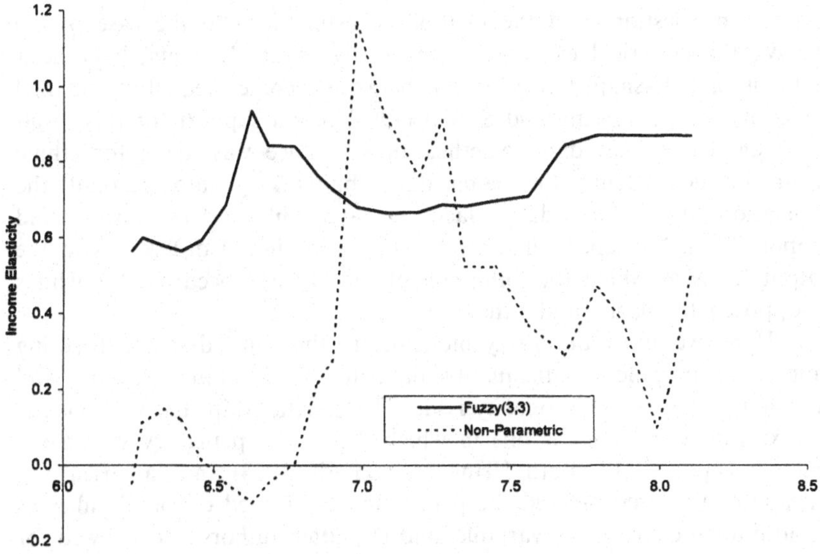

(a)

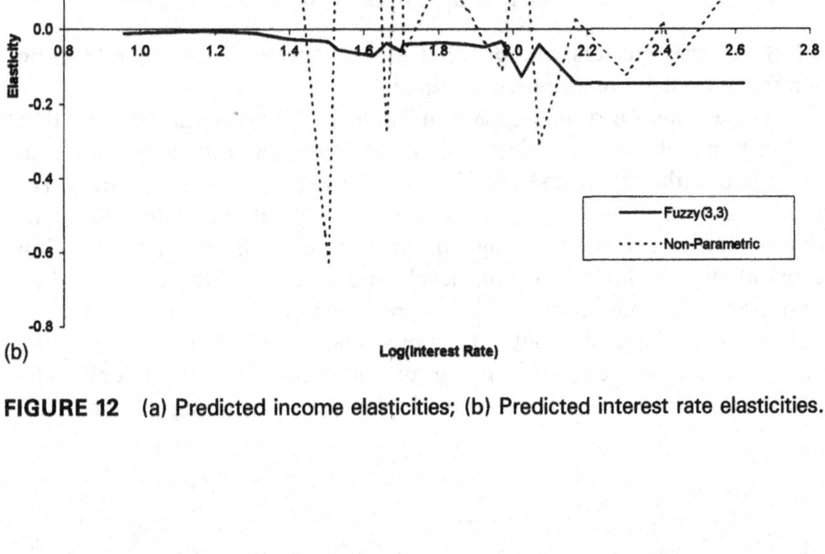

(b)

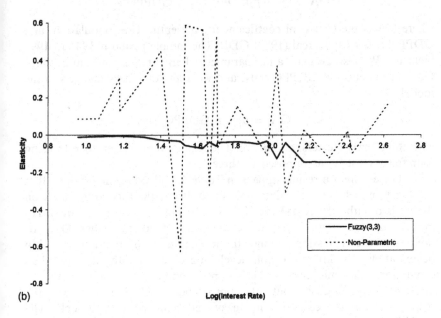

FIGURE 12 (a) Predicted income elasticities; (b) Predicted interest rate elasticities.

country in question, and the method of estimation. In the case of the U.S.A., the empirical evidence quite clearly rejects Kuznets' hypothesis in favor of a U-shaped relationship between income inequality and real per capita output. Hsing and Smyth (1994) found support for this result using aggregate U.S. data, whether an allowance was made for ethnic origin or not. Using the same data, but taking into account the nonstationarity of the data, Jacobsen and Giles (1998) also found support for a U-shaped relationship. They also found differences in the output levels at which the minimum of the U-curve occurred for whites as opposed to blacks and others.

Here, we apply our fuzzy modeling to this same data set, focusing simply on the income inequality data for "all" ethnic groups, and abstracting from nonstationarity issues. The relationship that we consider for expository purposes is one in which the only explanatory variable is real per capita GDP. Both Hsing and Smyth (1994) and Jacobsen and Giles (1998) also considered the percentage of married couple families as an additional explanatory variable, and the latter authors also allowed for dynamic effects in the model specification. The benchmark specification that we compare against is a simple but rigid quadratic relationship, estimated by OLS:

$$\mathrm{GA}_t = \beta_1 + \beta_2 \mathrm{GDPPC}_t + \beta_3 \mathrm{GDPPC}_t^2 + \varepsilon_t$$

where GA is the (%) Gini coefficient for the entire U.S. population, and GDPPC is per capita real (1987) GDP. The sample period is 1947 to 1991 inclusive. We also estimate a nonparametric kernel regression that explains GA as a function of GDPPC only, and we apply our fuzzy analysis to the model:

$$\mathrm{GA}_t = \beta_1 + \beta_2 \mathrm{GDPPC}_t + \varepsilon_t$$

with three fuzzy clusters, and $m = 2$. The results were not sensitive to using four fuzzy clusters, as opposed to three.

The estimation results appear in Table 4, and we again see there that the fuzzy model "fits" the data marginally better than its competitors on the basis of either % RMSE or % MAE. In Fig. 13 we see that the fuzzy model supports the U-shaped relationship found by either OLS or nonparametric regression, though the minimum of the fitted relationship occurs at slightly different output levels in each case. With quadratic OLS estimation this minimum is at a per capita real output level of approximately \$13,300, but at approximately \$12,700 and \$13,500, respectively, in the cases of the fuzzy and nonparametric models. The membership functions associated with the fuzzy clustering of the GDP

TABLE 4 Fuzzy Regression Results for Kuznets' U-Curve ($c=3$; $m=2$)

(a) SubModel Results

Fuzzy cluster	Observations	Cluster centre	Slope (t-value)	Intercept (t-value)
1	1–18	10421.87	-0.5140×10^{-5}	0.4190
	(1947–1964)		(−3.19)	(24.84)
2	19–32	14676.64	0.2956×10^{-5}	0.3126
	(1965–1978)		(3.00)	(21.75)
3	33–45	18256.73	0.9617×10^{-5}	0.2132
	(1979–1991)		(6.24)	(7.70)

(b) Comparative Model Performances

	Least squares	Nonparametric	Fuzzy
% RMSE[a]	1.574	1.890	1.570
% MAE[b]	1.288	1.595	1.196

[a]Percentage root mean squared error of fit.
[b]Percentage absolute error of fit.

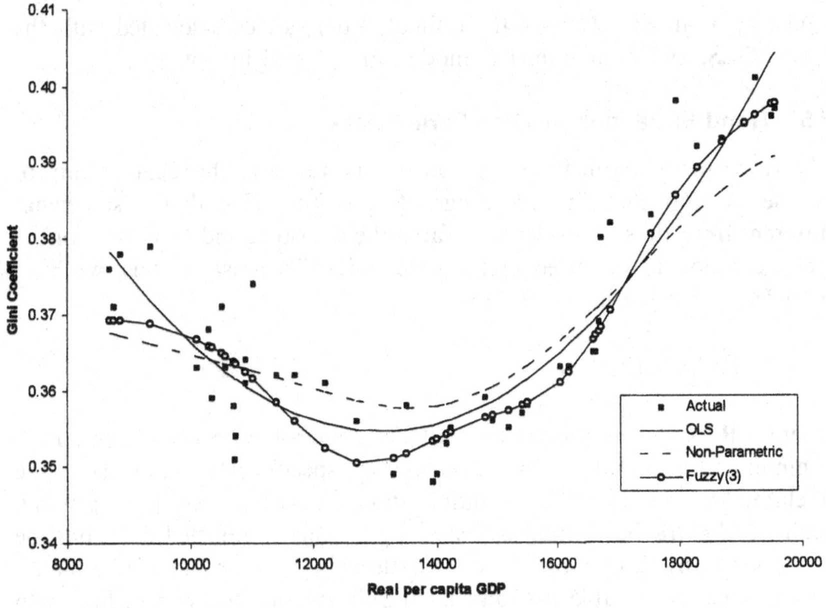

FIGURE 13 Kuznets' U-curve model.

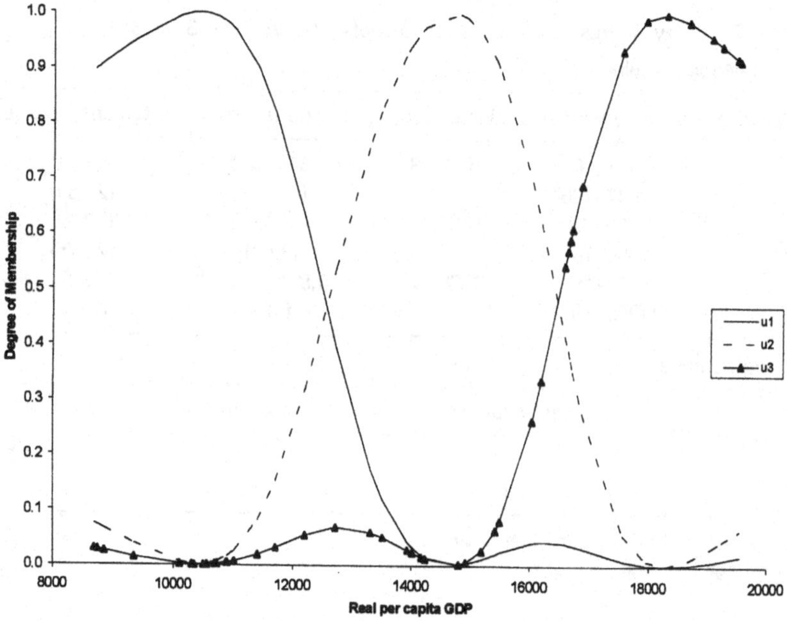

FIGURE 14 Membership functions of GDP.

data appear in Fig. 14, and the estimated derivatives associated with the fuzzy, OLS, and nonparametric models are plotted in Fig. 15.

5.5 Trend Extraction in Time-Series Data

The fuzzy modeling methodology under discussion in this chapter can, of course, be applied to a wide range of situations. One that is somewhat different from those considered so far is the detection and extraction of the trend component of an economic time series. To illustrate this, we have modeled a relationship of the form:

$$\log(\mathrm{ER}_t) = f(t) + \varepsilon_t$$

where ER is the monthly Canada/U.S. exchange rate (noon spot rate, beginning of month). The logarithmic specification recognizes the likelihood of an underlying multiplicative time-series model. In general, such trend extraction would be based on seasonally adjusted data, but the absence of any discernible seasonal pattern in this particular case makes this unnecessary. Table 5 shows the results of our fuzzy modeling with a seven-cluster specification.

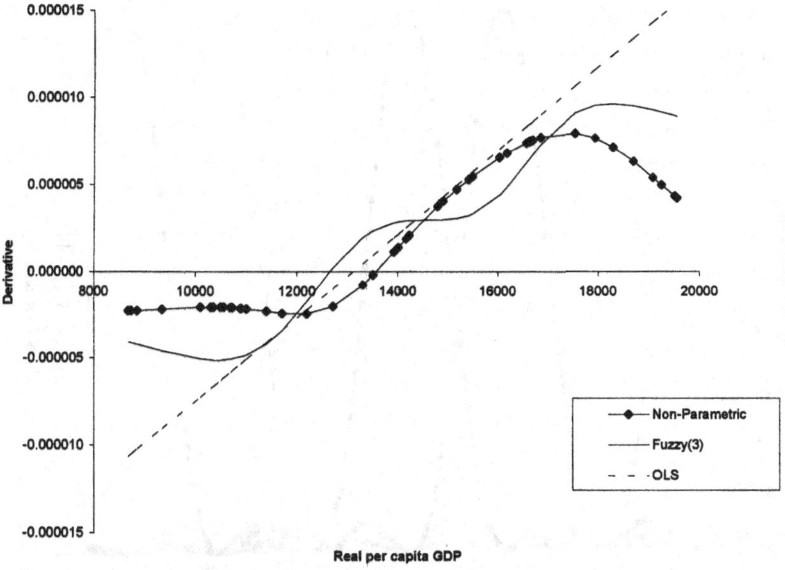

FIGURE 15 Estimated Gini-GDP derivatives.

TABLE 5 Fuzzy Regression Results for the Exchange Rate Trend ($c=7$; $m=2$)

Fuzzy cluster	Observations	Cluster centre	Slope (t-value)	Intercept (t-value)
1	1-50	24.07	−0.0010 (−8.38)	−0.0273 (7.70)
2	51-104	77.37535	0.0032 (11.51)	−0.2054 (−9.39)
3	105-158	131.5562	0.0014 (10.59)	0.0020 (0.12)
4	159-212	185.7503	0.0014 (5.16)	0.0171 (0.33)
5	213-266	239.6784	−0.0013 (−8.00)	0.4860 (12.16)
6	267-319	293.1247	0.0020 (8.37)	−0.3126 (−4.35)
7	320-368	345.4138	0.0020 (7.44)	−0.3027 (−3.31)

As in the previous examples, we compare the fuzzy modeling with nonparametric kernel estimation. Rather than also consider an OLS model where $f(t)$ is some polynomial in time, we have used the Hodrick–Prescott (H–P) (1980,1977) filter as a further benchmark for measuring the trend.

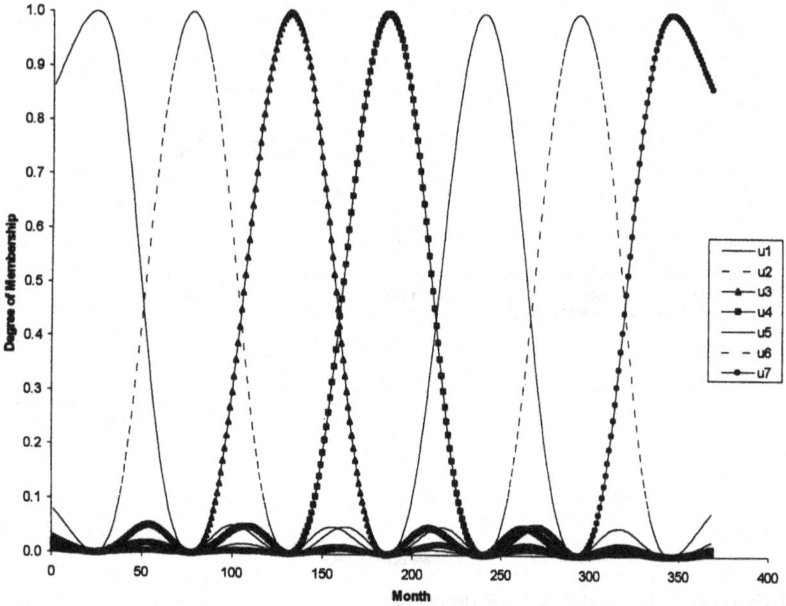

FIGURE 16 Membership functions for trend.

Our sample covers the period since the (re)floating of the Canadian dollar, and runs from May 1970 to December 2000. We follow Kydland and Prescott (1990, p. 9) and set the smoothing parameter for the H–P filter to $\lambda = 14{,}400$ (i.e., 100 times the square of the data frequency), as we have monthly data. The H–P filter was implemented with use of the TSP package (Hall, 1996) using code written by Cummins (1994).

The membership functions associated with this fuzzy modeling appear in Fig. 16, and Fig. 17 depicts the associated trend analysis. The corresponding nonparametric and H–P filter trend analyses appear in Fig. 18. In this application of H–P filter results are quite robust to the choice of the smoothing parameter. For instance, when $\lambda = 100$ (the usual choice of *annual* data) is used, the results are very similar to those in Fig. 18, but there is undue variation in the extracted trend. On the other hand, in the case of the fuzzy trend analysis a good deal of care has to be taken over the choice of the number of fuzzy clusters if a trend as "flexible" as that produced by the H–P filter is to be obtained. This can be seen in Fig. 19, though obviously this sensitivity is at least in part a function of the sample size and the nature of the trend itself.

In many respects, the trend of these exchange rate data that is identified by means of our fuzzy modeling appears to be somewhat more

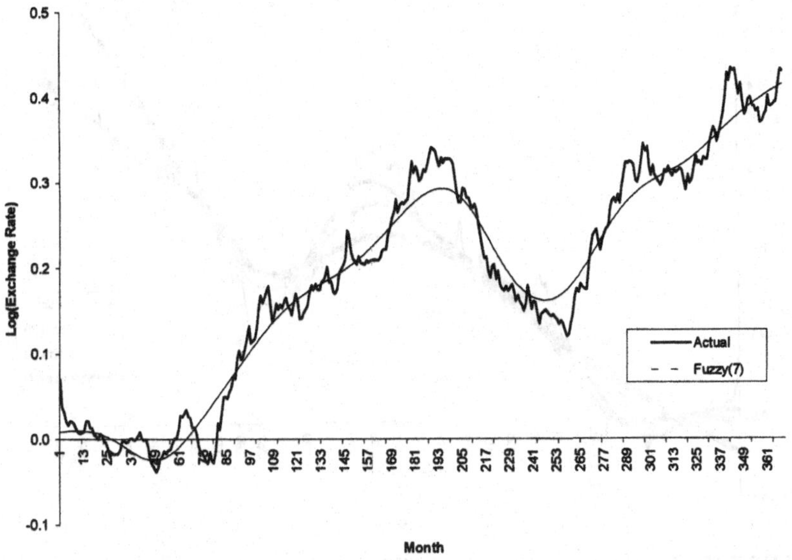

FIGURE 17 Exchange rate fuzzy trend analysis.

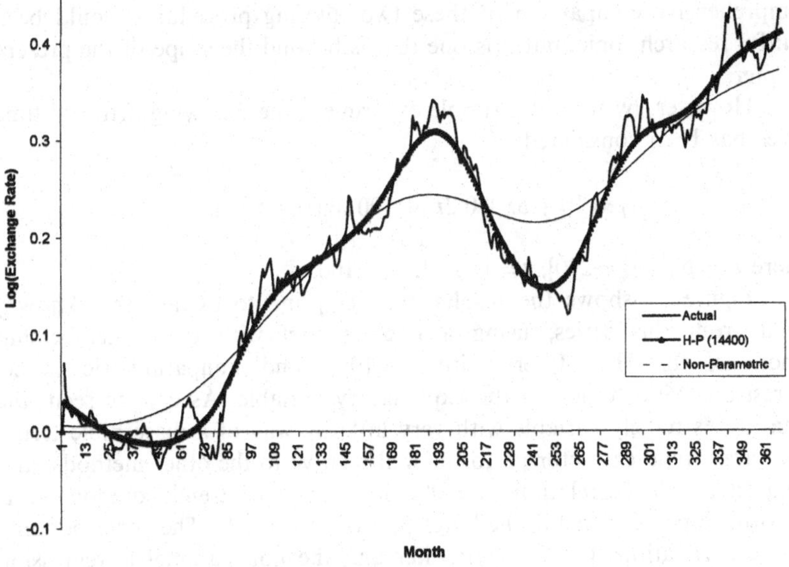

FIGURE 18 Nonparametric and H-P filter trend analysis.

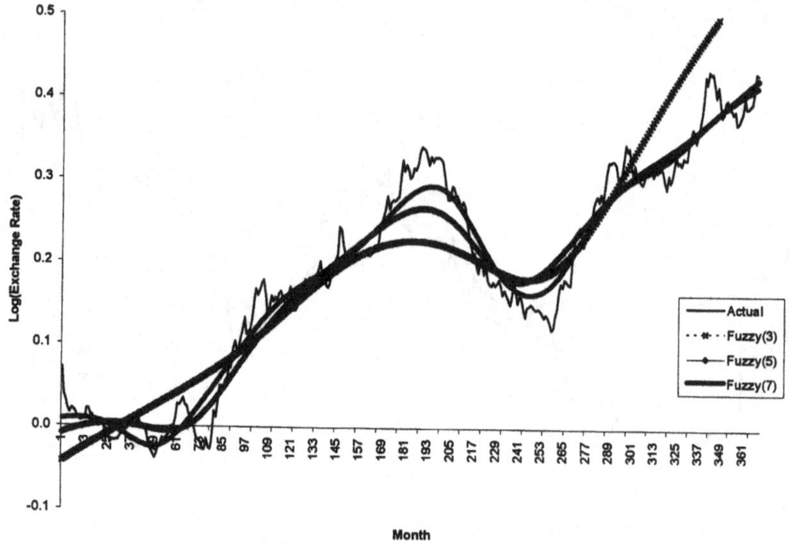

FIGURE 19 Sensitivity of the fuzzy modeling to the number of fuzzy clusters.

reasonable than that resulting from the application of the H–P filter. In particular, the latter seems to exhibit too much cyclical variation in this case, at least visually, as can be seen in Fig. 20. A much more comprehensive comparison of these two filtering procedures would be a fruitful research topic, but it is one that is beyond the scope of the present chapter.

However, by way of a simple experiment the following artificial time series has been considered:

$$y_t = 10 + 5t + 0.2t^2 + 100\sin(t) + 500\sin^2(t) + \varepsilon_t$$

where $\varepsilon_t \sim \mathrm{N}[0, \sigma = 200]$; $t = 1, 2, 3, \ldots, 100$.

Figure 21 shows the results of attempting to extract the (known) trend from this series, using a fuzzy model with three membership functions, the H–P filter (with $\lambda = 100$), and nonparametric kernel regression, with "time" as the explanatory variable. As can be seen, the data are extremely variable with very little by way of a discernible trend. The fuzzy trend extractor performs well relative to the other methods, and yields a simple correlation of 0.992 with the true trend component. It performs especially well in the latter part of the sample. The corresponding simple correlations for the H–P filter and the non parametric regression are 0.964 and 0.983, respectively.

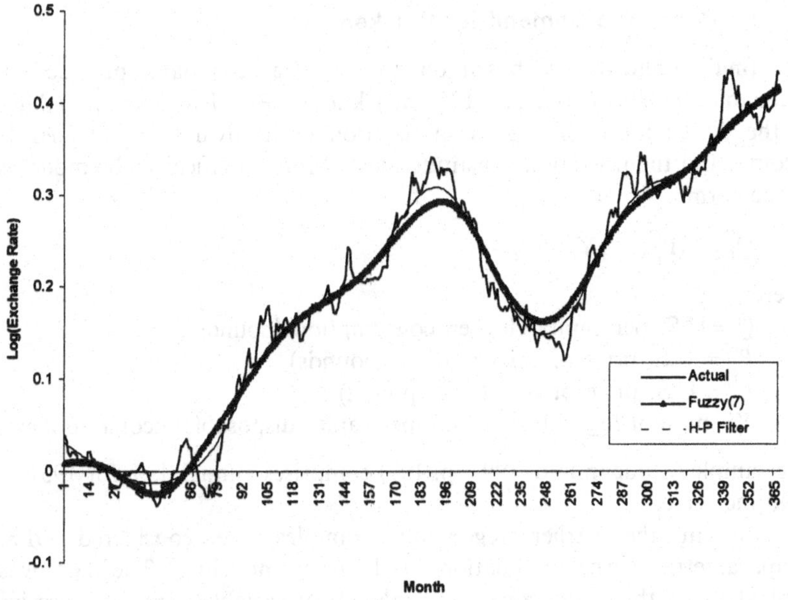

FIGURE 20 Fuzzy trend versus H-P trend.

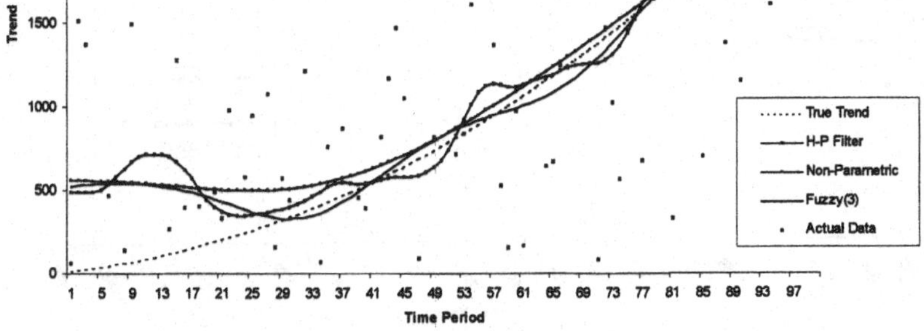

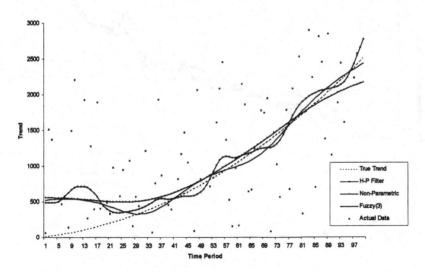

FIGURE 21 Artificial data-comparative trend analysis.

5.6 Modeling the Demand for Chicken

Our final application is based on an example, and data, provided in Studenmund's (1997, pp. 174–175) well known text. This example relates to the demand for chicken, as a function of its own price, disposable income, and the price of a substitute meat. More specifically, the model is of the form:

$$Q_t = f(\mathrm{P}_t^{\mathrm{C}}, \mathrm{P}_t^{\mathrm{B}}, \mathrm{Y}_t) + \varepsilon_t$$

where:

Q = U.S. per capita chicken consumption (pounds)
P^{C} = U.S. price of chicken (cents/pounds)
P^{B} = U.S. price of beef (cents/pound)
Y = natural logarithm of U.S. per capita disposable income (dollars)

The sample (Studenmund, 1997, p. 199) comprises annual data for 1951 to 1990 inclusive.

As in the earlier regression examples, we considered OLS, nonparametric kernel estimation, and fuzzy modeling. The last was applied to all three (nonconstant) explanatory variables. In each case we

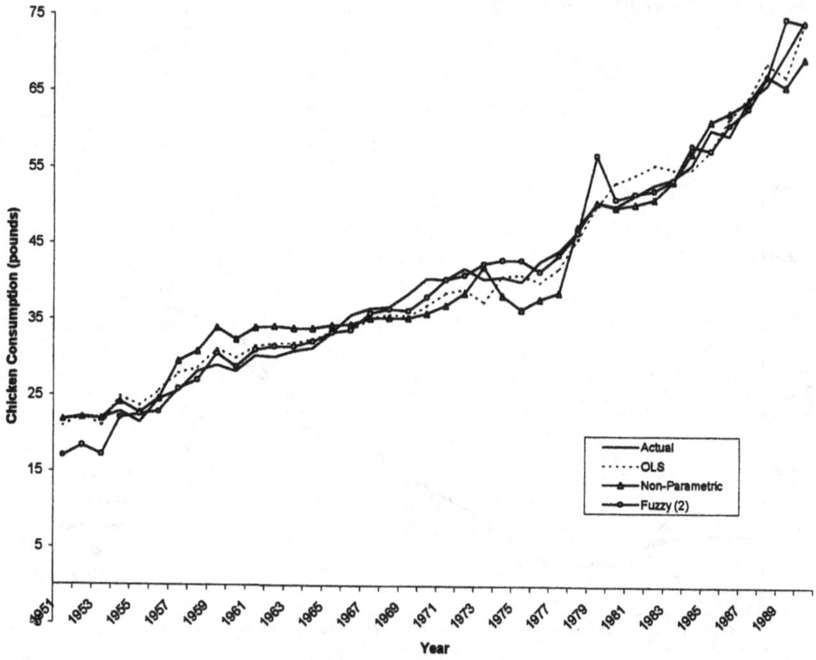

FIGURE 22 Chicken consumption model.

used two fuzzy clusters, resulting in eight potential subsamples, three of which had positive degrees of freedom. (Three fuzzy clusters implied 27 potential subsamples, and this resulted in only two with positive degrees of freedom. The resulting fuzzy prediction path exhibited erratic movement at several time points.)

The basic results are shown in Fig. 22 and Table 6. The membership functions were of the same general form as in the previous examples, and are not shown here to reduce space. As can be seen in Fig. 22, the fuzzy model performs very creditably, and this is reflected in the % RMSE and % MAE values shown in Table 6. Indeed, if it were not for the relatively poor predictive performance of the fuzzy model during the period 1974–1981, its overall performance would be exceptionally good. Our view is

TABLE 6 Fuzzy Regression Results for Chicken Consumption Model ($c_1 = c_2 = c_3 = 2$; $m = 2$)

(a) Submodel Results

Price of chicken		Price of beef		log(Income)	
Fuzzy cluster	Cluster centre	Fuzzy cluster	Cluster centre	Fuzzy cluster	Cluster centre
P^C1	10.420	P^B1	23.463	$Y1$	2875.560
P^C2	18.166	P^B2	60.183	$Y2$	12175.260

(b) Cluster Intersection Results[a]

Fuzzy intersection	Observations	β_1 (*t*-value)	β_2 (*t*-value)	β_3 (*t*-value)	β_4 (*t*-value)
$(P^C1 \cap P^B1 \cap Y1)$	7-22,	30.4720	−1.0523	0.2901	0.0026
	24-27	(9.90)	(−5.34)	(1.84)	(3.70)
$(P^C1 \cap P^B2 \cap Y2)$	29-33,	9.7358	−0.0243	0.3360	0.0024
	36-38, 40	(1.42)	(−0.07)	(4.29)	(13.36)
$(P^C2 \cap P^B2 \cap Y1)$	1-6	15.1140	−0.2595	0.1433	0.0061
		(0.70)	(−0.91)	(0.82)	(0.58)

(c) Comparative Model Performances

	Least squares	Nonparametric	Fuzzy
% RMSE[b]	5.130	7.692	6.879
% MAE[c]	4.407	6.001	4.559

[a] β_1, β_2, β_3, and β_4 are the coefficients of the intercepts, P^C, P^B, and Y respectively.
[b] Percentage root mean squared error of fit.
[c] Percentage mean absolute error of fit.

that the relatively small sample size disadvantages the fuzzy model in this example, and this may also be an instance where some experimentation is needed with the value for the exponent parameter, m, in the objective functional for the FCM algorithm.

Figure 23 displays the predicted own price, cross price, and income elasticities of demand (by year in the sample) implied by the OLS, nonparametric, and fuzzy models. Those relating to the nonparametric estimation are quite unsatisfactory. In particular, in each case there are values that have a sign opposite to that implied by the underlying economic theory. There are no such aberrations in the case of the fuzzy and OLS results. The latter generally appear somewhat more plausible as they exhibit fewer marked fluctuations than do the fuzzy elasticities. This weakness of the fuzzy elasticities is related to a similar feature in the predicted time paths, noted above in relation to Fig. 22. It is also interesting that the fuzzy income elasticities always exceed those from OLS estimation, and this is almost always the case with the cross-price elasticities.

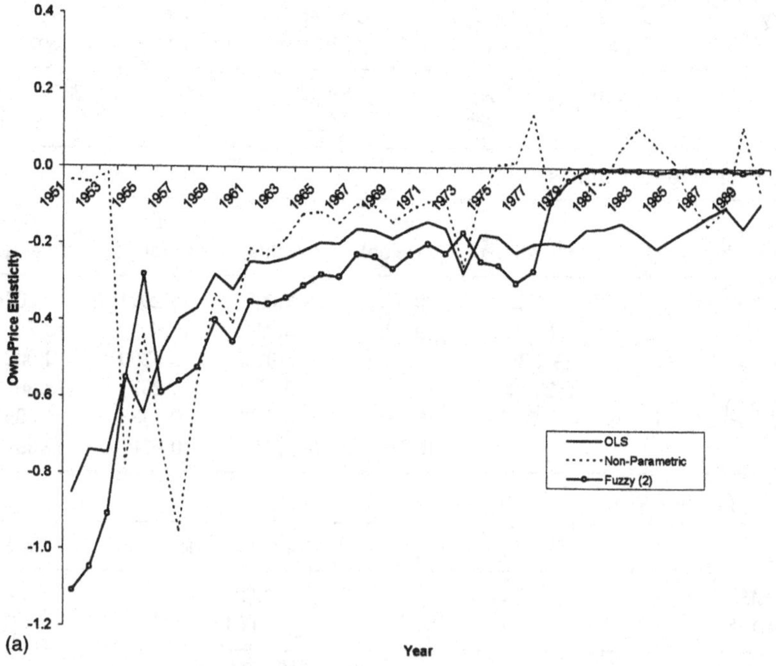

(a)

FIGURE 23 (a) Predicted own-price elasticities; (b) Predicted cross-price elasticities; (c) Predicted income elasticities (multiplied by 10,000).

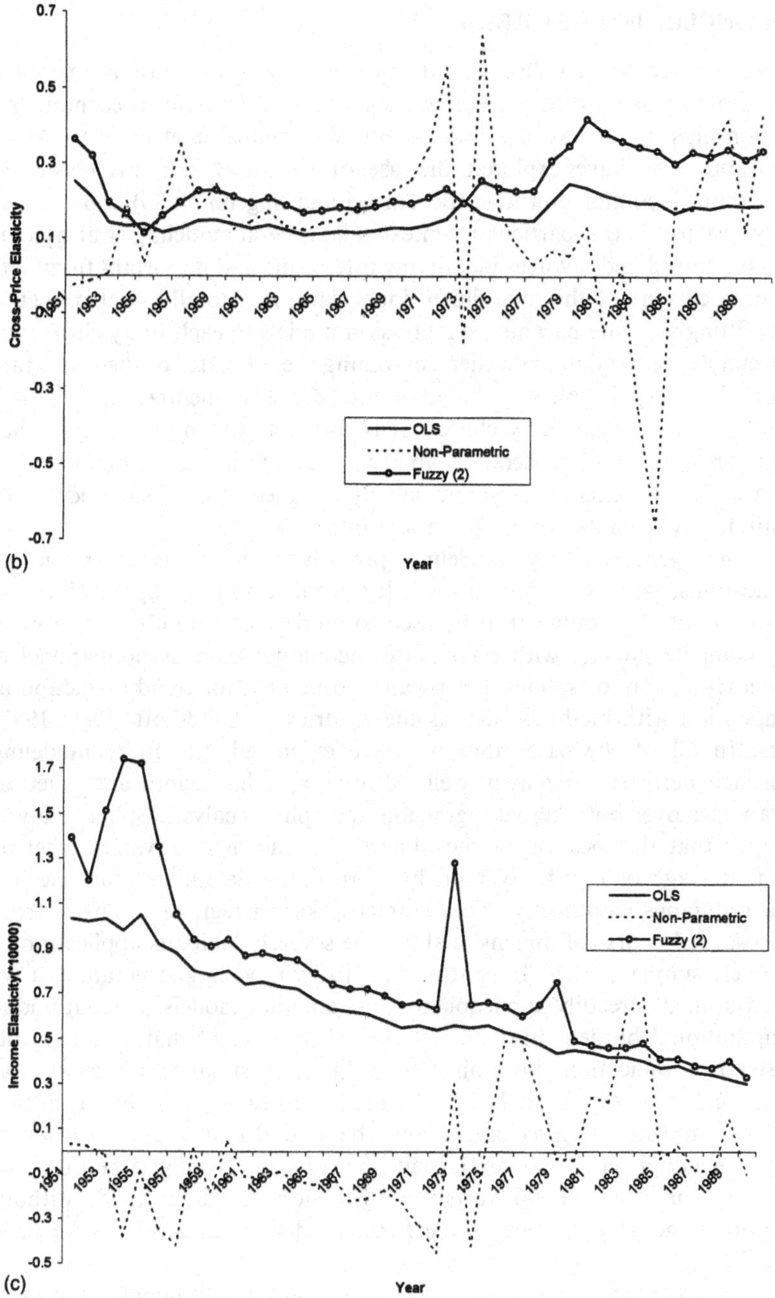

FIGURE 23 (Continued)

6 CONCLUDING REMARKS

In this chapter we have discussed the possibility of using various analytical tools from fuzzy set theory and fuzzy logic to model nonlinear econometric relationships in a flexible, and essentially semiparametric way. More specifically, we have explored the use of the fuzzy c-means clustering algorithm, in conjunction with the Takagi and Sugeno (1985) approach to fuzzy systems, in this particular context. The general modeling strategy that we have considered involves identifying interesting and important fuzzy sets or fuzzy clusters of the multidimensional data, in a totally nonparametric way; fitting separate parametric regression models to each fuzzy cluster (or sub-sample) of the data; and then combining the estimates of these separate models in a very flexible way, based on the "degree of membership" of each sample point to each fuzzy cluster. It is this last step in the analysis that facilitates the overall procedure's ability to capture intrinsic nonlinearities, because the "weights" that are effectively assigned to each submodel vary continuously from data point to data point.

This general fuzzy modeling procedures has a wide range of applications, and we have provided several empirical applications to illustrate this. Not only can it be used to model and smooth data in ways that compete directly with established techniques such as nonparametric kernel regression, or splines, but it can also be used for trend extraction in competition with methods such as the Hodrick and Prescott (1980, 1997) filter. In all of the cases that we have examined, the fuzzy modeling approach performs extremely well. Moreover, it has significant practical advantages over both kernel regression and spline analysis. Spline analysis requires that the location of the "knots" be known in advance, whereas the first stage of our fuzzy modeling procedure determines the relevant data groupings empirically. Nonparametric kernel regression suffers from the so-called "curse of dimensionality" the severely limits its application to relatively simple models. In contrast, the fuzzy modeling structure that we have outlined is readily applicable to quite complex models, without undue computational burden. Indeed, it should also be noted that although our illustrative applications have all used ordinary least squares regression at the second stage of the analysis to fit models to each fuzzy cluster, in fact any relevant estimation technique could be used at that stage. For instance (and depending on the context), logit or probit models could be fitted, or nonlinear or instrumental variables regression could be used, without disrupting the general style of the fuzzy modeling strategy that we have described.

We recognize that the bulk of the discussion in this chapter addresses issues relating to curve fitting and data smoothing, and relatively little attention has been paid to strictly inferential issues. It was argued heuristically at the end of Section 4 that the fuzzy predictor emerging from our modeling approach will be a (weakly) consistent estimator of the conditional mean of the dependent variable. While this is minimally helpful, it will hold only under suitable conditions, and our work in progress explores this issue (and other aspects of the sampling properties of the fuzzy predictor) more thoroughly.

In this context it is also worth noting that the various "fuzzy predictions" that we have presented in our examples are simply "point predictions," and the corresponding confidence intervals are also extremely important. While it is not immediately clear how straightforward it would be to construct exact finite-sample prediction intervals within this framework, they could certainly be approximated by means of bootstrap simulation. Asymptotic prediction intervals can be constructed relatively easily, as follows. It will be recalled that once the sample has been partitioned into fuzzy clusters (typically with a different number of data points in each cluster), our modeling procedure involves estimating a regression over each cluster, and then combining the results using weights based on the membership values. The precise details are given in Step 3 in Section 4. These weights vary continuously, but are exogenous if the regressors also satisfy this property. Rather than estimate each cluster regression separately, by least squares, another option is to estimate the

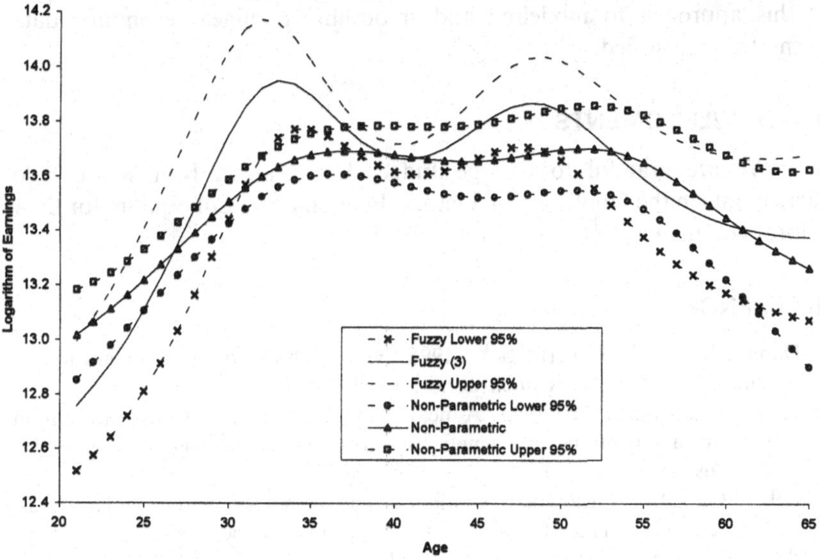

FIGURE 24 Confidence intervals for earning–age profile predictions.

group of cluster regressions as a "seemingly unrelated regression" (SUR) model. The only complication is that the SUR model is "unbalanced," in the sense that the samples associated with each equation are different from one another. However, unbalanced SUR models can be estimated quite readily (e.g., see Srivastava and Giles, 1987, pp. 339–346, for details). This then provides a complete estimated asymptotic covariance matrix for *all* of the estimated parameters in all the cluster regressions. This information (together with the membership values) can then be used to construct observation-by-observation asymptotic standard errors for the prediction of the conditional mean of the dependent variable.

By way of illustration we have undertaken this analysis in the case of the earnings–age profile data analyzed in Section 5.1, and the results appear in Fig. 24. SHAZAM code was written to produce these results, which are compared with their counterparts from nonparametric kernel regression. As can be seen, the fuzzy modeling procedure produces intervals that are very comparable to those from nonparametric regression, in terms of their shape and width. Given that our illustrative applications are based on relatively small samples, we have not provided corresponding asymptotic confidence intervals for our other "fuzzy predictions."

Clearly, much remains to be done in order to validate the real worth of the type of "fuzzy econometric modeling" that has been described and illustrated in this chapter. Nonetheless, we view the results presented here as being rather promising. Certainly, further exploration of this approach to modeling and smoothing nonlinear economic data seems to be justified.

ACKNOWLEDGMENTS

We are grateful to George Judge, Joris Pinkse, Ken White, and participants in the University of Victoria Econometrics Colloquium for their helpful comments.

REFERENCES

Al-Daoud, M.B. and Roberts, S.A., 1996. New methods for the initialisation of clusters, *Pattern Recognition Letters* **17**, 451–455.

Babu, G. and Murty, M., 1993. A near-optimal initial seed value selection in k-means algorithm using a genetic algorithm. *Pattern Recognition Letters* **14**, 763–769.

Bezdek, J.C., 1973. Fuzzy mathematics in pattern classification. Ph.D. thesis, Applied Mathematics Center, Cornell University, Ithaca, NY.

Bezdek, J.C., 1981. *Pattern Recognition With Fuzzy Objective Function Algorithms.* Plenum Press, New York.

Bishop, C.M., 1995. *Neural Networks for Pattern Recognition.* Oxford University Press, Oxford.

Chiu, S.L., 1994. Fuzzy model identification based on cluster estimation. *Journal of Fuzzy and Intelligent Systems* **2**, 267–278.

Coppejans, M., 2000. Breaking the curse of dimensionality. Unpublished manuscript, Department of Economics, Duke University, Raleigh, NC.

Cummins, C., 1994, TSP procedure hptrend44. TSP International, www.stanford.edu/~clint/tspex/hptrend44.tsp

Dasgupta, M. and Deb, R., 1996. Transitivity and fuzzy preferences. *Social Choice and Welfare* **13**, 305–318.

Davé, R.N., 1991. Characterization and detection of noise in clustering. *Pattern Recognition Letters* **12**, 657–664.

Draeseke, R. and Giles, D.E.A., 1999. A fuzzy logic approach to modelling the underground economy. In: L. Oxley, F. Scrimgeour and M. McAleer eds. *Proceedings of the International Conference on Modelling and Simulation (MODSIM 99)*, vol. 2. Modelling and Simulation Society of Australia and New Zealand, Hamilton, pp. 453–458.

Draeseke, R. and Giles, D.E.A., 2002. A fuzzy logic approach to modelling the New Zealand underground economy. *Mathematics and Computers in Simulation* **59**, 115–123.

Dunn, J.C., 1973. A fuzzy relative of the ISODATA process. *Journal of Cybernetics* **3**, 32–57.

Dunn, J.C., 1974. Well separated clusters and optimal fuzzy partitions. *Journal of Cybernetics* **4**, 95–104.

Dunn, J.C., 1977. Indices of partition fuzziness and the detection of clusters in large data sets. In: M. Gupta and G. Seridis, eds. *Fuzzy Automata and Decision Processes.* Elsevier, New York.

Frigui, H. and Krishnapuram, R., 1996. A robust algorithm for automatic extraction of an unknown number of clusters from noisy data. *Pattern Recognition Letters* **7**, 1223–1232.

Griffiths, W.E., Hill, R.C., and Judge, G.G., 1993. *Learning and Practicing Econometrics.* Wiley, New York.

Hall, B.H., 1996. Time Series Processor Version 4.3, *User's Guide.* TSP International, Palo Alto, CA.

Heckman, J. and Polchek, A., 1974. Empirical evidence on functional form of the earnings–schooling relationship. *Journal of the American Statistical Association* **69**, 350–354.

Hodrick, R.J. and Prescott, E.C., 1980. Postwar U.S. business cycles: An empirical investigation. Discussion Paper No. 451, Department of Economics, Carnegie-Mellon University, Pittsburgh, PA.

Hodrick, R.J. and Prescott, E.C., 1997. Postwar U.S. business cycles: An empirical investigation. *Journal of Money, Credit and Banking* **29**, 1–16.

Hsing, Y. and Smyth, D.J., 1994. Kuznets's inverted-U hypothesis revisited. *Applied Economics Letters* **1**, 111–113.

Huang, C. and Harris, R., 1993. A comparison of several vector quantization codebook generation approaches. *IEEE Transactions in Image Processing* **2**, 108–112.

Jacobsen, P.W.F. and Giles, D.E.A., 1998. Income distribution in the United States: Kuznets' inverted-U hypothesis and data non-stationarity. *Journal of International Trade & Economic Development* **7**, 405–423.

Jolion, J. and Rosenfeld, A., 1989. Cluster detection in background noise. *Pattern Recognition* **22**, 603–607.

Josef, S., Korosi, G., and Matyas, L., 1992. A possible new approach of panel modeling. *Structural Change and Economics Dynamics* **3**, 357–374.

Katsavounidis, I., Kuo, Jay C.-C. and Z. Zharg, 1994. A new initialization technique for generalized Lloyd iteration. *IEEE Signal Processing Letters* **1**, 144–146.

Keller, A., 2000. Fuzzy clustering with outliers, Unpublished manuscript, Institute of Flight Guidance, German Aerospace Center, Braunschweig, Germany.

Kolmogorov, A.N., 1957. On the representation of continuous functions of several variables by superpositions of continuous functions of one variable and addition. *Doklady Akademii Nauk* **114**, 674–681.

Kuznets, S., 1955. Economic growth and income inequality. *American Economic Review* **45**, 1–28.

Kydland, F. and Prescott, E.C., 1990. Business cycles: Real facts and a monetary myth. *Federal Reserve Bank of Minneapolis Quarterly Review*, Spring, 3–18.

Linde, Y., Buzo, A., and Gray, R., 1980. An algorithm for vector quantizer design, *IEEE Transactions in Communication* **28**, 84–95.

Lindström, T., 1998. A fuzzy design of the willingness to invest in Sweden. *Journal of Economic Behaviour and Organization* **36**, 1–17.

Lorentz, G.G., 1996. *Approximation of Functions*. Holt, Rinehart, and Winston, New York.

MacQueen, J., 1967. Some methods for classification and analysis of multivariate observations. In: J. M. Le Cam and J. Neyman, eds. *Proceedings of the 5th Berkeley Symposium in Mathematical Statistics and Probability*. University of California Press, Berkeley, CA, pp. 281–297.

Mincer, J., 1974. *Schooling, Experience and Earnings*. Columbia University Press for the National Bureau of Economic Research, New York.

Murphy, K.M. and Welsch, F., 1990. Empirical age–earnings profiles. *Journal of Labor Economics* **8**, 202–229.

Pagan, A. and Ullah, A., 1999. *Nonparametric Econometrics*. Cambridge University Press, Cambridge, UK.

Richardson, G., 1998. The structure of fuzzy preferences: Social choice implications. *Social Choice and Welfare* **15**, 359–369.

Ruspini, E., 1970. Numerical methods for fuzzy clustering. *Information Science* **2**, 319–350.

Sengupta, K., 1999. Choice rules with fuzzy preferences: Some characterizations. *Social Choice and Welfare* **16**, 259–272.

SHAZAM, 2000. *User's Reference Manual*, Version 9. SHAZAM Econometrics Software, Vancouver, BC.

Shepherd, D. and Shi, F.K.C., 1998. Economic modelling with fuzzy logic. *CEFES'98 Conference*, Cambridge, UK.

Silverman, B.W., 1986. *Density Estimation for Statistics and Data Analysis*. Chapman and Hall, London.

Singh, R.S., Ullah, A., and Carter, R.A.L., 1987. Nonparametric inference in econometrics: New applications. In: I.A. MacNeil and G.J. Umphrey, eds. *Time Series and Econometric Modelling*, vol. III. Reidell, Boston, MA.

Srivastava, V.K. and Giles, D.E.A., 1987. *Seemingly Unrelated Regression Equations Models: Estimation and Inference*. Marcel Dekker, New York.

Studenmund, A.H., 1997. *Using Econometrics: A Practical Guide*, 3rd ed. Addison-Wesley, Reading, MA.

Sugeno, M. and Kang, G.T., 1985. Structure identification of a fuzzy model. *Fuzzy Sets and Systems* **28**, 15–33.

Takagi, T. and Sugeno, M., 1985. Fuzzy identification of system and its application to modelling and control. *IEEE Transactions on Systems, Man and Cybernetics* **15**, 116–132.

Tou, J. and Gozales, R., 1974. *Pattern Recognition Principles*. Addison-Wesley, Reading, MA.

Ullah, A., 1985. Specification analysis of econometric models. *Journal of Quantitative Economics* **1**, 187–209.

U.S. Department of Commerce, 2000. U.S. real personal consumption expenditures, monthly, seasonally adjusted at annual rates, billions of chained 1996 dollars. Downloaded from FRED database at URL www.stls.frb.org/fred/data/gdp/pcec96 on 14 August 2000; U.S. real disposable personal income, monthly, seasonally adjusted at annual rates, billions of chained 1996 dollars, downloaded from FRED database at URL www.stls.frb.org/fred/data/gdp/dspic96 on 14 August 2000.

U.S. Department of Commerce, various years, Economic Report of the President.

Weiss, I., 1988. Straight line fitting in a noisy image. *Proceedings of the IEEE Computer Society Conference on Computer Vision and Pattern Recognition*, IEEE, New York, 647–652.

White, H., 1992. *Artificial Neural Networks: Approximation and Learning Theory*. Blackwell, Cambridge, MA.

White, H. and Gallant, A.R., 1992. On learning the derivatives of an unknown mapping with multiplayer feedforward networks. *Neural Networks* **5**, 129–138.

Yager, R.R., Filev, D.P., 1992. Learning of fuzzy rules by mountain clustering. *Proceedings of the SPIE Conference on Applications of Fuzzy Logic Technology*, Boston, MA, pp. 246–254.

Zadeh, L.A., 1965. Fuzzy sets. *Information and Control* **8**, 338–353.

Zadeh, L.A., 1987. *Fuzzy Sets and Applications: Selected Papers*. Wiley, New York.

15

Nonparametric Bootstrap Specification Testing in Econometric Models

Tae-Hwy Lee and Aman Ullah
University of California at Riverside, Riverside, California, U.S.A.

1 INTRODUCTION

Since the path-breaking work of Karl Pearson the 20th century saw significant advance in parametric statistical and econometric hypothesis-testing procedures [see Bera (2000) for an excellent survey]. A problem with the parametric testing procedures is that the tests may not be consistent under the mis-specified alternative hypotheses. In the last two decades a rich literature has developed on constructing consistent model specification tests using nonparametric estimation techniques. Bierens (1982) was the first to provide a consistent conditional moment test for model mis-specification. Ullah (1985) first suggested the construction of a model specification test using the nonparametric estimation technique. A nonparametric specification test for time-series data was first proposed by Robinson (1989).

Since the publication of these works various test statistics have been proposed for consistently testing parametric regression functional forms, e.g., Andrews (1997), Azzalini *et al.* (1989), Bierens (1982, 1990), Bierens and Ploberger (1997), Cai *et al.* (2000), De Jong (1996), Eubank and Hart (1992), Eubank and Spiegelman (1990), Fan and Li (1996), Fan *et al.* (2001), Gozalo (1993), Härdle and Mammen (1993), Hart

(1997), Hong and White (1995), Horowitz and Härdle (1994), Horowitz and Spokoiny (2000), Li and Wang (1998), Robinson (1991), Ullah (1985), Whang (2000), Wooldridge (1992), Yatchew (1992), and Zheng (1996), among others. Similarly, several papers have appeared on testing the significance of omitted or excluded variables from the model, e.g., Aït-Sahalia *et al.* (1994), Fan and Li (1996), Härdle and Mammen (1993), Li (1999), Linton and Gozalo (1997), Racine (1997), Ullah and Vinod (1993), and Whang and Andrews (1993), among others. Delgado and Stengos (1994), Lavergne and Vuong (1996), and Ullah and Singh (1989) explore non-nested hypothesis-testing problems. In addition to omitted variables and functional forms, there are many papers that look into the nonparametric approach to general hypothesis-testing problems encountered in econometrics, e.g., Cai *et al.* (2000), Fan *et al.* (2001), Hart (1997), Lewbel (1993, 1995), Robinson (1989), and Ullah and Singh (1989), among others. For details, see Pagan and Ullah (1999).

While most of the early developments in nonparametric hypothesis testing appeared for the independent data, except, e.g., Robinson (1989), in recent years the problem of hypothesis testing with the dependent time-series data has been addressed by many authors. For example, Berg and Li (1998), Fan and Li (1997), Hjellvik and Tjøstheim (1995, 1998), Hjellvik *et al.* (1998), Kreiss (1998), Lee (2001), and Lee and Ullah (2001), among others, have considered tests for functional form and omitted variables. In particular, for time series, testing for omitted variables is often to identify the number of lags. Nonparametric lag selection in nonlinear time-series models are studied by Auestad and Tjøstheim (1990), Cheng and Tong (1992), Fan and Li (1999a), Granger and Lin (1994), Granger *et al.* (2000), Hong and White (2001), Tjøstheim and Auestad (1994), Tschernig and Yang (2000), and Yao and Tong (1994). Chen and Fan (1999) provided consistent tests for time-series models, but the asymptotic distributions of their tests were nonstandard. In a major development, Fan and Li (1999b) developed the central limit theorems for the degenerate U-statistics for the weakly dependent data. This has provided a significant breakthrough and important contribution in Li (1999), who shows the asymptotic normality of Li and Wang (1998) type tests for a wide range of hypothesis-testing problems, with dependent data, e.g., parametric functional forms in regression, single index models, semiparametric regressions, variable selection, and mean–variance ratio hypothesis in finance.

We note here that the test statistics for most of the testing problems described above are based on the following alternative procedures: (1) Ullah (1985) type F or likelihood ratio procedure comparing the residual sum of squares under the null and alternative hypotheses (see also

Azzalini *et al.*, 1989; Cai *et al.*, 2000; Fan *et al.*, 2001); (2) the procedure comparing the sum of squares of the differences in the fitted values of the models under the null and alternative specifications, e.g., Härdle and Mammen (1993), Ullah and Vinod (1993), and Aït-Sahalia *et al.* (1994); and (3) the conditional moment procedure looking into the covariance between the residual under the null and the model specified under the alternative (e.g., Fan and Li, 1996; Zheng, 1996; Li and Wang, 1998). These three alternative procedures are equivalent in the sense that they conform to the same population value of the null hypothesis of no difference between the null and alternative specifications. However, sample statistics based on them are different and may give different results. The purpose here is not to introduce a new procedure of nonparametric testing, instead the modest aim is to explore the bootstrap simulation comparison of these three procedures with respect to size and power properties in small as well as large samples. In the earlier simulation studies usually the case of testing a parametric specification is considered by only one of the three procedures. This is perhaps the first study that considers all the three procedures and looks into not only the testing of a parametric specification but also the testing of varying parameters and omitted variables. We also consider both the independent cross-section and dependent time-series data models. Both the naive bootstrap and wild bootstrap procedures are used for out analysis.

The plan of the chapter is as follows. In Section 2, we present the nonparametric kernel regression estimator. Section 3 presents the three procedures of nonparametric hypothesis testing. Then, in Section 4 we give our simulation results. Finally, Section 5 gives conclusions.

2 NONPARAMETRIC REGRESSION

Let us consider the regression model:

$$y_t = m(x_t) + u_t = E(y_t|x_t) + u_t \tag{1}$$

where $t=1,\ldots,n$, y_t is a scalar dependent variable, $x_t=(x_{t1},\ldots,x_{tk})$ is an $1\times k$ vector, $m(x_t)=E(y_t|x_t)$ is the true but unknown regression function, and u_t is the error term such that $E(u_t|x_t)=0$. The model in Eq. (1) includes the autoregressive model as a special case in which x_t consists of lagged values of y_t. For the time-series case we assume that $\{y_t, x_t\}$ is a strictly stationary discrete-time stochastic process.

A parametric approach to estimating $m(x_t)$ in Eq. (1) may begin by fitting a linear parametric regression model through the data as

$$\begin{aligned} y_t &= \alpha + \beta x_t + u_t \\ &= X_t\delta + u_t \end{aligned} \tag{2}$$

or more generally a nonlinear parametric model $y_t = m(x_t, \delta) + u_t$, where $X_t = (1\ x_t)$ and $\delta = (\alpha\ \beta)'$. One can obtain a least squares (LS) estimator of $\hat{m}(x_t)$ by $m(x_t, \hat{\delta})$ where $\hat{\delta}$ is the LS estimator of δ obtained by minimizing the global LS objective function:

$$\sum_{t=1}^{n} u_t^2 = \sum_{t=1}^{n} (y - m(x_t, \delta))^2 \tag{3}$$

However, this global parametric LS estimator, based on global modeling, is inconsistent and biased at least in the regions of data where the a priori specified regression is not correctly specified.

An alternative improved approach is to use the nonparametric kernel regression estimation of the unknown $m(x_t)$. Essentially, the idea behind the kernel regression is to model the regression function $m(x_t)$ locally. For example, to obtain the regression function at a given point x, we apply the standard linear regression technique to the data in the interval of length h around x. That is, for the data in the interval of length h, we consider the linear model:

$$\begin{aligned} y_t &= \alpha(x) + x_t\beta(x) + u_t \\ &= X_t\delta(x) + u_t \end{aligned} \tag{4}$$

and then estimate $\delta(x)$ by minimizing the local LS or weighted LS errors:

$$\sum_{t=1}^{n} u_t^2 K_{tx} = \sum_{t=1}^{n} (y_t - X_t\delta(x))^2 K_{tx} \tag{5}$$

with respect to $\delta(x)$, where $K_{tx} = K((x_t - x)/h)$ is called a kernel (weight) function and $h \to 0$ as $n \to \infty$ is usually called the window width (smoothing parameter). Generally, the kernel function can be any probability density function having a finite second moment. The estimator so obtained is

$$\hat{\delta}(x) = (\mathbf{X}'\mathbf{K}(x)\mathbf{X})^{-1}\mathbf{X}'\mathbf{K}(x)\mathbf{y} \tag{6}$$

where $\mathbf{K}(x)$ is the $n \times n$ diagonal matrix with diagonal elements $K_{tx} (t = 1, \ldots, n)$, $\mathbf{X}$ is an $n \times (k+1)$ matrix with the tth row X_t, and $\mathbf{y}$ is an $n \times 1$ vector. The estimator of $m(x_t)$ is then given by $m(x_t) = X_t\delta(x_t)$. The approach in Eq. (4) is called local linear regression and the estimator in Eq. (6) is known as the local linear LS (LLLS) estimator. For details on the kernel regression estimators and the choices of $K(\cdot)$ and h, see Wand and Jones (1995), Fan and Gijbels (1996), and Pagan and Ullah (1999).

It is interesting to note the special cases and generalization of Eqs. (4) and (6). When $h = \infty$, the local linear regression modeling in Eq. (4) becomes global modeling in Eq. (2) and the LLLS estimator $\hat{\delta}(x)$ in Eq. (6)

becomes the global LS estimator of $\hat{\delta}$. This is because when $h = \infty$, $K_{tx} = K(0)$ and the minimization of $\Sigma(y_t - X_t\delta(x))^2$ becomes the minimization of $K(0)\Sigma(y_t - X_t\delta(x))^2 =$ minimization of $\Sigma(y_t - X_t\delta)^2$. Also note that when $X_t = 1$, the LLLS estimator $\hat{\delta}(x)$ reduces to

$$\begin{aligned}\hat{\delta}(x) &= \hat{\alpha}(x) = (\mathbf{i}'\mathbf{K}(x)\mathbf{i})^{-1}\mathbf{i}'\mathbf{K}(x)\mathbf{y} \\ &= \frac{\sum_{t=1}^{n} y_t K_{tx}}{\sum_{t=1}^{n} K_{tx}}\end{aligned} \tag{7}$$

which is the Nadaraya (1964) and Watson (1964) kernel regression estimator, where $\mathbf{i}$ is an $n \times 1$ vector of unit elements. The LLLS can be extended to the pth order local polynomial LS estimator where, for $k = 1$, $X_t = [1, x_t, \ldots, x_t^p]$ and δ is a $(p+1) \times 1$ vector (see Fan and Gijbels, 1996).

One advantage of the local estimators is that they can be viewed as the varying coefficient (functional coefficient) estimators. This is because $\hat{\delta}(x)$ may have varying values at different data points x_t. In this sense the local linear model $y_t = X_t\delta(x) + u_t$ is a varying coefficient model $y_t = m(x_t) + u_t$ where $m(x_t) = X_t\delta(x_t) \simeq X_t\delta(x)$ where $\delta(x_t) \simeq \delta(x)$ is the first term of the Taylor's approximation around x. This is in contrast to the global estimator $\hat{\delta}$, which is the estimator of δ in the constant coefficient model.

The above idea of a varying coefficients model can be extended to the situations where the coefficients are varying with respect to z_t, which may be a subset of x_t or something else, i.e.,

$$E(y_t|x_t) = m(x_t) = X_t\delta(z_t) \tag{8}$$

Examples of these include the functional coefficient autoregressive model (Chen and Tsay, 1993; Cai *et al.*, 2000), smooth coefficient model (Li *et al.*, 1997), random coefficient model (Raj and Ullah, 1981), smooth transition autoregressive model (Granger and Teräsvirta, 1993), exponential autoregressive model (Haggan and Ozaki, 1981), and threshold autoregressive model (Tong, 1990). Also see Section 4. To estimate $\delta(z_t)$ we can again do a local approximation $\delta(z_t) \simeq \delta(z)$ and then minimize $\Sigma[y_t - X_t\delta(z)]^2 K_{tz}$ with respect to $\delta(z)$, where $K_{tz} = K((z_t - z)/h)$. This gives the varying coefficient estimator:

$$\hat{\delta}(z) = (\mathbf{X}'\mathbf{K}(z)\mathbf{X})^{-1}\mathbf{X}'\mathbf{K}(z)\mathbf{y}$$

where $\mathbf{K}(z)$ is a diagonal matrix of K_{tz}, $t = 1, \ldots, n$. When $z_t = x_t$, this reduces to the LLLS estimator $\hat{\delta}(x)$ in Eq. (6).

Cai *et al.* (2000) consider a local linear approximation $\delta(z_t) \simeq \delta(z) + D(z)(z_t - z)'$ where $D(z) = (\partial\delta(z_t)/\partial z_t')$ evaluated at $z_t = z$. The LL varying coefficient (LLVC) estimator of Cai et al. (2000) is then obtained by minimizing

$$\sum_{t=1}^{n}[y_t - X_t\delta(z_t)]^2 K_{tz}$$
$$= \sum_{t=1}^{n}[y_t - X_t\delta(z) - [(z_t - z)\otimes X_t]\ \mathrm{vec}\ D(z)]^2 K_{tz} = \sum_{t=1}^{n}[y_t - X_t^z\delta^z(z)]^2 K_{tz}$$

with respect to $\delta^z(z) = [\delta(z)'\ (\mathrm{vec}\ D(z))']'$ where $X_t^z = [X_t\ (z_t - z)\otimes X_t]$. This gives

$$\ddot{\delta}^z(z) = (\mathbf{X}^{z\prime}\mathbf{K}(z)\mathbf{X}^z)^{-1}\mathbf{X}^{z\prime}\mathbf{K}(z)\mathbf{y} \tag{9}$$

and $\ddot{\delta}(z) = (\mathbf{I}\ 0)\ddot{\delta}^z(z)$. Hence,

$$\tilde{m}(x) = (1\ x\ 0)\ddot{\delta}^z(z) = (1\ x)\ddot{\delta}(z) \tag{10}$$

For the asymptotic properties of these varying coefficient estimators, see Cai *et al.* (2000).

3 NONPARAMETRIC BOOTSTRAP TESTS

We consider here two types of null hypotheses on $m(\cdot)$:

$$H_0 : m(x_t) = m(x_t, \delta) \tag{11}$$

$$H_0 : m(x_t) = m(x_{t1}) \tag{12}$$

where $x_t = (x_{t1}, x_{t2})$; x_{t1} is $1\times k_1$, x_{t2} is $1\times k_2$, and $k = k_1 + k_2$. The alternative hypothesis in each case is the unspecified nonparametric regression:

$$H_1 : m(x_t) = E(y_t|x_t) \tag{13}$$

The null hypothesis in Eq. (11) can be used as the null hypothesis for testing both for the functional form as well as the varying coefficient models. As a simple example of Eq. (11), we may consider testing for the functional form to be a linear regression, namely, $H_0 : m(x_t, \delta) = X_t\delta$ against H_1 in Eq. (13). However, if H_1 in Eq. (13) is specified to be local linear models $m(x_t) = X_t\delta(x_t)$ or $m(x_t) = X_t\delta(z_t)$ then the testing for the null $H_0 : m(x_t, \delta) = X_t\delta$ against $H_1 : m(x_t) = X_t\delta(z_t)$ is testing for the constant regression model against the varying coefficient regression (see Section 4 for more examples).

The null hypothesis in Eq. (12) is for testing the significance of the omitted variables x_{t2}, i.e., to test for selection of variables or lags. In this case both the null and alternative models are nonparametric. Though not considered here the test statistics considered below can also be used for testing situations where the null hypothesis is the partially linear model, $m(x_t) = x_{t1}\beta + m(x_{t2})$, single index models $m(x_t) = m(x_t, \delta)$, among others.

We will consider three main approaches for the above testing problems.

The first test procedure we consider is, as suggested in Ullah (1985), to compare the residual sum of squares RSS_0 under the null with the nonparametric residual sum of squares under the alternative, RSS_1. The test statistic is

$$T = \frac{(\text{RSS}_0 - \text{RSS}_1)}{\text{RSS}_1} \tag{14}$$

where for the null, Eq. (11), $\text{RSS}_0 = \Sigma\hat{u}_t^2, \hat{u}_t = y_t - m(x_t, \hat{\delta})$, and for the null, Eq. (12), $\text{RSS}_0 = \Sigma\tilde{u}_t^2, \tilde{u}_t = y_t - \tilde{m}(x_{t1})$, and $\tilde{m}(x_{t1})$ is given by Eq. (10) with $x_t = x_{t1}$. Further, $\text{RSS}_1 = \Sigma\tilde{u}_t^2, \tilde{u}_t = y_t - \tilde{m}(x_t)$. We reject the null hypothesis when T is large.

Fan and Li (1992) show the asymptotic normality of $nh^{k/2}T$ [see also Cai *et al.* (2000) and Fan *et al.* (2001, Theorem 5)].

Fan *et al.* (2001) further show that a suitably normalized T will have its asymptotic null distribution that is independent of nuisance parameters. They call this property the Wilks (1938) phenomenon. An important consequence of this result is that one does not have to derive theoretically the normalizing factors in order to be able to use the test. As long as the Wilks phenomenon holds, one can simply simulate the null distribution of the test statistic T. This is in stark contrast with some other tests whose asymptotic null distributions depend on nuisance parameters. Based on these Wilks results of Fan *et al.* (2001), Cai *et al.* (2000) suggest using the bootstrap method, which allows the implementation of Eq. (14). It involves the following steps to evaluate p values of T to test the null hypotheses in Eqs. (11) and (12):

1. Generate the bootstrap residual $\{\tilde{u}_t^*\}$ from the centered residuals from the nonparametric (NP) alternative model $(\tilde{u}_t - \bar{u})$ where $\bar{u} = n^{-1}\Sigma\tilde{u}_t$ and $\tilde{u}_t = y_t - \tilde{m}(x_t)$.
   a. For naive bootstrap, $\{\tilde{u}_t^*\}$ is obtained from randomly resampling $\{\tilde{u}_t - \bar{u}\}$ with replacement.
   b. For wild bootstrap, $\tilde{u}_t^* = a(\tilde{u}_t - \bar{u})$ with probability $r = (\sqrt{5}+1)/2\sqrt{5}$ and $\tilde{u}_t^* = b(\tilde{u}_t - \bar{u})$ with probability $1 - r \times (t = 1, \ldots, n)$, where $a = -(\sqrt{5}-1)/2$ and $b = (\sqrt{5}+1)/2$. See Li and Wang (1998, pp. 150–151).
2. Generate the bootstrap sample $\{y_t^*\}_{t=1}^n$ from the null model, from $y_t^* \equiv m(x_t, \hat{\delta}) + \tilde{u}_t^* (t = 1, \ldots, n)$ for the null in Eq. (11) to test for

parametric functional form, and from $y_t^* = \tilde{m}(x_{t1}) + \tilde{u}_t^*$ for the null in Eq. (12) to test for omitted variables x_{t2}.

3. Using the bootstrap sample $\{y_t^*, x_t, z_t\}_{t=1}^n$, calculate the bootstrap test statistic T^* using, for the sake of simplicity, the same h used in estimation with the original sample as done in Cai *et al.* (2000).
4. Repeat the above steps B times and use the empirical distribution of T^* as the conditional null distribution of T given $\{y_t^*, x_t, z_t\}_{t=1}^n$. We use $B = 500$. The bootstrap p value of the test T is simply the relative frequency of the event $\{T^* \geq T\}$ in the bootstrap resamples.

We use both naive bootstrap (Efron, 1979) and wild bootstrap (Wu, 1986; Liu, 1988). The wild bootstrap method preserves the conditional heteroskedasticity in the original residuals. For wild bootstrap, see also Härdle (1990, p. 247), Shao and Tu (1995, p. 292), or Li and Wang (1998, p. 150).

Two more versions of the T test in (14) can be considered:

$$S = \frac{\text{RSS}_0 - \text{RSS}_1}{\text{RSS}_0} \tag{15}$$

$$R = \frac{1}{n}(\text{RSS}_0 - \text{RSS}_1) \tag{16}$$

where S is the same as T with RSS_1 in the denominator replaced by RSS_0 in the spirit of Rao's score test, and R is essentially the numerator of T. In our Monte Carlo study in Section 4 the statistics T, S, and R are compared and calculated on the basis of weighted (trimmed) RSS to control the tail behavior of the nonparametric estimator. For example, weighted $\text{RSS}_1 = \Sigma \tilde{u}_t^2 w(z_t, a)$, and for the null in Eq. (11) weighted $\text{RSS}_0 = \Sigma \hat{u}_t^2 w(z_t, a)$, where $w(z_t, a) = \mathbf{1}(|z_t/\hat{\sigma}_z| < a)$, $\hat{\sigma}_z$ is the sample standard deviation, $a = \infty$, 2, 1.5, and $\mathbf{1}(\cdot)$ is the indicator function. The test statistics with weighted RSSs will be denoted as R_a, S_a, and T_a. Note that $w(z_t, \infty) = 1$ and thus $R_\infty, S_\infty, T_\infty$ are R, S, T in Eqs. (14)–(16) without weights. When the weight $w(\cdot, \cdot)$ is a known function so that it is not estimated, the Wilks phenomenon continues to hold as shown by Fan *et al.* (2001, Remark 4.2 and Theorem 9) and thus the Cai *et al.* (2000) bootstrap procedure can be applied to the statistics R_a, S_a, T_a with weighted RSSs.

The second test procedure we consider is to compare the fitted values from the null and alternative models as suggested in Härdle and Mammen

(1993), Ullah and Vinod (1993), and Aït-Sahalia *et al.* (1994). For the null in Eq. (11), this is given by

$$Q_a = \frac{1}{n}\sum_{t=1}^{n}(m(x_t, \hat{\delta}) - \tilde{m}(x_t))^2 w(z_t, a) \tag{17}$$

For the omitted variable testing in Eq. (12), $m(x_t, \hat{\delta})$ in Eq. (17) is replaced by the NP estimator $\tilde{m}(x_{t1})$ with x_{t2} omitted. The bootstrap procedure described above for R_a, S_a, T_a may also be applied to Q_a.

The third test procedure we consider is the conditional moment test for $E(u_t|x_t) = 0$, which is identical to testing

$$E[u_t E(u_t|x_t) f(x_t)] = 0 \tag{18}$$

where $f(x_t)$ is the density of x_t. A sample estimator of the left-hand side of Eq. (18) is

$$\begin{aligned} L' &= \frac{1}{n}\sum_{t=1}^{n}\hat{u}_t E(\hat{u}_t|x_t)\hat{f}(x_t) \\ &= \frac{1}{n(n-1)h^k}\sum_{t=1}^{n}\sum_{t'=1, t'\neq t}^{n}\hat{u}_t\hat{u}_{t'}K_{t't} \end{aligned} \tag{19}$$

where $\hat{u}_t = y_t - m(x_t, \hat{\delta}) = y_t - X_t\hat{\delta}$ to test for the null hypothesis in Eq. (11) or $\hat{u}_t = y_t - \tilde{m}(x_{t1})$ to test for the null hypothesis in Eq. (12), $E(\hat{u}_t|x_t) = \sum_{t'\neq t}\hat{u}_{t'}K_{t't}/\sum_{t'\neq t}K_{t't}$ from Eq. (7), and $\hat{f}(x_t) = [(n-1)h^k]^{-1}\sum_{t'\neq t}K_{t't}$ is the kernel density estimator, $K_{t't} = K((x_{t'} - x_t)/h)$. Note that we estimate the auxiliary regression function $E(\hat{u}_t|x_t)$ from the local constant LS estimator, Eq. (7), of Nadaraya (1964) and Watson (1964), not from the LLVC estimator of Cai *et al.* (2000) in Eq. (9) just to maintain the original formula of Li and Wang (1998) and Zheng (1996).

The asymptotic test statistic is then given by

$$L = nh^{k/2}\frac{L'}{\sqrt{\hat{\sigma}}} \xrightarrow{d} N(0, 1) \tag{20}$$

where $\hat{\sigma} = 2(n(n-1)h^k)^{-1}\Sigma_t\Sigma_{t'\neq t}\hat{u}_t^2\hat{u}_{t'}^2, K_{t't}^2$ is a consistent estimator of the asymptotic variance of $nh^{k/2}L'$; see Fan and Li (1996), Zheng (1996), Li and Wang (1998), Fan and Ullah (1999), and Rahman and Ullah (1999) for details. Also, see Pagan and Ullah (1999, Chap. 3) and Ullah (2001) for the relationship between R, Q, and L test statistics. Based on the asymptotic results of Fan and Li (1996, 1997, 1999b) and Li (1999) for dependent data, Berg and Li (1998) establish the asymptotic validity of using the wild bootstrap method for L for time series. The bootstrap p

values for L to test for the adequacy of the linear parametric model, $m(x_t, \delta) = X_t\delta$ in Eq. (11), can be computed as follows:

1. Generate the bootstrap residuals $\{\hat{u}_t^*\}$ from the residual from the null model $\hat{u}_t = y_t - X_t\hat{\delta}$:
   a. For naive bootstrap, $\{\hat{u}_t^*\}$ is obtained from randomly resampling $\{\hat{u}_t\}$ with replacement.
   b. For wild bootstrap, $\hat{u}_t^* = a\hat{u}_t$ with probability r and $\hat{u}_t^* = b\hat{u}_t$ with probability $1 - r$ as discussed above.
2. Generate the bootstrap sample $\{y_t^*\}_{t=1}^n$ from the null model $y_t^* \equiv m(x_t, \hat{\delta}) + \hat{u}_t^* = X_t\hat{\delta} + \hat{u}_t^*(t = 1, \ldots, n)$ for the null in Eq. (11) to test for neglected nonlinearity.
3. Using the bootstrap sample $\{y_t^*\}_{t=1}^n$, calculate the bootstrap test statistic L^*.
4. Repeat the above steps B times and use the empirical distribution of L^* as the null distribution of L. We use $B = 500$. The bootstrap p value of the test L is the relative frequency of the event $\{L^* \geq L\}$ in the bootstrap resamples.

For testing the omitted variables in the null, Eq. (12), we replace $X_t\hat{\delta}$ in the steps 1 and 2 above by the NP regression estimator $\tilde{m}(x_{t1})$ in Eq. (10) with $x_t = x_{t1}$ and with x_{t2} omitted, and use centered NP residuals. That is:

1. Generate the bootstrap residuals $\{\hat{u}_t^*\}$ from the centered residuals from the NP alternative model $(\hat{u}_t - \bar{u})$ where $\bar{u} = n^{-1}\Sigma\hat{u}_t$ and $\hat{u}_t = y_t - \tilde{m}(x_{t1})$.
2. Generate the bootstrap sample $\{y_t^*\}_{t=1}^n$ from the null model $y_t^* = \tilde{m}(x_{t1}) + \hat{u}_t^*$.

For parametric models, Davidson and MacKinnon (1999) show that the size distortion of a bootstrap test is at least of the order $n^{-1/2}$ smaller than that of the corresponding asymptotic test. For nonparametric models, h also enters in the order of refinement. Li and Wang (1998) show that if the distribution of L^j ($j = A$ for asymptotic, B for naive bootstrap, and W for wild bootstrap) admit an Edgeworth expansion then the bootstrap distribution approximates the null distribution of L with an error of order $n^{-1/2}h^{k/2}$, improving over the normal approximation. Since L is asymptotically normal under the null, the bootstrap tests L^B and L^W are more accurate than the asymptotic test L^A, as confirmed in the simulation of Section 4. See Hall (1992) for further discussion on Edgeworth expansions and the extent of the refinements in various contexts.

The above three different testing approaches are related. Under the null in Eq. (11) $H_0 : m(x_t) = m(x_t, \delta)$, the RSS-based test statistics R, S, T will be expected to be zero as $E(y_t - m(x_t, \delta))^2 = E(y_t - m(x_t))^2$. Also, by construction $E(u_t|x_t) = 0$ under the null, which implies $E[u_t E(u_t|x_t) f(x_t)] = 0$ for the L test. From this, we obtain the relationship used for the Q test because $E[u_t E(u_t|x_t) f(x_t)] = E[E(u_t|x_t)^2 f(x_t)] = E[\{E(y_t|x_t) - m(x_t, \delta)\}^2 f(x_t)] = 0$.

4 MONTE CARLO

In this section we examine the finite sample properties of the test statistics T, Q, L especially with the empirical null distributions being generated by the bootstrap method. Asymptotic critical values are also used for the L test. We consider four cases as indicated in "blocks" below. All of the error terms ε_t below are i.i.d. $N(0, 1)$.

Block 1

This block is to study the size and power of the tests for functional form or varying coefficients in time-series models. Let $x_t = y_{t-1}$. The following two data generating processes (DGP) are taken from Lee *et al.* (1993).

DGP 1 Linear AR (1)

$$y_t = 0.6 y_{t-1} + \varepsilon_t$$

DGP 2 Threshold Autoregressive [TAR(1)] (Nonlinear AR)

$$\begin{aligned} y_t &= 0.9 y_{t-1} + \varepsilon_t \quad |y_{t-1}| \le 1 \\ &= -0.3 y_{t-1} + \varepsilon_t \quad |y_{t-1}| > 1 \end{aligned}$$

Note that DGP 1 is a constant parameter model whereas the alternative DGP 2 is a varying parameter model $y_t = y_{t-1}\delta(y_{t-1}) + \varepsilon_t$, where $\delta(y_{t-1}) = 0.9$ or -0.3, depending on $x_t = z_t = y_{t-1}$. In this sense, testing for DGP 1 against DGP 2 is also a test for varying parameters.

Block 2

This block is to study the size and power of the tests for functional form in cross-sectional models. Let v_{t1} and v_{t2} be drawn from $IN(0, 1)$. Two regressors x_{t1} and x_{t2} are defined as $x_{t1} = v_{t1}$ and $x_{t2} = (v_{t1} + v_{t2})/\sqrt{2}$. Let $x_t = (x_{t1}\ x_{t2})$. The following two models are taken from Zheng (1996).

DGP 3

$$y_t = 1 + x_{t1} + x_{t2} + \varepsilon_t$$

DGP 4

$$y_t = |1 + x_{t1} + x_{t2}|^{5/3} + \varepsilon_t$$

Block 3

This block is to study the size and power of the tests for lag selection in time-series models. Let $x_t = (y_{t-1}\ y_{t-2})$. The alternative model DGP 6 is taken from Cai *et al.* (2000).

DGP 5 Exponential AR(1)

$$y_t = a_1(y_{t-1})y_{t-1} + 0.2\varepsilon_t$$
$$a_1(y_{t-1}) = 0.138 + (0.316 + 0.982y_{t-1})\exp(-3.89y_{t-1}^2)$$

DGP 6 Exponential AR(2)

$$y_t = a_1(y_{t-1})y_{t-1} + a_2(y_{t-1})y_{t-2} + 0.2\varepsilon_t$$
$$a_1(y_{t-1}) = 0.138 + (0.316 + 0.982y_{t-1})\exp(-3.89y_{t-1}^2)$$
$$a_2(y_{t-1}) = -0.437 - (0.659 + 1.260y_{t-1})\exp(-3.89y_{t-1}^2)$$

Block 4

This block is to study the size and power of the tests for variable selection in cross-sectional models. Let v_{t1} and v_{t2} be drawn from $IN(0, 1)$. Two regressors x_{t1} and x_{t2} are defined as $x_{t1} = v_{t1}$ and $x_{t2} = (v_{t1} + v_{t2})/\sqrt{2}$. Let $x_t = (x_{t1}\ x_{t2})$. The alternative model DGP 8 is taken from Zheng (1996).

DGP 7

$$y_t = |1 + x_{t1}|^{5/3} + \varepsilon_t$$

DGP 8

$$y_t = |1 + x_{t1} + x_{t2}|^{5/3} + \varepsilon_t$$

To estimate $\hat{u}_t$ for the null model and $\tilde{u}_t$ for the alternative model, the information sets used are $x_t = y_{t-1}$ for Block 1, $x_t = (y_{t-1}\ y_{t-2})$ for Block 3, and $x_t = (x_{t1}\ x_{t2})$ for Blocks 2 and 4. The omitted variable is y_{t-2} for Block 3 and x_{t2} for Block 4.

We use a scalar "threshold variable" z_t for all models: $z_t = y_{t-1}$ for Blocks 1 and 3, and $z_t = x_{t1}$ for Blocks 2 and 4.

For the Q_a, R_a, S_a, and T_a tests, as suggested by Cai *et al.* (2000), we select h using out-of-sample cross-validation. Let m and Q be two positive integers such that $n > mQ$. The basic idea is first to use Q subseries of lengths $n - qm$ $(q = 1, \ldots, Q)$ to estimate the coefficient functions $\delta_q(z_t)$ and then to compute the one-step forecast errors of the next segment of the time series of length m based on the estimated models. That is, to choose h minimizing the average of the mean square forecast errors:

$$AMS(h) = \sum_{q=1}^{Q} AMS_q(h) \tag{21}$$

$$AMS_q(h) = \frac{1}{m} \sum_{t=n-qm+1}^{n-qm+m} [y_t - X_t^z \ddot{\delta}_q^z(z)]^2 \tag{22}$$

and $\ddot{\delta}_q^z(\cdot)$ are computed from the sample $\{y_t\ x_t\}_{t=1}^{n-qm}$. We use $m = [0.1n]$, $Q = 4$, and the Epanechinikov kernel $K(z) = (3/4)(1 - z^2)1(|z| < 1)$.

For the L test, as in Li and Wang (1998, p. 154), we use a standard normal kernel. Note that x_t is an $1 \times k$ vector, and $k = 1$ for Block 1 and $k = 2$ for Blocks 2, 3, and 4. Thus, the smoothing parameter h is chosen as $h_i = c\hat{\sigma}_i n^{-1/5}$ $(i = 1)$ for the cases with $k = 1$, and $h_i = c\hat{\sigma}_i n^{-1/6}$ $(i = 1, 2)$ for the cases with $k = 2$, where $\hat{\sigma}_i$ is the sample standard deviation of ith element of x_t. The four values of $c = 0.1$, 0.5, 1, and 2 are used, and the corresponding estimated rejection probability will be denoted as L_c. In computing L, the h^k shown in Eqs. (19) and (20) is replaced with $\Pi_{i=1}^{k} h_i$.

Test statistics are denoted as Q_a^j, R_a^j, S_a^j, T_a^j, and L_c^j, with the superscripts $j = A$, B, W referring to the methods of obtaining the null distributions of the test statistics; asymptotics $(j = A)$, naive bootstrap $(j = B)$, and wild bootstrap $(j = W)$.

Monte Carlo experiments are conducted with 500 bootstrap resamples and 1000 Monte Carlo replications. The amount of computing time needed to obtain the size results (Panels A and B) of the 36 statistics shown in the tables for both 5 and 10% levels was as follows. It took a 600 MHz–256 MB Pentium III PC approximately 1 day for $n = 50$, 2–3 days for $n = 100$, and 5–7 days for $n = 200$. So, it took the PC roughly 7–10 days for the size results in each table. It took another 7–10 days for the power results (Panels C and D) in each table. It took less for Tables 1 and 2 where the null hypothesis is Eq. (11) than for Tables 3 and 4 where the null hypothesis is Eq. (12) and thus both the null and alternative models are nonparametric. For all the results in this chapter, it took the PC about 2 months. A GAUSS code for computing all the tests is available from the authors.

TABLE 1 Testing for Linearity in Time-Series Models (Block 1)

A. Size of tests at 5% level with DGP 1

n	Q^B_∞	Q^W_∞	Q^B_2	Q^W_2	$Q^B_{1.5}$	$Q^W_{1.5}$	R^B_∞	R^W_∞	R^B_2	R^W_2	$R^B_{1.5}$	$R^W_{1.5}$
50	0.089	0.125	0.089	0.099	0.081	0.075	0.102	0.135	0.101	0.117	0.097	0.089
100	0.041	0.072	0.029	0.036	0.030	0.024	0.049	0.087	0.039	0.050	0.036	0.036
200	0.033	0.059	0.024	0.026	0.021	0.019	0.037	0.066	0.033	0.036	0.027	0.024

n	S^B_∞	S^W_∞	S^B_2	S^W_2	$S^B_{1.5}$	$S^W_{1.5}$	T^B_∞	T^W_∞	T^B_2	T^W_2	$T^B_{1.5}$	$T^W_{1.5}$
50	0.022	0.092	0.025	0.073	0.020	0.055	0.022	0.092	0.025	0.073	0.020	0.055
100	0.018	0.062	0.015	0.031	0.019	0.020	0.018	0.062	0.015	0.031	0.019	0.020
200	0.021	0.057	0.024	0.028	0.015	0.018	0.021	0.057	0.024	0.028	0.015	0.018

n	$L^A_{0.1}$	$L^B_{0.1}$	$L^W_{0.1}$	$L^A_{0.5}$	$L^B_{0.5}$	$L^W_{0.5}$	L^A_1	L^B_1	L^W_1	L^A_2	L^B_2	L^W_2
50	0.047	0.059	0.055	0.013	0.049	0.048	0.001	0.046	0.042	0.000	0.039	0.039
100	0.031	0.041	0.042	0.015	0.044	0.047	0.006	0.049	0.046	0.000	0.043	0.047
200	0.038	0.045	0.044	0.027	0.052	0.057	0.008	0.047	0.048	0.002	0.050	0.041

B. Size of tests at 10% level with DGP 1

n	Q^B_∞	Q^W_∞	Q^B_2	Q^W_2	$Q^B_{1.5}$	$Q^W_{1.5}$	R^B_∞	R^W_∞	R^B_2	R^W_2	$R^B_{1.5}$	$R^W_{1.5}$
50	0.109	0.164	0.112	0.127	0.104	0.096	0.126	0.170	0.128	0.147	0.122	0.113
100	0.057	0.102	0.049	0.051	0.041	0.039	0.080	0.113	0.064	0.069	0.067	0.061
200	0.048	0.097	0.038	0.036	0.033	0.031	0.064	0.121	0.054	0.056	0.050	0.045

n	S^B_∞	S^W_∞	S^B_2	S^W_2	$S^B_{1.5}$	$S^W_{1.5}$	T^B_∞	T^W_∞	T^B_2	T^W_2	$T^B_{1.5}$	$T^W_{1.5}$
50	0.048	0.127	0.048	0.105	0.045	0.079	0.048	0.127	0.048	0.105	0.045	0.079
100	0.041	0.093	0.032	0.048	0.036	0.043	0.041	0.093	0.032	0.048	0.036	0.043
200	0.048	0.102	0.041	0.044	0.033	0.037	0.048	0.102	0.041	0.044	0.033	0.037

n	$L^A_{0.1}$	$L^B_{0.1}$	$L^W_{0.1}$	$L^A_{0.5}$	$L^B_{0.5}$	$L^W_{0.5}$	L^A_1	L^B_1	L^W_1	L^A_2	L^B_2	L^W_2
50	0.089	0.121	0.117	0.033	0.103	0.094	0.005	0.102	0.101	0.000	0.077	0.094
100	0.060	0.096	0.086	0.029	0.093	0.083	0.011	0.092	0.096	0.000	0.090	0.097
200	0.065	0.092	0.103	0.050	0.099	0.095	0.020	0.103	0.100	0.003	0.094	0.090

C. Power of tests at 5% level with DGP 2

n	Q^B_∞	Q^W_∞	Q^B_2	Q^W_2	$Q^B_{1.5}$	$Q^W_{1.5}$	R^B_∞	R^W_∞	R^B_2	R^W_2	$R^B_{1.5}$	$R^W_{1.5}$
50	0.422	0.505	0.422	0.452	0.432	0.414	0.457	0.533	0.473	0.488	0.489	0.461
100	0.649	0.731	0.677	0.674	0.684	0.663	0.717	0.776	0.732	0.741	0.744	0.727
200	0.922	0.946	0.908	0.907	0.906	0.905	0.954	0.965	0.947	0.948	0.959	0.956

(continued)

Table 1 Continued

n	S^B_∞	S^W_∞	S^B_2	S^W_2	$S^B_{1.5}$	$S^W_{1.5}$	T^B_∞	T^W_∞	T^B_2	T^W_2	$T^B_{1.5}$	$T^W_{1.5}$
50	0.232	0.454	0.250	0.397	0.271	0.362	0.232	0.454	0.250	0.397	0.271	0.362
100	0.628	0.739	0.658	0.694	0.675	0.688	0.628	0.739	0.658	0.694	0.675	0.688
200	0.942	0.962	0.945	0.947	0.950	0.956	0.942	0.962	0.945	0.947	0.950	0.956

n	$L^A_{0.1}$	$L^B_{0.1}$	$L^W_{0.1}$	$L^A_{0.5}$	$L^B_{0.5}$	$L^W_{0.5}$	L^A_1	L^B_1	L^W_1	L^A_2	L^B_2	L^W_2
50	0.304	0.331	0.325	0.408	0.567	0.549	0.228	0.575	0.569	0.001	0.335	0.388
100	0.656	0.696	0.706	0.857	0.928	0.922	0.773	0.943	0.943	0.086	0.825	0.840
200	0.967	0.973	0.972	0.997	0.999	0.999	0.996	0.998	0.998	0.842	0.998	0.999

D. Power of tests at 10% level with DGP 2

n	Q^B_∞	Q^W_∞	Q^B_2	Q^W_2	$Q^B_{1.5}$	$Q^W_{1.5}$	R^B_∞	R^W_∞	R^B_2	R^W_2	$R^B_{1.5}$	$R^W_{1.5}$
50	0.474	0.578	0.489	0.501	0.492	0.473	0.514	0.606	0.529	0.539	0.551	0.531
100	0.714	0.792	0.724	0.733	0.734	0.727	0.776	0.823	0.783	0.776	0.787	0.782
200	0.953	0.964	0.911	0.911	0.910	0.909	0.970	0.979	0.970	0.966	0.978	0.976

n	S^B_∞	S^W_∞	S^B_2	S^W_2	$S^B_{1.5}$	$S^W_{1.5}$	T^B_∞	T^W_∞	T^B_2	T^W_2	$T^B_{1.5}$	$T^W_{1.5}$
50	0.329	0.536	0.355	0.472	0.372	0.456	0.329	0.536	0.355	0.472	0.372	0.456
100	0.714	0.807	0.736	0.763	0.741	0.755	0.714	0.807	0.736	0.763	0.741	0.755
200	0.968	0.976	0.961	0.964	0.975	0.975	0.968	0.976	0.961	0.964	0.975	0.975

n	$L^A_{0.1}$	$L^B_{0.1}$	$L^W_{0.1}$	$L^A_{0.5}$	$L^B_{0.5}$	$L^W_{0.5}$	L^A_1	L^B_1	L^W_1	L^A_2	L^B_2	L^W_2
50	0.404	0.464	0.450	0.482	0.695	0.702	0.304	0.715	0.724	0.003	0.516	0.573
100	0.763	0.817	0.816	0.902	0.968	0.963	0.836	0.972	0.973	0.163	0.909	0.922
200	0.983	0.989	0.989	0.998	0.999	0.999	0.998	1.000	0.999	0.912	1.000	1.000

Test statistics are denoted as Q^j_a, R^j_a, S^j_a, T^j_a, and L^j_σ with the superscripts $j=A, B, W$ referring to the methods of obtaining the null distributions of the test statistics, using the asymptotics (A), naive bootstrap (B), and wild bootstrap (W). The number of bootstrap resamples = 500 and the number of Monte Carlo replications = 1000. The 95% asymptotic confidence interval of the estimated size is (0.036, 0.064) at 5% nominal level of significance and (0.081, 0.119) at 10% nominal level of significance.

Table 1 presents the empirical size (DGP 1) and power (DGP 2) of testing for neglected nonlinearity in time-series models in Block 1. We observe the following:

1. The L test using bootstrap (L^B and L^W) exhibits excellent size behavior and is better than all the other tests (L^A, Q^j, R^j, S^j, and T^j, $j=A$, B, W).

TABLE 2 Testing for Linearity in Cross-Sectional Models (Block 2)

A. Size of tests at 5% level with DGP 3

n	Q^B_∞	Q^W_∞	Q^B_2	Q^W_2	$Q^B_{1.5}$	$Q^W_{1.5}$	R^B_∞	R^W_∞	R^B_2	R^W_2	$R^B_{1.5}$	$R^W_{1.5}$
50	0.056	0.109	0.054	0.071	0.040	0.045	0.064	0.122	0.061	0.082	0.047	0.058
100	0.028	0.080	0.020	0.033	0.010	0.013	0.040	0.099	0.035	0.043	0.026	0.028
200	0.020	0.054	0.007	0.012	0.003	0.002	0.024	0.068	0.014	0.018	0.009	0.008

n	S^B_∞	S^W_∞	S^B_2	S^W_2	$S^B_{1.5}$	$S^W_{1.5}$	T^B_∞	T^W_∞	T^B_2	T^W_2	$T^B_{1.5}$	$T^W_{1.5}$
50	0.014	0.078	0.016	0.052	0.013	0.027	0.014	0.078	0.016	0.052	0.013	0.027
100	0.015	0.061	0.010	0.021	0.007	0.011	0.015	0.061	0.010	0.021	0.007	0.011
200	0.012	0.052	0.005	0.014	0.005	0.007	0.012	0.052	0.005	0.014	0.005	0.007

n	$L^A_{0.1}$	$L^B_{0.1}$	$L^W_{0.1}$	$L^A_{0.5}$	$L^B_{0.5}$	$L^W_{0.5}$	L^A_1	L^B_1	L^W_1	L^A_2	L^B_2	L^W_2
50	0.026	0.050	0.030	0.022	0.048	0.049	0.007	0.041	0.041	0.000	0.048	0.050
100	0.038	0.042	0.038	0.016	0.052	0.051	0.009	0.054	0.051	0.001	0.054	0.053
200	0.040	0.046	0.054	0.022	0.040	0.042	0.015	0.053	0.054	0.001	0.057	0.062

B. Size of tests at 10% level with DGP 3

n	Q^B_∞	Q^W_∞	Q^B_2	Q^W_2	$Q^B_{1.5}$	$Q^W_{1.5}$	R^B_∞	R^W_∞	R^B_2	R^W_2	$R^B_{1.5}$	$R^W_{1.5}$
50	0.076	0.139	0.077	0.093	0.054	0.058	0.097	0.154	0.095	0.116	0.076	0.085
100	0.054	0.116	0.038	0.051	0.022	0.024	0.075	0.132	0.056	0.075	0.042	0.049
200	0.035	0.089	0.016	0.023	0.008	0.008	0.052	0.103	0.030	0.032	0.021	0.023

n	S^B_∞	S^W_∞	S^B_2	S^W_2	$S^B_{1.5}$	$S^W_{1.5}$	T^B_∞	T^W_∞	T^B_2	T^W_2	$T^B_{1.5}$	$T^W_{1.5}$
50	0.030	0.114	0.033	0.080	0.027	0.044	0.030	0.114	0.033	0.080	0.027	0.044
100	0.032	0.106	0.026	0.047	0.022	0.034	0.032	0.106	0.026	0.047	0.022	0.034
200	0.024	0.086	0.017	0.025	0.016	0.014	0.024	0.086	0.017	0.025	0.016	0.014

n	$L^A_{0.1}$	$L^B_{0.1}$	$L^W_{0.1}$	$L^A_{0.5}$	$L^B_{0.5}$	$L^W_{0.5}$	L^A_1	L^B_1	L^W_1	L^A_2	L^B_2	L^W_2
50	0.098	0.105	0.084	0.038	0.098	0.093	0.012	0.098	0.101	0.000	0.100	0.105
100	0.087	0.093	0.082	0.049	0.109	0.101	0.020	0.107	0.108	0.002	0.114	0.108
200	0.091	0.103	0.104	0.046	0.086	0.088	0.021	0.094	0.096	0.003	0.112	0.112

C. Power of tests at 5% level with DGP 4

n	Q^B_∞	Q^W_∞	Q^B_2	Q^W_2	$Q^B_{1.5}$	$Q^W_{1.5}$	R^B_∞	R^W_∞	R^B_2	R^W_2	$R^B_{1.5}$	$R^W_{1.5}$
50	1.000	1.000	1.000	1.000	1.000	1.000	1.000	1.000	1.000	1.000	1.000	1.000
100	1.000	1.000	1.000	1.000	1.000	1.000	1.000	1.000	1.000	1.000	1.000	1.000
200	1.000	1.000	1.000	1.000	1.000	1.000	1.000	1.000	1.000	1.000	1.000	1.000

(continued)

TABLE 2 Continued

n	S^B_∞	S^W_∞	S^B_2	S^W_2	$S^B_{1.5}$	$S^W_{1.5}$	T^B_∞	T^W_∞	T^B_2	T^W_2	$T^B_{1.5}$	$T^W_{1.5}$
50	1.000	1.000	1.000	1.000	1.000	0.999	1.000	1.000	1.000	1.000	1.000	0.999
100	1.000	1.000	1.000	1.000	1.000	1.000	1.000	1.000	1.000	1.000	1.000	1.000
200	1.000	1.000	1.000	1.000	1.000	1.000	1.000	1.000	1.000	1.000	1.000	1.000

n	$L^A_{0.1}$	$L^B_{0.1}$	$L^W_{0.1}$	$L^A_{0.5}$	$L^B_{0.5}$	$L^W_{0.5}$	L^A_1	L^B_1	L^W_1	L^A_2	L^B_2	L^W_2
50	0.522	0.628	0.478	1.000	1.000	0.993	1.000	1.000	0.995	1.000	1.000	1.000
100	0.946	0.961	0.934	1.000	1.000	1.000	1.000	1.000	1.000	1.000	1.000	1.000
200	0.999	1.000	1.000	1.000	1.000	1.000	1.000	1.000	1.000	1.000	1.000	1.000

D. Power of tests at 10% level with DGP 4

n	Q^B_∞	Q^W_∞	Q^B_2	Q^W_2	$Q^B_{1.5}$	$Q^W_{1.5}$	R^B_∞	R^W_∞	R^B_2	R^W_2	$R^B_{1.5}$	$R^W_{1.5}$
50	1.000	1.000	1.000	1.000	1.000	1.000	1.000	1.000	1.000	1.000	1.000	1.000
100	1.000	1.000	1.000	1.000	1.000	1.000	1.000	1.000	1.000	1.000	1.000	1.000
200	1.000	1.000	1.000	1.000	1.000	1.000	1.000	1.000	1.000	1.000	1.000	1.000

n	S^B_∞	S^W_∞	S^B_2	S^W_2	$S^B_{1.5}$	$S^W_{1.5}$	T^B_∞	T^W_∞	T^B_2	T^W_2	$T^B_{1.5}$	$T^W_{1.5}$
50	1.000	1.000	1.000	1.000	1.000	1.000	1.000	1.000	1.000	1.000	1.000	1.000
100	1.000	1.000	1.000	1.000	1.000	1.000	1.000	1.000	1.000	1.000	1.000	1.000
200	1.000	1.000	1.000	1.000	1.000	1.000	1.000	1.000	1.000	1.000	1.000	1.000

n	$L^A_{0.1}$	$L^B_{0.1}$	$L^W_{0.1}$	$L^A_{0.5}$	$L^B_{0.5}$	$L^W_{0.5}$	L^A_1	L^B_1	L^W_1	L^A_2	L^B_2	L^W_2
50	0.792	0.791	0.672	1.000	1.000	0.998	1.000	1.000	0.999	1.000	1.000	1.000
100	0.993	0.994	0.986	1.000	1.000	1.000	1.000	1.000	1.000	1.000	1.000	1.000
200	1.000	1.000	1.000	1.000	1.000	1.000	1.000	1.000	1.000	1.000	1.000	1.000

2. R is better than S, and T, S, and T are identical; Q behaves similarly to R. The tests of Q, R, S, T tend to be oversized for $n=50$, and undersized with $n=100$ or 200, which is more apparent with the larger sample size.
3. Trimming for R_a, S_a, T_a, and Q_a is useful when n is small. For example, for $n=50$, T_2 works better than T_∞. However, the trimming makes the size worse when n is large (say, $n=200$).
4. The asymptotic test L^A works better with smaller c. L^A_c is not reliable for $c>0.1$ and gets worse as c gets larger. The bootstrap tests L^B and L^W work very well with all four values of c. The asymptotic L^A is sensitive to c while the bootstrap tests L^B and L^W are not sensitive to c. This is different from what is found by Lee and Ullah (2001) where the bootstrap tests L^B

TABLE 3 Lag Selection in Time-Series Models (Block 3)

A. Size of tests at 5% level with DGP 5

n	Q^B_∞	Q^W_∞	Q^B_2	Q^W_2	$Q^B_{1.5}$	$Q^W_{1.5}$	R^B_∞	R^W_∞	R^B_2	R^W_2	$R^B_{1.5}$	$R^W_{1.5}$
50	0.073	0.106	0.077	0.099	0.087	0.094	0.078	0.108	0.084	0.099	0.088	0.096
100	0.013	0.034	0.015	0.022	0.017	0.019	0.013	0.037	0.019	0.023	0.021	0.020
200	0.008	0.025	0.012	0.016	0.013	0.014	0.011	0.028	0.016	0.020	0.017	0.017

n	S^B_∞	S^W_∞	S^B_2	S^W_2	$S^B_{1.5}$	$S^W_{1.5}$	T^B_∞	T^W_∞	T^B_2	T^W_2	$T^B_{1.5}$	$T^W_{1.5}$
50	0.006	0.072	0.010	0.063	0.009	0.055	0.006	0.072	0.010	0.063	0.009	0.055
100	0.002	0.025	0.004	0.014	0.005	0.009	0.002	0.025	0.004	0.014	0.005	0.009
200	0.002	0.021	0.007	0.018	0.011	0.009	0.002	0.021	0.007	0.018	0.011	0.009

n	$L^A_{0.1}$	$L^B_{0.1}$	$L^W_{0.1}$	$L^A_{0.5}$	$L^B_{0.5}$	$L^W_{0.5}$	L^A_1	L^B_1	L^W_1	L^A_2	L^B_2	L^W_2
50	0.011	0.037	0.034	0.014	0.051	0.045	0.006	0.040	0.038	0.000	0.045	0.052
100	0.029	0.046	0.043	0.015	0.048	0.048	0.004	0.053	0.056	0.001	0.052	0.054
200	0.040	0.055	0.054	0.018	0.044	0.049	0.008	0.042	0.043	0.003	0.060	0.067

B. Size of tests at 10% level with DGP 5

n	Q^B_∞	Q^W_∞	Q^B_2	Q^W_2	$Q^B_{1.5}$	$Q^W_{1.5}$	R^B_∞	R^W_∞	R^B_2	R^W_2	$R^B_{1.5}$	$R^W_{1.5}$
50	0.085	0.124	0.098	0.121	0.106	0.119	0.090	0.123	0.100	0.121	0.108	0.114
100	0.022	0.051	0.030	0.044	0.031	0.031	0.023	0.061	0.032	0.043	0.035	0.034
200	0.014	0.034	0.021	0.028	0.023	0.023	0.018	0.039	0.028	0.033	0.036	0.032

n	S^B_∞	S^W_∞	S^B_2	S^W_2	$S^B_{1.5}$	$S^W_{1.5}$	T^B_∞	T^W_∞	T^B_2	T^W_2	$T^B_{1.5}$	$T^W_{1.5}$
50	0.018	0.090	0.022	0.080	0.026	0.072	0.018	0.090	0.022	0.080	0.026	0.072
100	0.006	0.035	0.013	0.028	0.013	0.027	0.006	0.035	0.013	0.028	0.013	0.027
200	0.011	0.030	0.017	0.028	0.023	0.027	0.011	0.030	0.017	0.028	0.023	0.027

n	$L^A_{0.1}$	$L^B_{0.1}$	$L^W_{0.1}$	$L^A_{0.5}$	$L^B_{0.5}$	$L^W_{0.5}$	L^A_1	L^B_1	L^W_1	L^A_2	L^B_2	L^W_2
50	0.069	0.093	0.075	0.037	0.088	0.092	0.011	0.090	0.084	0.001	0.094	0.092
100	0.090	0.101	0.100	0.031	0.101	0.101	0.010	0.104	0.102	0.003	0.102	0.105
200	0.086	0.096	0.096	0.038	0.098	0.091	0.017	0.096	0.089	0.004	0.110	0.100

C. Power of tests at 5% level with DGP 6

n	Q^B_∞	Q^W_∞	Q^B_2	Q^W_2	$Q^B_{1.5}$	$Q^W_{1.5}$	R^B_∞	R^W_∞	R^B_2	R^W_2	$R^B_{1.5}$	$R^W_{1.5}$
50	0.983	0.988	0.985	0.988	0.989	0.990	0.999	0.998	0.999	0.998	1.000	0.998
100	1.000	1.000	1.000	1.000	1.000	1.000	1.000	1.000	1.000	1.000	1.000	1.000
200	1.000	1.000	1.000	1.000	1.000	1.000	1.000	1.000	1.000	1.000	1.000	1.000

(continued)

Table 3 Continued

n	S^B_∞	S^W_∞	S^B_2	S^W_2	$S^B_{1.5}$	$S^W_{1.5}$	T^B_∞	T^W_∞	T^B_2	T^W_2	$T^B_{1.5}$	$T^W_{1.5}$
50	0.995	0.998	0.995	0.998	0.997	0.998	0.995	0.998	0.995	0.998	0.997	0.998
100	1.000	1.000	1.000	1.000	1.000	1.000	1.000	1.000	1.000	1.000	1.000	1.000
200	1.000	1.000	1.000	1.000	1.000	1.000	1.000	1.000	1.000	1.000	1.000	1.000

n	$L^A_{0.1}$	$L^B_{0.1}$	$L^W_{0.1}$	$L^A_{0.5}$	$L^B_{0.5}$	$L^W_{0.5}$	L^A_1	L^B_1	L^W_1	L^A_2	L^B_2	L^W_2
50	0.282	0.496	0.336	0.965	0.976	0.967	0.979	0.985	0.982	0.980	0.986	0.994
100	0.828	0.899	0.836	1.000	1.000	1.000	1.000	1.000	1.000	1.000	1.000	1.000
200	0.999	1.000	1.000	1.000	1.000	1.000	1.000	1.000	1.000	1.000	1.000	1.000

D. Power of tests at 10% level with DGP 6

n	Q^B_∞	Q^W_∞	Q^B_2	Q^W_2	$Q^B_{1.5}$	$Q^W_{1.5}$	R^B_∞	R^W_∞	R^B_2	R^W_2	$R^B_{1.5}$	$R^W_{1.5}$
50	0.986	0.990	0.987	0.991	0.991	0.991	1.000	0.999	1.000	0.998	1.000	1.000
100	1.000	1.000	1.000	1.000	1.000	1.000	1.000	1.000	1.000	1.000	1.000	1.000
200	1.000	1.000	1.000	1.000	1.000	1.000	1.000	1.000	1.000	1.000	1.000	1.000

n	S^B_∞	S^W_∞	S^B_2	S^W_2	$S^B_{1.5}$	$S^W_{1.5}$	T^B_∞	T^W_∞	T^B_2	T^W_2	$T^B_{1.5}$	$T^W_{1.5}$
50	0.997	0.998	0.997	0.998	0.998	0.999	0.997	0.998	0.997	0.998	0.998	0.999
100	1.000	1.000	1.000	1.000	1.000	1.000	1.000	1.000	1.000	1.000	1.000	1.000
200	1.000	1.000	1.000	1.000	1.000	1.000	1.000	1.000	1.000	1.000	1.000	1.000

n	$L^A_{0.1}$	$L^B_{0.1}$	$L^W_{0.1}$	$L^A_{0.5}$	$L^B_{0.5}$	$L^W_{0.5}$	L^A_1	L^B_1	L^W_1	L^A_2	L^B_2	L^W_2
50	0.571	0.680	0.550	0.979	0.978	0.978	0.985	0.987	0.990	0.984	0.991	0.996
100	0.946	0.961	0.931	1.000	1.000	1.000	1.000	1.000	1.000	1.000	1.000	1.000
200	1.000	1.000	1.000	1.000	1.000	1.000	1.000	1.000	1.000	1.000	1.000	1.000

and L^W are also sensitive to c. This is because $\{y_t^*\}$ in this chapter is not recursively generated (as described above) while Lee and Ullah (2001) generated $\{y_t^*\}$ recursively for time-series data. Note that in this chapter we generated the bootstrap data $\{y_t^*\}$ conditional on $\{x_t\}$ for both the cases when x_t are exogenous (Blocks 2 and 4) and the cases when x_t are lagged dependent variables (Blocks 1 and 3). The bootstrap method used in Lee and Ullah (2001) may be called the "recursive" bootstrap, while the bootstrap method used in this chapter may be called the "conditional" bootstrap. As discussed in Lee (2001), the bootstrap method treating x_t as given and generating $\{y_t^*\}$ conditional on x_t gives more robust size behavior than the recursive bootstrap even for the time-series data.

Table 4 Variable selection in Cross-Sectional Models (Block 4)

A. Size of tests at 5% level with DGP 7

n	Q^B_∞	Q^W_∞	Q^B_2	Q^W_2	$Q^B_{1.5}$	$Q^W_{1.5}$	R^B_∞	R^W_∞	R^B_2	R^W_2	$R^B_{1.5}$	$R^W_{1.5}$
50	0.025	0.055	0.032	0.043	0.032	0.036	0.029	0.048	0.035	0.042	0.036	0.036
100	0.011	0.022	0.014	0.021	0.011	0.013	0.013	0.028	0.019	0.024	0.021	0.018
200	0.003	0.014	0.011	0.015	0.008	0.010	0.004	0.013	0.007	0.014	0.008	0.009

n	S^B_∞	S^W_∞	S^B_2	S^W_2	$S^B_{1.5}$	$S^W_{1.5}$	T^B_∞	T^W_∞	T^B_2	T^W_2	$T^B_{1.5}$	$T^W_{1.5}$
50	0.003	0.024	0.007	0.020	0.010	0.022	0.003	0.024	0.007	0.020	0.010	0.022
100	0.006	0.016	0.007	0.013	0.008	0.011	0.006	0.016	0.007	0.013	0.008	0.011
200	0.002	0.010	0.005	0.011	0.005	0.007	0.002	0.010	0.005	0.011	0.005	0.007

n	$L^A_{0.1}$	$L^B_{0.1}$	$L^W_{0.1}$	$L^A_{0.5}$	$L^B_{0.5}$	$L^W_{0.5}$	L^A_1	L^B_1	L^W_1	L^A_2	L^B_2	L^W_2
50	0.027	0.057	0.041	0.018	0.051	0.050	0.005	0.058	0.055	0.000	0.045	0.054
100	0.032	0.041	0.038	0.009	0.039	0.041	0.003	0.043	0.042	0.000	0.039	0.040
200	0.041	0.049	0.053	0.020	0.047	0.047	0.004	0.042	0.041	0.000	0.035	0.035

B. Size of tests at 10% level with DGP 7

n	Q^B_∞	Q^W_∞	Q^B_2	Q^W_2	$Q^B_{1.5}$	$Q^W_{1.5}$	R^B_∞	R^W_∞	R^B_2	R^W_2	$R^B_{1.5}$	$R^W_{1.5}$
50	0.045	0.073	0.051	0.067	0.049	0.050	0.039	0.077	0.047	0.070	0.048	0.052
100	0.017	0.037	0.024	0.030	0.024	0.026	0.021	0.049	0.031	0.041	0.036	0.038
200	0.014	0.028	0.018	0.022	0.016	0.018	0.008	0.030	0.025	0.027	0.022	0.020

n	S^B_∞	S^W_∞	S^B_2	S^W_2	$S^B_{1.5}$	$S^W_{1.5}$	T^B_∞	T^W_∞	T^B_2	T^W_2	$T^B_{1.5}$	$T^W_{1.5}$
50	0.009	0.038	0.016	0.033	0.021	0.035	0.009	0.038	0.016	0.033	0.021	0.035
100	0.008	0.038	0.018	0.030	0.025	0.025	0.008	0.038	0.018	0.030	0.025	0.025
200	0.006	0.023	0.017	0.022	0.013	0.015	0.006	0.023	0.017	0.022	0.013	0.015

n	$L^A_{0.1}$	$L^B_{0.1}$	$L^W_{0.1}$	$L^A_{0.5}$	$L^B_{0.5}$	$L^W_{0.5}$	L^A_1	L^B_1	L^W_1	L^A_2	L^B_2	L^W_2
50	0.091	0.098	0.081	0.034	0.105	0.107	0.009	0.114	0.116	0.000	0.097	0.102
100	0.076	0.085	0.079	0.029	0.088	0.084	0.010	0.097	0.100	0.000	0.087	0.090
200	0.085	0.096	0.098	0.038	0.094	0.092	0.011	0.090	0.088	0.000	0.085	0.090

C. Power of tests at 5% level with DGP 8

n	Q^B_∞	Q^W_∞	Q^B_2	Q^W_2	$Q^B_{1.5}$	$Q^W_{1.5}$	R^B_∞	R^W_∞	R^B_2	R^W_2	$R^B_{1.5}$	$R^W_{1.5}$
50	0.993	0.993	0.992	0.992	0.992	0.991	0.995	0.995	0.994	0.994	0.996	0.995
100	1.000	1.000	1.000	1.000	1.000	1.000	1.000	1.000	1.000	1.000	1.000	1.000
200	1.000	1.000	1.000	1.000	1.000	1.000	1.000	1.000	1.000	1.000	1.000	1.000

(continued)

Table 4 Continued

n	S^B_∞	S^W_∞	S^B_2	S^W_2	$S^B_{1.5}$	$S^W_{1.5}$	T^B_∞	T^W_∞	T^B_2	T^W_2	$T^B_{1.5}$	$T^W_{1.5}$
50	0.993	0.994	0.993	0.993	0.994	0.995	0.993	0.994	0.993	0.993	0.994	0.995
100	1.000	1.000	1.000	1.000	1.000	1.000	1.000	1.000	1.000	1.000	1.000	1.000
200	1.000	1.000	1.000	1.000	1.000	1.000	1.000	1.000	1.000	1.000	1.000	1.000

n	$L^A_{0.1}$	$L^B_{0.1}$	$L^W_{0.1}$	$L^A_{0.5}$	$L^B_{0.5}$	$L^W_{0.5}$	L^A_1	L^B_1	L^W_1	L^A_2	L^B_2	L^W_2
50	0.154	0.259	0.182	0.907	0.964	0.913	0.952	0.994	0.982	0.777	0.993	0.994
100	0.527	0.589	0.526	1.000	1.000	1.000	1.000	1.000	1.000	1.000	1.000	1.000
200	0.941	0.962	0.948	1.000	1.000	1.000	1.000	1.000	1.000	1.000	1.000	1.000

D. Power of tests at 10% level with DGP 8

n	Q^B_∞	Q^W_∞	Q^B_2	Q^W_2	$Q^B_{1.5}$	$Q^W_{1.5}$	R^B_∞	R^W_∞	R^B_2	R^W_2	$R^B_{1.5}$	$R^W_{1.5}$
50	0.993	0.995	0.993	0.994	0.992	0.992	0.996	0.996	0.997	0.996	0.996	0.996
100	1.000	1.000	1.000	1.000	1.000	1.000	1.000	1.000	1.000	1.000	1.000	1.000
200	1.000	1.000	1.000	1.000	1.000	1.000	1.000	1.000	1.000	1.000	1.000	1.000

n	S^B_∞	S^W_∞	S^B_2	S^W_2	$S^B_{1.5}$	$S^W_{1.5}$	T^B_∞	T^W_∞	T^B_2	T^W_2	$T^B_{1.5}$	$T^W_{1.5}$
50	0.995	0.996	0.995	0.996	0.995	0.996	0.995	0.996	0.995	0.996	0.995	0.996
100	1.000	1.000	1.000	1.000	1.000	1.000	1.000	1.000	1.000	1.000	1.000	1.000
200	1.000	1.000	1.000	1.000	1.000	1.000	1.000	1.000	1.000	1.000	1.000	1.000

n	$L^A_{0.1}$	$L^B_{0.1}$	$L^W_{0.1}$	$L^A_{0.5}$	$L^B_{0.5}$	$L^W_{0.5}$	L^A_1	L^B_1	L^W_1	L^A_2	L^B_2	L^W_2
50	0.369	0.414	0.328	0.942	0.982	0.955	0.968	0.995	0.993	0.860	0.997	0.995
100	0.726	0.751	0.688	1.000	1.000	1.000	1.000	1.000	1.000	1.000	1.000	1.000
200	0.981	0.988	0.977	1.000	1.000	1.000	1.000	1.000	1.000	1.000	1.000	1.000

5. Turning to the power behavior, although the size of L^B and L^W are quite robust to c, the power of these tests can vary with c and is generally best with larger c. The tests Q, R, S, T have a similar power pattern but these are slightly worse than L.

Table 2 presents the size (DGP 3) and power (DGP 4) of testing for neglected nonlinearity in cross-sectional models in Block 2. The following observations are made. All the size results in Table 1 for time series (summarized above) hold here for Table 2 with cross-sectional data. While the size of L^B and L^W are quite robust to c, the power of these tests can vary with c and is higher with larger c.

Table 3 presents the size (DGP 5) and power (DGP 6) of testing for lag selection in time-series models in Block 3. The results are very similar to those in Tables 1 and 2. The tests L^B and L^W have good size and power in testing for lag selection in nonparametric time-series models as well as in testing for parametric functional forms.

Table 4 presents the size (DGP 7) and power (DGP 8) of testing for variable selection in cross-sectional models in Block 4. The results are again very similar to those in Tables 1–3. The tests L^B and L^W have very good size and power in testing for omitted variables in cross-sectional models.

5 CONCLUSIONS

We consider three nonparametric tests for functional form, varying parameters, and omitted variables in regression models both of time-series data and of cross-sectional data. The first approach (R, S, T) is to compare the sums of squared residuals from the null and the alternative models and the second test (Q) is to compare the fitted values of the null and alternative models. The third test (L) is the nonparametric conditional moment test, which is to see if the residuals from the null model are related to the conditioning variables in the alternative models. We find that the bootstrap tests of Li and Wang (1998) and Zheng (1996), L^B and L^W, have very good size and power properties in all the situations we considered.

One of the reasons for the better performance of these L tests compared to R, S, T, and Q tests may be because the asymptotic distribution of L is asymptotically normal with the mean zero under the null hypothesis, whereas this is not the case for R, S, and T tests. Therefore, it will be an interesting future study to compare the L test with the bias-adjusted R, S, T tests as described in Fan and Li (2001). It will also be useful to develop the theoretical power properties of the tests under various local alternatives, as studied in Hong and Lee (2001) and Tripathi and Kitamura (2000) in different but related contexts. Moreover, this chapter has considered the tests based on the kernel smoothing procedure only. It will be useful to study how our study compares with other specification testing procedures, especially using other smoothing procedures such as neural network, spline regression, and fuzzy c-means algorithm in Giles and Draeseke (this volume, Chap. 14). Finally, the issue of optimal choice of window-width for the tests considered here needs further future investigations.

ACKNOWLEDGMENTS

We thank Zongwu Cai, Jianqing Fan, and Qi Li for their programs and comments, and the UCR Academic Senate for their research support.

REFERENCES

Aït-Sahalia, Y., Bickel, P.J., and Stoker, T.M., 1994. Goodness-of-fit tests for regression using kernel models. Unpublished paper, Princeton (NJ), University of California (Berkeley), and MIT (Cambridge, MA).

Andrews, D.W.K., 1997. A conditional Kolmogorov test. *Econometrica* **65**, 1097–1128.

Auestad, B. and Tjøstheim, D., 1990. Identification of nonlinear time series: First order characterization and order determination. *Biometrika* **77**, 669–687.

Azzalini, A., Bowman, A., and Härdle, W.H., 1989. On the use of nonparametric regression for model checking. *Biometrika* **76**, 1–12.

Bera, A., 2000. Hypothesis testing in the 20th century with a special reference to testing with misspecified models. In: C.R. Rao and G.J. Szekely, eds. *Statistics for the 21st Century: Methodologies for Applications of the Future.* Marcel Dekker, New York.

Berg, M.D. and Li, D., 1998. A consistent bootstrap test for time series regression models. Unpublished manuscript, University of Guelph, Guelph, Ontario, Canada.

Bierens, H.J., 1982. Consistent model specification tests. *Journal of Econometrics* **20**, 105–134.

Bierens, H.J., 1990. A consistent conditional moment test of functional form. *Econometrica* **58**, 1443–1458.

Bierens, H.J. and Ploberger, W., 1997. Asymptotic theory of integrated conditional moment tests. *Econometrica* **65**, 1129–1151.

Bollerslev, T., 1986. Generalized autoregressive conditional heteroskedasticity. *Journal of Econometrics* **31**, 307–327.

Cai, Z., Fan, J., and Yao, Q., 2000. Functional-coefficient regression models for nonlinear time series. *Journal of the American Statistical Association* **95**, 941–956.

Chen, X. and Fan, Y., 1999. Consistent hypothesis testing in semiparametric and nonparametric models for econometric time series. *Journal of Econometrics* **91**, 373–401.

Chen, R. and Tsay, R.S., 1993. Functional-coefficient autoregressive models. *Journal of the American Statistical Association* **88**, 298–308.

Cheng, B. and Tong, H., 1992. On consistent nonparametric order determination and chaos. *Journal of Royal Statistical Society, Series B* **54**, 427–449.

Cleveland, W.S., 1979. Robust locally weighted regression and smoothing scatter plots. *Journal of the American Statistical Association* **74**, 829–836.

Davidson, R. and MacKinnon, J.G., 1999. The size distortion of bootstrap tests. *Econometric Theory* **15**, 361–376.

De Jong, R.M., 1996. On the Bierens test under data dependence. *Journal of Econometrics* **72**, 1–32.

Delgado, M.A. and Stengos, T., 1994. Semiparametric testing of non-nested econometric models. *Review of Economic Studies* **75**, 345–367.

Efron, B., 1979. Bootstrapping methods: Another look at the jackknife. *Annals of Statistics* **7**, 1–26.

Eubank, R.L. and Hart, J., 1992. Testing goodness-of-fit in regression via order selection criteria. *Annals of Statistics* **20**, 1412–1425.

Eubank, R.L. and Spiegelman, C.H., 1990. Testing the goodness-of-fit of linear models via nonparametric regression techniques. *Journal of the American Statistical Association* **85**, 387–392.

Fan, J. and Gijbels, I., 1996. *Local Polynomial Modelling and its Applications*. Chapman and Hall, London.

Fan, J. and Li, R., 1999a. Variable selection via penalized likelihood. Unpublished paper, University of North Carolina, Chapel Hill.

Fan, J., Zhang, C.M., and Zhang, J., 2001. Generalized likelihood ratio statistics and Wilks phenomenon. *Annals of Statistics*, **29**(1), 153–159.

Fan, Y., 1998. Goodness of fit tests based on kernel density estimation with fixed smoothing parameters. *Econometric Theory* **14**, 604–621.

Fan, Y. and Li, Q., 1992. Consistent model specification tests: Omitted variables and semiparametric functional forms. Manuscript, University of Windsor, Windsor, Ontario, Canada.

Fan, Y. and Li, Q., 1996. Consistent model specification tests: Omitted variables and semiparametric functional forms. *Econometrica* **64**, 865–890.

Fan, Y. and Li, Q., 1997. A consistent nonparametric test for linearity for AR(p) models. *Economics Letters* **55**, 53–59.

Fan, Y. and Li, Q., 1999b. Central limit theorem for degenerate U-statistics of absolutely regular processes with applications to model specification tests. *Journal of Nonparametric Statistics* **10**, 245–271.

Fan, Y. and Li, Q., 2002. A consistent model specification test based on the kernel sum of squares of residuals. *Econometric Reviews*, **21**(3), 337–352.

Fan, Y. and Ullah, A., 1999. Asymptotic normality of a combined regression estimator. *Journal of Multivariate Analysis* **71**, 191–240.

Gozalo, P.L., 1993. A consistent model specification test for nonparametric estimation of regression function models. *Econometric Theory* **9**, 451–477.

Granger, C.W.J. and Lin, J.-L., 1994. Using the mutual information coefficient to identify lags in nonlinear models. *Journal of Time Series Analysis* **15**, 371–384.

Granger, C.W.J. and Teräsvirta, T., 1993. *Modelling Nonlinear Economic Relationships*. Oxford University Press, New York.

Granger, C.W.J. and Teräsvirta, T., 1999. A simple nonlinear time series model with misleading linear properties. *Economics Letters* **62**, 161–165.

Granger, C.W.J., Maasoumi, E., and Racine, J., 2000. A dependence metric for nonlinear time series. Unpublished paper, University of California, San Diego.

Haggan, V. and Ozaki, T., 1981. Modeling nonlinear vibrations using an amplitude-dependent autoregressive time series model. *Biometrika* **68**, 189–196.

Hall, P., 1992. *The Bootstrap and Edgeworth Expansion*. Springer-Verlag, New York.

Härdle, W., 1990. *Applied Nonparametric Regression*. Cambridge University Press, New York.

Härdle, W. and Mammen, E., 1993. Comparing nonparametric versus parametric regression fits. *Annals of Statistics* **21**, 1926–1947.

Hart, J.D., 1997. *Nonparametric Smoothing and Lack-of-Fit Tests*. Springer-Verlag, New York.

Hjellvik, V. and Tjøstheim, D., 1995. Nonparametric tests of linearity for time series. *Biometrika* **2**, 351–368.

Hjellvik, V. and Tjøstheim, D., 1998. Nonparametric statistics for testing of linearity and serial independence. *Journal of Nonparametric Statistics* **6**, 223–251.

Hjellvik, V., Yao, Q., and Tjøstheim, D., 1998. Linearity testing using local polynomial approximation. *Journal of Statistical Planning and Inference* **68**, 295–321.

Hong, Y. and Lee, T-H., 2001. Diagnostic checking for adequacy of linear and nonlinear time series models. Unpublished manuscript, Cornell University and University of California, Riverside.

Hong, Y. and White, H., 1995. Consistent specification testing via nonparametric series regression. *Econometrica* **63**, 1133–1160.

Hong, Y. and White, H., 2001. Asymptotic distribution theory for nonparametric entropy measures for serial dependence. Unpublished manuscript, Cornell University and University of California, San Diego.

Horowitz, J.L. and Härdle, W.H., 1994. Testing a parametric model against a semiparametric alternative. *Econometric Theory* **10**, 821–848.

Horowitz, J.L. and Spokoiny, V.G., 2001. An adaptive, rate-optimal test of a parametric mean-regression model against a nonparametric alternative. *Econometrica*, **69**, 599–631.

Kreiss, J.-P., Neumann, M.H., and Yao, Q., 1998. Bootstrap tests for simple structures in nonparametric time series regression. Unpublished manuscript, University of Kent at Canterbury, UK.

Lavergne, P. and Vuong, Q., 1996. Nonparametric selection of regressors: The nonnested case. *Econometrica* **64**, 207–219.

Lee, B.-J., 1992. A heteroskedasticity test robust to conditional mean misspecification. *Econometrica* **60**, 159–171.

Lee, T.-H., 2001. Neural network test and nonparametric kernel test for neglected nonlinearity in regression models. *Studies in Nonlinear Dynamics and Econometrics*, **4**(4), 169–182.

Lee, T.-H. and Ullah, A., 2001. Nonparametric bootstrap tests for neglected nonlinearity in time series regression models. *Journal of Nonparametric Statistics*, forthcoming.

Lee, T.-H., White, H., and Granger, C.W.J., 1993. Testing for neglected nonlinearity in time series models: A comparison of neural network methods and alternative tests. *Journal of Econometrics* **56**, 269–290.

Lewbel, A., 1993. Consistent tests with nonparametric components with an application to Chinese production data. Unpublished manuscript, Brandeis University, Waltham, MA.

Lewbel, A., 1995. Consistent nonparametric testing with an application to testing Slutsky symmetry. *Journal of Econometrics* **67**, 379–401.

Li, Q., 1999. Consistent model specification tests for time series econometric models. *Journal of Econometrics* **92**, 101–147.

Li, Q. and Wang, S., 1998. A simple consistent bootstrap test for a parametric regression function. *Journal of Econometrics* **87**, 145–165.

Li, Q., Huang, C.J., and Fu, T-T., 1997. Semiparametric smooth coefficient models. Unpublished manuscript, Texas A&M University, College Station, TX.

Linton, O. and Gozalo, P., 1997. Consistent testing of additive models. Unpublished manuscript, Brown University, Providence, RI

Liu, R.Y., 1988. Bootstrap procedures under some non-iid models. *Annals of Statistics* **16**, 1697–1708.

Nadaraya, É.A., 1964. On estimating regression. *Theory of Probability and its Applications* **9**, 141–142.

Pagan, A.R. and Ullah, A., 1999. *Nonparametric Econometrics*. Cambridge University Press, Cambridge, UK.

Racine, J., 1997. Consistent significance testing for nonparametric regression. *Journal of Business and Economic Statistics* **15**, 369–379.

Rahman, M. and Ullah, A., 1999. Improved combined parametric and nonparametric regressions: Estimation and hypothesis testing. Unpublished manuscript, University of California, Riverside.

Raj, B. and Ullah, A., 1981. *Econometrics: A Varying Coefficients Approach*. Croom Helm, London.

Robinson, P.M., 1989. Hypothesis testing in semiparametric and nonparametric models for econometric time series. *Review of Economic Studies* **56**, 511–534.

Robinson, P.M., 1991. Consistent nonparametric entropy-based testing. *Review of Economic Studies* **58**, 437–453.

Ruppert, D. and Wand, M.P., 1994. Multivariate locally weighted least squares regression. *Annals of Statistics* **22**, 1346–1370.

Shao, J. and Tu, D., 1995. *The Jackknife and Bootstrap*. Springer-Verlag, New York.

Stone, C.J., 1977. Consistent nonparametric regression. *Annals of Statistics* **5**, 595–645.

Teräsvirta, T., Lin, C.-F., and Granger, C.W.J., 1993. Power of the neural network linearity test. *Journal of Time Series Analysis* **14**, 209–220.

Tong, H., 1990. *Nonlinear Time Series: A Dynamic System Approach*. Clarendon Press, Oxford, UK.

Tjøstheim, D. and Auestad, B.H., 1994. Nonparametric identification of nonlinear time series: Selecting significant lags. *Journal of the American Statistical Association* **89**, 1410–1419.

Tripathi, G. and Kitamura, Y., 2000. On testing conditional moment restrictions: The canonical case. Unpublished manuscript, University of Wisconsin, Madison, WI.

Tschernig, R. and Yang, L., 2000. Nonparametric lag selection for time series. *Journal of Time Series Analysis* **21**, 457–487.

Ullah, A., 1985. Specification analysis of econometric models. *Journal of Quantitative Economics* **1**, 187–209.

Ullah, A., 2001. Nonparametric kernel methods of estimation and hypothesis testing. In: B. Baltagi, ed. *Companion in Econometric Theory*. Blackwell, Oxford, UK.

Ullah, A. and Singh, R., 1989. Estimation of a probability density function with applications to nonparametric inference in econometrics. In: B. Raj, ed. *Advances in Econometrics and Modelling, Advanced Studies in Theoretical and Applied Econometrics*. Kluwer, Dordrecht, pp. 69–83.

Ullah, A. and Vinod, H.D., 1993. General nonparametric regression estimation and testing in econometrics. In: G.S. Maddala, C.R. Rao, and H.D. Vinod, eds. *Handbook of Statistics*, vol. 11. Elsevier, Amsterdam, pp. 85–116.

Wand, M.P. and Jones, M.C., 1995. *Kernel Smoothing*. Chapman and Hall, London.

Watson, G.S., 1964. Smooth regression analysis. *Sankhya, Series A* **26**, 359–372.

Whang, Y.-J., 2000. Consistent bootstrap tests of parametric regression functions. *Journal of Econometrics* **98**, 27–46.

Whang, Y.J. and Andrews, D.W.K., 1993. Testing of specification for parametric and semiparametric models. *Journal of Econometrics* **57**, 277–318.

Wilks, S.S., 1938. The large sample distribution of the likelihood ratio for testing composite hypotheses. *Annals of Mathematical Statistics* **9**, 60–62.

Wooldridge, J.M., 1992. A test for functional form against nonparametric alternatives. *Econometric Theory* **8**, 452–475.

Wu, C.F.J., 1986. Jackknife, bootstrap, and other resampling methods in regression analysis. *Annals of Statistics* **14**, 1261–1350.

Yao, Q. and Tong, H., 1994. On subset selection in non-parametric stochastic regression. *Statistica Sinica* **4**, 51–70.

Yatchew, A.J., 1992. Nonparametric regression tests based on least squares. *Econometric Thoery* **8**, 435–451.

Zheng, J.X., 1996. A consistent test of functional form via nonparametric estimation techniques. *Journal of Econometrics* **75**, 263–289.

16

The Effect of Economic Growth on Standard of Living: A Semiparametric Analysis

Nilanjana Roy
University of Victoria, Victoria, British Columbia, Canada

1 INTRODUCTION

Up until the late 1960s, development was synonymous with economic growth, the idea being that the increase in national income would eventually "trickle down" to the masses and the standard of living of the people would improve. However, the experience of the developing countries in the 1970s did not lend empirical support to this view. So the view started changing in the 1970s and the idea of development became more people oriented than goods oriented. Development policies started focusing directly on the well-being of the people or on raising the standard of living of the people rather than on increasing the national income of the country.

The question for development economists and policymakers then became: "How to improve the standard of living of the people?" and a related question that was raised was "Does economic growth positively affect the standard of living, and if so, how strong is the influence?" The latter is an important question since an affirmative answer to that question would provide support for policies that are aimed at increasing income in developing countries. However, the answer to the latter question was not easy to find and, in fact, the debate still goes on. According to some researchers (Iseman, 1980; Sen, 1981; Anand and Kanbur, 1991; Anand and Ravallion, 1993; among others), it is not income but public provision of services that plays an important role in raising the standard of living. While others, such as Bhalla (1988), Kakwani (1993), Pritchett and Summers (1996), and Chakraborty (1999),* have concluded from their analyses that there exists a positive and statistically significant relationship between *per capita* income and standard of living. It should be noted here, that standard of living has been measured by various indicators in the literature. Some researchers have used life expectancy at birth, others have used infant mortality, still others have used composite indexes combining health indicators and indicators for literacy, among others.

One of the contributions of Pritchett and Summers (1996) is that they controlled for unobserved country-specific effects such as good government or cultural values that could affect the standard of living indicator positively. However, they did not pay much attention to the issue of functional form specifications and stated "*We didn't consider the issue of functional form in the context of the present work to be sufficiently important to merit formal specification tests*".[†] However, it is well known in the econometrics literature that mis-specification of functional form can lead to inconsistent and inefficient estimates and suboptimal tests.

In fact, the nonlinearity in the relationship between *per capita* income and various indicators of standard of living has been recognized by many researchers and some of the functional forms used in this literature are logistic, log-linear, and semilog, among others. However, as Chakraborty (1999) correctly points out, no formal specification testing was done in any of these studies and the various specifications were chosen arbitrarily. Chakraborty (1999) contributes to the literature by using an alternative methodology, namely, the nonparametric approach, to study the relation between standard of living and *per capita* income. This data-dependent approach does not need the specification of any functional form and hence avoid the problem of functional form mis-specification.

However, the rate of convergence of nonparametric estimators are slower than that of their parametric counterparts. So it is meaningful to do specification testing to see if the data supports the parametric model, before we move to the more computationally involved nonparametric estimator. Chakraborty (1999) does not undertake any such specification testing to provide support for her preferred nonparametric model. Given

---

*This paper has a very nice survey of the empirical literature.

[†]See Prichett and Summers, (1996, p. 847, footnote 4).

that one of the aims of this chapter is to undertake such specification testing to see if the use of the nonparametric models are appropriate in studying the relationship between standard of living and *per capita* income, we now briefly discuss some of the issues dealt with specifically in Chakraborty (1999).

Chakraborty (1999) estimated a nonparametric regression of the "achievement function" on the logarithm of *per capita* income and a nonparametric regression of the "improvement function" on the logarithm of *per capita* income and its change. The concepts of the "achievement function" and the "improvement function" were developed by Kakwani (1993) using an axiomatic approach. The achievement function is a transformation of some standard of living indicator such as life expectancy and was constructed such that it gives different weights to equal increments in the indicator, depending on the initial level of the indicator. More specifically, the function gives higher weights to those countries that have already achieved a higher level of longevity, say, since it is relatively more difficult to increase longevity from a higher level than from a lower level. The achievement function is given as

$$f(x, M_0, M) = \frac{\ln(M - M_0) - \ln(M - x)}{\ln(M - M_0)} \tag{1}$$

where x denotes the value of the positive standard of living indicator for the country in question, e.g., it can be the country's life expectancy, while M and M_0 denote the upper and lower limits of the indicator. For life expectancy, the limits are generally set at 80 and 30 respectively. Also, note that for an indicator such as infant mortality, which is a negative measure of the standard of living, the above formula needs to be adjusted as

$$f(I, m_0, m) = \frac{\ln(m - m_0) - \ln(I - m_0)}{\ln(m - m_0)} \tag{2}$$

where I is the infant mortality rate and m and m_0 are its upper and lower limits, set generally at 300 and 5, respectively. See Kakwani (1993) for details on the achievement function. The improvement functions corresponding to Eqs. (1) and (2) are defined as the difference in the achievement function in two given periods. According to Kakwani (1993), the nonlinearity in the relationship between a standard of living indicator and the *per capita* income is captured adequately by specifying a linear relationship between the achievement function and the logarithm of *per capita* income. Hence, he estimated linear models using achievement function as the dependent variable and logarithm of *per capita* income as the independent variable. Chakraborty (1999) claimed that this model was mis-specified and used a nonparametric specification instead. She also detected nonlinearity in the relationship between the improvement function and the level of *per capita* income and its difference and again proposed using a purely nonparametric specification. The results from her specifications were contradictory to Kakwani's (1993) result that a country's level of economic welfare has a positive and statistically significant impact on its improvement in the living standard. In this chapter, using the same data set, we formally test Kakwani's linear specification against Chakraborty's nonparametric specification using Li and Wang's test (1998).

Neither Kakwani (1993) nor Chakraborty (1999) considered the effect of including public health expenditure per head as a regressor in the model. The second aim of this chapter is to include the new variable in the regression to examine its effect on the estimated coefficient of the income variable. As we discussed before, some researchers claim that it is public provision of services that plays an important role in raising the living standard and the income coefficient will lose its strength and statistical significance once the other variable has been accounted for (Anand and Ravallion, 1993). We have tested a parametric specification against a semiparametric one using Li and Wang's test (1998) and based on the results from the test, we have presented a semiparametric regression model with the logarithm of the public health expenditure per head appearing nonlinearly in the model. Our results show that irrespective of the specification, the income coefficient continues to be statistically significant. The semiparametric specification allows us to provide a $\sqrt{n}$ consistent estimator of the income coefficient and also allows us to offer a graphical representation of the nonlinear behavior of the logarithm of the health expenditure variable.

Section 2.1 offers a brief discussion of the semiparametric estimation method (Robinson, 1988), Section 2.2 provides an outline of the specification test of Li and Wang (1998), while Section 2.3 contains a discussion of some of the computational issues related to the estimation methods and the test procedures. Section 3.1 gives a brief description of the data while Section 3.2 has the empirical results. Section 4 concludes.

2 METHODOLOGY

2.1 Semiparametric Model

The semiparametric model can be written as

$$y_i = X_i'\beta + m(Z_i) + u_i \quad i = 1, \ldots, n \tag{3}$$

where X_i contains q regressors, which enter the specification linearly, β is a $q \times 1$ vector of parameters, Z_i contains p continuous variables, and $m(\cdot)$ is an unknown function; u_i is the error term and satisfies the standard assumptions about the error term. Following Robinson (1988), we can obtain a $\sqrt{n}$ consistent estimator of β by first taking an expectation of Eq. (3) conditional on Z_i and then subtracting that from Eq. (3) to remove the effect of $m(Z_i)$, and finally estimating the parameter β in the second step by using the transformed variables where the expectation terms are replaced by their kernel estimates.

We can then rewrite Eq. (3) as

$$y_i - X_i'\hat{\beta} = m(Z_i) + (u_i + X_i'(\beta - \hat{\beta})) \tag{4}$$

where the term in the parenthesis on the right-hand side is the new composite error term. Given that in our case Z_i contains only one variable, we can estimate Eq. (4) using standard nonparametric methods to obtain an estimator for $m(Z_i)$. If Z_i contained more than one variable, then to obtain the individual effects of the variables, one can follow the additive partially linear approach and use marginal integration to estimate the components. See Linton and Nielsen (1995) and Fan and Li (1996), among others, for more details on that.

2.2 Specification Test

The two kinds of specification tests that we have used in this chapter have been proposed by Li and Wang (1998). The first one is to test the null of a linear regression model against a purely nonparametric specification. The null hypothesis is given as

$$H_0 : y_i = g(X_i, \beta) + u_i \quad i = 1, \ldots, n$$

and the alternative is given as

$$H_1 : y_i = m(X_i) + v_i \quad i = 1, \ldots, n$$

where u_i and v_i are assumed to have conditional mean zero and conditional variance $\sigma_u^2(X_i)$ and $\sigma_v^2(X_i)$, respectively. For our case, $g(X_i, \beta) = X_i'\beta$, but the test holds as long as $g(\cdot)$ is some known function.

The test statistic is given by

$$J_n = nh^{q/2} I_n / \sqrt{\hat{\Omega}} \tag{5}$$

where $I_n = (1/n^2h^q)\sum_{i=1}^{n}\sum_{j\neq i,j=1}^{n}\hat{u}_i\hat{u}_jK_{ij}$, $K_{ij} = K((X_i - X_j)/h)$, is the kernel function, h is the smoothing parameter, $\hat{u}_i$ is the ith least squares residual, and

$$\hat{\Omega} = \frac{2}{n^2h^q}\sum_{i=1}^{n}\sum_{j\neq i,j=1}^{n}\hat{u}_i^2\hat{u}_j^2K_{ij}^2$$

Li and Wang (1998) show that the test statistic has an asymptotic standard normal distribution under the null hypothesis. However, in small samples they suggest a wild bootstrap method to obtain the critical values. The rejection region for the test at the level α is $J_n > c$ where c is the critical value. In this chapter, we have chosen the kernel function to be the normal kernel and have reported the bootstrapped critical values.

The second test that we have used is to test the null of a linear regression model augmented by a square term against a semiparametric alternative. The null hypothesis is given as

$$H_0 : \ y_i = X_i'\beta + g(Z_i, \gamma) + u_i \quad i = 1, \ldots, n$$

and the alternative is given as

$$H_1 : \ y_i = X_i'\beta + m(Z_i) + v_i \quad i = 1, \ldots, n$$

where in our particular case, $g(Z_i, \gamma) = \gamma_1 Z_i + \gamma_2 Z_i^2$ and the dimension of Z_i is $p = 1$. The test statistic is the same as in Eq. (5) except that the q is replaced by p and the $\hat{u}_i$ is replaced by u_{ni} where $u_{ni} = y_i - X_i'\tilde{\beta} - g(Z_i, \hat{\gamma})$ is the residual from the "mixed" regression, $\tilde{\beta}$ is any semiparametric $\sqrt{n}$ consistent estimator of β based on the alternative model,* and $\hat{\gamma}$ is the least-squares estimator based on the null model. The test statistic is again asymptotically distributed as standard normal but Li and Wang (1998), based on their Monte Carlo Work, suggest the use of bootstrapped critical values.

2.3 Computational Issues

The kernel functions used throughout the chapter have been the normal kernel. The bandwidths in this chapter have been chosen using the formula $h = c \times \text{sd} \times n^{-1/(q+4)}$, where "sd" is the estimated standard deviation of the relevant regressor, q is the number of regressors appearing nonparametrically in the model, and c is a constant. The results for the different tests have been presented for a range of c values† to show the

*See Heckman (1986) and Robinson (1988), among others, for such a choice. We used the Robinson estimator for this chapter.
†The c values used were 0.8, 1.0, 1.2, and 1.4.

robustness of the results to the choice of the bandwidth. The semiparametric estimations were also undertaken for the same range of c values, but the results and figures presented here are for only $c = 1$, partly because visually this bandwidth seemed to offer the correct degree of smoothness and partly due to space constraint. In other words, the smaller bandwidths offered undersmoothing while the larger ones offered oversmoothing. One can also use some computer-intensive approach to calculate the bandwidth, such as the cross-validation method, but we did not use any such method here since the finite sample performances of such approaches are often not very good and our data set has less than 100 observations.

The semiparametric estimations and all the Li and Wang tests were coded and calculated using GAUSS Version 3.2.34. The computation is not time consuming; in fact, none of them required more than 5 min to run. The parametric estimation results and the scatter plots were all calculated using STATA Version 5.0, because of the ease with which panel data can be handled in STATA.

3 EMPIRICAL RESULTS

3.1 Description of the Data

The first part of our analysis uses the data set used by Kakwani (1993). We use the same data set since we want formally to test Chakraborty's (1999) claim that the parametric models used by Kakwani are misspecified and that a nonparametric model would be a better alternative even for that particular data set.

The data are based on 80 developing countries and contain information on the population weighted average of the achievement index for each country and the population weighted average GDP *per capita* in purchasing power parity (PPP) U.S. dollar for the two subperiods, 1971–1980 and 1981–1990. Kakwani used the logarithm of the *per capita* GDP in PPP U.S. dollar of a country as a measure of its economic welfare. Kakwani (1993) also provides data on the improvement index of each country measured as the difference in the achievement index in the two subperiods. We will focus on the achievement and improvement indices for life expectancy since Chakraborty (1999) focused on that particular variable. For details on the data, see Kakwani (1993, p. 316).

The second part of our analysis extends to a trivariate model similar to that of Anand and Kanbur (1993) but unlike them, we use a semiparametric approach and use data for more than one year. The latter allows us to estimate panel data models and to examine Prichett and

Summers' (1966, p. 847) observation that good health might be "*associated with some other country-level factor such as 'cultural values' or 'good government,' that is either intrinsically unobservable or for some other reason excluded from the estimation.*"

The data used in this part of the analysis have been obtained from the Global Development Network (GDN) growth database.* This database, put together by William Easterly and Hairong Yu, contains panel data for 122 countries on more than 100 variables. The data on the different variables have been collected from many different reliable data sources and the source for each variable is clearly indicated in this database. From the GDN growth database, we collected data on infant mortality rate (per 1000 live births), real gross domestic product *per capita* in constant dollars (international prices, base year 1985), and health expenditure as percentage of GDP. The original source of the infant mortality data is Global Development Finance and World Development Indicators, that for the real GDP data is the Penn World Table 5.6 and that for the health expenditure data is the IMF: Government Financial Statistics. The data used for our extended model constitute a balanced panel for 26 countries for 3 different years (1982, 1987, and 1992). We have focused our analysis on developing countries only, in keeping with this literature. We chose countries with real *per capita* GDP in constant dollars less than $6000,† as in Pritchett and Summers (1996).

3.2 Results

Kakwani (1993) estimated a linear regression model with the achievement index as the dependent variable and the welfare of a country as the independent variable for each subperiod. He also estimated a linear regression model with the improvement index as the dependent variable and the welfare level and the change in welfare level of a country between the two subperiods as the independent variables. Chakraborty (1999), using a nonparametric framework of analysis and focusing on the achievement index and improvement index related to life expectancy only, concluded that the functional forms of Kakwani's models were misspecified and detected nolinearity in the relationship. However,

---

*The database can be found at www.worldbank.org/research/growth/GDNdata.htm

†In fact, there are four countries in our data set that have real *per capita* GDP greater than $6000 but less than $8000 for the year 1992. We included these countries (Korea, Malta, Mauritius, and Mexico) in our data set since their GDP figures in the other 2 years are less than $6000.

TABLE 1 Tests of Linear Parametric Models Against Nonparametric Models

	$c=0.8$	$c=1.0$	$c=1.2$	$c=1.4$
1971-1980 (*Bivariate model—Kakwani's data*)				
Test statistic	−0.406	−0.439	−0.494	−0.549
c.v.[a]	0.653	0.478	0.239	0.101
1980-1990 (*Bivariate model—Kakwani's data*)				
Test statistic	0.015	0.055	0.087	0.084
c.v	0.839	0.652	0.418	0.273
1971-1990 (*Bivariate model—Kakwani's data*)				
Test statistic	0.843	0.914	0.904	0.847
c.v	0.896	0.778	0.483	0.257
1971-1990 (*Trivariate model—Kakwani's data*)				
Test statistic	−0.776	−0.800	−0.870	−0.953
c.v	1.239	0.926	0.762	0.622
1982, 1987, 1992 (*Bivariate model—GDN data, transformed variables*)				
Test statistic	−0.698	−0.640	−0.589	−0.554
c.v	0.899	0.681	0.414	0.292
1982, 1987, 1992 (*Bivariate model—GDN data, untransformed variables*)				
Test statistic	2.732	2.392	2.086	1.800
c.v	−0.390	−0.422	−0.469	−0.493

[a]c.v. is the bootstrapped 5% critical value based on 399 replications.

Chakraborty's conclusions were based on a visual inspection of the nonparametric regression plots, using Kakwani's data set.*

One of the aims of this study is to test formally for this suspected mis-specification of functional form by using the same data set. Table 1 presents the results from using the nonparametric specification test by Li and Wang (1998), which tests a simple linear model against a purely nonparametric one. The first nine rows of results relate to Kakwani's bivariate model with achievement index as the dependent variable and the welfare level as the independent variable. The results have been presented for various bandwidths since the choice of bandwidth is usually very important in the nonparametric framework. We have presented bootstrapped critical values since Li and Wang's (1998) Monte Carlo exercise showed that the wild bootstrap method provided a much better

*It should also be noted that the confidence intervals for the regression curves were not presented.

null approximation for the test than the asymptotic normality result. For both subperiods, we fail to reject the null hypothesis of the linear model at both the 5 and 10% levels of significance for all the different bandwidths. This result agrees with Li and Wang's (1998) finding that, due to the unbiasedness of the relevant test statistic, this testing procedure is not as sensitive to the choice of h as many other kernel smoothing procedures. The results lend strong support to Kakwani's model in the bivariate setup and contradict Chakraborty's claim that Kakwani's model relating the achievement index to the welfare level suffered from functional form mis-specification. The test results under the heading 1971–1990 (bivariate model) are for the pooled data set and they show that there is not much support for the null model at the 5% level. However, at the 1% level of significance we fail to reject the model even with the pooled data.* It should be noted though that neither Kakwani (1993) nor Chakraborty (1999) were analysing a pooled model.

The results in Table 1 under the heading 1971–1990 (trivariate model) present the results from Li and Wang's (1998) test where the null model now is the linear model with the improvement index as the dependent variable and the regressors are the welfare level and the change in welfare level. The alternative model is a purely nonparametric model; *i.e.*, both the regressors appear as unknown functions. The results from the table show that we fail to reject the null model at the 5% level of significance for all c values considered. In fact, the result holds also at the 10% level of significance. This means that Chakraborty's (1999) claim that this model of Kakwani's suffers from functional form mis-specification is not supported by the data either. So the conclusion from Kakwani (1993, p. 335) that *"the level of economic welfare (lagged) and change in it have a positive and significant influence on the improvement in the standard of living"* still holds.

The above analysis shows that the use of the achievement index adequately captures the nonlinearity in the relationship between the life expectancy at birth and the welfare level and that there is no further need to model achievement index as an unknown function of the welfare level as suggested by Chakraborty (1999). However, both Kakwani (1993) and Chakraborty (1999) used bivariate models even though both mentioned factors other than income that many have a positive effect on the standard of living over time. One such factor is the public provision of health services as indicated by Anand and Ravallion (1993).

---

*The figures for the 1% critical value are available from the author on request. In fact, all the results that are mentioned but not shown here are available from the author on request.

Other factors may include unobservable country-specific factors such as good government or cultural values, as indicated by Pritchett and Summers (1996), which would affect health positively. Given this scenario, we use the second data set and examine the role, if any, played by public expenditure on health services and other unobserved country-specific factors in enhancing the health status of a country. We will use infant mortality rate as an indicator of standard of living for this study. The results using life expectancy index can be obtained similarly, but we choose to focus our attention on infant mortality rate mainly since the data on it are more reliable. The life expectancy figures usually are not observed directly from death registrations but are derived by using infant or child mortality figures together with the model life table. This means that they do not contain any more information than the infant or child mortality figures, and in fact suffer from more measurement errors. For further discussion of this issue, see Pritchett and Summers (1996).

Kakwani (1993) and Anand and Ravallion (1993) use slightly different nonlinear transformations of the relevant health indicator as the dependent variable in their analysis. While Anand and Ravallion (1993) do not offer much discussion about their choice of the transformation function, a major focus of Kakwani's (1993) paper was on the derivation of the achievement index based on an axiomatic approach. The tests we performed so far indicated that this transformation seems to capture adequately the nonlinearity in the relationship. Kakwani (1993, p. 324, footnote 11) while referring to Anand and Ravallion's (1993) paper noted that their transformation function is quite similar to his. We have also used Kakwani's longevity data and an Anand and Ravallion (1993) type bivariate model and found that the estimated levels of longevity from the two models are very similar. Also, Li and Wang's (1998) tests of the preceding kind used with this dependent variable yielded extremely similar results to those in Table 1. Given this and the slightly easier formula for Anand and Ravallion's index, we chose their index rather than Kakwani's for our analysis. So the infant mortality index in our study is given by the logarithm of the variable (IMR-5) where IMR denotes infant mortality rate and 5 refers to the "best" level of infant mortality.

Before going into regression analysis, first we present in Fig. 1 the scatter plot matrices of the variable involved in our regression analysis. The upper scatter plot is based on the untransformed variable (infant mortality rate, real GDP *per capita*, and health expenditure per head) while those in the lower scatter plot are based on the transformed variables (achievement index of infant mortality, welfare, and logarithm of health expenditure per head). The plots indicate some nonlinearity between

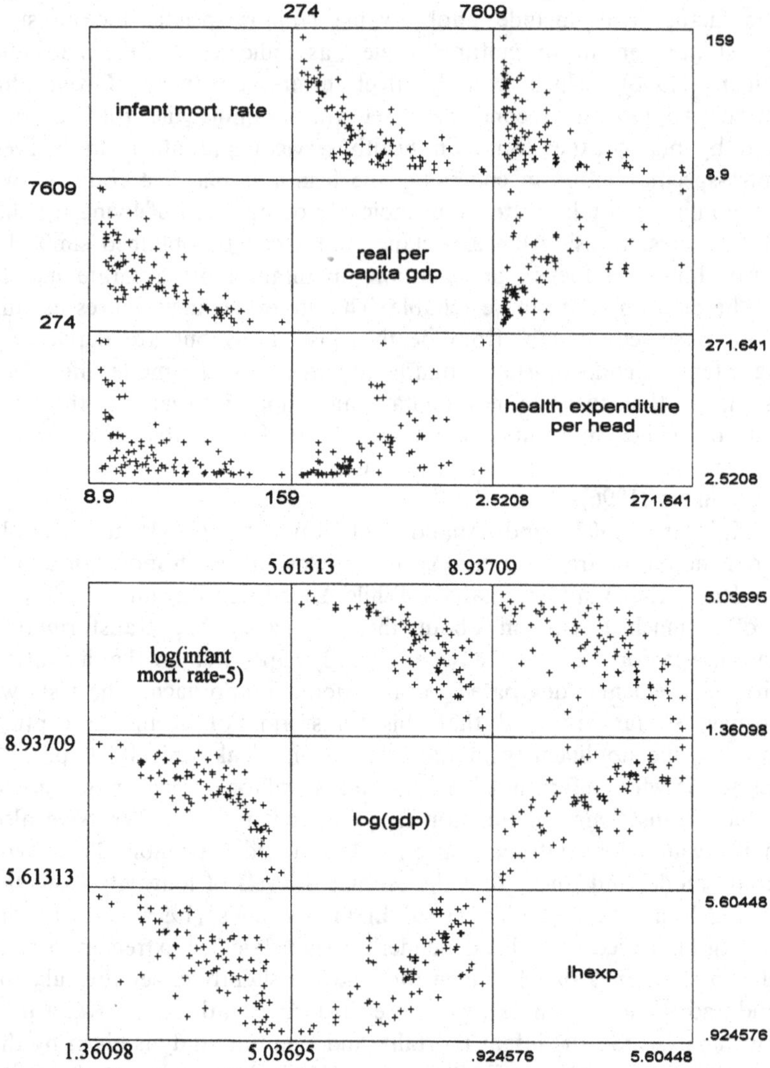

FIGURE 1 Scatter plots of variables using Global Development Network data (26 developing countries).

the index and the health expenditure variable, in both cases. We formally tested a linear relationship between the infant mortality index and the logarithm of GDP *per captia* against a purely nonparametric specification for this data set, using once again Li and Wang's test (1998), and found that we failed to reject the linear model at the 5% level of significance. We

TABLE 2 Regression Results for Infant Mortality Index ($n=26$, $t=3$)

	OLS(1)	OLS(2)	OLS(3)	OLS(4)	SPP($c=1$)	RE	SPRE($c=1$)
Constant	10.542	9.707	9.194	10.067		9.276	
	(0.000)	(0.000)	(0.000)	(0.000)		(0.000)	
Welfare	−0.880	−0.711	−0.836	−0.806	−0.766	−0.798	−0.752
	(0.000)	(0.000)	(0.000)	(0.000)	(0.000)	(0.000)	(0.000)
Dummy 87	−0.171	−0.169	−0.164	−0.157	−0.249		
	(0.270)	(0.266)	(0.231)	(0.247)	(0.075)		
Dummy 92	−0.334	−0.317	−0.296	−0.283	−0.371		
	(0.033)	(0.040)	(0.033)	(0.040)	(0.009)		
ln(hexp)		−0.142	0.861	−0.441		0.543	
		(0.049)	(0.001)	(0.605)		(0.023)	
$(\ln(\text{hexp}))^2$			−0.147	0.292		−0.106	
			(0.000)	(0.296)		(0.005)	
$(\ln(\text{hexp}))^3$				−0.045			
				(0.114)			
Adjusted R^2	0.613	0.628	0.698	0.704			

also tested a linear relationship between infant mortality and GDP *per capita* and, in fact, rejected the linear model. The results from the tests are presented in the last six rows of Table 1. The results from these tests confirm our a priori belief that there exists a nonlinear relationship between the transformed variables. Given the scatter plots and this result, we are then going to assume a linear relationship between the infant mortality index and the welfare variable and are going to be concerned about possible nonlinearity in the public health expenditure variable.

Table 2 reports the regression results based on data on 26 countries in 3 different years from various specifications with infant mortality index as the dependent variable. The first four columns present results from various linear regression models while the fifth column presents the results from a semiparametric pooled model. From column one to column two the fit improves slightly as the logarithm of the variable public health expenditure per head, ln(hexp), is added to the model. The signs are according to our a priori expectation, namely, everything else remaining the same as welfare increases infant mortality goes down and similarly as public health expenditure increases infant mortality decreases, all else being equal. Both variables are statistically significantly different from zero at 5% level of significance. The fit improves further as the model in the second column is augmented by a squared term in order to capture possible nonlinearity coming in through the health expenditure variable. The squared term is strongly statistically significant. The next specification has a cubic term in the health expenditure variable added to it, but now none of the terms involving the public health expenditure variable are statistically significant.*

The above specifications assumed that the welfare and the health expenditure variables are exogenous to the model; otherwise, the OLS estimator will be biased and inconsistent. There is a possibility that the welfare variable may be correlated with the error term since it may not be exogenous to the model, i.e., the infant mortality index is not only affected by the logarithm of income but the latter is also affected by the former. This is a possibility and so we need to test for the exogeneity of the welfare variable in our model. In order to perform the Hausman (1978) test to check for such correlation, we require instruments for the income variable. Following Pritchett and Summers (1996) and the discussion therein, we have chosen the logarithm of the ratio of the gross domestic public investment to the GDP of a country, and the logarithm of one plus the black market premium,† as possible instruments for the logarithm of income. We performed the Hausman test using Wu's (1973) statistic, by including the predicted value from a regression of the logarithm of income on a constant and the instrument(s)‡ in our linear or quadratic specification, described above, and then checking the statistical significance of the predicted value. In all the cases considered, the estimated coefficient on the predicted value was not statistically significant, and we failed to reject the null hypothesis of no correlation at any desired significance level. This provided some support for the assumption of exogeneity of the income variable in our regressions above.

Given that mis-specification of functional form can have an impact on our results, we next estimate a semiparametric (SP) model with ln(hexp) modeled as an unknown function. We first test the null hypothesis of a parametric model with welfare, ln(hexp), $(\ln(\text{hexp}))^2$, dummy variable for 1987 (dummy87), and dummy variable for 1992 (dummy92) as the independent variables against a SP specification with ln(hexp) appearing

*The dummy variable for the year 1987 is not statistically significant in any of the OLS specifications. We have also estimated the same specifications excluding the dummy variable for 1987. The results for the estimated coefficients on the welfare and health expenditure variables remain almost identical to the results in Table 2.

†The black market premium is measured as (black market exchange rate/official exchange rate)−1. The data on black market premium (%) and gross domestic public investment as % of GDP are available in the Global Development Network Growth data base.

‡We calculated the predicted value by running regressions with each of the instruments separately as well as with both instruments together.

TABLE 3 Tests of Quadratic Parametric Model Against Semiparametric Model

	$c=0.8$	$c=1.0$	$c=1.2$	$c=1.4$
Test statistic	16.101	16.851	16.124	14.063
c.v.[a]	−0.038	−0.271	−0.481	−0.610

[a]c.v. is the bootstrapped 5% critical value based on 399 replications.

as an unknown function using the test by Li and Wang (1998). The results are given in Table 3 and they show that for all the bandwidths considered we can reject the null model at the 5% level of significance. This led us to choose the SP specification over the linear model augmented by a squared term as our preferred model.

Given that the data are panel in nature, we have also estimated a standard random effects (RE) model and a semiparametric random effects (SPRE) model (Li and Ullah, 1998). The estimates from the RE model are given in column 7 of Table 2 and show that the welfare variable continues to be statistically significant, and the result is quite similar to that of the pooled data case [OLS(3) in Table 2]. We carried out the Breusch and Pagan (1980) Lagrange multiplier test for RE and it strongly rejected the null hypothesis of the variance of the country-specific error term being zero. Given that the Hausman test (1978) failed to reject the null hypothesis of no correction between the country-specific error term and the regressors, we prefer the RE model over a fixed effects model. However, as in the pooled model case, the issue of mis-specification of functional form can still be a problem in this case. To address that issue, we estimated a SPRE model following Li and Ullah (1998).

The estimated nonlinear relationship between infant mortality index and ln(hexp) from the SP model along with its bootstrapped confidence interval* is presented in Fig. 2, while that from the SPRE model is presented in Fig. 3. By providing a confidence interval over the entire range of the logarithm of the health expenditure variable, we are able to offer a better representation of the nonlinearity in the relationship relative to the simple linear benchmark. The shape of the plot of the relationship between the index and the logarithm of the health expenditure from the SPRE model is very similar to that from the pooled SP model. The

*We used the wild bootstrap to obtain the standard errors of the estimate and then used the normal approximation method to calculate the confidence bands. We chose the normal approximation method since the latter does not perform well when the sample size is small because of the importance this method attaches to the tails of the distribution (DiCiccio and Romano, 1988).

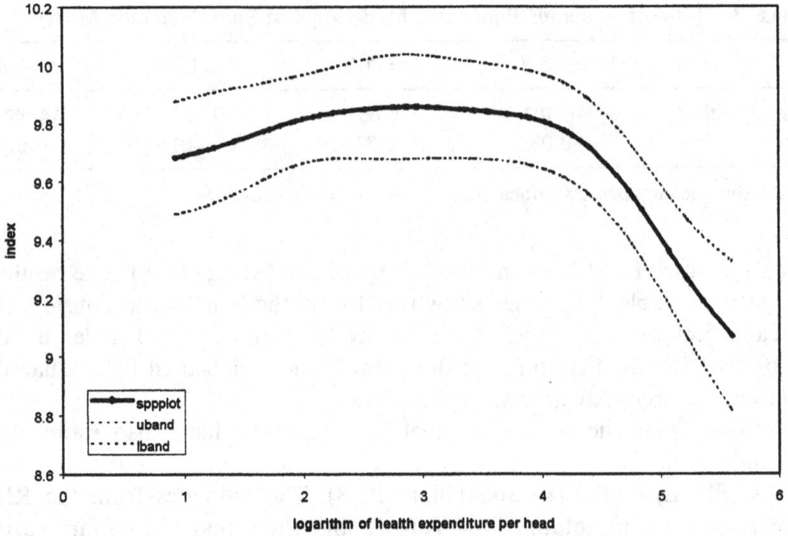

FIGURE 2 Infant mortality index against health expenditure from a semiparametric pooled model.

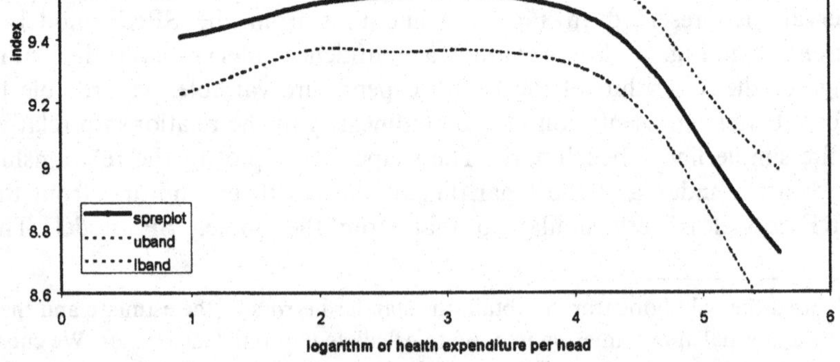

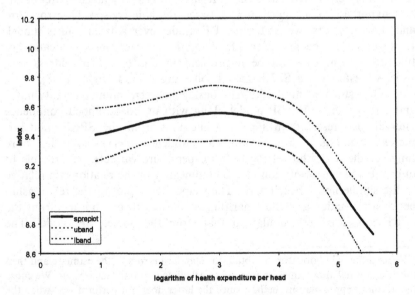

FIGURE 3 Infant mortaility index against health expenditure from a semiparametric random effects model.

estimated coefficient of the welfare variable is also very similar for the two models. These results indicate that the unobserved country-specific factors are not very important in affecting the infant mortality index since their inclusion did not change the results much. So given that it is relatively easier to calculate the pooled models and the results are similar across the different specifications, we prefer the pooled models in this case.

These results show that the public health expenditure has an effect on the standard of living as measured by the infant mortality index and that the relationship may not be linear as considered in the existing literature. However, the importance of income in influencing standard of living does not diminish much as the public health expenditure variable is included in the model. The results show strong and statistically significant effects of income on the infant mortality index even after we have taken into account the nonlinearity in the public health expenditure variable. This result is in contrast to Anand and Ravallion's (1993) result that the effect of income on the standard of living diminishes substantially and may even become statistically insignificant once other variables like public health expenditure per head is added to the regression. It should be noted that Anand and Ravallion also had a poverty measure as an additional regressor in their models, involving life expectancy and literacy indices, but not in the model with the infant mortality index. The latter was due to the lack of data on the poverty measure for the case considered.

We faced a similar problem and were unable to obtain reliable data on the poverty measure for the 3 years without losing more countries from our sample. So we are unable to control for this particular variable. We recognize that this has the potential of biasing our estimate of the income coefficient. However, two things are to be noted in this respect. First, the inclusion of the poverty variable seems somewhat ad hoc in Anand and Ravallion's (1993) paper, so it is not certain whether this is a relevant regressor or not. Second and more importantly, the purpose of this study was to analyze the effect of the inclusion of a variable representing public provision of health services on the estimated income coefficient. We found that the income variable continued to be highly statistically significant and that the inclusion of the health expenditure variable did not reduce its importance much,* contrary to what some researchers have suggested.†

---

*The coefficient estimate changes from 0.880 to −0.711 with the inclusion of the public health expenditure variable.

†It should be noted that the issues relating to possible nonstationarity of the data have not been taken into account in this study. However, none of the existing studies with which we are trying to compare our results have done so either. This issue remains an area of future research.

One other thing to note is that, like Anand and Ravallion (1993), we chose public health expenditure per head to represent public provision of health services. However, as Chakraborty (1999) discussed in her paper, what might be more important in developing countries are the coverage of public services and the nature of the orientation of the programs rather than the public health expenditure figure in itself. Also, private health expenditure seems to be relatively more important in developing countries as compared to developed countries. Unfortunately, reliable data on some of these variables are hard to obtain and hence a study involving these other variables remains an area of future research as better data become available.

4 CONCLUSION

We first tested for mis-specification in the functional forms of the linear models proposed by Kakwani (1993), who studied the effect of income on the achievement index and improvement index of life expectancy as an indicator of the standard of living of a country. Using a test by Li and Wang (1998), we found that the data supported the linear specifications as opposed to the nonparametric specifications put forth by Chakraborty (1999). This highlights, in general, the usefulness of performing a test such as the one proposed by Li and Wang before deciding on a nonparametric specification as the preferred model. The nonparametric estimator is biased in a finite sample and its attractiveness comes from its asymptotic properties. So given that we normally deal with relatively small-sized samples in cross-country studies, it is preferable to use a parametric model, if it is supported by the data, since the estimators will then have known finite sample properties and much faster speed of convergence.

We were also interested in the claim by certain researchers (Iseman, 1980; Sen, 1981; Anand and Ravallion, 1993; among others) that it is not economic growth but public provision of services which is more important in influencing the standard of living of a country, and that the strong and statistically significant effect of income vanishes once public provision of services has been accounted for. So we studied the effect of including another explanatory variable representing public provision of health services on the estimated coefficient of the income variable. The issue of nonlinearity in the relationship of this variable to the health index used was tested and we analyzed a semiparametric model that avoids the mis-specification of the functional form. Irrespective of the specification used, income continued to have a strong and statistically significant effect on the health indicator. Further, controlling for unobserved country-specific effects did not seem to impact the results much.

In general, the usefulness of the semiparametric model is that it allows us to provide a pictorial representation of the nonlinearity in the relation between the dependent variable and the variable* that appears as an unknown function in the model, along with its confidence interval. The methodology allows the data to decide the functional form and does not impose any ad hoc functional form a priori, hence avoiding the issue of functional form mis-specification. It is a very useful tool when one has a number of regressors and the prior knowledge that some of them are linearly related to the dependent variable. This kind of model avoids the problem of the "curse of dimensionality," which makes the purely nonparametric estimators to be of little use if there are more than three or four regressors in the model. On the computational side, these estimators are not very burdensome either. Even though there are not many software packages that would automatically perform a semiparametric estimation, as long as those programs can do nonparametric estimation, they can be quite easily manipulated to calculate semiparametric estimators as well.

ACKNOWLEDGMENTS

I thank the seminar participants at the University of Victoria and an anonymous referee for their helpful comments on an earlier draft of this chapter. I am very grateful to Thanasis Stengos for his help with some of the GAUSS codes used for this study. The financial support from the internal grant provided by the University of Victoria is gratefully acknowledged.

REFERENCES

Anand, S. and Kanbur, S.M.R., 1991. Public policy and basic needs provision: Intervention and achievement in Sri Lanka. In: J.P. Dreze and A.K. Sen, eds. *The Political Economy of Hunger*, vol. 3. Clarendon Press, Oxford, pp. 59–92

Anand, S. and Ravallion, M., 1993. Human development in poor countries: On the role of private incomes and public services. *Journal of Economic Perspectives* 7, 133–150.

---

*More than one variable can appear as the unknown function in the model, so to obtain the individual effects one can follow the additive partially linear approach and use marginal integration. See Section 2.1 for references.

Bhalla, S.S., 1988. Sri Lanka's achievement: Fact and fancy In: T.N. Srinivasan and P.K. Bardhan, eds. *Rural Poverty in South Asia*. Columbia University Press, New York, 557–565.

Breusch, T. and Pagan, A., 1980. The LM test and its application to model specification in econometrics. *Review of Economic Studies* **47**, 239–254.

Chakraborty, I., 1999. Living standard and economic growth: A fresh look at the relationship through the nonparametric approach. *Journal of Quantitative Economics* **15**, 39–66.

DiCiccio, T.J. and Ramano, J.P., 1988. A review of bootstrap confidence intervals (with discussions). *Journal of the Royal Statistical Society, Series B* **50**, 338–370.

Fan, Y. and Li, Q., 1996. On estimating additive partially linear models. Unpublished manuscript. University of Guelph, Guelph, Ontario, Canada.

Hausman, J., 1978. Specification tests in econometrics. *Econometrica* **46**, 1251–1271.

Heckman, N.E., 1986. Spline smoothing in a partly linear model. *Journal of the Royal Statistical Society, Series B* **48**, 244–248.

Iseman, P., 1980. Basic needs: The case of Sri Lanka. *World Development* **8**, 237–258.

Kakwani, N., 1993. Performance in living standards: An international comparison. *Journal of Development Economics* **41**, 307–336.

Li, Q. and Ullah, A., 1998. Estimating partially linear panel data model with one-way error components. *Econometric Reviews* **17**, 145–166.

Li, Q. and Wang, S., 1998. A simple consistent bootstrap test for a parametric regression function. *Journal of Econometrics* **87**, 145–165.

Linton, O.B. and Nielson, J.P., 1995. A kernel method of estimating structural nonparametric regression based on marginal integration. *Biometrika* **82**, 93–100.

Pritchett, L. and Summers, L.H., 1996. Wealthier is healthier. *Journal of Human Resources* **31**, 841–868.

Robinson, P., 1988. Root-N-consistent semiparametric regression. *Econometrica* **56**, 931–954.

Sen, A., 1981. Public action and the quality of life in developing countries. *Oxford Bulletin of Economics and Statistics* **43**, 287–319.

Wu, D., 1973. Alternative tests of independence between stochastic regressors disturbance. *Econometrica* **41**, 733–750.

Index

Achievement index, 481, 486, 488
AIC (*see* Information criterion, Akaike's)
Animation, 6
ARCH, 326–327, 346–347
Australian Bureau of Statistics, 43
Autocorrelation (*see also* Autoregressive process)
  positive, 44
  of residuals, 44, 326n
Autoregressive (AR) process (*see also* Autocorrelation), 38, 212, 213, 328, 360
  coefficient in, 327, 347
  exponential, 462
  filter using, 131
  functional coefficient, 455
  integrated moving average, 4
  model for, 4, 176
  moving average (ARMA), 333, 346
  parameter in, 329n,
  threshold (TAR), 356, 455, 461

Bartlett correction, 12
Bayes factor, 219, 220, 244, 250, 251
Bayes' rule (*see* Bayes' theorem)
Bayes' theorem, 219, 266, 278
Bayesian inference
  communication of results of, 251–258
  computer-aided, 264
  with conjugate prior distribution, 217–218
  with hierarchical prior distribution, 220–222
  intercept correction in, 389
  and large data sets, 2
  and model comparison, 218–220, 243–251
  and model prior probabilities, 218–219
  and model selection, 41
  posterior density (distribution) for, 212, 214–215, 224–225, 232, 243, 247, 258, 263, 266–270, 274–276, 278–279, 284
    tractable, 217
  posterior odds ratio in, 219
  and posterior simulation methods, 223–243
  predictive density (distribution) for, 283–285
  prior density (distribution) for, 2, 41, 212, 215, 218, 237, 250, 258, 385–386
    conjugate, 216, 217–218, 227
    construction of, 243
    improper, 253

[Bayesian inference]
inequality-restricted, 274
Litterman, 386
noninformative, 253, 264, 266, 276
subjective, 253
tightness of, 390
two-stage hierarchical, 221
robustness of, 253
simulation methods for, 209–258
software for, 210, 237–240, 249–250, 254
Bias (*see* Estimator)
BIC (*see* Information criterion, Bayesian)
Bootstrap
block, 70, 82
conditional, 469
confidence interval from, 493, 497
corrected, 71
coverage probability for, 69–90
critical values from, 484
method of, 25, 445
naive, 453, 458, 460, 463
parametric, 13, 21, 23n
recursive, 469
sieve, 82
test, 69–82
wild, 453, 458–460, 463, 487, 493n
Brownian motion, 98, 179
Bubble, price (*see also* Volatility, excess), 316–317, 318, 323–324, 327, 336, 344, 347

Calculus, fractional matrix, 11
CAPM model, 8
Causal ordering, 91
Causality
Granger, 91–145
long-run, 96
short-run, 96
Central Bank
European, 372
Norwegian, 387
Central limit theorem, 225, 235, 253
Chaos, 352, 361–362
Characteristic function, 73n
Cointegration (*see also* Data, cointegrated), 175, 391
and causality testing, 91–145
and stock prices, 323
Engle and Granger test for, 386
Johansen test for, 386
Collinearity (*see also* Data, ill-conditioned), 301
Computation (*see* Computing)
Computing
accuracy of, 248, 250
burden of, 245, 247, 325, 415, 418, 480, 497
difficulty with, 307
efficiency of,
execution time for, 241, 243, 250, 258, 335, 403, 463, 485
and FCM algorithm, 414–415
intensive nature of, 316
power of, 9
problems with, 165
rounding error, 308
speed of, 1, 9
Conditioning number (*see also* Data, ill-conditioned), 301
Cornish-Fisher expansion, 74, 75, 77
Cp (*see* Mallows' Cp)
Cross validation (*see* Generalized cross validation)
Cumulant, 73, 74, 77, 78, 79
Curse of dimensionality, 408, 444

Data
achievement index, 485
animated (*see* Animation)
availability of, 1, 9
beef price, 440
Canadian earnings, 306–309, 419–422
cell phone, 200
chicken consumption, 440
chicken price, 440
cointegrated, 91–145
concrete price, 200
consumer price index, 372
consumption expenditure, 423
cross-section(al), 3, 7–8, 428, 471
dependent, 70, 72, 453
weakly, 452
differenced, 94, 144
exchange rate, 359, 369–372, 434, 436

[Data]
financial market, 317
gas, 200
Global Development Network (GDN), 486
gross domestic product (GDP) per capita, 485
ill-conditioned (*see also* Collinearity, Conditioning Number, and Multicollinearity), 291, 293, 306, 312
improvement index, 485
income inequality (*see also* Gini coefficient), 432
inflation rate, 372
integrated, 91–145
seasonally, 175–203
interest rate, 427
knitted fabric sales, 199
large sets of, 1–10
lot size, 240
missing, 6
money stock, 427
motorcycle, 301–304
nonannual, 175
non-stationary, 175, 312
orbit, 366, 368
output, 427
panel (*see also* Panel data), 2, 3
Panel Survey of Income Dynamics, 242
quality of, 6
real disposable personal income, 423
per capita, 440
real estate price, 240
retail trade, 43
reliability of, 495–496
scanner, 359
seasonal, 179
seasonally adjusted, 422–423, 434
semiannual, 175–203
share price (*see* Stocks)
stationary, 91, 176, 203
stock price (*see* Stocks)
tick-by-tick, 5
time-series (*see* Time-series)
unemployment rate, 372
wheat yield, 280
world tables, 43
Delta method, 78, 80, 246
Difference equation, 363
Differential equation, 223n
Distribution
asymptotic (*see also* Distribution, limit, non-standard), 179, 253, 347, 472, 484
chi square (*see also* Test, chi square), 25, 97–98, 131, 153, 254, 257
empirical, 74n
exact, 20
gamma, 224
inverted, 42
Gaussian (*see also* Distribution, normal), 7, 12, 15, 16, 17, 24
joint, 222
high frequency, 5
invariant, 231, 233, 234, 236
limit, non-standard (*see also* Distribution, asymptotic), 98, 180
lognormal, 334n
multinomial, 151
multivariate, 301
non-Gaussian, 12
non-standard, 95
normal (*see also* Distribution, Gaussian), 224, 250, 253, 254, 270, 285, 334n, 472
bivariate, 213
independent, 243
multivariate, 42, 154–155, 243, 265, 267, 275, 279, 290
truncated, 278–279, 286
standard, 484
univariate, 223
posterior (*see* Bayesian inference)
prior (*see* Bayesian inference)
reference, 295
t- (*see also* Test, t-), 163
multivariate, 268, 270, 279, 284
univariate, 270
two-sided, 75
univariate, 224, 295
Wishart, inverted, 267, 269, 279, 284, 285
Drift, 178, 179, 184, 185, 201, 386–387, 388, 403
Dummy variable, 389–391, 492

[Dummy variable]
seasonal, 179, 180, 181, 184, 185, 201

Edgeworth expansion, 71–72, 74, 75, 77, 78, 80, 460
Equivalence scale, 282
Equity premium puzzle, 343
Error correction model, 96
vector (VEC), 102, 105–107, 129, 145, 386–387, 390–391, 403
Estimating function, 294
Estimation (*see* Estimator)
Estimator (*see also* Regression)
asymptotic properties of, 456
Bayes (*see* Bayesian inference)
biased, 150, 154, 157, 159, 160, 297, 311, 316, 322, 329n, 492
canonical modulation, 299
consistent, 2, 152, 236, 296, 386
weakly, 418, 445
data-based information theoretic (DBIT), 293–294, 300, 301–302, 304–306, 307–309, 310–312, 419, 422
efficiency of, 96, 297
efficient, asymptotically, 152
extremum, 297
fuzzy (*see* Fuzzy econometric modeling)
inconsistent, 480, 492
inefficient, 322, 480
kernel (*see* Estimator, nonparametric, and Regression)
least absolute deviations, 67, 415
least squares (LS) (*see also* Regression, linear), 44, 95, 99, 156, 157, 166, 177, 216, 241, 274, 290, 294, 295–296, 297, 300, 301–302, 303–304, 306, 308–309, 312, 385, 420, 425, 427, 432, 435, 442, 454–455, 459, 492
generalized, 25–26, 31, 267, 390
indirect, 389
local linear (LLLS), 454–455
local limit varying coefficient (LLVC), 455–456
local polynomial, 455
ordinary (OLS) (*see* Estimator, least squares)
trimmed, 415
M-, 415
maximum empirical likelihood (EL), 149–166
maximum likelihood (MLE), 13, 16, 24, 25–26, 40, 106, 243, 258, 267, 274, 292, 312
quasi-, 26
method of moments (MOM), 150
method of regularization (MOR), 297, 312–313
minimax, 292
mixed regression, 386, 387, 484
Nadaraya-Watson, 311, 459
nonparametric, 300, 407, 419–420, 422, 425, 451, 458–459, 480
normal, asymptotically, 152, 296
panel (*see* Panel data)
parametric, 407
precision, 300, 301
principal components (PC), 300, 302–304, 359
quantile, 4
ridge, 297, 299, 300, 302, 304
iterative, 297
robust, 415
semiparametric, 482–483, 485, 492
single-equation, 280
Stein, 292, 293, 313
modified, 293
positive-rule, 292–293
superconsistent, 388
SUR (*see* Seemingly unrelated regressions)
three stage least squares (3SLS), 389
iterated, 389
two-stage least squares (2SLS), 150, 154, 156, 158–166, 388, 390
two-step, 267
unbiased, 294, 298, 299, 308, 312
variance of, 157, 160, 306, 311
European Monetary Union, 372
Exogeneity, 215nFederal Reserve Board, 315–316

Filter
autoregressive (*see* Autoregressive process, filter using)
Hodrick-Prescott (HP) (*see* Time-series, trend in)

[Hodrick-Prescott (HP)]
moving average, 103
stationarity, 175
Forecasting
Bayesian, 266, 283–285
conditional distribution for, 4
ex ante, 92, 101–103, 107–108
out-of-sample, 5, 92, 102
real-time, 102, 385–404
Fuzzy clustering, 407–446
c-means (FCM) algorithm for, 412–415, 417, 419–420, 423, 442, 444, 472
binary splitting (BS) variant of, 414
binary splitting direct search variant of, 414
KKZ variant of, 415
mountain method variant of, 415
simple cluster-seeking (SCS) variant of, 414
subtractive clustering variant of, 415
Fuzzy econometric modeling, 415–418
Fuzzy logic, 410–412, 444
Fuzzy set, 410–412, 444
degree of membership for, 410, 411, 412, 413, 444
membership function for, 420, 425, 427, 432, 436, 438

Gamma function, 270
General to simple methodology, 7
Generalized cross validation (GCV), 38
Gibbs sampler (*see* Markov-chain Monte Carlo)
Gini coefficient (*see* also, Data, income inequality), 428
Granger causality (*see* Causality)

Hadamard product, 295
Harmonic mean, 246–249
Hodrick-Prescott (HP) filter (*see* Time-series, trend in)
Hyperparameter, 221

Improvement index, 481, 486
Income inequality (*see also* Gini coefficient and Data, income inequality), 428
Indicator function, 213n, 458
Indirect least squares (*see* Estimator, least squares)
Inference
Bayesian (*see* Bayesian inference)
ill-posed, 291, 312
Information (*see* Kullback-Leibler information)
Information criterion
Akaike's (AIC), 38, 40, 44, 101, 107–108, 130, 143–145
Bayesian (BIC), 38, 40, 44, 67, 326, 333
Hannan and Quinn's (HQ), 38, 40, 44
Schwartz's (SIC), 92, 101, 107–108, 129–130, 143–145
Input-output relationship, 416
Instrument (*see also* Instrumental variable), 155, 492
Instrumental variable (IV) (*see also* Instrument), 149, 151, 156, 164
Integration
analytical, 249, 267
bivariate, 223
marginal, 483
Monte Carlo, 325,
numerical, 284, 320, 325, 330, 335
by simulation, 222
Intercept correction, 387, 388–391
Inverse problem, 294, 297

Kolmogorov's representation theorem, 356
Kullback-Leibler (KL) information, 295
Kurtosis, excess, 338, 343
Kuznets' U-curve, 428–432
dynamic effects and, 432
nonstationarity data and, 432

Lagrange multiplier test (*see* Test)
Latent variable, 220–222
Law of the excluded middle, 411
Least squares (*see* Estimator, least squares)
Likelihood function, 211, 221, 251, 254, 265, 276, 278

[Likelihood function]
  marginal, 219, 220, 237, 243–244, 245–246, 248–249, 250, 251, 266
  for probit model, 251
  unbounded, 212
Likelihood ratio test (*see* Test, likelihood ratio)
Lindeberg-Levy central limit theorem (*see* Central limit theorem)
Logistic function, 354–355, 357
Loss
  balanced, 293, 298, 299
  squared error (SEL), 154, 157, 292, 297, 298
  squared error prediction (SEPL), 299
Lyapunov exponent, 352, 361–364, 366, 368–371, 381

M-estimator (*see* Estimator)
Mallows' Cp, 308, 311
Marginal propensity to consume (m.p.c.), 425
Markov model, 319–320, 332
Markov-chain Monte Carlo (MCMC)
  and Bayesian inference, 263
  burn-in iterations for, 240, 273
  central limit theorem for, 236
Gibbs sampler for, 224, 227–229, 231, 244, 245–246, 248, 264, 269, 273–274, 276–277, 279, 280, 285
    convergence of, 234–235, 273
    two-block, 231
  Hastings-Metropolis algorithm for, 224, 229–232, 244, 264, 274–277, 279, 281, 283
    convergence of, 235
  methods for, 224, 263–264, 273–274, 286
  Metropolis-within-Gibbs algorithm for, 231–232
    numerical accuracy of, 235–237
Maximum empirical likelihood (*see* Estimator, maximum empirical likelihood)
Maximum likelihood (*see* Estimator, maximum likelihood)
Mean absolute error (MAE), 422, 423, 427, 432, 441
  of forecast, 102
Mean squared error
  of estimation, 157
  of forecast, 102, 393, 396–397, 463
  of prediction, 157
  root (RMSE), 422, 423, 427, 432, 441
Method of moments (*see* Estimator, method of moments)
Missing value, 6
MLE (*see* Estimator)
Model comparison
  Bayesian approach to, 218–220, 243–251
Model selection
  backward, 308
  Bayesian approach to, 41
  for causal orderings, 91
  general to specific (*see* General to simple methodology)
  for linear regressions, 37–67
  for VAR models, 92, 100, 145
MOM (*see* Estimator, method of moments)
Moment conditions, 149, 150, 153
Monte Carlo
  experiment, 24–30, 33, 39, 43–67, 71, 74n, 92, 94, 103, 153, 154, 185, 325n, 327, 328, 335–336, 403, 487
  integration (*see* Integration, Monte Carlo)
  method, 45, 300, 306, 317, 324, 458
  p-value, 21, 23
  procedure (*see* Monte Carlo, method)
  sampling, 149–150, 154, 165
  simulation, 41, 184, 325, 333–335
    error of, 325n
  test, 11–33
  variance reduction in, 330n
Moving average (MA)
  in errors, 108, 14–144, 181
  model for, 2, 326
  process for, 38, 92, 105, 130, 145, 328
Multicollinearity (*see* Collinearity, and Data, ill-conditioned)

Neyman factorization criterion, 217n
Neural network, 351–382, 472
activation function in, 353–354
artificial, 352
layers of, 353, 364–365, 373, 376, 378
node reduction in, 358
pruned, 360, 369–370
pruning methods for, 357–358
reduced (*see* Neural network pruned)
three-layer feed forward, 352, 356
two-stage, 352
weight reduction in, 358
Noncausality (*see* Causality)
Nuisance parameter, 13, 18, 97

OLS (*see* Estimator, least squares)
Optimization
difficulty with, 45, 403
numerical, 45, 351
Ordinary least squares (*see* Estimator, least squares)
Outlier, 4, 6, 415
Overfitting, 312, 357

P value
local Monte Carlo (LMC), 21
maximized Monte Carlo (MMC), 21
simulated, 23
Panel data (*see* lso Data, panel)
balanced, 486
estimation with, 2, 3, 6, 479–497
Petroleum Monitoring Agency Canada, 200
Phillips curve, 372–381
Pivotal statistic (*see* Statistic,pivotal)
Posterior odds ratio (*see* Bayesian inference)
Power (*see* Test, power of)
Prediction (*see* orecasting)
Preliminary test, 387
bias from, 38
Pricing model
capital asset (*see* CAPM)
fundamental, 318–322
Donaldson and Kamstra (DK), 320–322, 323, 332–333, 346
Gordon, 319, 332, 346
Markov, 319–320, 332
Prior (*see* Bayesian inference)
Probit model, 242–243, 251, 286
Purchasing power parity (PPP), 371, 485

Quadrature (*see* Integration)

Random walk, 178, 185, 385
Regression (*see also* Estimator)
auxiliary, 184, 185, 459
kernel, 311, 432, 444, 483
multivariate, 407
linear (*see also* Estimator, least squares), 38, 39, 43, 215, 240, 310, 483, 491
multivariate linear (MLR), 12, 18
nonparametric, 310, 408, 427, 438, 490
parametric, 451
polynomial, 307, 310
ridge (*see* Estimator, ridge)
semiparametric, 452
with principal components, 303
spline (*see* Spline)
varying coefficient, 456
Reversibility condition, 230
Ridge estimator (*see* Estimator, ridge)
Risk, 155, 156, 158,161, 290, 292, 297, 304, 313
empirical, 303
Risk premium, 322n

Sampling
acceptance, 224–225
importance, 224, 226–227, 238, 244
Seemingly unrelated regressions (SUR, SURE)
Bayesian inference for, 263–287
inequality restrictions in, 278–279
nonlinear, 275–276
restrictions on parameters of, 11–33, 277–279
testing restrictions in, 11–33
unbalanced, 446
Serial correlation (*see also* Autoregressive process, and Moving Average), 238, 316n, 326

Serial dependence (*see* Serial correlation)
Shares (*see* Stocks)
SIC (*see* Information criterion, Schwartz's)
Simulated annealing (SA), 45, 50, 52, 66
Simulation (*see also* Monte Carlo)
  of economies, 315–348
  experiment (*see* Monte Carlo, experiment)
Simultaneous equations model, 286
Size (*see* Test, size of)
Skew, 338, 343
Spline, 310, 311, 444, 472
Standard of living, 479–497
Statistic, pivotal, 12, 18, 23n
  boundedly, 12
  sufficient (*see* Sufficient statistic)
Statistics Canada, 200
Statistics New Zealand, 199
Stein estimator (*see* Estimator)
Stocks
  data for, 359
  market for, 9
  on New York Stock Exchange, 8
  prices of, 315–348
  ex post rational, 322, 332
  S&P 500, 317, 326, 328–329, 331, 334n, 336–340, 343–344
Structural break, 4
  in DVAR model, 390, 392–393
  in MDVAR model, 389
  testing for, 3, 387
  and seasonal unit root tests, 203
Structural equations, 149–166
Sufficient statistic, 217, 218
SUR (*see* Seemingly unrelated regressions)
SURE (*see* Seemingly unrelated regressions)

Taylor's theorem, 75n
  approximation, 455
  series expansion from, 76, 81
Test
  Anderson-Darling, 326n
  asymptotic, 26n, 33, 460
  bias of, 29, 181
  bootstrap, 69–82
    nonparametric, 451–472
  bounds, 13
Monte Carlo (BMC), 12
  chi-square (*see also* Distribution, chi square), 26, 29, 31, 129, 240
  for bubbles, 324
  Camerer, 323, 346
  Campbell-Shiller, 323, 346
  conditional moment, 459
    nonparametric, 472
  consistent, 452
  Cramer-von Mises, 326n
  Diebold-Mariano, 102
  empirical likelihood ratio (ELR), 162
  Engle and Granger, 386
  ex ante, 129
  F-, 11, 26, 29, 106, 129, 201n, 452
    nonstandard, 177, 201n
  feasibility of, 33
  based on generalized method of moments (GMM), 72
  Hausman, 492
    Wu's version of, 492
  Johansen, 386
  Kolmogorov-Smirnov, 326n
  L, 459–460, 461
  Lagrange multiplier (LM), 12, 162
  with large samples, 2
  Li and Wang, 482–483, 485, 489
  likelihood ratio (LR), 12–13, 16, 18–19, 21, 452
    quasi- (QLR), 25, 31
  Monte Carlo, 12
    bounds, 22
    local (LMC), 26–27, 29
  MRS1, 324
  MRS2, 324
  nonparametric, 452, 472
  pairwise, 37
  parametric, 451
preliminary (*see* Preliminary test)
  power of, 26, 29, 33, 181, 185, 461–462, 465, 471
Q, 459, 461
  score (Rao's), 458
  Shapiro-Wilks, 326n
  simulation-based, 11–33
  size of, 26, 27, 33, 95, 129, 143, 181, 185, 461–462, 465
  specification, 451–472, 480, 482, 483–484

[Q]
t- (*see also* Distribution, t-), 179, 181, 201n
nonstandard, 177, 201n
T, 457, 460, 461
S version of, 458, 460
R version of, 458, 460
unit root, 4, 371
Dickey-Fuller (DF), 176, 181, 203, 323, 346
augmented (ADF), 178, 199, 201
F-, 179, 201n
t-, 179, 181, 201n
Hylleberg-Engle-Granger-Yoo (HEGY), 176–178, 181, 203
with monthly data, 176
seasonal, 175–203
with semi-annual data, 175–203
with quarterly data, 176
Wald
with empirical likelihood, 153
for Granger noncausality, 94–99, 101, 105–107, 129, 131
sequential, 96–97, 143
in SUR model, 12, 24–26, 31
with two-stage least squares, 163
Three stage least squares (*see* Estimator, three stage least squares)
3SLS (*see* Estimator, three stage least squares)
Time-series, 469
artificial, 438
estimation with, 8–9
long, 3
mixing conditions for, 236
multiplicative model for, 434
nonlinear, 452
seasonal pattern in, 434
seasonally integrated, 175–203
trend in, 178, 179, 184, 185, 306
cubic, 278
extraction of, 434–439
Hodrick-Prescott (HP) filter for, 435–436, 438, 444
Tobit model, 285, 286
Trend (*see* Time-series)
Trimming parameter, 291–313
TSSS methodology (*see* Fuzzy econometric modeling)
Two-stage least squares (*see* Estimator, two stage least squares)
2SLS (*see* Estimator, two stage least squares)

U-curve (*see* Kuznets' U-curve)
Unit root (*see* Data, integrated)
test (*see* Test, unit root)

VAR (*see* Vector autoregression model)
VEC (*see* Error correction model, vector)
Vector autoregression (VAR) model, 4,8, 92–93, 94, 96–97, 102–107, 130–131, 143–145, 385–404
Bayesian (BVAR), 386, 390
in differences (DVAR), 387, 390–393, 396–397
Bayesian, 396–397
mixed drift (MDVAR), 389–391, 393, 396–397, 403
Bayesian, 396–397
Volatility, excess (*see also* Bubble), 316–317, 323–324, 336

Wald test (*see* Test, Wald)
Wilks phenomenon, 457
World Bank, 43

Printed in the United States
by Baker & Taylor Publisher Services